EX LIBRIS

FUNDAMENTALS OF MATHEMATICS

VOLUME III

Analysis

Fundamentals of Mathematics

Volume I

Foundations of Mathematics
The Real Number System and Algebra

Volume II

Geometry

Volume III

Analysis

FUNDAMENTALS OF MATHEMATICS

VOLUME III

Analysis

Edited by
H. Behnke
F. Bachmann
K. Fladt
W. Süss

with the assistance of
H. Gerike
F. Hohenberg
G. Pickert
H. Rau
H. Tietz

Translated by
S. H. Gould

The MIT Press Cambridge, Massachusetts, and London, England

Originally published by Vandenhoeck & Ruprecht, Göttingen, Germany, under the title *Grundzüge der Mathematik*. The publication was sponsored by the German section of the International Commission for Mathematical Instruction. The translation of this volume is based upon the German edition of 1962.

Printed and bound in the United States of America.

ISBN 0 262 02049 1(hardcover)
Library of Congress catalog card number: 68-14446

Contents

Translator's Foreword

The pleasant task of translating this unique work has now extended over several years, in the course of which I have received invaluable assistance from many sources. Fortunately I had the opportunity, in personal conversation or in correspondence, of discussing the entire translation with the original authors, many of whom suggested improvements, supplied exercises, or made changes and additions in the German text, wherever they seemed desirable to bring the discussion up to date, for example, on the continuum hypothesis, Zorn's lemma, or groups of odd order. To all these authors I express my gratitude.

For technical and clerical help I am especially indebted to Linda Shepard, of the Law School at the University of Utah, for her expert typing and discriminating knowledge of English; to Diane Houle, supervisor of the Varitype Section of the American Mathematical Society, for her unrivaled skill and experience in the typing of mathematical translations; to Linda Rinaldi and Ingeborg Menz, secretaries, respectively, of the Translations Department of the Society and the firm Vandenhoeck and Ruprecht, for keeping straight a long and complicated correspondence; to the staff of The MIT Press for their customary technical expertness; and to my wife, Katherine Gould, for help too varied and too substantial to be readily described.

S. H. Gould
Institute of Mathematics
Academia Sinica
Taipei, Taiwan
Republic of China

September 1973

Preface

The many branches of analysis—the infinitesimal calculus, the theories of measure and integration, differential and integral equations, the calculus of variations, and the theory of functions of a complex variable—form the most comprehensive part of mathematics. The present volume, however, is not an encyclopedia or textbook of these mathematical disciplines; it is addressed to readers who are making practical use of mathematics in their daily lives—in particular, young instructors in colleges—and who wish to orient themselves quickly in the foundations and recent developments of their science before they consult specialized works in a literature which is by now voluminous and often almost inaccessible.

Consequently we are concerned here with presenting the essential features of the subject in a reasonably ample form.

The first chapter deals with convergence. This subject is fundamental in all branches of analysis (in fact, analysis can be described as that branch of mathematics in which the concept of convergence is indispensable); at the same time it forms a bridge to topology. In recent times the concept of convergence has been made more profound through the introduction of Moore-Smith sequences, filters (introduced by Henri Cartan and Bourbaki), and uniform structures. Chapter 1 presents these concepts to the reader. Chapter 2 deals with the concept of a function. A real function is a mapping of a point set in the space R^n into the space R^m, and in particular the "classical" function is a mapping from R^1 into R^1 (in other words, a mapping of real numbers into real numbers). In this chapter we discuss the concepts of continuity and particularly of differentiability, which for a mapping of R^n into R^m amount, in coordinate-free terms, to linear approximation; and then by introducing coordinate axes we arrive at the concept of partial differentiation. The third chapter introduces the Riemann integral in R^n as a preparation for abstract measure theory. By specialization of abstract measures we arrive at the theory of the Lebesgue integral, which in modern times is more and more replacing the Riemann integral. In chapter 3*a* the results of the preceding chapter are utilized to present a modern theory of probability. Chapter 4 deals with the transformation of multiple integrals and thus leads directly to the theory of alternating differential forms. In chapter 5 we turn to complex analysis. Here we begin anew with elementary matters. The rules for operating with complex numbers are given, and complex differentiability and integrability are discussed. With the concept of holomorphy we pass to chapter 6, which presents the fundamental principles of the classical theory of functions of a complex variable. The identity theorems and the introduction of Riemann surfaces as the domains of definition of "many-valued" functions form the main part of this chapter. The starting point of the following chapter is the fact that in projective geometry the infinitely distant points are not introduced in the same way as in theory of functions of a complex variable. Are we then to conclude that these points at infinity have a type of reality somehow different from that of finite points? This chapter also closes with a look at present-day research (the theory of modifications).

Now follow those parts of analysis that are most important for the applications. Chapters 8 and 9, on real analysis, deal with ordinary and partial differential equations. Chapter 10, based on function-theoretic methods, deals with difference equations and definite integrals, in particular the Γ-function. Chapter 11 is an introduction to the far-reaching theory of Hilbert and Banach spaces, with applications to the theory of integral equations. Chapter 12, with the title "Real Functions," describes various connections between the properties of point sets and functions. Chapter 13 is an introduction to the productive and invariably surprising applications of analysis to questions in the theory of numbers. Chapter 14 has the title "The Changing Structure of Modern Mathematics." This chapter represents our summarizing remarks on the subject of pure mathematics. The rapidly increasing number of specialized results threatens to divide mathematics into many separate special disciplines. Opposing this tendency are the modern efforts to create a unified treatment of the whole of mathematics, depending chiefly on a theory of structure and based, of course, on elements that are common to many different branches of mathematics. These efforts have been incorporated, in a form that has by now become historic, in the work of N. Bourbaki, whose great and often surprising success will not be denied by anyone well acquainted with the situation. The last chapter of the present volume will discuss this modern tendency. We recommend that our readers peruse it carefully since it gives articulate expression to the purposes we have had in mind throughout.

It was difficult to coordinate the various parts of the present volume. From this point of view we are particularly grateful to Professor H. Tietz and Dr. H. Rau, as well as to many other colleagues, among them H. Suchlandt, of Munich, who has greatly assisted Dr. Rau in the preparation of the manuscript. We are also grateful to Dr. H. Liermann and Dr. Kuhlmann, who have worked, respectively, on chapter 13 (analysis and number theory) and chapter 12 (real functions). Special gratitude is due to the association of sponsors, in particular to Dr. Fritz Gummert, for their munificent financial support, without which the volume could never have been written. We also express our gratitude to the publishers, who have carried out our requests with great patience and have given the volume a magnificent format.

A staff of more than one hundred authors have been working on this set of volumes for more than four years. During that period we have been bereaved by the death of Messrs. Dreetz, Lohmeyer, Reimann, Süss, and Zühlke. To all these colleagues we remain enduringly grateful.

F. Bachmann H. Behnke K. Fladt

Spring 1962

Münster (Westf.)

CHAPTER I

Convergence

1. Introduction

Analysis is essentially the mathematics of approximation. Mathematical entities that are difficult to deal with directly are approximated by others that are more convenient, and an attempt is made to apply to the original entities the results obtained for the approximate ones. Our first task will be to illustrate this statement for sequences, which involve the most elementary but also the most basic concepts of analysis. Many of the theorems about sequences can be generalized in far-reaching ways and can be set in a much broader framework.

It is not our intention here to repeat well-known details. Rather we are interested in bringing to the foreground certain dominating aspects of the subject. Thus from the very beginning the relevant concepts will be presented in a somewhat more general (and therefore more abstract) form than is customary in textbooks. From the pedagogical point of view it would be desirable to make as much use as possible of concrete examples, but in a systematic treatment of the subject we must not lose sight of the advantages offered by abstraction; it enables us to break up the longer proofs into shorter steps and thereby gain a better understanding of their logical structure.

Let us give a brief summary of the organization of the present chapter.

For the description of limit processes in analysis it is customary to introduce sequences. Our purpose is to analyze the resulting concepts and theorems from several points of view with the idea in mind that in mathematics such a procedure always leads to valuable insights into the structure of the subject and to productive generalizations.

The concept of *convergence* does not in itself require the whole structure of the real or complex numbers but can already be formulated in arbitrary *topological spaces* (§2.1). But even the condition that a convergent sequence shall have only one limit already forces us to introduce our first restriction: this condition holds only in so-called *Hausdorff spaces* (§2.2).

We are accustomed to the idea of *calculating* with the limits of numerical sequences, but in a general topological space such calculations may not be possible, since none of the relevant operations are necessarily defined in such a space. But if they are defined, then calculation with limit values has a meaning, and the question arises: under what circumstances can we prove the well-known theorems about interchange of a limit process with addition or multiplication? In other words, when are addition and multiplication consistent with the topology of the space? This question leads to the concept of a *topological group* and to related concepts (§§2.3 to 2.6).

The order structure of the real numbers makes them into a topological space with striking properties; consequently the sequences of real numbers (and also, to some extent, of complex numbers, which can be reduced to them) have properties that require the special structure of this topological space for their very formulation (§3).

The *fundamental Cauchy theorem* for numerical sequences, which is proved in §3, does not make use of the full order structure of the real numbers but only of its weaker metric structure; but even the latter is definitely stronger (that is, more special) than a merely topological structure, in which the *Cauchy condition* cannot even be formulated. In a metric space, however, this condition has meaning and the validity of the fundamental Cauchy theorem is an additional property, called *completeness*. An important criterion for completeness is given at the close of this section (§4).

In the following section (§5) we focus our attention, not on the set of values of a sequence, but on its index set. For a sequence in the classical sense this index set is the set of natural numbers, which proves for many purposes to be far too "thin"; for example, in the symbol $\lim_{x \to a} f(x)$ the "index" x runs through all the numbers in a neighborhood of a. Of course this particular example of a "continuous" limit process can be reduced to the convergence of sequences in the classical sense, but that fact does not make it any more natural to insist that index sets must be countable. The natural generalization is provided by the *Moore-Smith sequences*, or by the equivalent, and often more convenient, concept of a *filter*.

From the theoretical point of view the most important application of filters is the following: just as the topological properties of a space can be described without reference to any metric but merely in terms of subsets of the space, we may also introduce a general structure in which the *Cauchy condition* can be formulated without reference to a metric. Such a "*uniform structure*" induces a natural topology and allows us, in particular, to give the most general formulation (§6) of the extremely important concept of uniform convergence.

2. Sequences

2.1. *The Limit Concept*

The definition of a *sequence* involves three entities: the index set, namely (in the classical case) the set N of non-negative whole numbers 0, 1, 2, . . . ; the set of values M, which may be any non-empty set, and finally a mapping of N into M.[1] Thus to each number n from N there corresponds a well-defined element of M, denoted by a_n. A sequence is not merely a subset; it is an indexed (enumerated) subset. If the elements of M are numbers, we speak of a numerical sequence, if they are functions, we speak of a sequence of functions, and so forth. If no further properties are assigned to the set M, the sequences are of no great interest, but the situation is quite different if M is a *topological space*, a concept that arises in the following way. Certain subsets of M are distinguished, in the manner described below, as *neighborhoods*, whereupon M is said to have a *topological structure*; or equivalently, the set M, along with the topological structure thus defined for it, is said to be a topological space.[2] The elements of such a set are usually called *points*, so that these "points" may, for example, be functions. In the set M we distinguish certain subsets which, as already mentioned, are called neighborhoods. They are required to have the following properties.

H 1) *Every point $a \in M$ has at least one neighborhood $U(a)$ and is contained in each of its neighborhoods: $a \in U(a)$.*

H 2) *For any two neighborhoods $U(a)$ and $V(a)$ there exists a neighborhood $W(a)$ which is contained in the intersection of $U(a)$ and $V(a)$: $W(a) \subset U(a) \cap V(a)$.*

H 3) *For every point b in the neighborhood $U(a)$ there exists at least one neighborhood $U(b)$ which is contained in $U(a)$.*

Thus a set, together with a system of subsets, which satisfies H 1 *through* H 3 *is called a topological space.*

It is now possible to introduce one of the most important concepts of analysis, namely that of a *limit*. It may happen that the elements of the sequence (which, as we have seen, are points of the topological space M) are ultimately contained in an arbitrarily chosen neighborhood of a point a of the space. We then say that *the sequence converges to the point a* or that a is a *limit* of the sequence. The situation can be described more precisely as follows: we say that *the sequence $a_n (a_n \in M, n \in N)$ converges to the point $a \in M$ if for every neighborhood $U(a)$ of a there exists a number n_0 such that $a_n \in U(a)$ for all $n \geqq n_0$.*

Thus convergence is a so-called infinitary property of the sequence. In other words, any valid statement about the sequence remains valid if only finitely many elements are changed or omitted. We shall often regard a given property as valid for all the elements of a sequence, even though there may be finitely many exceptions, since it is usually irksome to keep mentioning the exceptional cases.

[1] For the concept of "mapping" see IA, §8.4.

[2] A set can be given a topological structure in many different ways, so that one and the same set can be interpreted as a topological space in many ways.

2.2 *Uniqueness of the Limit*

In a completely general topological space it is possible for a sequence to have several limits, an undesirable feature of the space which can be eliminated only if we specialize further to a so-called *Hausdorff space*, that is, if we also require the following *Hausdorff separation axiom*.

H 4) *For any two distinct points a and b there exist mutually exclusive neighborhoods $U(a)$, $U(b)$, with $U(a) \cap U(b) = \varnothing$.*[3]

A topological space with the properties H 1) *through* H 4) *is called a Hausdorff space.* As mentioned above, the following theorem holds in such a space:

Theorem 1. *In a Hausdorff space any sequence has at most one limit.*

The proof is very easy. If a and a' are limits of a sequence a_n, $n \in N$, and $U(a)$ and $U(a')$ are arbitrary neighborhoods of a and a', there exist numbers n_1, n_2 such that $a_n \in U(a)$ for all $n \geqq n_1$, and $a_n \in U(a')$ for all $n \geqq n_2$. If n_0 is the greater of the two numbers n_1, n_2, then a_n is in the intersection $U(a) \cap U(a')$, for all $n \geqq n_0$. Thus no two neighborhoods have an empty intersection, which by H 4) is possible only if a and a' coincide.

In a Hausdorff space[4] the uniquely determined limit of a sequence is denoted by

(1) $$a = \lim_{n \to \infty} a_n.$$

An alternative notation is

(2) $$a_n \to a \qquad \text{for} \qquad n \to \infty.$$

As a final remark we note that the elements of a sequence may be considered as approximations to the limit, so that in the sense of the topology of M the limit can be approximated with arbitrary exactness.

2.3. *Topological Groups*

A great deal more can be said about sequences if the space M, in addition to being a Hausdorff space, has an algebraic structure that is "consistent" with its topology. A good example is given by the real numbers, which form not only a Hausdorff space but also a field, so that we can speak of a *topological field.* In the same way we can also speak of *topological groups*, *topological rings*, and so forth.

Before giving an exact definition of a topological group, let us introduce the following useful notation:

If A and B are subsets of the group[5] M, we denote by AB the set of

[3] The symbol $\varnothing$ denotes the empty set.

[4] For simplicity we shall assume below (often tacitly) that the space is Hausdorff.

[5] For the axioms defining a group see IB2, §1.1.

"points" ab with $a \in A$, $b \in B$. Similarly, $1/A = A^{-1}$ denotes the set of "points" $1/a = a^{-1}$ inverse to a, with $a \in A$. Compare the rules for calculation with complexes in IB2, §3.1.

When we say that the group property is "*consistent*" with the topology of M we mean that the group operations are continuous mappings of the space into itself, in the following precise sense:

G 1) *If $c = ab$ and $U(c)$ is an arbitrary neighborhood of c, there exist neighborhoods $U(a)$, $U(b)$ of a and b such that $U(a)U(b) \subset U(c)$.*

G 2) *For an arbitrary neighborhood $U(1/a)$* of $1/a$ *there exists a neighborhood $U(a)$ of a such that* $1/U(a) \subset U(1/a)$.

In more intuitive language the meaning of G 1) and G 2) is as follows: the product $z = xy$ is arbitrarily close to c if x is sufficiently close to a and y is sufficiently close to b; and the inverse $1/x$ is arbitrarily close to $1/a$ if x is sufficiently close to a.

Definition 1. *A topological space which is also a group and which satisfies the axioms* G 1) *and* G 2) *is called a topological group.*

Note. These remarks are independent of the notation used for the operation of the group. For example, in Abelian groups it is customary to regard the operation as addition and to denote it by $+$, and consequently (cf. IB2, §1) to write $-a$ for a^{-1}.

2.4. *Calculation with Limits*

Theorem 2. *Let M be a topological group, the group operation being written multiplicatively. If the sequences a_n and b_n (a_n, $b_n \in M$, $n \in N$) are convergent, then the sequence $c_n = a_n b_n$ is also convergent and has the limit $c = ab$; or*: $\lim c_n = \lim a_n \lim b_n$.

Proof. Let $U(c)$ be an arbitrary neighborhood of c. In view of the continuity of the product operation, there exist neighborhoods $U(a)$, $U(b)$ of a and b such that $U(a)U(b) \subset U(c)$; that is, $xy \in U(c)$, if $x \in U(a)$, $y \in U(b)$. By hypothesis, there exist numbers n_1, n_2 such that $a_n \in U(a)$ for $n > n_1$ and $b_n \in U(b)$ for $n > n_2$. For all $n > n_0 = \max(n_1, n_2)$ we then have $c_n = a_n b_n \in U(c)$, as was to be proved.

Remark. If the group operation is written additively, the theorem becomes:

Theorem 3. $\lim (a_n + b_n) = \lim a_n + \lim b_n$.

Theorem 4. *Let M be a topological group. If the sequence a_n converges to a (a_n, $a \in M$; $n \in N$), then the sequence $1/a_n$ is also convergent and has the limit $1/a_n$ that is,* $\lim (1/a_n) = 1/\lim a_n$.

Proof. For an arbitrary neighborhood $U(1/a)$ of $1/a$ there exists a neighborhood $U(a)$ of a such that $1/U(a) \subset U(1/a)$; that is, $1/x \in U(1/a)$, if $x \in U(a)$. Moreover, there exists a number n_0 such that $a_n \in U(a)$ for $n > n_0$ and also such that $1/a_n \in U(1/a)$.

Since $\frac{a_n}{b_n}$ can be written in the form $a_n \frac{1}{b_n}$, theorems 2 and 4 imply the following theorem for a topological group M with multiplicatively written group operation:

Theorem 5. $\lim \frac{a_n}{b_n} = \frac{\lim a_n}{\lim b_n}$ *if the right side exists.*

2.5. *Topological Fields*

If we proceed still further with this specialization of the space M and require that it be a *topological field*, we mean thereby that the following conditions are satisfied:

1. *M is a Hausdorff space;*
2. *M is a field* (cf. IB5, §§1.2 *and* 1.10); *in other words, M is an additive group, $M^* = M - 0$ is a multiplicative group, and the two group operations are connected by the distributive laws;*
3. *The consistency conditions* G 1) *and* G 2) *are satisfied, both for addition and for multiplication, where in* G 2) *we must, of course, assume $a \neq 0$.*

In a topological field we have the formulas

$$\lim (a_n + b_n) = \lim a_n + \lim b_n,$$
$$\lim a_n b_n = \lim a_n \lim b_n,$$
$$\lim \frac{a_n}{b_n} = \frac{\lim a_n}{\lim b_n}, \qquad b_n \neq 0, \quad b = \lim b_n \neq 0, \tag{3}$$

if the sequences a_n and b_n converge; and in each case the limit is uniquely determined.

2.6. *Application to the Field C of Complex Numbers*

It is natural to ask why the proofs in the textbooks are so much more complicated than the ones given here. The explanation is that in the proofs of the theorems about limits it is customary to include, in implicit form, a proof of the continuity of the two operations, addition and multiplication. Let us illustrate in more detail for the case of the field C of complex numbers. It is clear that the limit theorems from §2.5 will have been proved if we show that this field is a topological field. Regarded as a metric space it is certainly also a Hausdorff space, and thus we need only prove that the addition and multiplication are continuous.

Continuity of addition. Let $c = a + b$ and let ϵ be a given positive number. The ϵ-neighborhood of c is then the set $U(c)$ of numbers z with $|z - c| < \epsilon$. For $U(a)$ we take the set of numbers x with $|x - a| < \frac{\epsilon}{2}$ and for $U(b)$ the set of numbers y with $|y - b| < \frac{\epsilon}{2}$. From $|(x + y) - (a + b)| = |(x - a) +$

$(y - b)| \leqq |x - a| + |y - b| < \frac{\epsilon}{2} + \frac{\epsilon}{2} = \epsilon$ it follows at once that $z = x + y$ is contained in the ϵ-neighborhood of $c = a + b$; that is,

$$U(a) + U(b) \in U(a + b).$$

Continuity of multiplication. After choice of the arbitrary positive number ϵ we choose two other positive numbers ϵ_1 and ϵ_2 in the manner described below. We note that $|xy - ab| = |b(x - a) + a(y - b) + (x - a)(y - b)| \leqq |b|\,|x - a| + |a|\,|y - b| + |x - a|\,|y - b|$. Let $U(a)$ be an ϵ_1-neighborhood of a and $U(b)$ an ϵ_2-neighborhood of b. For $x \in U(a)$ and $y \in U(b)$ we have $|b|\,|x - a| \leqq |b|\epsilon_1$, $|a|\,|y - b| \leqq |a|\epsilon_2$ (with equality if either a or b is equal to 0) and $|x - a|\,|y - b| < \epsilon_1\epsilon_2$. If we take $\epsilon_1 = \frac{1}{3}\epsilon/(|b| + 1)$, $\epsilon_2 = \frac{1}{3}\epsilon/(|a| + 1)$ with $\epsilon_2 < 1$, then obviously $|xy - ab| < \frac{1}{3}\epsilon + \frac{1}{3}\epsilon + \frac{1}{3}\epsilon$, which completes the proof.

Continuity of the operation of forming the inverse. Assume $a \neq 0$ and let η be a positive number to be determined below. If we first assume $\eta < \frac{1}{2}|a|$, then for every number x in the η-neighborhood $U(a)$ of a we have the inequality $|x - a| < \frac{1}{2}|a|$, and consequently $|x| > \frac{1}{2}|a|$. Thus the numbers in $U(a)$ are different from 0. Furthermore,

$$\left|\frac{1}{x} - \frac{1}{a}\right| = \frac{|x - a|}{|x|\,|a|} < \frac{\eta}{\frac{1}{2}|a|^2}.$$

If we now take $\eta < \frac{1}{2}|a|^2\epsilon$, the proof is complete.

The above proof is a good illustration of the remark in §1 to the effect that a proof can often be divided up into shorter steps.

3. Monotone Sequences and Limits of Indeterminacy

3.1. *Monotone Sequences*

We shall make use of the fact (see IB1, §4.4) that in the field of real numbers every set A that is bounded above has a least upper bound

(4) $$\mu = \sup_{x \in A} x,$$

and every set B that is bounded below has a greatest lower bound

(5) $$\lambda = \inf_{x \in B} x.$$

If the sets are unbounded above or below, then for μ, λ it is convenient to take the numbers $+\infty$ and $-\infty$, which are adjoined to the field as improper elements.

For monotone sequences it is easy to establish convergence a priori, without knowing the value of the limit, where, as usual, we say that a sequence a_n, $n \in N$ is *monotone increasing* if $a_n \leqq a_{n+1}$ for all n, and *monotone decreasing* if $a_n \geqq a_{n+1}$ for all n (where it is obviously convenient to allow a finite number of exceptions to these inequalities).

The fundamental theorem on monotone sequences runs as follows:

Theorem 6. *If a monotone increasing sequence is bounded above, then the sequence is convergent, and similarly for a monotone decreasing sequence bounded below.*

It will be sufficient to prove the first part of the theorem. Since the set of numbers a_n, $n \in N$ is bounded above, it has a least upper bound

$$a = \sup_{n \in N} a_n.$$

Thus for all n we have $a_n < a + \epsilon$, with arbitrary positive ϵ, since it is true that $a_n \leqq a$. Furthermore, there exists a number n_0 with $a_{n_0} > a - \epsilon$ and thus, since the sequence is monotone, $a_n > a - \epsilon$ for all $n \geqq n_0$.

3.2. *Limits of Indeterminacy*

All sequences of real numbers, including those that do not converge, display certain regularities closely connected with monotone sequences.

Let μ_k be the least upper bound of the set of terms a_n, $n \geqq k$ in a given sequence a_n, $n \in N$; that is,

$$\mu_k = \sup_{n \geqq k} a_n, \qquad k = 0, 1, 2, \ldots. \tag{6}$$

If the sequence is not bounded above, we set $\mu_k = +\infty$. Otherwise the numbers μ_k are the terms of a monotone decreasing sequence, the *majorizing sequence* of the given sequence. The number

$$\mu = \inf_{k \in N} \mu_k \tag{7}$$

is either $-\infty$ or the limit of the sequence μ_k. This number is called the *upper limit* (limes superior) of the sequence a_n. We write

$$\mu = \limsup_{n \to \infty} a_n. \tag{8}$$

If one of the $\mu_k = +\infty$, we write $\mu = +\infty$.

Correspondingly we can form

$$\lambda_k = \inf_{n \geqq k} a_n, \qquad k = 0, 1, 2, \ldots \tag{9}$$

and

$$\lambda = \sup_{k \in N} \lambda_k. \tag{10}$$

The sequence λ_k, $k \in N$ is called the *minorizing sequence* of the given sequence if the λ_k are finite. The number (10) is equal to $-\infty$ if a $\lambda_k = -\infty$, is equal to $+\infty$ if the sequence λ_k is unbounded above, and is otherwise the limit of the monotone sequence λ_k. This number is called the *lower limit* (limes inferior) of the sequence a_n. We write

$$\lambda = \liminf_{n \to \infty} a_n. \tag{11}$$

It is easy to show that

$$\liminf_{n\to\infty} a_n = -\limsup_{n\to\infty} (-a_n). \tag{12}$$

Since $$\lambda_0 \leqq \lambda_k \leqq \mu_k \leqq \mu_0, \qquad k \in N,$$

we have $$\lambda_0 \leqq \lambda \leqq \mu \leqq \mu_0,$$

and thus

$$\inf_{n\in N} a_n \leqq \liminf_{n\to\infty} a_n \leqq \limsup_{n\to\infty} a_n \leqq \sup_{n\in N} a_n. \tag{13}$$

The upper and lower limits of a sequence are called its *limits of indeterminacy*. They are not necessarily identical with the least upper and greatest lower bounds.

The following theorem is useful in many applications:

Theorem 7. *If a number β is such that ultimately*

$$a_n < \beta + \epsilon,$$

for every fixed choice of the positive number ϵ, then

$$\limsup_{n\to\infty} a_n \leqq \beta.$$

If a number α is such that ultimately

$$a_n > \alpha - \epsilon,$$

then $$\liminf_{n\to\infty} a_n \geqq \alpha.$$

It is sufficient to prove the first part of the theorem. If $\beta = +\infty$, the assertion is trivial. But if β is finite, then by hypothesis the sequence is bounded above and thus the majorizing sequence exists. We can find a number n_0 with $a_n < \beta + \epsilon$ for all $n \geqq n_0$. Thus $\beta + \epsilon$ is an upper bound for the set of numbers a_n, $n \geqq n_0$, and consequently $\mu_k \leqq \beta + \epsilon$, for $k \geqq n_0$, and a fortiori $\mu \leqq \beta + \epsilon$. But since ϵ is arbitrarily small, we have $\mu \leqq \beta$. The proof requires only slight changes for the case $\beta = -\infty$, since the hypothesis then states that ultimately the numbers are less than any preassigned bound and consequently are unbounded below.

It is easy to show that λ is the smallest and μ is the largest number with the property stated in the hypothesis of the preceding theorem. Equivalently we have

Theorem 8. *Let ϵ be an arbitrary positive number. If λ and μ are the limits of indeterminacy of the sequence a_n, $n \in N$, then ultimately*

$$a_n > \lambda - \epsilon, \qquad a_n < \mu + \epsilon$$

and infinitely often

$$a_n < \lambda + \epsilon, \qquad a_n > \mu - \epsilon.$$

Again it is sufficient to consider the case of the upper limit. Since μ is the greatest lower bound of the numbers μ_k, there exists a number k with $\mu_k < \mu + \epsilon$, so that certainly $a_n < \mu + \epsilon$ for $n \geqq k$. If it were not true that $a_n > \mu - \epsilon$ infinitely often, then ultimately we would have $a_n \leqq \mu - \epsilon$, or in view of the preceding theorem $\mu \leqq \mu - \epsilon$, which is impossible.

The connection between limits of indeterminacy and convergence can be stated as follows.

Theorem 9. *A sequence is convergent if and only if its limits of indeterminacy are finite and are equal to each other. Then the sequence converges to this common value.*

The condition stated in the theorem is necessary, since if a_n converges to a, then ultimately $a_n > a + \epsilon$, so that $\mu \leqq a$ and $a_n > a - \epsilon$, and thus $\lambda \geqq a$. Since $\lambda \leqq \mu$, we thus have $\lambda = \mu = a$. Conversely, if $\lambda = \mu = a$, then ultimately $a_n > \lambda - \epsilon = a - \epsilon$ and ultimately $a_n < \mu + \epsilon = a + \epsilon$, so that a is the limit of the sequence.

3.3. *The Fundamental Theorem of Cauchy*

This theorem states a criterion for the convergence of a sequence of complex[6] numbers.

Let c_n, $n \in N$, be a sequence of complex numbers. We first formulate a necessary condition for convergence. Let c denote the limit of the sequence. Then if ϵ is a preassigned positive number, there exists a number n_0 such that $|c_n - c| < \frac{1}{2}\epsilon$ for $n \geqq n_0$. Thus

$$|c_m - c_n| \leqq |c_m - c| + |c_n - c| < \tfrac{1}{2}\epsilon + \tfrac{1}{2}\epsilon = \epsilon$$

for all $m, n \geqq n_0$.

But this necessary condition, to the effect that ultimately the terms must remain arbitrarily close to one another, is also sufficient. The two assertions constitute the *fundamental theorem of Cauchy*:

Theorem 10. *For the convergence of a sequence c_n, $n \in N$, of complex numbers it is necessary and sufficient that for an arbitrarily preassigned positive ϵ there exists a number n_0 such that*

$$|c_m - c_n| < \epsilon \tag{14}$$

for all $m, n \geqq n_0$.

It remains to prove the sufficiency of the condition. If we write $c_n = a_n + ib_n$, with real a_n and b_n, then obviously c_n, $n \in N$, is convergent if both the sequences a_n and b_n, $n \in N$, are convergent. From $|c_m - c_n| \leqq |a_m - a_n| + |b_m - b_n|$ and $|a_m - a_n| \leqq |c_m - c_n|$, $|b_m - b_n| \leqq |c_m - c_n|$ we see that the Cauchy condition for c_n implies the same condition for a_n and b_n, and conversely. Thus it is sufficient to prove the theorem for real sequences.

[6] And thus, in particular, of real numbers.

Let a_n, $n \in N$, be a real sequence satisfying the Cauchy condition. If ϵ is an arbitrary positive number, there exists a number n_0 such that for some $m \geqq n_0$ and every $n \geqq n_0$ we have $|a_m - a_n| < \epsilon$ or $a_m - \epsilon < a_n < a_m + \epsilon$. If m is regarded as fixed, these inequalities hold for all numbers a_n with finitely many exceptions. But this statement means that the sequence is bounded and that the two limits of indeterminacy are finite. By §3.2, theorem 8, we have $a_m - \epsilon \leqq \liminf_{n\to\infty} a_n$ and $\limsup_{n\to\infty} a_n \leqq a_m + \epsilon$, so that

$$\limsup_{n\to\infty} a_n - \liminf_{n\to\infty} a_n \leqq 2\epsilon.$$

But since the positive number ϵ is arbitrary, the two limits of indeterminacy must be equal, which completes the proof of convergence.

4. Metric Spaces

4.1. *Axiomatic Definition*

In III 2, §1.2 we shall be making use of special topological spaces, namely metric spaces. They are characterized by the fact that for every two points there is defined a real non-negative number $d(a, b)$, called the "*distance*" between a and b, with the following properties:

M 1) $d(a, b) = 0$ *if and only if* $a = b$,
M 2) $d(a, b) = d(b, a)$,
M 3) *For every choice of elements* a, b, c *we have*

$$d(a, b) \leqq d(a, c) + d(c, b) \qquad \textit{(triangle inequality)}.$$

The connection here with the classical arguments of analysis is obvious.

As an example of a metric space let us take the set of rational numbers, or the set of real numbers, with the distance function $d(a, b) = |a - b|$. The axioms M 1) to M 3) can be verified at once. Thus theorem 1 in §2.2 is valid if we are dealing with sequences of rational or real numbers.

4.2. *Complete Metric Spaces*

If a sequence a_n converges to a point a in a metric space, then by the triangle inequality (cf. §4.1, M 3)) and the definition of convergence we have, as in §3.3, the following *necessary condition for convergence*:

$$(15)\quad d(a_n, a_m) \leqq d(a_n, a) + d(a, a_m) \leqq \epsilon \qquad \text{for} \qquad n, m > n_0, \quad \epsilon > 0.$$

If the space is such that (15) is also a *sufficient* condition for convergence of sequences, it is said to be *complete*. Thus a sufficient condition for the completeness of a space is that in it every bounded infinite subset has at least one *limit point*, i.e., a point such that every neighborhood of it contains infinitely many points of the set.

We can now prove the following theorem:

Theorem 11. *If in a metric space every bounded infinite subset has at least one limit point, then the space is complete.*

The proof is as follows. For a sequence a_n let $d(a_m, a_n) < \frac{\epsilon}{2}$ for arbitrary $\epsilon > 0$ and $n, m > n_0(\epsilon)$. Then it follows that the sequence is bounded and thus has a limit point a. Consequently, $d(a, a_m) < \frac{\epsilon}{2}$ for infinitely many m; and by hypothesis $d(a_n, a_m) < \frac{\epsilon}{2}$ for $n, m > n_0$. Consequently $d(a_n, a) \leqq d(a_n, a_m) + d(a_m, a) < \epsilon$ for all $n > n_0$, which means that the sequence a_n converges to a.[7]

This criterion for convergence in an arbitrary metric space is obviously a *generalization of the Cauchy criterion for convergence* in classical analysis.

5. Filters

5.1. *Moore-Smith Sequences*

A sequence is a mapping of the set N of natural numbers into a set of values M. This mapping determines a structure in M, which can be described in terms of the set of segments a_n, $n \geqq n_0$, of M, where n_0 is an arbitrary non-negative number.

But in the above discussion the order properties of the set N have not been used to their full extent. Our statements remain correct if for the index set N we take a set in which there is defined a *relation* ρ with the following properties:

gM 1) *If $p \rho q$ and $q \rho r$, then $p \rho r$; namely, the relation is transitive.*

gM 2) *For any two elements p and q in N there exists in N an element r with $p \rho r$ and $q \rho r$.*

gM 3) We have $p \rho p$ for all p.

Such a set is called a *directed set*. For the non-negative integers the relation "ρ" usually means "$\leqq$," but for the same set of numbers we can easily define other relations with the above properties, e.g., the relation "divisible," in which $p \rho q$ means that q is divisible by p.

A set M, a directed set N and a mapping of N into M define a *Moore-Smith sequence*. If for M we take a topological space R, the sequence is said to be *convergent to a point a if for every neighborhood $U(a)$ of a there exists an element n_0 in the index set such that $a_n \in U(a)$ for all n with $n_0 \rho n$.*

The following examples show that we are dealing here with an essential extension of the concept of a sequence.

[7] G. Aumann, Reelle Funktionen. Springer, Berlin-Göttingen-Heidelberg 1954, p. 126.

Let the index set be a (non-empty) set X of real numbers with a limit point a. The set becomes a directed set if we agree that $p \rho q$ means $|p - a| \geqq |q - a|$. The above conditions are then easily verified. A mapping f of the set X into another set Y of real numbers, or in other words a real function, determines a Moore-Smith sequence. The terms of the sequence are the values $f(x)$ of the function. In the sense of Moore-Smith convergence this sequence has a limit b if for every assigned number ϵ there exists a number $x_0 \in X$ such that $|f(x) - b| < \epsilon$ for all x with $x_0 \rho x$, which means that $|x_0 - a| \geqq |x - a|$, or in other words $|x - a| \leqq \delta$, with $\delta = |x_0 - a|$. Since the definition would otherwise be trivial, we assume that a is not an element of X. Then $\delta > 0$, and we have the usual definition of the limit of a function for $x \to a$. Of course it is also possible that $f(a)$ exists. If the value of the function at this point coincides with the limit, then $f(x)$ is said to be *continuous* for $x = a$.

Another striking example[8] is provided by the *Riemann integral.* Here we let the function f be defined in the interval $a \leqq x \leqq b$ and for a partition $\mathfrak{z}$ of the interval defined by the intermediate points

$$a = x_0 \leqq x_1^* \leqq x_1 \leqq x_2^* \leqq x_2 \leqq \cdots \leqq x_{n-1} \leqq x_n^* \leqq x_n = b$$

we consider the Riemann sum

$$S(f, \mathfrak{z}) = \sum_{i=1}^{n} f(x_i^*)(x_i - x_{i-1}).$$

Then the set $\mathfrak{Z}$ of all partitions $\mathfrak{z}$ forms a directed set if by the norm of the partition $\mathfrak{z}$ we mean

$$d_{\mathfrak{z}} = \max_{1 \leqq i \leqq n} |x_i - x_{i-1}|$$

and define the relation $\mathfrak{z} \rho \mathfrak{z}'$ by $d_{\mathfrak{z}} \geqq d_{\mathfrak{z}'}$. Consequently $S(f, \mathfrak{z})$ is a Moore-Smith sequence with the index set $\mathfrak{Z}$ and the image set R of real numbers. If this sequence converges, the function f is said to be integrable in the sense of Riemann.

5.2. *Filters*

The concept of a Moore-Smith sequence can be translated into purely set-theoretic language, since the ordering relation ρ can be replaced by relations between sets.

Let there be given a sequence a_n, $n \in N$, where N is a directed set. In the image set M we consider a nonempty system $\mathfrak{F}$ of subsets with the following property: a subset U of M is an element of the system $\mathfrak{F}$ if and only if the

[8] Cf. G. Pickert, Folgen und Filter in der Infinitesimalrechnung, Der mathematische und naturwissenschaftliche Unterricht **13**, pp. 150–153 (1960).

elements of the sequence "ultimately" belong to U, e.g., if in the index set there exists an element n_0 such that $a_n \in U$ for all n with $n_0 \rho n$.

The system $\mathfrak{F}$ obviously has the following properties:

F 1) *The empty set is not an element of* $\mathfrak{F}$.
F 2) *Every set that includes an element of* $\mathfrak{F}$ *is an element of* $\mathfrak{F}$.
F 3) *The intersection of two elements of* $\mathfrak{F}$ *is an element of* $\mathfrak{F}$.

Such a system is called a *filter*. Let us give some examples.

In the set of non-negative integers we consider the subsets whose complements have only finitely many elements. This filter is called a *Fréchet filter* and is denoted symbolically by $n \to \infty$. A similar filter arises in any set M in which a sequence a_n, $n \in N$, is defined if as subsets we admit those subsets of n that contain "almost" all terms of the sequence, i.e., all terms with the exception of at most finitely many.

In a topological space we may consider the system of subsets that include a neighborhood of a given point a. This filter is called the neighborhood filter of a and is denoted symbolically by $x \to a$.

We now wish to show conversely that for a given filter $\mathfrak{F}$ in a set M we can always construct a Moore-Smith sequence related to $\mathfrak{F}$ in the above way.

Thus our first task is to construct from $\mathfrak{F}$ a suitable index set N. We construct this set in the following manner, which is natural enough but may seem somewhat artificial at first sight.

Let the elements of N be the pairs (a, A), where A is an element of $\mathfrak{F}$ (and is thus a subset of M) and a is an element of M that is contained in A. In N we define an order as follows:

$$(a, A) \, \rho \, (b, B) \quad \text{means} \quad A \supset B,$$

so that N is now a directed set. Since the properties gM 1) and gM 3) are immediately clear, we need only verify gM 2): for $A \in \mathfrak{F}$, $B \in \mathfrak{F}$ we see by F 3) that $C = A \cap B \in \mathfrak{F}$, and by F 1) the set C contains an element c; thus if $n = (a, A)$ and $m = (b, B)$ are two elements of N, then $p = (c, C)$ is also an element of N and we have a $n \rho p$ and $m \rho p$. We now obtain a Moore-Smith sequence by means of the mapping

$$n = (a, A) \to a \text{ of } N \text{ into } M.$$

It remains to show that the filter consisting of those subsets U of N which "ultimately" contain all the elements of the sequence is identical with the given filter $\mathfrak{F}$. For such a U there exists a $n_0 \in N$ such that $a_n \in U$ follows from $n_0 \rho n$. If n_0 is the pair (a_0, A_0) and if $a \in A_0$, then in particular we have $n_0 \rho n$ for $n = (a, A_0)$; consequently $a \in U$ for every $a \in A_0$, so that U contains A_0. Since $A_0 \in \mathfrak{F}$ we see from F 2) that $U \in \mathfrak{F}$. Conversely, every element A of the filter is a subset of M "ultimately" containing all the elements of the sequence, and for $a_0 \in A$ and $n_0 = (a_0, A)$ we see that for $n = (b, B)$ the relation $n_0 \rho n$ means precisely that $B \subset A$, so that in fact $a_n = b \in B \subset A$ for $n_0 \rho n$.

Two Moore-Smith sequences corresponding to the same filter obviously behave in the same way with respect to convergence, so that it must be possible to express convergence or nonconvergence solely in terms of filters.

In fact, a Moore-Smith sequence converges to the limit a if and only if every neighborhood $U(a)$ of a "ultimately" contains all the elements of the sequence; in other words, if and only if every $U(a)$ belongs to the filter $\mathfrak{F}$; or expressed still differently, if and only if $\mathfrak{F}$ includes the neighborhood filter $\mathfrak{U}(a)$ of a. This statement remains meaningful if $\mathfrak{F}$ is an arbitrary filter, not necessarily arising from a sequence. Thus it is natural to regard convergence not as a property of sequences but of filters and to make the following definition.

A filter $\mathfrak{F}$ in a topological space M is said to be convergent to $a \in M$ if it is finer[9] *than the neighborhood filter* $\mathfrak{U}(a)$.[10]

Thus a sequence $f : N \to M$ is convergent if and only if the corresponding filter converges.

We have interpreted the behavior of a sequence in terms of filters in the image space M. But the structure of the index set N can also be defined by means of a filter: in N the subsets that "ultimately" contain all elements of N form a filter $\mathfrak{F}_0$ which is obviously a natural generalization of the Fréchet filter for the set of natural numbers.

The *image filter* $\mathfrak{F}$ in M generated by the mapping $f : N \to M$ of the "direction filter" $\mathfrak{F}_0$ in N can now be described as the set of all subsets of M that contain images of sets in $\mathfrak{F}_0$. These image sets themselves form a system of sets $\mathfrak{B}$ in M with the properties

FB 1) *The empty set does not belong to* $\mathfrak{B}$,
FB 2) *The intersection of any two sets in* $\mathfrak{B}$ *contains a set in* $\mathfrak{B}$.

Such a system of sets $\mathfrak{B}$ is called a *filter basis*. Thus the mapping $f : N \to M$ maps the sets of a filter $\mathfrak{F}_0$ in N onto a filter basis $\mathfrak{B}$ in M, and the image filter $\mathfrak{F}$ arises from $\mathfrak{B}$ by the adjunction of all the sets that include sets in $\mathfrak{B}$.[11]

[9] A filter $\mathfrak{F}_1$ is said to be *finer* than the filter $\mathfrak{F}_2$ if it contains more sets, i.e., if $\mathfrak{F}_1 \supset \mathfrak{F}_2$.

[10] If the space M is a Hausdorff space (cf. §2.2) the filter $\mathfrak{F}$ has at most one limit; for then $\mathfrak{U}(a) \subset \mathfrak{F}$, $\mathfrak{U}(b) \subset \mathfrak{F}$ with $b \neq a$ is impossible, since for the two disjoint neighborhoods $U(a)$ and $U(b)$ it would follow from F 3 that $\varnothing = U(a) \cap U(b) \in \mathfrak{F}$, in contradiction to F 1.

[11] A filter is itself a filter basis and as such generates itself.

Since a sequence is a mapping $f: N \to M$ in which a filter $\mathfrak{F}_0$ is distinguished in N, the concept of convergence in the sense of Moore-Smith is subsumed in the following definition:

The mapping $f: N \to M$ is said to be convergent on the filter $\mathfrak{F}_0$ in N if the image filter $\mathfrak{F}$ of $\mathfrak{F}_0$ under f is convergent.[12]

Finally, the concept of *limits of indeterminacy of sequences* can be taken over for *filters of real numbers*:

If $\mathfrak{F}$ is a filter in the set R of real numbers and if

$$\mu_A \quad \text{and} \quad \lambda_A$$

denote, respectively, the least upper and greatest lower bounds of a set $A \subset R$, we define

$$\mu = \limsup \mathfrak{F} = \inf_{U \in \mathfrak{F}} \mu_U, \qquad \lambda = \liminf \mathfrak{F} = \sup_{U \in \mathfrak{F}} \lambda_U.$$

It is obvious that a real filter is convergent if and only if $\mu = \lambda$.

6. Uniform Spaces

6.1. Our discussion up to now has served the purpose of generalizing the elementary concept of convergence in such a way that it can be defined, not only for metric spaces, but in purely topological terms.

However, the concept of convergence as defined in terms of a metric accomplishes somewhat more than the purely topological definition, as we may show by the example of sequences of real numbers: the *fundamental Cauchy theorem* (§3.3) allows us to express the convergence of a sequence in such a way that the limit itself is not mentioned.

The statement contained in formula (14) in §3.3, namely that "ultimately" two elements of the sequence in question are arbitrarily close to each other, is immediately expressible in terms of a metric but cannot be formulated in terms of convergence alone: for neighborhoods of *one* point it is meaningful to assert that one neighborhood is greater than another (namely, if one of them includes the other), but we cannot make any such comparison between neighborhoods of *distinct* points. A comparison of this sort, as it occurs in the formulation of the fundamental Cauchy theorem, can nevertheless be made without any reference to a metric, provided that the space M has a so-called uniform structure, which means simply that we have some way of measuring the "nearness to each other" of two points.

In order to distinguish the various concepts as clearly as possible, we at first pay no attention to the topology of M, regarding it merely as a set

[12] Thus the definition assumes that M is a topological space, which is not necessarily true for N.

without structure. The definition we now proceed to develop for the concept of a *uniform structure as the totality of nearness relations* in M will then define a topology in M in a natural way. Only those topologies that arise from a uniform structure, in a sense to be made more precise below, allow us to formulate the fundamental Cauchy theorem.

In order to construct the desired definition of nearness we allow ourselves to be guided by the special case of a metric space M. The statements of interest are those that refer to pairs of points, so that it is convenient to formulate them as statements concerning the product space $M \times M$ consisting of the ordered[13] pairs (a, b) of points a and b in M. If $d(a, b)$ is the distance function that makes M into a metric space, the pairs (a, b) with $d(a, b) < \epsilon$ for a preassigned $\epsilon > 0$ form a set V_ϵ in $M \times M$. The system of these sets V_ϵ has the following properties, which are only restatements of the properties of the distance function:

Since $d(a, a) = 0$ for every point $a \in M$, every set V_ϵ contains the set Δ of all pairs (a, a); this set is called the "diagonal."

The equality $d(a, b) = d(b, a)$ states that every set V_ϵ is symmetric: $V_\epsilon^{-1} = V_\epsilon$, where V^{-1} is the set arising from V by reflection in the diagonal:

$$(a, b) \in V^{-1} \text{ if and only if } (b, a) \in V.$$

Finally, the triangle inequality $d(a, c) \leqq d(a, b) + d(b, c)$ states that $(a, c) \in V_\epsilon$ if $(a, b) \in V_{\epsilon/2}$ and $(b, c) \in V_{\epsilon/2}$. This latter statement suggests that we should introduce the following operation (called *composition*) for the subsets of $M \times M$:

For $A \subset M \times M$ and $B \subset M \times M$ let $A \circ B$ be the subset of $M \times M$ consisting of those pairs (a, c) for which there exists a *$b \in M$ such that $(a, b) \in A$ and $(b, c) \in B$.*

The above assertion then has the following form:

$$V_{\epsilon/2} \circ V_{\epsilon/2} \subset V_\epsilon.$$

The system of sets V_ϵ obviously forms a filter basis, since in view of $V_\epsilon \supset \Delta$ no set V_ϵ is empty and also $V_{\epsilon 1} \cap V_{\epsilon 2} = V_{\min(\epsilon_1, \epsilon_2)}$. Thus if to every V_ϵ we adjoin all the sets that include it, we obtain a filter $\mathfrak{G}$ of subsets U from $M \times M$ with the following properties:

F* 1) *Every $U \in \mathfrak{G}$ contains the diagonal Δ.*
F* 2) *If $U \in \mathfrak{G}$, then also $U^{-1} \in \mathfrak{G}$.*
F* 3) *For every $U \in \mathfrak{G}$ there exists a $V \in \mathfrak{G}$ with $V \circ V \subset U$.*

In this way we have gained a vantage point from which we can make the following definition, without reference to any metric:

Let M be a set and let $\mathfrak{G}$ be a filter in $M \times M$ with the properties F* 1)

[13] I.e., the pairs (a, b) and (b, a) are distinct if $a \neq b$.

to F* 3); *then a uniform structure is thereby defined in M; a set U in* $\mathfrak{G}$ *is called an entourage.*

If a uniform structure has been defined on a set M, we can formulate the Cauchy condition (as was our original purpose) for sequences, or more generally for filters:

A filter $\mathfrak{F}$ *in M is called a Cauchy filter if it contains arbitrarily "small" sets, i.e., if for every entourage U there exists a set* $A \in \mathfrak{F}$ *with* $A \times A \subset U$.

Finally, in order to state the desired generalization of the fundamental Cauchy theorem we must define what we mean by saying that a Cauchy filter converges; in other words, we must convert the set M into a topological space. As was pointed out above, the given uniform structure will enable us to define the desired topology in a natural way:

For every point $a \in M$ and every entourage $U \in \mathfrak{G}$ let $U(a)$ be the set of those $x \in M$ for which $(a, x) \in U$.[14] These sets $U(a)$ will then form the system of neighborhood filters for the desired topology on M. Thus we make the following definition:

A set $A \subset M$ *is said to be open if for every* $a \in A$ *there exists a* $U(a)$ *with* $U(a) \subset A$; *the empty set is also said to be open.*

This definition satisfies the axioms under which the system of open sets on M defines a topology, namely:

The entire space M is open, since $M \times M \in \mathfrak{G}$ and $M = (M \times M)(a)$.

The union of arbitrarily many open sets is obviously open; and finally, the intersection of two open sets A and B is also open, since for $a \in A \cap B$ and $U_1(a) \subset A$, $U_2(a) \subset B$ we have

$$A \cap B \supset U_1(a) \cap U_2(a) = (U_1 \cap U_2)(a),$$

and with U_1 and U_2 their intersection $U_1 \cap U_2$ also belongs to $\mathfrak{G}$.

Thus from the uniform structure we have constructed on M the topology known as the *uniform topology*, under which the space M itself becomes a *uniform space.*

The question of deciding when a given topological space is uniform will not be discussed here. Let us simply state that every metric space, and thus in particular the space R of real numbers, is uniform; moreover, we shall later construct important examples of function spaces that are uniform and shall use them to illustrate the far-reaching importance of these concepts.

For the present, let us fix our attention again on the *fundamental Cauchy theorem*, which for certain uniform spaces can now be stated in the following way:

Every Cauchy filter converges.

[14] When the uniform structure arises from a metric $d(x, y)$ on M and $U = V_\epsilon$ is an ϵ-entourage, this definition of $U(a)$ obviously means that $U(a)$ is an ϵ-neighborhood of a.

This theorem establishes a particular property[15] of those uniform spaces in which it is valid, namely their *completeness*. Here again we shall not investigate the criteria for completeness in a uniform space but shall content ourselves with the most important example:

Theorem 12. *The space R of real numbers, regarded as a uniform space, is complete.*[16]

Proof. Let $\mathfrak{F}$ be a real Cauchy filter. For $\epsilon > 0$ let V_ϵ again be the ϵ-entourage in $R \times R$. Since $\mathfrak{F}$ is a Cauchy filter, there exists for every $\epsilon > 0$ a $U_\epsilon \subset \mathfrak{F}$ with $U_\epsilon \times U_\epsilon \subset V_\epsilon$, i.e., $|x - y| < \epsilon$ for $x, y \in U_\epsilon$. If y is kept fixed, we have $y - \epsilon < x < y + \epsilon$ for all $x \in U_\epsilon$. Consequently,

$$y - \epsilon \leqq \lambda_{U_\epsilon} \leqq \lambda \leqq \mu \leqq \mu_{U_\epsilon} \leqq y + \epsilon \quad \text{and thus} \quad 0 \leqq \mu - \lambda \leqq 2\epsilon.$$

Since this result holds for arbitrary $\epsilon > 0$, it follows that $\mu = \lambda$, so that the filter is convergent.

Thus we have laid the basis for all the general applications of the fundamental Cauchy theorem in analysis. As an example let us again consider a real function f defined on a set of real numbers with the limit point a. The images of neighborhoods of a and sets including them in R are the elements of a filter, which is a Cauchy filter if for preassigned $\epsilon > 0$ we can always find a δ-neighborhood $U_\delta(a)$ of a such that $f(U_\delta(a)) \times f(U_\delta(a)) \subset V_\epsilon$, where V_ϵ is the set of pairs of numbers $(f(x), f(y))$ with $|f(x) - f(y)| < \epsilon$. But this condition obviously means that $|f(x) - f(y)| < \epsilon$ for all x, y for which $|x - a| < \delta$, $|y - a| < \delta$. Thus under these conditions $\lim_{x \to a} f(x)$ exists.

Although the Cauchy condition is not sufficient for convergence in all uniform spaces, still it is always necessary:

A convergence filter $\mathfrak{F}$ in a uniform space M is a Cauchy filter.

Proof. Let G be the entourage filter in $M \times M$ and let a be the limit of $\mathfrak{F}$. Then $\mathfrak{F}$ is finer than the neighborhood filter $\mathfrak{U}(a)$, i.e., for every neighborhood $V(a)$ of a there exists a set $A \in \mathfrak{F}$ with $A \subset V(a)$. Let $U \in \mathfrak{G}$ be arbitrary and let $V \in \mathfrak{G}$ be such that $V \circ V \subset U$. Here we may take $V^{-1} = V$, since with V the inverse V^{-1} and consequently the intersection $V \cap V^{-1}$ also belong to $\mathfrak{G}$. Then if $A \in \mathfrak{F}$ and $A \subset V(a)$, it follows that $A \times A \subset U$, since for arbitrary elements $x \in A$, $y \in A$ we have $(a, x) \in V$ and $(a, y) \in V$, and thus $(x, a) \in V^{-1} = V$, so that $(x, y) \in V \circ V \subset U$.

[15] For example, in the space of rational numbers with the usual metric, the Cauchy sequence $\left(1 + \frac{1}{n}\right)^n$—and consequently the Cauchy filter generated by it—is not convergent.

[16] It would be easy to reduce this theorem to theorem 11 by showing that every metrically complete space is also complete when regarded as a uniform space.

6.2. *Uniform Convergence*

Let f_n, $n \in N$, be a sequence of real functions of a real variable x. Let the sequence be convergent for every value of x in a certain domain. Then the limit values are the values $g(x)$ of a well-defined function g. For given x and given positive ϵ we can find a number n_0 such that $|g(x) - f_n(x)| < \epsilon$ for $n \geqq n_0$. But these inequalities are not necessarily satisfied for a different value of x. Thus to every value of x there corresponds a suitable number n_0. If all these numbers have a (finite) upper bound m, then $|g(x) - f_n(x)| < \epsilon$ for all $n \geqq m$, where m no longer depends on x. In this case we say that the sequence f_n converges *uniformly* to g.

This situation can be put in a general setting by means of the concepts introduced in the preceding sections.

Let X be an arbitrary set and Y a uniform space. The mappings f of X in Y will again simply be called *functions*. They form the elements of a new set F. We shall show that F can be converted into a uniform space in a natural way. For this purpose we must make a sharp distinction between functions and their values. A function is an element of F, but the value of a function is an element of Y. To an element x of X and an element f of F there is assigned an element $f(x)$ of Y, namely the value of the function for the argument x.

We again let $\mathfrak{G}$ denote the filter of the uniform structure of the space Y. If U is a given subset of $Y \times Y$, we let U^* denote the subset of $F \times F$ such that (f, g) is an element of U^* if and only if the element $(f(x), g(x))$ belongs to U for every choice of x from X. It is clear that $U \subset V$ implies $U^* \subset V^*$. We now let U run through the filter $\mathfrak{G}$. Then U^* runs through a system of subsets in $F \times F$ which together with the sets that include them form a filter $\mathfrak{G}^*$. We now prove that $\mathfrak{G}^*$ defines a uniform structure in F.

Every set U^* obviously includes the diagonal in $F \times F$, since $(f(x), f(x))$ is always an element of U, independently of the choice of x.

Moreover, $(U^{-1})^* = U^{*-1}$ and $(U \circ V)^* = U^* \circ V^*$, since U^{*-1} consists of the pairs (f, g) with $(g, f) \in U^*$, i.e., $(g(x), f(x)) \in U$ or $(f(x), g(x)) \in U^{-1}$ for all values of x. Then (f, g) belongs to $U^* \circ V^*$ if and only if there exists an h with $(f, h) \in U^*$ and $(h, g) \in V^*$, which means that $(f(x), h(x)) \in U$ and $(h(x), g(x)) \in V$ for every x, i.e., $(f(x), g(x)) \in U \circ V$ or $(f, g) \in (U \circ V)^*$.

The uniform structure defined by $\mathfrak{G}^*$ defines a uniform topology for the function space F. Thus it is now possible to define the concept of *uniform convergence*, since it can now be reduced to ordinary convergence:

A filter Φ in the function space F converges uniformly to a function g if Φ converges in the uniform topology of F.

Let us analyze this definition. A neighborhood $U^*(g)$ of g is the set of functions f with $(f, g) \in U^*$, and thus $(f(x), g(x)) \in U$ for all x, where U denotes an element in the filter of the uniform structure of Y. Convergence means that in the filter Φ there exists an element A such that for all $f \in A$

$$(f(x), g(x)) \in U \qquad \text{for all} \qquad x \in X.$$

Let us now suppose that we are dealing with a sequence, in other words with a mapping of the set N of natural numbers into the function space F. The function corresponding to the number n will be denoted, as usual, by f_n. We obtain a filter Φ if we take those subsets of the sequence which contain all but finitely many of its elements. If X and Y denote the set of real numbers, the above definition of uniform convergence means, in terms of the usual uniform structure, that a sequence f_n, $n \in N$, has the limit g if for every positive number ϵ there exists a number n_0 such that

$$|f_n(x) - g(x)| < \epsilon$$

for $x \in X$ and $n \geqq n_0$.

But now the set A in the above discussion is exactly the set of functions f_n with $n \geqq n_0$, so that this specialization to the case of a sequence of real functions is identical with the classical definition with which we began our investigation of uniform convergence.

CHAPTER 2

Functions

Introduction

In §§1.1 to 1.8 we describe the most important facts about functions and continuity. Then in §§2.1 to 2.5 we introduce the differential quotient and show how difference quotients can be estimated by means of differential quotients. Such estimates are used below, in place of the customary use of the mean-value theorem, in our results about the growth of functions (§2.6), in our justification of the interchange of differentiation and passage to a limit (§2.7), in our applications to power series in §2.8 and finally, after the introduction of higher derivatives in §3.1 and §3.2, in our discussion of the remainder term in the Taylor formula (§3.3). The exponential function is defined by its functional equation and is then expressed in the form of its Taylor series (§4.1). The logarithm is obtained (§4.2) as its inverse function (in the real field). Then in §§4.3 to 4.5 we discuss the mapping $x \to e^x$ in the complex field, introduce sines and cosines, and prove the existence of $\frac{\pi}{2}$ as a zero of the cosine.

The remainder of the article deals with mappings of n-dimensional spaces. A curve is a mapping from the 1-dimensional space into a space of higher dimensionality; here differentiation leads to the tangent vector. For a function of several variables the most important concept is that of the total derivative, which allows us to approximate a function at a point by means of a linear function, or more generally (when we are dealing with mappings from one Cartesian space into another) by a linear mapping (§§5.1 to 5.4). If such a mapping is to have a unique inverse, the approximating linear mapping must likewise have a unique inverse, and this condition is also sufficient for unique invertibility, at least in the small (§§5.5 and 5.6). This important inversion theorem, which is proved by iterative methods, can be used to solve systems of equations, to make functions "explicit" when they are defined implicitly, and to prove theorems about dependence and independence of systems of functions (§§5.7 and 5.8).

In its mathematical meaning the word "function" is found for the first time in correspondence between Leibniz and Johann Bernoulli, where a function was considered to be an algebraic or analytical expression in which one or more magnitudes were allowed to vary. But then the 18th-century mathematicians encountered difficulties and paradoxes of every sort, because they could not decide exactly which operations they should allow in the construction of a function. For example, could curves drawn mechanically or by hand be regarded as the graphical images of functions? Or must they restrict themselves to functions that they could expand in a Taylor series?

In 1807 it was discovered by Fourier that functions of a highly arbitrary character can be expanded in a "Fourier series,"

$$f(x) = a_0 + \sum (a_n \cos nx + b_n \sin nx).$$

(This fact was already clear to Daniel Bernoulli on physical grounds, but Fourier showed how the coefficients could be determined from the given function by integration, an achievement which lent much greater weight to the theorem.) For example, the function f may have a corner, which is a very remarkable phenomenon, since the sines and cosines on the right-hand side of the equation are smooth functions.

Clarification began in 1829, with Dirichlet. The genetic definition of a function, whereby it was regarded as the result of some operation, was given up, and a function was understood to be any correspondence between numbers, quite independently of the way in which the correspondence arose. This point of view did not carry the day completely until the present century, and at the same time a process of generalization was taking place, in which the concept of a function was extended to that of a mapping.

In this chapter we oscillate between two extremes. In the sections on measure and integration we deal with functions of an extremely general kind. On the other hand, the concept of an analytic function in the theory of functions of a complex variable is a very specialized notion, whose theory depends on what is concisely called the infinitesimal calculus.

1. Continuity

1.1. We speak of a *mapping* f of a set M into a set N if to every element x of M there is assigned exactly one element $y = f(x)$ of N (cf. IA, §8.4). (It is possible to consider many-valued mappings, in which more than one element of N is assigned to an element of M, but at least for the time being we shall not do so.)

Example. Let M and N be the set of human beings (living and dead) and let f be the function to which each human being assigns his father.

1.2. We now restrict the concept of a mapping to that of a *continuous mapping*, where "continuous" is taken to mean: if x changes only slightly, then y also changes only slightly. Of course this remark is meaningful only if we have defined a concept of "slightly" in M and N. In other words, we must define a topology in M and N. Now the topology can be defined in many different ways. Let us consider two of them.

1. In M and N the concept of a limit is already defined, and we say that f is continuous at a if

$$\lim a_n = a$$

implies

$$\lim f(a_n) = f(a).$$

In words: to a sequence converging to a the mapping f assigns a convergent sequence, and the limit of the one sequence corresponds to the limit of the other. The mapping f is said to be continuous (without further qualification) if f is continuous at every a.

Alternatively we may say: f is continuous if the symbols f and lim are permutable

$$f(\lim a_n) = \lim f(a_n), \tag{1}$$

in the sense that if the left-hand side exists, then the right-hand side also exists and is equal to the left-hand side.

2. Let M and N be metric spaces, i.e., in M and in N let there exist a concept of distance ($\rho\,(a, b)$ = the distance between the elements a and b) with the properties

$$\begin{aligned} &\rho(a, b) > 0 \quad \text{for} \quad a \neq b, \\ &\rho(a, b) = 0 \quad \text{for} \quad a = b, \\ &\rho(a, b) = \rho(b, a), \\ &\rho(a, b) + \rho(b, c) \geqq \rho(a, c). \end{aligned}$$

A mapping f of M into N is said to be continuous at a if the image of a sufficiently small neighborhood of a falls within any preassigned neighborhood of $f(a)$; more precisely:

For every $\epsilon > 0$ there exists a $\delta > 0$ such that for all x with $\rho(a, x) < \delta$, we have

$$\rho(f(a), f(x)) < \epsilon. \tag{2}$$

Again f is said to be continuous (without further qualification) if f is continuous at every a.

The spaces to be discussed below are subsets of the m-dimensional (real or complex) Euclidean space R_m.

The elements of R_m are the systems of m numbers (real or complex), also called vectors,

$$\mathfrak{x} = (x_1, \ldots, x_m)$$

with the definitions

$$|\mathfrak{x}| = \sqrt{|x_1|^2 + \cdots + |x_m|^2}$$

and

$$\rho(\mathfrak{x}, \mathfrak{y}) = |\mathfrak{x} - \mathfrak{y}|.$$

By convergence in such a space we mean coordinate-wise convergence,

$$\lim_{n\to\infty} \mathfrak{x}^{(n)} = \mathfrak{x} \text{ if and only if } \lim_{n\to\infty} x_i^{(n)} = x_i \quad (i = 1, \ldots, m).$$

The special case $m = 1$ is the set of numbers regarded as a space.

1.3. The two definitions of continuity turn out to be equivalent as soon as we have set up a certain connection between the concept of limit and the concept of distance. In a metric space we agree that

$$\lim a_n = a$$

if and only if

$$\lim \rho(a, a_n) = 0.$$

We now demonstrate the equivalence in two steps:

I ϵ-δ-continuity at $a \to$ limit-continuity at a.

Let there be given a sequence a_n with $\lim a_n = a$. Then we must show: $\lim f(a_n) = f(a)$, and $\lim \rho f(a), f(a_n)) = 0$. But this latter equality means that for preassigned $\epsilon > 0$ we have

$$\rho(f(a), f(a_n)) < \epsilon \qquad \text{for almost all } n.$$

(Strictly speaking, we should write the symbols for absolute value, but since ρ is non-negative, this may be disregarded.) Thus we must prove the above inequality for arbitrary preassigned ϵ.

For the given ϵ we determine $\delta > 0$ in accordance with (2). By hypothesis we have $\lim a_n = a$, and consequently $\lim \rho(a, a_n) = 0$, so that

$$\rho(a, a_n) < \delta \qquad \text{for almost all } n.$$

Thus by (2)

$$\rho(f(a), f(a_n)) < \epsilon \qquad \text{for almost all } n,$$

as was to be proved.

II Limit-continuity at $a \to \epsilon$-δ-continuity at a.

Here we shall prove the equivalent statement:

If f is not ϵ-δ-continuous at a, then it is not limit-continuous at a.

Thus we assume that f is not ϵ-δ-continuous at a: then we can find an ϵ, say $\epsilon_0 > 0$, such that for every $\delta > 0$ there exists an x for which

$$\rho(a, x) < \delta$$

is true but

$$\rho(f(a), f(x)) < \epsilon_0$$

is false.

For δ let us successively take the numbers $\frac{1}{n}$ $(n = 1, 2, \ldots)$. Then for each of these numbers there exists an x, call it a_n, for which

$$\rho(a, a_n) < \frac{1}{n},$$

but for which

$$\rho(f(a), f(a_n)) \geqq \epsilon_0 > 0$$

also holds. In other words

$$\lim a_n = a,$$

whereas $f(a_n)$ either has no limit at all or at any rate not the limit $f(a)$. Thus f is not limit-continuous at a.

1.4. The following situation often arises. A mapping f is defined on a subset of M for which the point a is a limit point; the images lie in N and the mapping f can be extended to a mapping f^* which is also defined at a and is continuous there; $f^*(a) = b \in N$. Then we say that

$$\lim_{x \to a} f(x) = b.$$

Example. $f(x) = \dfrac{\sin x}{x}$, $M =$ the set of $x \neq 0$, $a = 0, f^*(0) = 1$.

1.5. The word *function* is customarily used for mappings in which the set of images is a set of numbers (real or complex). But the terminology varies. For example, it is also possible to speak of functions if the images are vectors. Then we speak of *vector functions*, in contrast to *scalar functions*, the values of which are numbers.

Let M be a compact space (e.g., a bounded closed set in the Cartesian R_n); and let f be a continuous function. Then the image $f(M)$ of M is again compact, and thus is a bounded closed set of numbers (see 5, §4.3) In particular, for real functions f we have:

A real continuous function on a compact space has a maximum and a minimum.

Let M be connected (e.g., an interval). *Then $f(M)$ is also connected.* In particular:

A real continuous function on a connected space has an interval (possibly semi-open) *for its image and thus, between any two of its values* α *and* β, *it assumes every intermediate value* γ $(\alpha < \gamma < \beta)$.

1.6. From the definition of a continuous mapping it follows immediately that *successive application* of continuous mappings again produces a continuous mapping.

The following functions are continuous in R_2:

$$(x_1, x_2) \to x_1 + x_2, \qquad (x_1, x_2) \to x_1 - x_2$$

$$(x_1, x_2) \to x_1 x_2, \qquad (x_1, x_2) \to \frac{x_1}{x_2} \quad (\text{for } x_2 \neq 0).$$

By repeated application of these principles we obtain:

Every rational expression in a variable x,

$$f(x) = \frac{P(x)}{Q(x)},$$

($P(x)$, $Q(x)$ *entire rational*), *is continuous provided* $Q(x) \neq 0$.

The same remark holds for rational expressions in $x_1, \ldots, x_n$ as functions defined in R_n.

1.7. The distance between two bounded mappings f and g of the space M into the space N can be measured by

$$\rho(f, g) = \sup_{x \in M} \rho(f(x), g(x)).$$

The functions f_n,

$$f_n(x) = x^n \qquad (0 \leqq x \leqq 1),$$

converge for every x to the function f:

$$f(x) = \begin{matrix} 0 \\ 1 \end{matrix} \quad \text{for} \quad \begin{matrix} 0 \leqq x < 1 \\ x = 1 \end{matrix}.$$

Yet the distances $\rho(f_n, f)$ are all equal to 1.

If the *distances* converge to 0 we speak of *uniform convergence*.

The limit of a uniformly convergent sequence of functions that are continuous at a in a metric space M is again continuous at a.

For let $\lim f_n(x) = f(x)$ (uniformly). Let $a \in M$ and let $\delta > 0$ be given. Then there exists an n_0 such that

$$|f(x) - f_{n_0}(x)| < \tfrac{1}{3}\epsilon \qquad \text{for all} \qquad x \in M.$$

In particular,

$$|f(a) - f_{n_0}(a)| < \tfrac{1}{3}\epsilon.$$

Also there exists a $\delta > 0$ such that

$$|f_{n_0}(a) - f_{n_0}(x)| < \tfrac{1}{3}\epsilon \qquad \text{for all} \qquad x \in M \text{ with } \rho(a, x) < \delta.$$

From these three inequalities it follows that:

$$|f(x) - f(a)| < \epsilon \qquad \text{for all} \qquad x \in M \text{ with } \rho(a, x) < \delta,$$

as was to be proved.

1.8. Let M be the set of real numbers $\alpha \leqq x \leqq \beta$. Every real continuous function in M can be generated by the processes described in §§1.6 and 1.7. More precisely:

Weierstrass approximation theorem. *Every neighborhood of a function f that is continuous in the interval $\alpha \leqq x \leqq \beta$ contains a polynomial* (we omit the proof).

2. Differentiability

2.1. First we consider only functions f that map a set of numbers M into a set of numbers N, real or complex. If a is kept fixed, the *difference quotient*

$$\varphi(x, a) = \frac{f(x) - f(a)}{x - a} \qquad (x \neq a)$$

can be regarded as a function of x. Let us ask whether $\varphi(x, a)$ can be extended to a function that is defined for $x = a$ and is continuous at a. Such a definition is certainly possible if a is an isolated point of M; for then we may assign to $\varphi(x, a)$ any value we like for $x = a$. We exclude this trivial case, so that a is a limit point of M.

The function $\varphi(x, a)$ can be extended in the desired way if there exists a number l with the property:

$$\lim x_n = a \qquad (x_n \neq a;\ x_n \in M)$$

implies

$$\lim \frac{f(x_n) - f(a)}{x - a} = l;$$

or alternatively: for every $\epsilon > 0$ there exists a $\delta > 0$ such that

$$\left|\frac{f(x) - f(a)}{x - a} - l\right| < \epsilon \tag{3}$$

for all $x \in M$, $x \neq a$ with $|x - a| < \delta$.

More concisely

$$\lim_{x \to a} \frac{f(x) - f(a)}{x - a} = l.$$

Such an l, if it exists, is uniquely determined and is called the *differential quotient* of f at a:

$$l = f'(a) = \left(\frac{df(x)}{dx}\right)_{x=a}.$$

In this case f is said to be *differentiable* at a.

The inequality (3) can also be written in the form:

$$(4) \qquad f(x) = f(a) + (x - a)f'(a) + |x - a|\epsilon(x \to a),$$

where $\epsilon(x \to a)$ denotes a variable quantity that converges to 0 for $x \to a$; i.e., for every $\epsilon > 0$ there exists a $\delta > 0$ such that

$$|\epsilon(x \to a)| < \epsilon \qquad \text{for} \qquad |x - a| < \delta, \qquad x \neq a.$$

Formula (4) provides us with an approximation to a differentiable function in the neighborhood of a by means of a linear function; the remainder approaches 0 faster than the difference between the values of the independent variable.

2.2. If the differential quotient of $f(x)$ exists for all $x = a$ in a set N, we can consider the function f' which for every $x = a \in N$ assumes the value $f'(a)$. This function is called the *derivative* of f (in N).

The derivative of a constant function is 0.

$$(f + g)' = f' + g', \qquad (fg)' = fg' + f'g, \qquad \left(\frac{f}{g}\right)' = \frac{gf' - fg'}{g^2}.$$

In rules of this sort we always mean that if the right-hand side exists, then the left-hand side also exists and is equal to the right-hand side.

If f has an inverse g and if $f(a) = b, f'(a) = 0$, then $g'(b)$ also exists and $g'(b) = f'(a)^{-1}$.

If $f(a) = b$, $g(b) = c$, $h(x) = g(f(x))$ and if $f'(a)$ and $g'(b)$ exist, then $h'(a)$ also exists and $h'(a) = f'(a)g'(b)$. (Chain rule.)

For the proofs of such theorems it is usually most convenient to make use of (4). For example, for the product rule we have

$$\begin{aligned} f(x) &= f(a) + (x - a)f'(a) + |x - a|\epsilon_1(x \to a), \\ g(x) &= g(a) + (x - a)g'(a) + |x - a|\epsilon_2(x \to a), \\ f(x)g(x) &= f(a)g(a) + (x - a)(f(a)g'(a) + f'(a)g(a)) \\ &\quad + |x - a|\epsilon_3(x \to a), \end{aligned}$$

where it is easy to show that $\epsilon_3(x \to a)$ has the desired property.

Again, for the chain rule we have

$$\begin{aligned} f(x) &= f(a) + (x - a)f'(a) + |x - a|\epsilon_1(x \to a), \\ g(y) &= g(b) + (y - b)g'(b) + |y - b|\epsilon_2(y \to b), \end{aligned}$$

or, setting $y = f(x)$ and $b = f(a)$,

$$g(f(x)) = g(f(a)) + (f(x) - f(a))g'(b) + |f(x) - f(a)|\epsilon_2(f(x) \to f(a)),$$

so that if we replace $f(x) - f(a)$ by its value from the first formula above, we obtain

$$g(f(x)) = g(f(a)) + (x - a)f'(a)g'(b) + |x - a|\epsilon_3(x \to a).$$

2.3. *Estimation of Difference Quotients by Differential Quotients*

Let M be a real interval, and let f be a real function defined and differentiable at every point of M; $a, b \in M$, $a < b$. Then

$$\inf_{a \leqq x \leqq b} f'(x) \leqq \frac{f(b) - f(a)}{b - a} \leqq \sup_{a \leqq x \leqq b} f'(x). \tag{5}$$

This estimate of the difference quotient in terms of the differential quotient is of basic importance. We shall use it here in place of the usual mean-value theorem, which states: there exists a ξ with

$$a \leqq \xi \leqq b \quad \text{and} \quad \frac{f(b) - f(a)}{b - a} = f'(\xi).$$

We do this because the existence of such a ξ is never of any importance; in every case we are interested only in the estimate (5).

It will be enough to prove half of the theorem. Let $a < c < b$.

$$\frac{f(b) - f(a)}{b - a} = \frac{c - a}{b - a}\frac{f(c) - f(a)}{c - a} + \frac{b - c}{b - a}\frac{f(b) - f(c)}{b - c}.$$

We now note that

$$\frac{c - a}{b - a} \quad \text{and} \quad \frac{b - c}{b - a}$$

are positive and that their sum is 1. If we had

$$\frac{f(b) - f(a)}{b - a} = C > \sup_{a \leqq x \leqq b} f'(x).$$

then one of the two difference quotients

$$\frac{f(c) - f(a)}{c - a}, \quad \frac{f(b) - f(c)}{b - c},$$

would also have to be $\geqq C$. By successive bisection we would obtain a sequence of intervals I_n with endpoints d_n and e_n $(> d_n)$ such that $I_{n+1} \subset I_n$ and

$$\frac{f(e_n) - f(d_n)}{e_n - d_n} \geqq C.$$

But then d_n and e_n would have a common limit c (it may be that almost all the d_n, or almost all the e_n, are $= c$). Then we would have $d_n \leqq c \leqq e_n$, so that for every n it would follow that

$$\frac{f(c) - f(d_n)}{c - d_n} \quad \text{or} \quad \frac{f(e_n) - f(c)}{e_n - c} \geqq C.$$

In the limit we would then have

$$f'(c) \geqq C > \sup f'(x),$$

which is a contradiction.

2.4. The inequalities (5) can be taken over to the complex field: by the segment (a, b) we mean the set of all

$$c = (1 - s)a + sb \qquad \text{with} \qquad 0 \leqq s \leqq 1,$$

or in other words, all c with

$$|a - c| + |c - b| = |a - b|.$$

Let f be differentiable at every point of M and let the segment (a, b) lie entirely in M. Then

(6) $$\left|\frac{f(b) - f(a)}{b - a}\right| \leqq \sup_{x \in M} |f'(x)|.$$

For the proof we choose c on the segment (a, b) and note that

$$\frac{c - a}{b - a} \quad \text{and} \quad \frac{b - c}{b - a}$$

are again positive and their sum is equal to 1. Then in the same way as in §2.3 we see that

$$\left|\frac{f(b) - f(a)}{b - a}\right| \geqq C$$

implies

$$\left|\frac{f(c) - f(a)}{c - a}\right| \quad \text{or} \quad \left|\frac{f(b) - f(c)}{b - c}\right| \geqq C$$

and we proceed as before.

If in (6) we replace $f(x)$ by $f(x) - \mu x$, we have

(7) $$\left|\frac{f(b) - f(a)}{b - a} - \mu\right| \leqq \sup_{x \in M} |f'(x) - \mu|.$$

If M is the union of segments with endpoint x_0, we can write (6) (with x_0 in place of a, and x in place of b) in the form

(8) $$f(x) = f(x_0) + \text{rem}\,(\leqq |x - x_0| \sup_{\xi \in M} |f'(\xi)|).$$

2.5. a) A simple consequence of §2.4, (8) (or (7)) is: *let the derivative of $f(x)$ be equal to 0 at every point of M and let it be possible to join all the points of M to a point $x_0 \in M$ by means of segments lying entirely in M. Then f is constant in M.*

For $\sup |f'(x)| = 0$, so that by (8): $f(x) - f(x_0) = 0$.

Of course, the same result follows if we assume only that all the points of M can be joined by polygonal lines.

b) A remarkable consequence is the following:

Let M be the union of segments with end point x_0, where x_0 itself does not necessarily belong to M. Let the derivative f' exist everywhere in M. Let f be continuous at x_0. Let it be possible to extend f' to a function g that is continuous at x_0. Then f' also exists at x_0 and is continuous there and equal to $g(x_0)$.

The proof runs as follows. For $\epsilon > 0$ we determine $\delta > 0$ such that

$$|f'(x) - g(x_0)| < \tfrac{1}{2}\epsilon \quad \text{for} \quad |x - x_0| < \delta, \quad x \in M.$$

Then for $x \in M$ we choose y on the segment (x_0, x). By (7) we have

$$\left|\frac{f(x) - f(y)}{x - y} - g(x_0)\right| \leqq \sup_{\xi \in M, |\xi - x_0| < \delta} |f'(\xi) - g(x_0)| \leqq \tfrac{1}{2}\epsilon$$

for $|x - x_0| < \delta$, $x \in M$. Since f is continuous, the same inequality holds with x_0 in place of y. Consequently

$$\left|\frac{f(x) - f(x_0)}{x - x_0} - g(x_0)\right| < \epsilon \quad \text{for} \quad |x - x_0| < \delta, \quad x \in M,$$

as was to be proved.

c) An important application is the following:

Let M be convex, i.e., let it contain the entire line segment joining any two points in M. Let f be defined in M and everywhere differentiable, and let f' be continuous at x_0. Then

$$\lim_{x \to x_0, x^* \to x_0, x^* \neq x} \frac{f(x^*) - f(x)}{x^* - x} = f'(x_0),$$

i.e., the function

$$\varphi(x, x^*) = \frac{f(x^*) - f(x)}{x^* - x}$$

can be extended to $x = x^* = x_0$ in such a way as to be continuous there.

For we have

$$\left|\frac{f(x^*) - f(x)}{x^* - x} - f'(x_0)\right| \leqq \sup_{y \in M_\delta} |f'(y) - f'(x_0)|,$$

where M_δ is the set of $z \in M$ with $|z - x_0| < \delta$ and $x, x^* \in M_\delta$. Since f' is continuous at x_0, the right-hand side approaches 0 as $\delta \to 0$, which implies the desired assertion.

2.6. Another application is as follows: Let us show that

$$\log x \leqq \epsilon x$$

for arbitrary $\epsilon > 0$ and all sufficiently large x.

From (5) it follows that for $x > c > 1$ we have

$$\frac{\log x - \log c}{x - c} \leqq \sup_{\xi \geqq c} \frac{1}{\xi} \leqq \frac{1}{c},$$

and thus

$$\log x \leqq \log c + \frac{x}{c}.$$

Taking $c = \dfrac{2}{\epsilon}$ and $A = \dfrac{2 \log c}{\epsilon}$ we obtain

$$\log x \leqq \frac{\epsilon}{2} A + \frac{\epsilon}{2} x,$$

so that for $x \geqq \max(c, A)$

$$\log x \leqq \epsilon x.$$

2.7. How do matters stand with respect to *differentiation of convergent sequences of functions*? When can we interchange the order of differentiation and passage to the limit?

Let M be convex. Let the functions $f_n(x)$ be everywhere differentiable in M. Let $\lim f'_n(x) = g(x)$ *exist uniformly; let* $\lim f_n(a)$ *exist for a certain $a \in M$. Then*

$$\lim f_n(x) = f(x) \qquad (x \in M)$$

exists (and if M is bounded, exists uniformly); the function $f(x)$ is differentiable everywhere in M and

$$\lim f'_n(x) = f'(x).$$

Proof. For $\epsilon > 0$ there exists an n_0 such that

$$(9) \qquad |f'_n(x) - g(x)| < \tfrac{1}{6}\epsilon \qquad \text{for} \qquad n \geqq n_0, \quad x \in M.$$

Consequently

$$(10) \qquad |f'_m(x) - f'_n(x)| < \tfrac{1}{3}\epsilon \qquad \text{for} \qquad m, n \geqq n_0, \quad x \in M.$$

By (8) we have

$$|f_m(x) - f_n(x)| \leqq |f_m(a) - f_n(a)| + |x - a| \sup_{\xi \in M} |f'_m(\xi) - f'_n(\xi)|.$$

If $\eta > 0$ is given, the first summand is $< \frac{1}{2}\eta$ for $m, n > n_1$ on account of the convergence of $f_n(a)$, and the same inequality holds (for fixed x) for the second summand if we make the appropriate choice of $\epsilon = \frac{1}{2}\eta/|x - a|$. For $n > \max(n_0, n_1)$ we then have $|f_m(x) - f_n(x)| < \eta$. Consequently $f_n(x)$ is convergent. For $|x - a| \leqq k$ we can choose $\epsilon = \frac{1}{2}\eta/k$, a choice which is independent of x, so that the convergence is uniform.

For $x_0, x \in M$ and $m, n \geqq n_0$ the inequalities (7) and (10) give

$$\left|\frac{f_m(x) - f_m(x_0)}{x - x_0} - \frac{f_n(x) - f_n(x_0)}{x - x_0}\right| \leqq \frac{1}{3}\epsilon.$$

We now pass to the limit $m \to \infty$ and replace n by n_0. Then we also have

$$\left|\frac{f(x) - f(x_0)}{x - x_0} - \frac{f_{n_0}(x) - f_{n_0}(x_0)}{x - x_0}\right| \leqq \frac{1}{3}\epsilon.$$

Now we determine $\delta > 0$ in such a way that for $|x - x_0| < \delta$

$$\left|\frac{f_{n_0}(x) - f_{n_0}(x_0)}{x - x_0} - f'_{n_0}(x_0)\right| < \frac{1}{3}\epsilon$$

(as is possible, since f_{n_0} is differentiable); further, by (9) we have

$$|f'_{n_0}(x_0) - g(x_0)| < \tfrac{1}{6}\epsilon.$$

Consequently

$$\left|\frac{f(x) - f(x_0)}{x - x_0} - g(x_0)\right| < \epsilon \qquad \text{for} \qquad |x - x_0| < \delta,$$

as was to be proved.

2.8. If the power series

$$\sum_{n=0}^{\infty} a_n x^n$$

converges for $x = \gamma$, then for $|\gamma| = r$ we have

$$|a_n r^n| \leqq C,$$

for all the coefficients a_n, and consequently for $|x| \leqq \rho = \vartheta r$ $(0 \leqq \vartheta < 1)$ it follows that

$$|a_n x^n| \leqq C\vartheta^n.$$

But this inequality implies the convergence of $\sum a_n x^n$ for all x with $|x| < r$, and this convergence is uniform for all $|x| \leqq \rho$. *The power series represents a continuous function $f(x)$ for all $|x| < r$.*

But can we differentiate termwise? For this purpose we must show that the sequence of derivatives of partial sums converges uniformly for $|x| \leqq \rho$ and every $\rho < r$, i.e., that the series

$$\sum_{n=1}^{\infty} na_n x^{n-1}$$

converges uniformly. Now we have $\lim \sqrt[n-1]{n} = 1$, so that

$$n < \eta^{n-1}$$

for arbitrary $\eta > 1$ and all sufficiently large n. Thus

$$|na_n x^{n-1}| \leqq \eta^{n-1} C \vartheta^{n-1}$$

for $|x| \leqq \rho$ and large n. If we choose η so that $\vartheta_1 = \eta\vartheta < 1$, we see that the uniform convergence of

$$\sum na_n x^{n-1}$$

follows for $|x| \leqq \rho$. Here ρ is arbitrary $< r$. Consequently we can differentiate $\sum a_n x^n$ termwise for all $|x| < r$, i.e., $\sum na_n x^{n-1}$ represents the derivative of $\sum a_n x^n$.

3. Higher Derivatives

3.1. By repeated differentiation of a function f we obtain the derivatives $f', f'', \ldots, f^{(n)}, \ldots$ (if they exist).

Let M be convex, let f be $(k-1)$-times differentiable for $x \in M$ and k-times differentiable at $x_0 \in M$, and let $f(x_0) = f'(x_0) = \cdots = f^{(k)}(x_0) = 0$. Then in M

$$f(x) = \epsilon(x \to x_0)|x - x_0|^k.$$

For $k = 0$ this equation merely asserts the continuity of f at x_0, and for $k = 1$ it asserts the differentiability of f. The general assertion is proved by induction. We assume that it is correct for $k = n - 1$ $(\geqq 1)$ and for arbitrary functions f satisfying the hypothesis; and then we conclude that the assertion is true for $k = n$ as follows: for a given f with

$$f(x_0) = \cdots = f^{(n)}(x_0) = 0$$

we consider the function $g = f'$, for which

$$g(x_0) = \cdots = g^{(n-1)}(x_0) = 0$$

and consequently, by the induction hypothesis,

$$g(x) = \epsilon_1(x \to x_0)|x - x_0|^{n-1}.$$

But now by (6) we have

$$|f(x) - f(x_0)| \leqq |x - x_0| \sup_{\xi \in (x_0, x)} |f'(\xi)|;$$

moreover,

$$f'(\xi) = g(\xi), \qquad f(x_0) = 0,$$

$$\begin{aligned}\sup_{\xi \in (x_0, x)} |\epsilon_1(\xi \to x_0)|\, |\xi - x_0|^{n-1} &\leqq \sup |\epsilon_1(\xi \to x_0)| \sup |\xi - x_0|^{n-1} \\ &= \epsilon_2(x \to x_0)|x - x_0|^{n-1},\end{aligned}$$

from which it follows that

$$f(x) = \epsilon(x \to x_0)|x - x_0|^n.$$

3.2. Let M be convex and let f be $(n - 1)$-times differentiable for $x \in M$ and n-times differentiable at x_0. Then at x_0 the polynomial

$$g(x) = f(x_0) + \frac{x - x_0}{1!} f'(x_0) + \frac{(x - x_0)^2}{2!} f''(x_0) + \cdots + \frac{(x - x_0)^n}{n!} f^{(n)}(x_0)$$

has the same derivatives, up to the nth inclusive, as $f(x)$. To the remainder $f(x) - g(x)$ we apply §3.1:

$$f(x) - g(x) = \epsilon(x \to x_0)|x - x_0|^n.$$

Thus $f(x)$ is equal to a polynomial, up to a remainder which for $x \to x_0$ approaches 0 faster than $|x - x_0|^n$ (approaches 0 even after being divided by $|x - x_0|^n$):

$$f(x) = \sum_{p=0}^{n} \frac{(x - x_0)^p}{p!} f^{(p)}(x_0) + \epsilon(x \to x_0)|x - x_0|^n.$$

3.3. *Let M be convex and let f be n-times differentiable for $x \in M$. Then for x_0, $x \in M$ we have the Taylor formula*

$$\begin{aligned} f(x) &= \sum_{p=0}^{n-1} \frac{(x - x_0)^p}{p!} f^{(p)}(x_0) \\ &\quad + \text{rem}\left(\leqq |x - x_0| \sup_{\xi \in (x_0, x)} \frac{|x - \xi|^{n-1}}{(n - 1)!} |f^{(n)}(\xi)|\right). \end{aligned} \tag{11}$$

For the proof we denote the remainder, which is a function of x and x_0, by $g(x, x_0)$ and differentiate it with respect to x_0:

$$\begin{aligned}\frac{dg(x, x_0)}{dx_0} &= \sum_{p=1}^{n-1} \frac{(x - x_0)^{p-1}}{(p - 1)!} f^{(p)}(x_0) - \sum_{p=0}^{n-1} \frac{(x - x_0)^p}{p!} f^{(p+1)}(x_0) \\ &= -\frac{(x - x_0)^{n-1}}{(n - 1)!} f^{(n)}(x_0),\end{aligned}$$

so that with ξ in place of x_0:

$$\frac{dg(x, \xi)}{d\xi} = -\frac{(x - \xi)^{n-1}}{(n-1)!} f^{(n)}(\xi). \tag{12}$$

Application of the inequality (6) for x and x_0 leads, in view of the fact that $g(x, x) = 0$, to the desired estimate for the remainder, which is named after Cauchy.

Somewhat less important is the estimate of Lagrange

$$\text{rem} \leqq \frac{|x - x_0|^n}{n!} \sup_{\xi \in (x_0, x)} |f^{(n)}(\xi)|, \tag{13}$$

which is obtained from (12) if the expression

$$\left|\frac{dg(x, \xi)}{d\xi}\right| \leqq \frac{|x - \xi|^{n-1}}{(n-1)!} \sup_{\xi \in (x_0, x)} |f^{(n)}(\xi)|$$

is integrable with respect to ξ from x to x_0.

3.4. The most important application is obtained when f is arbitrarily often differentiable and the remainder for $n \to \infty$ approaches zero. Then we have the *Taylor series*

$$f(x) = \sum_{p=0}^{\infty} \frac{(x - x_0)^p}{p!} f^{(p)}(x_0). \tag{14}$$

Example (see the remark at the end of §4.2):

$$f(x) = (1 + x)^\mu, \qquad x_0 = 0, \qquad \mu \quad \text{not an integer},$$
$$f^{(p)}(x) = \mu(\mu - 1)\cdots(\mu - p + 1)(1 + x)^{\mu - p}.$$

The remainder after the $(n - 1)$-th derivative becomes (with $\xi = tx$)

$$\leqq |x| \left|\mu\binom{\mu - 1}{n - 1}\right| \sup_{0 \leqq t \leqq 1} \left|\frac{(1 - t)x}{1 + tx}\right|^{n-1} |(1 + tx)^{\mu - 1}|,$$

so that, since

$$\lim_{n \to \infty} \binom{\mu - 1}{n - 1} = 1,$$

and for $|x| < 1$,

$$(1 - t)|x| = |x| - t|x| < 1 - t|x| \leqq |1 + tx|,$$

we have

$$\lim \left|\frac{(1 - t)x}{1 + tx}\right|^{n-1} = 0.$$

Thus the remainder approaches zero and we may write the Taylor series

$$(1 + x)^\mu = \sum_{n=0}^{\infty} \binom{\mu}{n} x^n \qquad \text{for} \qquad |x| < 1.$$

4. Exponential Functions

4.1. The (natural) logarithms were invented as an instrument for changing products into sums. Let us first consider functions f that do the opposite, i.e., satisfy the equation

(15) $$f(x + y) = f(x)f(y)$$

for all x and y. Assuming that f is differentiable, we form the difference quotient

$$\frac{f(x + y) - f(x + 0)}{y} = \frac{f(y) - f(0)}{y} f(x),$$

from which, by letting y approach zero we have:

(16) $$f'(x) = f'(0)f(x).$$

From (15) it follows for $x = y = 0$ that $f(0)^2 = f(0)$, so that $f(0) = 0$ or 1. If f vanishes at any point, it vanishes identically. Thus we shall assume that

(17) $$f(0) = 1.$$

Then f is everywhere positive.

If we also require that

(18) $$f'(0) = 1,$$

it follows that

(19) $$f'(x) = f(x)$$

and by successive steps

$$f^{(n)}(x) = f(x).$$

Since $f' = f$ is everywhere positive, f is monotone increasing, and thus in every interval it attains its maximum at the right-hand endpoint. So in (13) the remainder term is $\frac{|x|^n}{n!} \max(f(x), 1)$ and therefore converges to 0 for $n \to \infty$. Consequently, by §3.4 we have the expansion:

(20) $$f(x) = 1 + \frac{x}{1!} + \frac{x^2}{2!} + \cdots = \sum_{n=0}^{\infty} \frac{x^n}{n!}.$$

Conversely, let $f(x)$ be defined by (20). Then it is easy to see that $f(x)$ satisfies conditions (17) and (19). Let $g(x)$ be a second function satisfying (17) and (19). Then

$$\left(\frac{f}{g}\right)' = \frac{gf' - fg'}{g^2} = 0,$$

so that

$$g = \text{const}\cdot f,$$

where by (17) the constant is seen to be 1. Thus f is uniquely determined by (17) and (19). But f also satisfies (15) and (17); since

$$h(x) = f(x + c)f(c)^{-1}$$

satisfies (17) and (19) and is therefore identical with $f(x)$, which is exactly the relation (15). Again, f is the only differentiable function that satisfies (15), (17), and (18).

If we are given a differentiable function h that satisfies (15) and (17), and thus also (16), we write $h'(0) = c$ and form the function

$$g(x) = h\left(\frac{x}{c}\right),$$

which again satisfies (15), (17), and (18), as is easily shown. Thus $g(x) = f(x)$ and therefore

$$h(x) = f(cx) \qquad (c = \text{const}) \tag{21}$$

is the most general function satisfying (15) and (17). (If $c = 0$, then by (16) the function h is constant and is thus $= 1$.)

From (20) it follows that $f(x) > 0$ for $x \geqq 0$, so that $f(x)f(-x) = 1$ implies $f(x) > 0$ for all x. By (19) we also have $f'(x) > 0$, so that the function $f(x)$ is monotone increasing. For large positive x it becomes arbitrarily large, and for large negative x, in view of the fact that $f(x)f(-x) = f(0) = 1$, arbitrarily small and positive; thus $f(x)$ assumes all positive values for real x and therefore maps the set of real numbers one-to-one onto the set of positive numbers.

Let us further examine the functions $h(x)$ (see (21)), for which $h(1)$ is positive. By the preceding argument there exists a uniquely determined *real* c such that

$$h(1) = f(c). \tag{22}$$

Then

$$h(x) = f(cx) \qquad (c \text{ real}). \tag{23}$$

Such functions $h(x)$ are called *exponential functions* and are written in the form

$$h(x) = a^x \qquad \text{with} \qquad a = h(1).$$

To justify this notation we must show that for every natural number n we have

$$h(n) = a^n$$

in the elementary sense. For $n = 1$ the statement is true by definition, and if $h(n - 1) = a^{n-1}$, then (15) gives $h(n) = h(n - 1)h(1) = a^{n-1}a$, which is exactly the definition of a^n (in the elementary sense).

4.2. In particular the function f (see (20)) is called e^x, so that

$$e = 1 + \frac{1}{1!} + \frac{1}{2!} + \cdots.$$

Then, as we have seen above, the mapping $x \to e^x$, considered for real x only, has a uniquely determined inverse, which is called $\log x$ and is defined for $x > 0$. By (15) it follows that

$$\log(xy) = \log x + \log y;$$

by (19)

$$\frac{d}{dx} e^x = e^x,$$

so that

$$\frac{d}{dx} \log x = \frac{1}{x};$$

and by (22) and (23)

$$\frac{d}{dx} a^x = a^x \log a.$$

The inverse of a^x is called ${}^a\log x$. From (22) and (23) we have

$${}^a\log x = \frac{\log x}{\log a},$$

$$\frac{d}{dx} {}^a\log x = \frac{1}{x \log a}.$$

The function x^μ was used above for arbitrary μ (x real and positive). By the present definition we have

$$x^\mu = e^{\mu \log x},$$

from which it follows that

$$\frac{d}{dx} x^\mu = \mu x^{\mu - 1}.$$

4.3. Let us now discuss the behavior of e^x for complex x. From (20) we see that

$$\overline{e^x} = e^{\bar{x}}.$$

Thus

$$|e^x|^2 = e^x e^{\bar{x}} = e^{2\,\mathrm{Re}\,x}.$$

Since for real z the equation $e^z = 1$ has the unique solution $z = 0$, all the solutions x of

$$e^x = 1$$

have $\mathrm{Re}\,x = 0$, and thus are purely imaginary. These solutions form a group with respect to addition; for if $e^x = e^y = 1$, then also $e^{x-y} = 1$. If they had a limit point, say $e^{x_n} = 1$, $\lim x_n = x_0$, then by continuity we would also have $e^{x_0} = 1$, so that $e^{x_n - x_0} = 1$, and the solutions would have the limit point 0; for $x'_n = x_n - x_0$ we would have $e^{x'_n} = 1$, $\lim x'_n = 0$. But then it would follow that $\lim \dfrac{e^{x'_n} - 1}{x'_n} = 0$, in contradiction to the fact that $\dfrac{d}{dx} e^x = 1$ for $x = 0$. Thus the solutions cannot have any limit point. If there are any solutions other than $x = 0$, there exists a solution of the form si, with minimal $s > 0$. Assuming its existence for the moment, we call it $2\pi i$. Then it is easily shown that all the solutions of $e^x = 1$ have the form $x = 2\pi i \cdot n$ (n an integer). If $e^x = e^y$, then $e^{x-y} = 1$, so that x and y differ by an integral multiple of $2\pi i$ (and conversely).

The mapping $y = e^x$ maps the real axis one-to-one onto the positive axis, and the imaginary axis into the unit circle $|y| = 1$ (since for real β: $|e^{i\beta}|^2 = e^{i\beta}e^{-i\beta} = 1$); this mapping is actually "onto" (cf. IA, §8.4), as will be shown below; for fixed β the quantity $e^{\alpha + i\beta}$ (α, β real) runs through the set $e^{\alpha}e^{i\beta}$, or in other words along the ray issuing from 0 (but not including 0) obtained from the positive axis by multiplication with $e^{i\beta}$; for fixed α the real axis is mapped onto the circle with center 0 obtained from the unit circle by multiplication with e^{α}. If we shift the real axis parallel to itself, its image rotates about the origin and the individual points trace out circles.

4.4. We now prove the existence, assumed above, of a nontrivial solution of $e^x = 1$. If we can show the existence of a (real) α with $e^{i\alpha} = \pm i$, the desired proof is complete, since $e^{4i\alpha} = (e^{i\alpha})^4 = 1$. To do this, we introduce the functions

$$\cos x = \tfrac{1}{2}(e^{ix} + e^{-ix}),$$

$$\sin x = \frac{1}{2i}(e^{ix} - e^{-ix}).$$

For real x these functions are real

$$e^{ix} = \cos x + i \sin x,$$
$$\sin^2 x + \cos^2 x = 1.$$

For $e^{i\alpha} = \pm i$ we have $e^{-i\alpha} = \mp i$, so that $\cos \alpha = 0$ (and conversely). Thus we now seek such an α. By (20) we have

$$\cos x = 1 - \frac{x^2}{2!} \quad \frac{x^4}{4!} - \cdots = 1 - \frac{x^2}{2!}\left(1 - \frac{x^2}{3\cdot 4}\right) - \frac{x^6}{6!}\left(1 - \frac{x^2}{7\cdot 8}\right) - \cdots.$$

For $x^2 < 12$ the quantities in parentheses are positive, so that

$$\cos x < 1 - \frac{x^2}{2}\left(1 - \frac{x^2}{12}\right) \qquad \text{for} \qquad x^2 < 12,$$

$$\cos \sqrt{3} < 1 - \tfrac{3}{2}(1 - \tfrac{1}{4}) = -\tfrac{1}{8} < 0.$$

Consequently there must exist a solution of $\cos \alpha = 0$ between 0 and $\sqrt{3}$; we could even show that this solution is the smallest positive solution.

4.5. We now show that e^x maps the imaginary axis *onto* the unit circle.

From $e^{2\pi i} = 1$ it follows that $e^{\pi i} = -1$, $\cos \pi = -1$. Furthermore, $\cos 0 = 1$. Thus between 0 and π the function $\cos \alpha$ assumes all values from 1 to -1. If y is given with $|y| = 1$, then $|\text{Re } y| \leqq 1$; we choose α' real and such that $\cos \alpha' = \text{Re } y$. Then

$$\sin \alpha' = \pm\sqrt{1 - \cos^2 \alpha'} = \pm\sqrt{1 - (\text{Re } y)^2} = \pm \text{Im } y.$$

According to whether

$$\sin \alpha' = \text{Im } y \qquad \text{or} \qquad = -\text{Im } y$$

we set $\alpha = \alpha'$ or $\alpha = -\alpha'$.

In any case $e^{ix} = \cos x + i \sin x = \text{Re } y + i \,\text{Im } y = y$, which proves the assertion.

The equation $e^x = y$ ($\neq 0$) has a solution, namely $x_0 \log |y|$, where x_0 is a solution of $e^{x_0} = \dfrac{y}{|y|}\cdot$ Thus e^x maps the totality of complex numbers onto the totality of complex numbers $\neq 0$. Under this mapping two numbers have the same image if and only if they differ from each other by an integral multiple of $2\pi i$.

5. Functions in *n*-Dimensional Space

5.1. We deal here with mappings of sets that lie in (real or complex) Cartesian spaces R or S. Let us first discuss the case that R (the space containing the original set) is one-dimensional. The mapping

$$t \to \mathfrak{x}(t), \qquad a \leqq t \leqq b,$$

is called a *curve* (in S); in *kinematic* terms we also speak of a *motion*.

The derivative or *velocity* at the time t_0 is defined as the limit of the velocity between t_0 and t ($\lim t = t_0$) (if it exists):

$$\left(\frac{d\mathfrak{x}}{dt}\right)_{t=t_0} = \mathfrak{x}'(t_0) = \lim_{t\to t_0} \frac{\mathfrak{x}(t) - \mathfrak{x}(t_0)}{t - t_0}.$$

Thus the velocity is a vector, in this case a tangent vector. If $\mathfrak{x}(t)$ has the coordinates $x_\nu(t)$, then $x_\nu'(t)$ are the coordinates of $\mathfrak{x}'(t)$. The tangent is represented by

$$\mathfrak{x}(t_0) + (t - t_0)\mathfrak{x}'(t_0).$$

If we compare this expression with $\mathfrak{x}(t)$, the remainder is again of the form

$$\epsilon(t \to t_0)|t - t_0|$$

(of course, ϵ is here a vector).

5.2. Now let us consider the case that S (the space containing the set of images) is of dimension 1. Then we are dealing with the function

$$f(\mathfrak{x}) = f(x_1, \ldots, x_r),$$

which is defined for certain points in the r-dimensional space R. (For $r = 2$ this function may be visualized as a mountain on the $x_1 - x_2$-plane.) If they exist, we may form the derivatives of f with respect to the individual variables, the so-called *partial derivatives*, for which it is customary to use the symbol ∂:

$$\frac{\partial f}{\partial x_\nu}, \qquad \text{also denoted by } f_{x_\nu}.$$

We now ask: if we traverse a curve $\mathfrak{x}(t)$ in R, how will the function f behave? (In the situation visualized above, we will then be proceeding along a curved path on the mountain that represents the function.) In particular we may ask about the derivative of

$$g(t) = f(\mathfrak{x}(t))$$

(i.e., we may investigate the tangent vector of the curve on the mountain, which is also one of the tangent vectors of the surface represented by the mountain).

We assume that in a neighborhood of $\mathfrak{x}^0$ the function f is defined and has partial derivatives that are continuous at $\mathfrak{x}^0$. For simplicity we write the formulas for $r = 2$:

$$f(x_1, x_2) - f(x_1^0, x_2) = (x_1 - x_1^0)f_{x_1}(x_1^0, x_2^0) + |x_1 - x_1^0|\epsilon_1(\mathfrak{x} \to \mathfrak{x}^0),$$
$$f(x_1^0, x_2) - f(x_1^0, x_2^0) = (x_2 - x_2^0)f_{x_2}(x_1^0, x_2^0) + |x_2 - x_2^0|\epsilon_2(\mathfrak{x} \to \mathfrak{x}^0).$$

Adding these equalities, and setting

$$\frac{|x_1 - x_1^0|}{|\mathfrak{x} - \mathfrak{x}^0|}\,\epsilon_1 + \frac{|x_2 - x_2^0|}{|\mathfrak{x} - \mathfrak{x}^0|}\,\epsilon_2 = \epsilon(\mathfrak{x} \to \mathfrak{x}^0),$$

we obtain

$$f(\mathfrak{x}) - f(\mathfrak{x}^0) = \sum (x_\nu - x_\nu^0) f_{x_\nu}(\mathfrak{x}^0) + |\mathfrak{x} - \mathfrak{x}^0|\epsilon(\mathfrak{x} \to \mathfrak{x}^0). \tag{24}$$

Thus f is linear "to the first approximation," i.e., except for a quantity that approaches 0 faster than $\mathfrak{x} - \mathfrak{x}^0$. We write this result in the form

$$df = \sum \frac{\partial f}{\partial x_\nu}\, dx_\nu, \tag{25}$$

where we have replaced $f(\mathfrak{x}) - f(\mathfrak{x}^0)$ by the "infinitesimal" increment df and $x_\nu - x_\nu^0$ by dx_ν. Then df is called the *total differential* of f.

If we now let $\mathfrak{x}$ depend differentiably on a parameter t and insert

$$\mathfrak{x}(t) - \mathfrak{x}(t^0) = \mathfrak{x}'(t^0)(t - t^0) + |t - t^0|\epsilon(t \to t^0)$$

into the formula (24), and if we then set

$$g(t) = f(\mathfrak{x}(t))$$

and

$$g(t) - g(t^0) = \sum f_{x_\nu}(\mathfrak{x}(t^0)) x_\nu'(t^0)(t - t^0) + |t - t^0|\epsilon_1(t \to t^0),$$

we obtain the formula

$$g'(t^0) = \sum f_{x_\nu}(\mathfrak{x}(t^0)) x_\nu'(t^0), \tag{26}$$

which is a generalization of the chain rule to the case of several variables.

The formula (24) can also be generalized in the sense of §2.5, c):

$$f(\mathfrak{x}^*) - f(\mathfrak{x}) = \sum (x_\nu^* - x_\nu) f(\mathfrak{x}^0) + |\mathfrak{x}^* - \mathfrak{x}|\epsilon(\mathfrak{x}^* \to \mathfrak{x}^0, \mathfrak{x} \to \mathfrak{x}^0). \tag{27}$$

If such a representation of f is possible, we say that f is *totally differentiable* at $\mathfrak{x}^0$. The function f is totally differentiable at $\mathfrak{x}^0$ provided that the partial derivatives exist in the neighborhood of $\mathfrak{x}^0$ and are continuous at $\mathfrak{x}^0$.

5.3. We can also form partial derivatives of higher order, provided they exist. Here it is to be noted that

$$\frac{\partial}{\partial x_2}\frac{\partial f}{\partial x_1} = f_{x_1 x_2}$$

is not necessarily the same as $f_{x_2 x_1}$. But it is true that if we assume, say, the existence of f_{x_1} and f_{x_2} in a neighborhood of $\mathfrak{x}^0$ and the existence and continuity of $f_{x_1 x_2}$ at $\mathfrak{x}^0$, then $f_{x_1 x_2}(\mathfrak{x}^0)$ *exists and is equal to* $f_{x_1 x_2}(\mathfrak{x}^0)$. We shall not give the proof here, since it is similar to proofs given above.

5.4. We now discuss the general case, in which a set of the r-dimensional Cartesian space R is mapped into an s-dimensional Cartesian space S:

$$\mathfrak{y} = f(\mathfrak{x}).$$

Here f consists of s scalar functions

$$y_\mu = f_\mu(\mathfrak{x}), \qquad \mu = 1, \ldots, s.$$

We assume that these functions are defined in a neighborhood of $\mathfrak{x}^0$ and that they have continuous partial derivatives at $\mathfrak{x}^0$. By (24) we have

$$f_\mu(\mathfrak{x}) - f_\mu(\mathfrak{x}^0) = \sum_\nu \frac{\partial f_\mu}{\partial x_\nu}(\mathfrak{x}^0)(x_\nu - x_\nu^0) + |\mathfrak{x} - \mathfrak{x}^0|\epsilon_\mu(\mathfrak{x} \to \mathfrak{x}^0).$$

This system of s equations can again be combined into one equation: the first term on the right-hand side is the result of a homogeneous linear mapping A, applied to the vector $\mathfrak{x} - \mathfrak{x}^0$. Consequently

$$(28) \qquad f(\mathfrak{x}) - f(\mathfrak{x}^0) = A(\mathfrak{x} - \mathfrak{x}^0) + |\mathfrak{x} - \mathfrak{x}^0|\epsilon(\mathfrak{x} \to \mathfrak{x}^0).$$

The difference between $f(\mathfrak{x})$ and $f(\mathfrak{x}^0)$ is linear in $\mathfrak{x} - \mathfrak{x}^0$ up to a remainder which approaches 0 faster than $|\mathfrak{x} - \mathfrak{x}^0|$. Of course, A depends on $\mathfrak{x}^0$, and its matrix

$$(29) \qquad \left(\frac{\partial f_\mu}{\partial x_\nu}\right)$$

is called the *functional matrix at* $\mathfrak{x}^0$. In analogy with (25) we also write

$$(29a) \qquad d\mathfrak{y} = A\,d\mathfrak{x};$$

here the "infinitesimal" increments of $\mathfrak{y}$ are homogeneous linear combinations of the increments of $\mathfrak{x}$. In analogy with (27) we again have

$$(30) \qquad f(\mathfrak{x}^*) - f(\mathfrak{x}) = A(\mathfrak{x}^* - \mathfrak{x}) + |\mathfrak{x}^* - \mathfrak{x}|\epsilon(\mathfrak{x} \to \mathfrak{x}^0, \mathfrak{x}^* \to \mathfrak{x}^0).$$

If the function f can be represented in this way, we say that f is *totally differentiable* at $\mathfrak{x}^0$.

If we combine two such mappings

$$\mathfrak{y} = f(\mathfrak{x}),$$
$$\mathfrak{z} = g(\mathfrak{y}),$$

to form

$$\mathfrak{z} = h(\mathfrak{x}),$$

we see from

$$\mathfrak{y}^* - \mathfrak{y} = A(\mathfrak{x}^* - \mathfrak{x}) + |\mathfrak{x}^* - \mathfrak{x}|\epsilon_1(\mathfrak{x} \to \mathfrak{x}^0, \mathfrak{x}^* \to \mathfrak{x}^0),$$
$$\mathfrak{z}^* - \mathfrak{z} = B(\mathfrak{y}^* - \mathfrak{y}) + |\mathfrak{y}^* - \mathfrak{y}|\epsilon_2(\mathfrak{y} \to \mathfrak{y}^0, \mathfrak{y}^* \to \mathfrak{y}^0)$$

that

$$\mathfrak{z}^* - \mathfrak{z} = BA(\mathfrak{x}^* - \mathfrak{x}) + |\mathfrak{x}^* - \mathfrak{x}|\epsilon(\mathfrak{x} \to \mathfrak{x}^0, \mathfrak{x}^* \to \mathfrak{x}^0),$$

i.e., *when the mappings f and g are combined, the corresponding linear mappings A and B are multiplied.*

5.5. If f is defined as in §5.4 in a neighborhood of $\mathfrak{x}^0$ and if f has an inverse g which is totally differentiable in a neighborhood of $\mathfrak{y}^0 = f(\mathfrak{x}^0)$ (in other words, gf = identity), then for the corresponding linear mappings A and B it must also be true that BA = identity. Consequently the dimensions of R and S must coincide and for the "functional determinant" we must have $\det A \neq 0$ at $\mathfrak{x}^0$. If $\det A = 0$, then f cannot have an inverse of this sort.

Now let f be totally differentiable with $\det A \neq 0$ at $\mathfrak{x}^0$, and let there exist a continuous inverse g of f in the neighborhood of $\mathfrak{y}^0 = f(\mathfrak{x}^0)$. Then g is also totally differentiable at $\mathfrak{x}^0$. For we have

$$\mathfrak{y}^* - \mathfrak{y} = A(\mathfrak{x}^* - \mathfrak{x}) + |\mathfrak{x}^* - \mathfrak{x}|\epsilon(\mathfrak{x} \to \mathfrak{x}^0, \mathfrak{x}^* \to \mathfrak{x}^0),$$

and thus

$$(31) \qquad A^{-1}(\mathfrak{y}^* - \mathfrak{y}) = (\mathfrak{x}^* - \mathfrak{x}) + |\mathfrak{x}^* - \mathfrak{x}|A^{-1}\epsilon(\mathfrak{x} \to \mathfrak{x}^0, \mathfrak{x}^* \to \mathfrak{x}^0).$$

Since A^{-1} is linear, it follows that $|A^{-1}\mathfrak{z}| \leqq \alpha|\mathfrak{z}|$ for all $\mathfrak{z}$. Consequently, if $\mathfrak{x}$, $\mathfrak{x}^*$ are so close to $\mathfrak{x}^0$ that

$$|\epsilon(\mathfrak{x} \to \mathfrak{x}^0, \mathfrak{x}^* \to \mathfrak{x}^0)| \leqq \frac{1}{2\alpha},$$

then

$$|A^{-1}(\mathfrak{y}^* - \mathfrak{y})| \geqq \tfrac{1}{2}|\mathfrak{x}^* - \mathfrak{x}|.$$

Thus $|\mathfrak{x}^* - \mathfrak{x}| \leqq 2\alpha|\mathfrak{y}^* - \mathfrak{y}|$. Inserting this result in (31), we obtain

$$\mathfrak{x}^* - \mathfrak{x} = A^{-1}(\mathfrak{y}^* - \mathfrak{y}) + |\mathfrak{y}^* - \mathfrak{y}|\epsilon_1(\mathfrak{y} \to \mathfrak{y}^0, \mathfrak{y}^* \to \mathfrak{y}^0),$$

in view of the fact that, after the substitution of $g(\mathfrak{y})$, $g(\mathfrak{y}^*)$, and $g(\mathfrak{y}^0)$ for $\mathfrak{x}$, $\mathfrak{x}^*$, and $\mathfrak{x}^0$, we may consider the $\epsilon(\mathfrak{x} \to \mathfrak{x}^0, \mathfrak{x}^* \to \mathfrak{x}^0)$ as an $\epsilon(\mathfrak{y} \to \mathfrak{y}^0, \mathfrak{y}^* \to \mathfrak{y}^0)$, since g is continuous.

5.6. We now come to the most important theorem in this part of the subject.

Let f map a set of points in the r-dimensional space R into a set of points in the r-dimensional space S. Let f be defined in a neighborhood of $\mathfrak{x}^0$ *and be totally differentiable at* $\mathfrak{x}^0$. *Assume also that the functional determinant does not vanish at* $\mathfrak{x}^0$. *Then there exists a neighborhood U of* $\mathfrak{x}^0$ *which is mapped one-to-one by f, has an image f(U) containing a neighborhood V of* $\mathfrak{y}^0 = f(\mathfrak{x}^0)$, *and is such that the inverse g of f defined in V (with images in U) is again totally differentiable at* $\mathfrak{x}^0$.

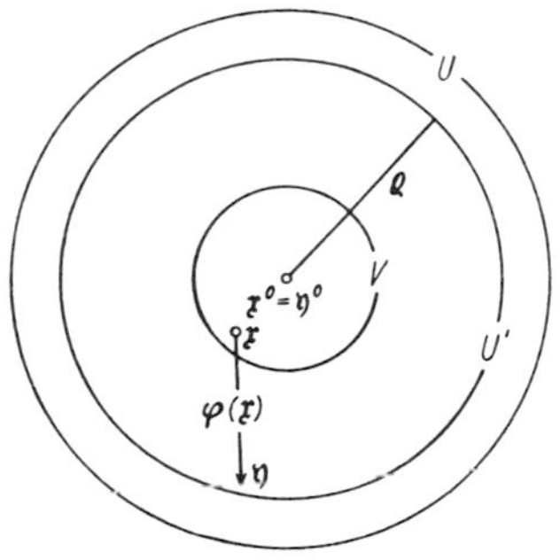

Fig. 1

For the proof (cf. Figure 1) we first assume:

$$R = S, \qquad \mathfrak{x}^0 = \mathfrak{y}^0 = 0, \qquad A = \text{identity},$$

so that

$$\mathfrak{y} = \mathfrak{x} + \varphi(\mathfrak{x}), \tag{32}$$

$$\begin{aligned} \varphi(0) &= 0, \\ |\varphi(\mathfrak{x}^*) - \varphi(\mathfrak{x})| &= |\mathfrak{x}^* - \mathfrak{x}|\epsilon(\mathfrak{x} \to 0, \mathfrak{x}^* \to 0). \end{aligned} \tag{33}$$

Thus there exists a (spherical) neighborhood U of $\mathfrak{x}^0$ in which

$$|\varphi(\mathfrak{x}^*) - \varphi(\mathfrak{x})| \leqq \vartheta|\mathfrak{x}^* - \mathfrak{x}|, \qquad \mathfrak{x}, \mathfrak{x}^* \in U, \qquad \vartheta \text{ fixed, } < 1. \tag{34}$$

We now choose a somewhat smaller sphere U' (with $\overline{U}' \subset U$). Then if ρ is the radius of U, we have from (34) that

$$|\varphi(\mathfrak{x})| \leqq \vartheta\rho \qquad \text{for} \qquad \mathfrak{x} \in U'. \tag{35}$$

We let V consist of all those points of U' that are at a distance $> \vartheta\rho$ from the boundary of U'.

The inverse g of f will be obtained by iteration. We wish to solve the equation

$$\mathfrak{x} = \mathfrak{y} - \varphi(\mathfrak{x}) \tag{36}$$

by means of a mapping $\mathfrak{x} = g(\mathfrak{y})$ and we begin with a "zeroth approximation" g_0 (say $g_0(\mathfrak{y}) \equiv \mathfrak{y}$). If we substitute $g_0(\mathfrak{y})$ into (36), we do not obtain an equality, in general. Thus we set

$$g_1(\mathfrak{y}) = \mathfrak{y} - \varphi(g_0(\mathfrak{y})) \tag{37, 1}$$

as a new approximation and proceed in exactly the same way.

In general, we set

$$g_{n+1}(\mathfrak{y}) = \mathfrak{y} - \varphi(g_n(\mathfrak{y})). \tag{37, n + 1}$$

If the g_n have a limit

$$\lim g_n = g, \tag{38}$$

it follows from the continuity of φ and from $(37, n + 1)$ by passage to the limit that

$$g(\mathfrak{y}) = \mathfrak{y} - \varphi(g(\mathfrak{y})).$$

But then we have found a solution $\mathfrak{x} = g(\mathfrak{y})$ of (36). The convergence is seen as follows: by (34) and (35)

$$|g_{n+2}(\mathfrak{y}) - g_{n+1}(\mathfrak{y})| = |\varphi(g_{n+1}(\mathfrak{y})) - \varphi(g_n(\mathfrak{y}))| \tag{39}$$

$$\leqq \vartheta|g_{n+1}(\mathfrak{y}) - g_n(\mathfrak{y})| \leqq \cdots \leqq \vartheta^{n+1}|g_1(\mathfrak{y}) - g_0(\mathfrak{y})| = \vartheta^{n+1}|\varphi(\mathfrak{y})| \leqq \vartheta^{n+1}\rho.$$

Thus there exists uniformly

$$\lim g_n = \lim\left(g_0 + \sum_{\nu=1}^{n}(g_\nu - g_{\nu-1})\right) = g_0 + \sum_{\nu=1}^{\infty}(g_\nu - g_{\nu-1}),$$

since $\sum \rho\vartheta^n$ is a majorant of this series and $0 \leqq \vartheta < 1$.

But the proof is not yet completely in order, since in (34) and (35) we made the assumption that the values substituted into φ lie in U', whereas we do not yet even know whether the values of a $g_{n+1}(\mathfrak{y})$ may not lie outside the domain of definition of φ.

We now stipulate that $g_0(\mathfrak{y}) = \mathfrak{y}$ is to be defined only in V (see the definition). Then g_1 is defined in V by (37, 1) and we have

$$g_0(\mathfrak{y}) \in U' \qquad \text{for} \qquad \mathfrak{y} \in V,$$
$$|g_1(\mathfrak{y}) - \mathfrak{y}| = |\varphi(g_0(\mathfrak{y}))| \leqq \vartheta\rho$$

in view of (37, 1) and (35); thus

$$g_1(\mathfrak{y}) \in U' \qquad \text{for} \qquad \mathfrak{y} \in V$$

by the definition of V.

Further, if we have already proved that

$$g_n(\mathfrak{y}) \in U' \qquad \text{for} \qquad \mathfrak{y} \in V, \tag{40}$$

then by $(37, n + 1)$ and (35)

$$|g_{n+1}(\mathfrak{y}) - \mathfrak{y}| = |\varphi(g_n(\mathfrak{y}))| \leqq \vartheta\rho,$$

and thus

$$g_{n+1}(\mathfrak{y}) \in U' \qquad \text{for} \qquad \mathfrak{y} \in V.$$

But then, by induction, (40) follows in general, and we have justified the application in (39) of the estimates (34) and (35). Consequently, $\lim g_n = g$ exists, and $\mathfrak{x} = g(\mathfrak{y})$ satisfies (36), so that

$$f(g(\mathfrak{y})) = \mathfrak{y} \qquad \text{for} \qquad \mathfrak{y} \in V. \tag{41}$$

Further, for $\mathfrak{y} \in V$ we have from (40) that

$$g(\mathfrak{y}) = \lim g_n(\mathfrak{y}) \in \bar{U}' \subset U,$$
$$g(V) \subset U.$$

From (41) we also have

$$V = f(g(V)) \subset f(U).$$

Consequently the f-image of U contains the set V.

The fact that f is one-to-one in U can be seen as follows. Let

$$f(\mathfrak{x}_1) = f(\mathfrak{x}_2) \qquad \text{for} \qquad \mathfrak{x}_1, \mathfrak{x}_2 \in U'.$$

Then

$$\varphi(\mathfrak{x}_1) - \varphi(\mathfrak{x}_2) = \mathfrak{x}_1 - \mathfrak{x}_2$$

and by (34)

$$|\varphi(\mathfrak{x}_1) - \varphi(\mathfrak{x}_2)| < \vartheta|\mathfrak{x}_1 - \mathfrak{x}_2|$$

with $|\vartheta| < 1$, so that

$$\mathfrak{x}_1 = \mathfrak{x}_2.$$

The inverse g of f is defined in V and is continuous there, since it is the uniform limit of continuous g_n. Thus the total differentiability of g follows from §5.5. We have obtained g constructively by iteration.

We now drop the special assumptions (32) and (33), so that f is arbitrary, $f(\mathfrak{x}^0) = \mathfrak{y}^0$,

$$f(\mathfrak{x}) - f(\mathfrak{x}^0) = A(\mathfrak{x} - \mathfrak{x}^0) + |\mathfrak{x} - \mathfrak{x}^0|\epsilon(\mathfrak{x} \to \mathfrak{x}^0).$$

We set

$$\mathfrak{X} = \mathfrak{x} - \mathfrak{x}^0, \qquad \mathfrak{Y} = A^{-1}(\mathfrak{y} - \mathfrak{y}^0).$$
$$\mathfrak{Y} = F(\mathfrak{X}) = A^{-1}(f(\mathfrak{x}) - f(\mathfrak{x}^0)).$$

Then $F(0) = 0$ and the linear mapping that corresponds to F is the identity. Thus the previous case applies and we can find neighborhoods, which we now call U^* and V^*, such that in U^* the mapping F is one-to-one, $F(U^*) \supset V^*$, and G, the inverse of F in V^*, is continuous and totally differentiable. Then we can set $U = \mathfrak{x}^0 + U^*$, $V = \mathfrak{y}^0 + AV^*$, where U, V are neighborhoods of $\mathfrak{x}^0$ and $\mathfrak{y}^0$. If $\mathfrak{y} \in V$, then $\mathfrak{y} - \mathfrak{y}^0 \in AU^*$, $\mathfrak{Y} = A^{-1}(\mathfrak{y} - \mathfrak{y}^0) \in V^*$, so that for the corresponding $\mathfrak{X}$ with $\mathfrak{Y} = F(\mathfrak{X})$ we have $\mathfrak{X} \in U^*$, and thus $\mathfrak{x} \in \mathfrak{x}^0 + U^* = U$. Consequently $f(U) \supset V$, from which the rest of the proof is clear.

5.7. The theorem of §5.6 can be used, for example, for the solution of systems of equations. Let there be given m equations with $n\,(>m)$ unknowns, where we combine the unknowns into a vector $\mathfrak{x}$ of the n-dimensional Cartesian space:

$$(42) \qquad f_i(\mathfrak{x}) = 0, \qquad i = 1, \ldots, m.$$

Let $\mathfrak{x}^0$ be a solution of this system of equations; in a neighborhood of $\mathfrak{x}^0$ assume that the partial derivatives of f_i exist and that they are continuous at $\mathfrak{x}^0$. Also let the functional matrix be such that

$$\text{rank}\left(\frac{\partial f_i}{\partial x_j}\right) = m \text{ at } \mathfrak{x}^0.$$

We may assume that among the $\binom{n}{m}$ subdeterminants of order m the first m are such that

$$\det\left(\frac{\partial f_i}{\partial x_j};\quad i = 1, \ldots, m;\quad j = 1, \ldots, m\right) \neq 0 \text{ at } \mathfrak{x}^0. \tag{43}$$

We now consider the mapping from n dimensions into n dimensions

$$\mathfrak{y} = f(\mathfrak{x}), \tag{44}$$

where the first m coordinates of f are the given f_i ($i = 1, \ldots, m$) and for the other $n - m$ coordinates we set

$$f_i(\mathfrak{x}) \equiv \mathfrak{x}_i, \qquad i = m + 1, \ldots, n. \tag{45}$$

The functional determinant of this mapping is equal to (43) and is thus $\neq 0$. We apply §5.6 and thus, in a neighborhood of $\mathfrak{x}^0$, obtain an inverse

$$\mathfrak{x} = g(\mathfrak{y})$$

of (44) with

$$f(g(\mathfrak{y})) \equiv \mathfrak{y}. \tag{46}$$

We set

$$g(0, \ldots, 0, y_{m+1}, \ldots, y_n) = h(y_{m+1}, \ldots, y_n).$$

Then by (46)

$$f_i(h(y_{m+1}, \ldots, y_n)) \equiv 0, \qquad i = 1, \ldots, m. \tag{47}$$

Setting $g(\mathfrak{y})$ in (45) we obtain

$$f_i(g(\mathfrak{y})) \equiv g_i(\mathfrak{y}), \qquad i = m + 1, \ldots, n,$$

and thus, in view of (46),

$$g_i(\mathfrak{y}) \equiv y_i, \qquad i = m + 1, \ldots, n,$$

and therefore

$$h_i(\mathfrak{y}) \equiv y_i, \qquad i = m + 1, \ldots, n. \tag{48}$$

From (47) and (48) we have

$$f_i(h_1(y_{m+1}, \ldots, y_n), \ldots, h_m(y_{m+1}, \ldots, y_n), y_{m+1}, \ldots, y_n) \equiv 0,$$
$$i = 1, \ldots, m.$$

If we now replace the y by the x, we obtain functions

$$h_1(x_{m+1}, \ldots, x_n), \ldots, h_m(x_{m+1}, \ldots, x_n),$$

which when substituted for

$$x_1, \ldots, x_m$$

in (42) satisfy the system of equations in a neighborhood of $\mathfrak{x}^0$. Consequently we have solved this system in a neighborhood of $\mathfrak{x}^0$.

5.8. But what can we say in case the functional matrix is of lower order? In general, nothing. But if this lowering of the order takes place not only at one point but on an entire open set, we can determine whether a given function is dependent on the functions in (42). More precisely, we have the following situation.

In a neighborhood of the point $\mathfrak{x}^0$ of the n-dimensional Cartesian space let the functions

$$\varphi(\mathfrak{x}) \qquad \text{and} \qquad f_i(\mathfrak{x}) \qquad i = 1, \ldots, p$$

be defined and have continuous partial derivatives. Also let

$$\text{rank}\left(\frac{\partial f_i}{\partial x_j}\right) = p. \tag{49}$$

We say that φ is dependent on the f_i in the neighborhood of $\mathfrak{x}^0$ if there exists a function F with continuous partial derivatives in the p-dimensional Cartesian space such that

$$F(f_1(\mathfrak{x}), \ldots, f_p(\mathfrak{x})) \equiv \varphi(\mathfrak{x}) \tag{50}$$

in a suitable neighborhood of $\mathfrak{x}^0$ (of course, the left-hand side is required to be meaningful in this neighborhood). We now prove the following theorem.

The function φ is dependent on the f_i in the neighborhood of $\mathfrak{x}^0$ if and only if there exists a neighborhood of $\mathfrak{x}^0$ in which the matrix (49) *preserves the same rank φ when extended by the row $\dfrac{\partial \varphi}{\partial x_j}$.*

For the proof we note that when we make a substitution $\mathfrak{x} = g(\mathfrak{y})$ with continuous partial derivatives and nonvanishing functional determinant in a neighborhood of $\mathfrak{x}^0$, dependence, independence, and rank of the functional matrix remain unchanged (the functional matrix will not change its rank, since it is multiplied by a nonsingular matrix).

We now consider again the mapping

$$\mathfrak{y} = f(\mathfrak{x}),$$

where the coordinates f_i of f for $i = 1, \ldots, p$ are the given coordinates and $f_i(\mathfrak{x}) \equiv x_i$ for $i > p$. We again invert the mapping: $\mathfrak{x} = g(\mathfrak{y})$. If we replace $\mathfrak{x}$ in the given $f_1, \ldots, f_p$ in accordance with this mapping, we obtain functions of $\mathfrak{y}$ which are simply the $y_1, \ldots, y_p$ respectively. Thus by the preceding section we may assume that the $f_1, \ldots, f_p$ have been so chosen in advance. Then the matrix (49) is the unit matrix and the condition on the rank of the extended matrix means simply that $\frac{\partial \varphi}{\partial x_j} = 0$ for $j > p$, so that φ depends only on the $y_1, \ldots, y_p$. But then the same situation holds in (50) (namely, $F(y_1, \ldots, y_p) \equiv \varphi(g(\mathfrak{y}))$).

CHAPTER 3

Integral and Measure

Introduction

Since the first appearance of the famous Lebesgue theory of measure and integral it has been known that "integral" and "measure" are very closely related; they are among the most important concepts of analysis.

In the six decades since Lebesgue constructed his theory the two concepts have undergone a vigorous development. It has been realized that the classical Lebesgue procedure can be replaced by less laborious methods, and it has become clear that the central theorems on integration remain valid, with the necessary changes, under essentially more general assumptions than in the original theory of Lebesgue.

In the present discussion we take account of these and other developments. The first section deals essentially with the *Riemann integral.* Since we may assume that the reader is already well acquainted with this integral, we have emphasized certain aspects of its theory that are not included in the standard textbooks on calculus. Our purpose here is to prepare the way for an abstract presentation of integral and measure in §2. By suitable specialization this abstract theory also gives rise to the *Lebesgue integral* and the *Lebesgue measure*, though naturally in a far different way from the classical Lebesgue theory. In §3 we give some discussion of the *theory of distributions* of L. Schwartz. The content of §3 is closely related to the exposition in §§1 and 2.

1. Elementary Theory of Integration

1.0. *Preliminary Remarks*

By R we shall always mean the field of real numbers. For an arbitrary fixed integer $n \geqq 1$ we denote by R^n the set of all n-tuples $x = (x_\nu) = (x_1, \ldots, x_n)$ of real numbers $x_1, \ldots, x_n$. The symbol $\alpha \,_{(}\!\leqq\!_{)}\, \xi \,_{(}\!\leqq\!_{)}\, \beta$ (α, β, ξ real, $\alpha \leqq \beta$) is used as a common notation for the four intervals $\alpha < \xi < \beta$,

$\alpha \leqq \xi < \beta$, $\alpha < \xi \leqq \beta$, $\alpha \leqq \xi \leqq \beta$ of R^1, in situations where it makes no difference whether the endpoints α, β belong to the interval or not. By *intervals in* R^n we mean sets of points $J \subset R^n$ of the form

$$J = \{x : x = (x_\nu),\ a_\nu \ (\leqq)\ x_\nu \ (\leqq)\ b_\nu\} \qquad (a_\nu), (b_\nu), (x_\nu) \in R^n,\ a_\nu \leqq b_\nu.$$

For a given J, a set of finitely many intervals $J^{(1)}, \ldots, J^{(k)}$ of R^n with

$$J = \bigcup_{\kappa=1}^{k} J^{(\kappa)} \qquad \text{and} \qquad J^{(\kappa)} \cap J^{(\lambda)} = \varnothing \qquad \text{for} \qquad \kappa \neq \lambda,$$

will be called a *partition* $\tau = (J^{(1)}, \ldots, J^{(k)}$ of $J)$.

1.1. *Step Functions, Vector Lattices (Riesz Spaces)*

Let J be a fixed closed interval $\{x : x = (x_\nu),\ a_\nu \leqq x_\nu \leqq b_\nu,\ a_\nu < b_\nu\}$ of R^n. With certain real-valued functions f[1] defined on J we now wish to associate an integral in some useful way; to do this, we may proceed, for example, in the following manner. We first define the integral only for "particularly simple" functions, and then attempt to extend the definition to "more complicated" ones.

It has turned out that in this sense the simplest functions are step functions. A real-valued function t defined on J is called a *step function* (*on* J) if for the function t there exist a partition $\tau = (J^{(1)}, \ldots, J^{(k)})$ of J and real numbers $c_1, \ldots, c_k$ such that for every $x \in J$

$$t(x) = \sum_{\kappa=1}^{k} c^{\kappa} \chi_{J^{(\kappa)}}(x) \qquad x \in J. \tag{1}$$

Here $\chi_{J^{(\kappa)}}$ denotes the characteristic function of $J^{(\kappa)}$ in J.[2]

We note that a step function t can be written in the form (1) in various ways (since for a given t there are various partitions τ with the property (1)).

The sum of two step functions and the product of a real number with a step function are again step functions. The maximum and the minimum of two step functions are step functions.[3]

These easily proved theorems state the *fundamental properties* of the

[1] As in III 2, functions will be denoted by $f, g, \ldots, t, \ldots$. On the other hand a symbol like $g(y)$ refers to the value of the function g at a fixed point y of the domain of definition E of g. Here E can be any set $\neq \varnothing$, and g may be defined by assigning to each $y \in E$ a unique value $g(y)$. In textbooks on calculus, the function g is often denoted by $g(y)$ with variable y, but at least in the present chapter this notation is quite unsuitable and can easily lead to misunderstanding.

[2] If $E \neq \varnothing$ and $A \subseteq E$ are sets, the *characteristic function* χ_A *of* A *in* E is defined as $\chi_A(y) = 1$ for $y \in A$, $\chi_A(y) = 0$ for $y \in (E - A)$.

[3] If f, g are arbitrary real-valued functions defined on a set $E \neq \varnothing$, their sum $f + g$, their maximum $\sup(f, g)$ and their minimum $\inf(f, g)$ are defined by $(f + g)(y) = f(y) + g(y)$, $(\sup(f, g))(y) = \max(f(y), g(y))$ and $(\inf(f, g))(y) = \min(f(y), g(y))$ respectively, for every $y \in E$, and the product αf of f with the real

functions t, but it is convenient for us to interpret these properties in a somewhat different way: we consider the set $T = T(J)$ of step functions t on J as a space whose points are the functions t. In terms of the space T the fundamental properties of t can be formulated thus:

(2) From $t_1, t_2 \in T$, $\alpha \in R$ follows $t_1 + t_2 \in T$ and $\alpha t_1 \in T$.
(3) From $t_1, t_2 \in T$ follows $\sup(t_1, t_2) \in T$ and $\inf(t_1, t_2) \in T$.

(Cf. footnote 3.) We then bring (2) and (3) into relation with general concepts of modern mathematics. For (2) this is easily done: from the usual rules for calculation with real-valued functions it follows from (2) that T is a vector space over R.[4] In contrast to the vector spaces of analytic geometry the present vector space is not of finite dimension.

The condition (3) expresses an order property of a special kind for the vector space T. If for $t_1, t_2 \in T$ we set

(4) $t_1 \leqq t_2$ if and only if for every $x \in J$
$t_1(x) \leqq t_2(x)$,

we thereby introduce into the space T the so-called natural (partial) order $\leqq$.[5] This relation is "compatible" with the structure of T as a vector space; in other words, we have

(5) From $t_1 \leqq t_2$ it follows that $t_1 + t \leqq t_2 + t$ for arbitrary $t \in T$
$t_1, t_2 \in T$;
(6) From $t \geqq 0$, $\alpha > 0$ it follows that $\alpha t \geqq 0$ $\quad \alpha \in R$, $t \in T$.

(Cf. footnotes 5 and 6.)
It is easy to see that the maximum and the minimum of two step functions coincide with their supremum and infimum in T in the sense of the order

---

number α is defined by $(\alpha f)(y) = \alpha f(y)$ for $y \in E$. Later we shall need the absolute value $|f|$, the positive part f^+, and the negative part f^- of a function f, which are defined as follows: $|f|(y) = |f(y)|$ for every $y \in E$; and if for the function N "identically zero on E": $N(y) = 0$ for $y \subset E$ we write simply 0, then f^+ and f^- are defined by $f^+ = \sup(f, 0)$ and $f^- = \inf(f, 0)$, respectively.

[4] A set $V \neq \varnothing$ of elements $u, v, \ldots$ is called a *vector space over* R if for every pair $u, v \in V$ there is defined a sum $u + v \in V$ and for every $\alpha \in R$ and every $u \subset V$ a (scalar) product $\alpha u \in V$ with the following properties: (I) the set V is an *Abelian group* with respect to addition; (II) $\alpha(u + v) = \alpha u + \alpha v$, $(\alpha + \beta)u = \alpha u + \beta u$; (III) $(\alpha\beta)u = \alpha(\beta u)$, $1u = u$. The conditions (II), (III) must be satisfied for arbitrary $\alpha, \beta \in R$ and arbitrary $u, v \in V$. The elements of a vector space are called *vectors*. (Cf. IB3, §1.)

[5] On order cf. IA, §8.3 and IB9, §1. A set $V \neq 0$ is (*partially*) *ordered* by a relation $\leqq$ if for certain u, v in V the relation $u \leqq v$ is so defined that the following axioms are satisfied: (I) $u \leqq u$ for every $u \in V$; (II) from $u \leqq v$ and $v \leqq w$ follows $u \leqq w$; (III) from $u \leqq v$ and $v \leqq u$ follows $u = v$. We agree that the notations $u \leqq v$ and $v \geqq u$ have the same meaning. If for $u, v \in V$ there exists a $w \in V$ with $u \leqq w$, $v \leqq w$, such that for every $z \in V$ with $u \leqq z$, $v \leqq z$ we have $w \leqq z$, then w is called the *supremum* (*least upper bound*) *of* u *and* v *in* V. The infimum (greatest lower bound) of u and v is defined analogously.

$\leqq$.[5] Thus (3) states that the vector space T, ordered by (4), *contains the supremum and the infimum of every pair of elements in T.* In general a vector space T over R with an order relation $\leqq$ and with properties (5), (6), (3) is called a *vector lattice* (or a *Riesz space*). The importance of this concept is due to the fact that practically all the spaces of real-valued functions f to which it is possible to assign an integral $L(f)$ in any usual way are *vector lattices* or closely related to them. Examples of *vector lattices*, in addition to the $T(J)$ considered above, are the space $C(J)$ of continuous functions on J and the space $\mathfrak{R}(J)$ of bounded *Riemann integrable functions* over J.

It is obvious that every vector lattice of real-valued functions that contains a function f also contains its positive part $f^+ = \sup(f, 0)$, its negative part $f^- = \inf(f, 0)$ and its absolute value $|f| = f^+ - f^-$.[6]

1.2. *Integration of Step Functions, Linear Functionals*

We continue with the assumptions and notation of 1.1. Given an arbitrary step function t on J, which we shall write as

$$(1') \qquad t = \sum_{\kappa=1}^{k} c_\kappa \chi_{J^{(\kappa)}} \qquad c_1, \ldots, c_k \in R; \qquad \text{where} \qquad (J^{(1)}, \ldots, J^{(k)}) \text{ is a partition of } J,$$

we now wish to associate with it an integral $I(t)$. It is natural to begin by defining $I(\chi_{J'})$ for the subintervals J' of J and then attempt to define $I(t)$ by setting

$$(7) \qquad I(t) = \sum_{\kappa=1}^{k} c_\kappa I(\chi_{J^{(\kappa)}}) \qquad t, c_\kappa, J^{(\kappa)} \qquad \text{as in } (1').$$

Under quite general assumptions on $I(\chi_{J'})$ a usable definition of an integral for step functions can in fact be constructed in this way.[7] Here we shall restrict ourselves to the following definition of $I(\chi_{J'})$, which naturally suggests itself,[8] if only for historical reasons,

$$(8) \quad I(\chi_{J'}) = \prod_{i=1}^{n} (b_i' - a_i') \quad \text{for} \quad J' = \{x : x = (x_\nu), \quad a_\nu' \,_{(\leqq)}\, x_\nu \,_{(\leqq)}\, b_\nu'\}.$$

[6] In $t \geqq 0$ and $\alpha t \geqq 0$ the symbol 0 denotes the function "identically zero on J" (see footnote 3); we shall continue to use this notation below, since it can hardly lead to confusion with the number zero.

[7] As is shown by MacShane [8], Chap. II, V, it is sufficient for many purposes to require that (I) $0 \leqq I(\chi_{J'}) = I(\chi_{\bar{J}'}) < \infty$, where $\bar{J}'$ is the closed hull (see footnote 19) of J'; and (II) $I(\chi_{J' \cup J''}) = I(\chi_{J'}) + I(\chi_{J''})$ for arbitrary subintervals J', J'' of J with $J' \cap J'' = \varnothing$.

[8] For $n = 1$, $t \geqq 0$ the equations (7), (8) mean that $I(t)$ is equal to the area of the point set $\{(x, y) : x \in J, 0 \leqq y \leqq t(x)\}$ in R^2. The connection between area and integral has played an extremely important part in the development of the integral calculus.

Here $I(\chi_{J'})$ is equal to the *n-dimensional elementary-geometric content* $|J'|$ of the interval J'. Since it is easy to see that the expression $\sum_{\kappa=1}^{k} c_\kappa |J^{(\kappa)}|$ is independent of the particular choice of representation (1′) for t, we can define $I(t)$ uniquely by setting

$$(9) \qquad I(t) = \sum_{\kappa=1}^{k} c_\kappa |J^{(\kappa)}| \qquad t, c_\kappa, J^{(\kappa)} \quad \text{as in } (1').$$

This value $I(t)$ is called the (*definite*) *integral of the step function t over J*; various notations for the integral are

$$(10) \quad I(t) = \int_J t\,dx = \int_J t(x)dx = \int_{a_1}^{b_1} \cdots \int_{a_n}^{b_n} t(x_1, \ldots, x_n)dx_1 \cdots dx_n.$$

Among the properties of the "step integral" $I(t) = \int_J t\,dx$ the following are almost self-evident:

$$(11) \qquad I(t_1 + t_2) = I(t_1) + I(t_2) \qquad t_1, t_2 \in T;$$

$$(12) \qquad I(\alpha t) = \alpha I(t) \qquad \alpha \in R,\ t \in T;$$

$$(13) \qquad I(t) \geqq 0 \quad \text{for} \quad t \geqq 0 \qquad t \in T.$$

For the interpretation of such relations we may, if we wish, take the point of view customary in the integral calculus. Then the definite integral $I(t) = \int_J t\,dx$ appears as a real number assigned to each fixed step function t, and (11), (12), (13) are rules for calculation. For example, (12) states that a constant factor can be put in front of the "integral sign" I.

But it is advantageous to make a different interpretation. The correspondence

$$(14) \qquad t \to I(t) = \int_J t\,dx \qquad t \in T$$

defines a real-valued function I, whose domain of definition is the space T of step functions on J (cf. footnote 1). For arbitrary fixed $t \in T$ the integral $\int_J t\,dx$ is the *value of this function I at the point t*. Since the function I assigns numerical values to certain functions, it is called a *functional* (a function of functions). More generally, any function L which maps a set $E \neq \varnothing$ into the field of real (or complex) numbers is often called a *functional on E*. A functional L on a vector space E over R is said to be *linear* (or to be a *linear form*) if the following equations are satisfied.

$$(15) \qquad L(y_1 + y_2) = L(y_1) + L(y_2) \qquad y_1, y_2 \in E$$

and

$$(16) \qquad L(\alpha y) = \alpha L(y) \qquad \alpha \in R, \quad y \in E.$$

In particular, if E is a *vector lattice* and if

(17) $$L(y) \geqq 0 \qquad \text{for} \qquad y \geqq 0 \qquad y \in E,$$

then L is said to be *positive. Consequently, I is a positive linear functional on the vector lattice of step functions on J.* Every positive linear functional L on a *vector lattice* E (say of real functions) satisfies the condition

$$L(|y| \pm y) = L(|y|) \pm L(y) \geqq 0,$$

i.e.,

(18) $$|L(y)| \leqq L(|y|) \qquad y \in E,$$

which for $L = I$ is equivalent to the inequality $\left|\int_J t dx\right| \leqq \int_J |t| dx$.

Finally, let us state one more property of the functional I:

$$\text{From} \quad t_1 \geqq t_2 \geqq \cdots \geqq 0 \qquad \text{and} \qquad \lim_{m \to \infty} t_m(x) = 0$$

for every $x \in J$ it follows that

(19) $$\lim_{m \to \infty} I(t_m) = I(\lim_{m \to \infty} t_m) = 0 \qquad t_1, t_2, \ldots \in T.$$[9]

We say that I is *continuous with respect to monotone limits.* The importance of the property (19) of continuity appears to have been first realized by Daniell in 1917. It will play an important role in the construction of the theory of *Lebesgue integration* and its generalizations (§2).

1.3. *Extension of the Step Integral to the Riemann Integral*

We now wish to extend the definition of an integral (9) to more complicated functions, and in the first place to functions that are *Riemann integrable.* This extension, and similar extensions below, of the original concept of an integral is best regarded as an extension of the positive linear functional I to a space of real functions that is an extension of $T = T(J)$.

If on spaces $E \neq \varnothing$ and $E' \neq \varnothing$ with $E \subseteqq E'$ there are defined functionals L and L' respectively and if for every $y \in E$

(20) $$L'(y) = L(y) \qquad y \in E,$$

then L' is said to be an *extension* (*continuation*) *of L to E'*. In particular, if E, E' are *vector lattices* and if L is positive and linear on E, then our chief attention will naturally be given to extensions L' of L that are positive and linear on E'.

Now let $E = T$ and $L = I$. In addition to T we consider, in the first place, the *vector lattice* $B = B(J)$ of all *bounded* real-valued functions on J. Then we have $T \subset B$. We now define a subspace $E' = \mathfrak{R}$ of B as follows:

[9] For the proof (for $n = 1$) see Riesz and Nagy [9], 16.

$f \in B$ belongs to $\Re$ if and only if for given f and arbitrary $\epsilon > 0$ there exist step functions $\underline{t}, \bar{t} \in T$ with $\underline{t} \leqq f \leqq \bar{t}$ and $I(\bar{t}) - I(\underline{t}) < \epsilon$. It is obvious that $T \subseteq \Re \subseteq B$ and that $\Re$ *is a vector lattice*. Moreover, for every $f \in \Re$ we have the relation

$$(21) \qquad \sup_{\underline{t} \leq f} I(\underline{t}) = \inf_{f \leq \bar{t}} I(\bar{t}) \qquad \underline{t}, \bar{t} \in T;$$

for $f \in T$ these expressions are equal to $I(f)$. Thus if for arbitrary fixed $f \in \Re$ we set

$$(22) \qquad I'(f) = \sup_{\underline{t} \leq f} I(\underline{t}) = \inf_{f \leq \bar{t}} I(\bar{t}) \qquad \underline{t}, \bar{t} \in T,$$

we have extended I to a functional I' (defined by $f \to I'(f)$ for $f \in \Re$) on $\Re$. The functional I' *is positive and linear on* $\Re$. The functions $f \in \Re = \Re(J)$ are the (*properly*)[10] *Riemann integrable* functions *over* J; the functional $I'(f)$ is called the (*definite*)[11] *Riemann integral of* f *over* J. For the case $f \in \Re$ various notations are customary, for example

$$(23) \quad I'(f) = \int_J f dx = \int_J f(x) dx = \int_{a_1}^{b_1} \cdots \int_{a_n}^{b_n} f(x_1, \ldots, x_n) dx_1 \cdots dx_n.$$

It is easy to establish the connection between these somewhat unusual definitions of *Riemann integrability* and *Riemann integral* with the definitions ordinarily found in the textbooks on integral calculus: for every $g \in B$ and every partition $\tau = (J^{(1)}, \ldots, J^{(k)})$ we define lower and upper sums as follows:

$$(24) \qquad \underline{I}(g; \tau) = \sum_{\kappa=1}^{k} \underline{g}(J^{(\kappa)}) |J^{(\kappa)}| \qquad \underline{g}(J') = \inf_{x \in J'} g(x)$$

and

$$(24') \qquad \bar{I}(g; \tau) = \sum_{\kappa=1}^{k} \bar{g}(J^{(\kappa)}) |J^{(\kappa)}| \qquad \bar{g}(J') = \sup_{x \in J'} g(x).$$

Then the supremum and the infimum for all τ and fixed g,

$$(25) \qquad \underline{I}(g) = \sup_{\tau} \underline{I}(g; \tau) \qquad g \in B$$

and

$$(25') \qquad \bar{I}(g) = \inf_{\tau} \bar{I}(g; \tau) \qquad g \in B,$$

[10] In distinction to the *improperly Riemann integrable* functions, which are not necessarily bounded or whose domain of definition is not a bounded closed set in R^n.

[11] In distinction to the *indefinite* integral Φ of f (not dealt with here), which for continuous $f \in \Re(J)$ could in the present case be defined by

$$\Phi(x) = \int_{a_1}^{x_1} \cdots \int_{a_n}^{x_n} f(y_1, \ldots, y_n) dy_1 \cdots dy_n, \qquad x = (x_\nu) \in J.$$

are respectively the *lower* and *upper Darboux integral of g over J*. It is easy to see that $g \in B$ is *Riemann integrable* over J if and only if $\underline{I}(g) = \bar{I}(g)$. In this case we also have $I'(g) = \underline{I}(g) = \bar{I}(g)$.

We assume that the reader is familiar, at least in outline, with the theory of the *Riemann integral* as presented in detail in the textbooks on integral calculus. In the present context we shall mention only the properties of $\mathfrak{R}$ that are of general interest. One such property is the fact that the *vector lattice* $C = C(J)$ *of continuous functions on* J *is contained in* $\mathfrak{R} = \mathfrak{R}(J)$. (The *Riemann integrability* of functions continuous on J is essentially a consequence of the uniform continuity of functions continuous on bounded closed subsets of R^n.) In addition to the continuous functions, the space $\mathfrak{R}$ obviously contains certain discontinuous functions, e.g., the non-constant step functions. The extent to which a *Riemann integrable function* can be discontinuous was already discussed by Riemann in his inaugural dissertation. The fact is that *a function* $g \in B(J)$ *is Riemann integrable over* J *if and only if the set of points of discontinuity of* g *in* J *is of* L_n*-measure zero*, where a subset of R^n is said to be of L_n-measure zero (n-dimensional Lebesgue measure zero) if it can be covered by an at most countable set of open intervals $J^{(1)}, J^{(2)}, \ldots$ in R^n whose total content $\sum_\kappa |J^{(\kappa)}|$ (cf. §1.2) is smaller than an arbitrarily prescribed number $\epsilon > 0$.[12]

In order to integrate even more strongly discontinuous functions we might try to repeat, with $\mathfrak{R}$ and I' in place of T and I, the procedure whereby we extended T and I to $\mathfrak{R}$ and I'. But it turns out that a function $g \in B$ already belongs to $\mathfrak{R}$: $g \in \mathfrak{R}$ if for arbitrary $\epsilon > 0$ there exist functions $\underline{f}, \bar{f} \in \mathfrak{R}$ with $\underline{f} \leqq g \leqq \bar{f}$ and $I'(\bar{f}) - I'(\underline{f}) < \epsilon$. This method of extension is "closed." Other methods are required if we wish to extend the concept of the *Riemann integral* in any significant way; one of them will be discussed in §2.

1.4. *Definition of the Riemann Integral by Means of Sums*

The definitions of *Riemann integrability* and *Riemann integral* given in §1.3 are essentially different from the sum-definition with which Riemann first introduced his concept of the integral.

In the extension of T and I to $\mathfrak{R}$ and I' in §1.3 we made use of the following properties only: the space T is a *vector lattice* of bounded real-valued functions on some set $A \neq \varnothing$, and I is a positive linear functional on T. No use was made of the topology of the domain of definition $A = J$ of the functions in T (nor was any use made of the continuity (19) of I). But the situation is quite different in the sum-definition of the *Riemann*

[12] Proofs of this theorem are to be found, e.g., in Haupt-Aumann-Pauc [5], 8.4.3.1 Theorem and in Kamke [6], 4.2, Theorem 4.

integral, in which $I'(f) = \int_J f dx$ with $f \in \mathfrak{R}$ is defined as the limiting value of the *Riemann sums*

$$(26) \qquad F(\tau; x^{(\lambda)}) = \sum_{\kappa=1}^{k} f(x^{(\kappa)}) |J^{(\kappa)}|,$$

where $\tau = (J^{(1)}, \ldots, J^{(k)})$ is a partition of J and $x^{(\kappa)} \in J^{(\kappa)}$ is arbitrary. Here the limit is taken with respect to a *filter basis* $\mathfrak{E}$ (cf. III 1, §5.2), and the definition of $\mathfrak{E}$ involves not only the topology but also the metric of R^n.

If $d(M)$ is the diameter of M[13] with $M \subset R^n$, we may call

$$(27) \qquad d(\tau) = \max_{\kappa=1,\ldots,k} d(J^{(\kappa)}) \qquad \tau = (J^{(1)}, \ldots, J^{(k)})$$

the diameter of the partition τ of J. On the set $\mathbf{T}$ of all partitions τ of J we now define $\mathfrak{E}$ as follows: the elements of $\mathfrak{E}$ are the subsets E_δ of $\mathbf{T}$ with

$$(28) \qquad E_\delta = \{\tau : \tau \in \mathbf{T}, \quad 0 < d(\tau) \leqq \delta\} \qquad \delta \text{ arbitrary } > 0.$$

It is easy to show that $\mathfrak{E} = \{E_\delta : \delta > 0\}$ is a filter basis on $\mathbf{T}$. Following Haupt [14], 1.11 we call it the *Riemann filter basis* on the set of all partitions of J. The filter generated by $\mathfrak{E}$ on $\mathbf{T}$ has a countable basis, say $\mathfrak{B} = \{E_\delta\}$, where $\delta_1 > \delta_2 > \cdots > 0$ and $\lim_{m \to \infty} \delta_m = 0$.

The sum-definition of the *Riemann integral* and its equivalence with the definition given in §1.3 now result from the following theorem: *let* $f \in \mathfrak{R}(J)$. *If to each* $\tau = (J^{(1)}, \ldots, J^{(k)})$ *we assign a Riemann sum* $F(\tau) = F(\tau; x^{(\lambda)})$ *by means of f as in* (26), *where* $x^{(\kappa)}$ *is chosen arbitrarily in* $J^{(\kappa)}$, *then*

$$(29) \qquad I'(f) = \int_J f dx = \lim_{\mathfrak{E}} F(\tau).$$

On the other hand, if we form sums $F(\tau) = F(\tau; x^{(\lambda)})$ by means of an $f \in B(J)$ and if for every choice of intermediate points $x^{(\kappa)}$ the limiting value $\lim_{\mathfrak{E}} F(\tau)$ exists, then $f \in \mathfrak{R}(J)$, and $I'(f)$ is given by (29).

The textbooks on calculus prove this theorem somewhat differently: a partition is not defined in quite the same way and the theory of filters is usually not introduced. Instead, the proof is based on sequences

$$F(\tau_m) = F(\tau_m; x_m^{(\lambda_m)}), \qquad m = 1, 2, \ldots,$$

of *Riemann* sums. In (29) $\lim_{\mathfrak{E}} F(\tau)$ is replaced by $\lim_{m \to \infty} F(\tau_m)$, where τ_m, $m = 1, 2, \ldots$, is any sequence of partitions with $\lim_{m \to \infty} d(\tau_m) = 0$ and the

[13] Let $|x, y| = (\sum_{i=1}^n (x_i - y_i)^2)^{1/2}$ be the distance between $x = (x_\nu)$ and $y = (y_\nu) \in R^n$; then we set

$$d(\varnothing) = 0, \qquad d(M) = \sup_{x,y \in M} |x, y| \qquad \text{for} \qquad \varnothing \neq M \subseteq R^n.$$

intermediate $x_m^{(\kappa_m)}$ are chosen arbitrarily for τ_m.[14] The equivalence of the two limit relations rests on the fact that the filter generated by $\mathfrak{E}$ has a countable basis. For in this case the general theory of convergence in terms of filters can be reduced to the classical theory in terms of sequences (see III 1, §5).

A proof of the equivalence of the sum-definition with other definitions of the *Riemann integral* cannot be given here. Let us merely mention the essential point: the *Darboux integrals* $\underline{I}(g)$, $\bar{I}(g)$, which exist for arbitrary $g \in B(J)$, can be defined as the supremum of the lower sums $\underline{I}(g; \tau)$ and the infimum of the upper sums $\bar{I}(g; \tau)$ ((25) and (25′)), and can also be represented as limits with respect to the filter basis $\mathfrak{E}$

$$\underline{I}(g) = \lim_{\mathfrak{E}} \underline{I}(g; \tau) \qquad g \text{ fixed} \in B, \tag{30}$$

$$\bar{I}(g) = \lim_{\mathfrak{E}} \bar{I}(g; \tau) \qquad g \text{ fixed} \in B.^{15} \tag{30'}$$

1.5. *Content as Associated with the Riemann Integral*

On the basis of the definition (9) the integral $I(t)$ of step functions t is connected with the elementary geometric content $|J'|$ of intervals $J' \subseteqq J$. In particular $|J'|$ is equal to the integral $I(\chi_{J'})$ of the characteristic function of J' in J. In the extension of I to I' this connection is extended, with the necessary changes, to a connection between the *Riemann integral* and the *(Peano-)Jordan content* about to be described. The relation between integral and content is completely symmetric in the sense that *the theory of the Riemann integral can be based on that of the Jordan content or, just as well, conversely*. In the first procedure the concept of content is fundamental and the theory takes on a predominantly geometric character.[16] But if the concept of integral is the primary one, as in the present discussion,[17] then the theory takes on a more analytic character.

A bounded point set A of R^n is said to be *Jordan-measurable* if there exists a (bounded) closed interval J of R^n which contains A and is such that the characteristic function $\chi_A = \chi_{A,J}$ of A in J is *Riemann integrable* over J: $\chi_{A,J} \in \mathfrak{R}(J)$. If this condition is satisfied for *one* interval J, it is also satisfied for every other bounded closed interval J' of R^n that includes A,

[14] See for example Mangoldt-Knopp, Einführung in die höhere Mathematik III, 9. ed., Hirzel, Leipzig 1950, in particular Sections 35 and 94. For the *Riemann integral* in terms of the theory of filters see Haupt-Aumann-Pauc [5], 8.4; compare also Haupt [14], 1.

[15] Cf. the references cited in footnote 14.

[16] For detailed accounts of this procedure see the references in footnote 14, Haupt-Aumann-Pauc [5], Sections II and III and from Mangoldt-Knopp, Section III.

[17] See also A. Ostrowski, Vorlesungen über Differential- und Integralrechnung III. Birkhäuser, Basel 1954, Chapter II, and Haupt [14], 1.13.

and we have $\int_J \chi_{A,J}dx = \int_{J'} \chi_{A,J'}dx$. This number is called the (*n-dimensional*) *Jordan content* $i(A) = i_n(A)$ of A

(31) $$i(A) = \int_J \chi_{A,J}dx \qquad A \subseteqq J,\ \chi_{A,J} \in \mathfrak{R}(J).$$

If we let Q denote the set of all measurable subsets of R^n, the Jordan content defined by (31) can be regarded as a function i defined on Q. Since the $A \in Q$ are sets, i is called a *set function.* The fundamental properties of Q and i follow at once from the corresponding properties of $R(J)$ and I'.

The domain of definition Q of i is a field of sets, i.e.,

(32) From $A, B \in Q$ follows $A \cup B \in Q$ and $A \cap B \in Q$,
(33) From $A, B \in Q$ follows $A - B = A - (A \cap B) \in Q$.[18]

The function i is non-negative, finite, and additive on Q:

(34) From $A \in Q$ follows $0 \leqq i(A) < \infty$;
(35) From $A, B \in Q$, $A \cap B = \varnothing$ follows $i(A \cup B) = i(A) + i(B)$.

The relations (32), . . ., (35) are essentially identical with the so-called *content axioms.* Their significance is obvious: if two sets have a content, then their union, their intersection, and the complement of one of them in the other also has a content, which is a certain non-negative finite number; and the content of the union of two disjoint sets is equal to the sum of their contents. The fact that only *certain* sets (namely the $A \in Q$) have a content is perhaps less obvious. It turns out that the decision as to exactly which sets are to be considered as measurable is of basic importance in the construction of a satisfactory theory of content.

Further properties of Q and i are as follows: Q contains the (bounded) intervals of R^n, and i assigns to them their n-dimensional elementary geometric content. Also, *i is complete on Q*: i.e., every bounded subset B of R^n for which, given an arbitrary $\epsilon > 0$, there exist measurable sets $\underline{A}, \overline{A} \in Q$ with $\underline{A} \subseteqq B C \overline{A}$ and $i(\overline{A} - \underline{A}) < \epsilon$ is itself measurable.[19] *Every congruent mapping*

(36) $$x = (x_\nu) \to x' = (x'_\nu) = \left(\sum_{\lambda=1}^{n} a_{\nu\lambda}x_\lambda + c_\nu\right)$$

$(a_{\nu\lambda})$ n-rowed orthogonal matrix

[18] We speak here of a field (in the sense of the theory of sets) because the set-theoretical operations $\cup$, $\cap$ (the latter defined by $A - B = \{y : y \in A, y \notin B\}$) do not lead out of Q. The concept of a field in this sense is to be distinguished from a field in algebra.

[19] Compare the fact that the method for extending T, I to $\mathfrak{R}$, I' is closed (see end of §1.3).

of R^n onto itself carries each measurable set $A \subset R^n$ into a measurable set $A' \subset R^n$ with the same content,

$$i(A') = i(A). \tag{37}$$

In particular, i is therefore independent of the system of coordinates defining the space R^n.

An important criterion for measurability is provided by the following theorem. *A bounded subset A of R^n is measurable if and only if $A_g = \tilde{A} - \mathring{A}$ (where $\tilde{A}$ is the closed hull, and $\mathring{A}$ is the open kernel of A in R^n)*[20] *of the boundary points of A can be covered by finitely many intervals $J^{(1)}, \ldots, J^{(k)}$ such that the sum of their contents $\sum_{\kappa=1}^{k} |J^{(\kappa)}|$ is less than an arbitrarily preassigned $\epsilon > 0$.*

For bounded real-valued functions f whose domain of definition is a *measurable set A in R^n* the *Riemann integral* $\int_A fdx$ can be defined in the following way: the function f is extended by

$$f_J(x) = f(x) \quad \text{for} \quad x \in A, \quad f_J(x) = 0 \quad \text{for} \quad x \in (J - A) \tag{38}$$

to a bounded function f_J on some bounded closed interval J in R^n that contains A. Then f is said to be *Riemann integrable over A* if $f_J \in \mathfrak{R}(J)$; and the (*definite*) *Riemann integral* $\int_A fdx$ *of f over A* is defined, in case $f_J \in \mathfrak{R}(J)$, as

$$\int_A fdx = \int_J f_J dx \qquad f_J \in \mathfrak{R}(J). \tag{39}$$

(Both these concepts are independent of the subsidiary interval J.)

1.6. *The Stieltjes Integral*

By §1.4 the *Riemann integral* $I'(f) = \int_J fdx$ of every function f that is *Riemann integrable* over J can be represented as the limiting value of certain (*Riemann*) sums. This fact is sometimes stated in the following way: the positive linear functional I' defined on the *vector lattice* $\mathfrak{R}(J)$ consisting of functions that are *Riemann integrable* over J admits an "*integral representation*." It is natural to ask whether other positive linear functionals L, whose domain of definition is a *vector lattice* E of bounded real functions allow an integral representation in such a sense. For the case that E is the space $C = C(\langle a, b\rangle)$ of continuous functions on the closed interval $J = \langle a, b\rangle$, $a < b$, of R^1 the complete answer to this question was first given by F. Riesz.

A class of positive linear functionals on C can be expressed by means of

[20] By the closed hull $\tilde{A}$ and the open kernel $\mathring{A}$ of A in R^n we mean the smallest closed set in R^n that contains A and the greatest open set in R^n in which A is contained.

the *Stieltjes integral.*[21] Let g be an arbitrary monotone non-decreasing (bounded) function defined on $\langle a, b\rangle$. For arbitrary $f \in C$ we form, in analogy with (26), the *Stieltjes sums*

$$F_g(\tau) = F_g(\tau; x^{(\lambda)}) = \sum_{\kappa=1}^{k} f(x^{(\kappa)})(g(c_\kappa) - g(c_{\kappa-1})); \tag{40}$$

here τ is the partition of $\langle a, b\rangle$ defined by the points $a = c_0 < c_1 < \cdots < c_k = b$ and the $x^{(\kappa)}$ are intermediate points with $c_{\kappa-1} \leqq x^{(\kappa)} \leqq c_\kappa$, which are not varied for the given partition. Since f is uniformly continuous on $\langle a, b\rangle$, this sum converges, with respect to the *Riemann filter basis* $\mathfrak{F}$ on the set of all partitions τ of $J = \langle a, b\rangle$ (cf. §1.4), to a number $I_g(f)$, no matter how the intermediate points $x^{(\kappa)}$ are chosen for τ. *The number $I_g(f)$ is called the Stieltjes integral of f over $\langle a, b\rangle$ with respect to the integrator g.* Various notations are

$$I_g(f) = \int_a^b f dg = \int_a^b f(x) dg(x) = \lim_{\mathfrak{F}} F_g(\tau; x^{(\lambda)}). \tag{41}$$

In the special case $g(x) = x$ for $a \leqq x \leqq b$ we obviously obtain the *Riemann integral* of $f \in C$ over $\langle a, b\rangle$.

For fixed g let us now consider the correspondence

$$f \to I_g(f) = \int_a^b f dg \qquad f \in C. \tag{42}$$

This correspondence defines a positive linear functional I_g on C. It was recognized by F. Riesz in 1909 that these I_g represent the totality of positive linear functionals on C: *for every positive linear functional L on the vector lattice C there exists a monotone nondecreasing bounded function g such that $L(f) = I_g(f) = \lim_{\mathfrak{F}} F_g(\tau; x^{(\lambda)})$ for every $f \in C$* (cf. footnote 21). Consequently, every positive linear functional on C can be represented as the limiting value of *Stieltjes sums.*

The concept of a *Stieltjes integral* can be generalized by taking g to be a function of bounded variation on $\langle a, b\rangle$.[22] Every such function can be

[21] The *Stieltjes integral* and its generalizations are presented in detail by Aumann [1] 8.5.12, . . . , 8.5.12.9 and Riesz and Nagy [9], Chapter III. Proofs of the Riesz theorem can be found in [1], 8.5.12.5, 8.5.12.6 and in [9], 50. The point of view in [9] is somewhat different from our present one.

[22] The function g is said to be *of bounded variation* on $\langle a, b\rangle$ if for every partition τ of $\langle a, b\rangle$, defined say by $a = c_0 < c_1 < \cdots < c_k = b$, the values of $T_g(\tau) = T_g(\tau; a, b) = \sum_{\kappa=1}^{k} |g(c_\kappa) - g(c_{\kappa-1})|$ remain less than a fixed number independent of τ. The number $T_g = T_g(a, b,) = \sup_\tau T_g(\tau; a, b,)$ is called the *total variation of g in* $\langle a, b\rangle$.

represented as the difference[23] of two nondecreasing bounded functions. Then the representation of g as the difference between the positive variation p and the negative variation n of g (cf. footnote 23) is particularly convenient. We set

$$I_g(f) = \int_a^b f dg = \int_a^b f dp - \int_a^b f dn \qquad f \in C. \tag{43}$$

The correspondence (42) with a g of bounded variation also defines a linear functional I_g on C, which is no longer necessarily positive. Nevertheless it can be written as the *difference of positive linear functionals*, e.g., in the form $I_g = I_p - I_n$.

2. Abstract Measure and Its Extension

2.0. *Historical Remarks*

Next to the *Riemann integral*, the concept of an integral due to Lebesgue has played the most important role in pure and applied mathematics. The procedure followed by Lebesgue in 1901 consists of three steps. In the first step the Jordan content is used to assign to certain bounded point sets in R^1 (or R^n) a more general content, namely a *Lebesgue measure*. The second step deals with the *measurability of functions*, which is reduced to the measure of point sets. Finally, in the third step it is shown how an integral, the *Lebesgue integral*, can be assigned to measurable functions.[24] It is closely related to the *Riemann integral*; for example, every function that is *properly Riemann integrable* is also *Lebesgue integrable*, and to the same value. *In this sense the Lebesgue integral is a generalization of the Riemann integral.* With respect to limit processes the *Lebesgue integral* behaves much more satisfactorily and, as a result of these and other advantages, has almost completely replaced the *Riemann integral* in many branches of mathematics (for example, in the theory of real functions, functional analysis, integral equations, and probability).

In many respects the classical Lebesgue construction appears to be a detour, since it operates with the theory of measure (i.e., measure of point sets, see, e.g., [7]), and we can arrive more rapidly at the central theorems of the Lebesgue theory by considering an integral from the point of view of positive linear functionals and attempting to obtain the *Lebesgue integral* as an appropriate extension of some suitably chosen basic functional. It was in this way that the *Riemann integral* was constructed in §1,

[23] Set $T(x) = T_g(a, x)$ for $a \leqq x \leqq b$; the *positive variation p and the negative variation n of g in $\langle a, b\rangle$* are defined by $2p(x) = T(x) + g(x) - g(a)$ and $2n(x) = T(x) - g(x) + g(a)$, $a \leqq x \leqq b$. Obviously $g = p - n$.

[24] This procedure corresponds to the method of basing the *Riemann integral* on the theory of *Jordan content* (cf. §1.5).

and in fact we shall see that this fundamental idea is of very wide application; it can be used to construct much more general concepts of integration than the *Lebesgue integral.*

2.1. *Abstract Measures, Examples*

For the construction of a theory of integration that has many of the properties of the Lebesgue theory we require very few assumptions, for example: let there be given an arbitrary set $A \neq \varnothing$, a *vector lattice* E of real-valued (finite) functions t defined on the set A, and a positive linear functional μ on E that is continuous with respect to monotone passage to the limit (§1.2). Thus in particular:

$$\text{From } t_1 \geqq t_2 \geqq \cdots \geqq 0 \text{ and } \lim_{m \to \infty} t_m(x) = 0$$

(44) it follows for every $x \in A$ that

$$\lim_{m \to \infty} \mu(t_m) = \mu(\lim_{m \to \infty} t_m) = 0 \qquad t_1, t_2, \ldots \in E.$$

As in Bourbaki [2], Chap. IV, §1.5 we call μ *an abstract measure on A*. The $t \in E$ are called *elementary functions* and the $\mu(t)$ are *elementary integrals.*

Let us give some simple examples for these concepts:

a) Let $A = J$ be a bounded closed interval of R^n, let $E = T(J)$ be the space of step functions on J, and let $\mu = I$ be the functional defined in (14). Here the elementary functions are the step functions on J and the elementary integrals are the step integrals over J.

b) Let $A = \langle a, b \rangle$ be a closed bounded interval in R^1, let $E = C$ be the space of continuous functions on $\langle a, b \rangle$, and let $\mu = I_g$ be the functional introduced in (42) by means of the monotone nondecreasing function g. The elementary functions are the continuous f on $\langle a, b \rangle$ and the elementary integrals are the *Stieltjes integrals* of these f with respect to the integrator g.[25]

c) Let $A = R^n$, $E = K(R^n)$ be the space of functions f that are continuous on R^n and vanish outside of some bounded closed interval J_f (which depends in each case on the $f \subset K(R^n)$). Then we can define the integral

$$\int_{R^n} f dx = \int_{-\infty}^{\infty} \cdots \int_{-\infty}^{\infty} f(x_1, \ldots, x_n) dx_1 \cdots dx_n,$$

$f \in K(R^n)$ uniquely by setting

$$\int_{R^n} f dx = \int_{-\infty}^{\infty} \cdots \int_{-\infty}^{\infty} f(x_1, \ldots, x_n) dx_1 \cdots dx_n = \int_{J_f} f dx.$$

[25] The proof that the condition of continuity (44) is satisfied here will be found, for example, in Haupt-Aumann-Pauc [5], 6.2.3, III.

The functional defined by $f \to \int_{R^n} f dx$, $f \in K(R^n)$, is an abstract measure on R^n (cf. footnote 1). The $f \in K(R^n)$ are identical with *the functions that are continuous in* R^n *and have compact support*. (By the support of a function f defined in R^n we mean the smallest closed set $S_f \subseteqq R^n$ on whose complement $R^n - S_f$ the function f vanishes.)

The extension of an abstract measure μ, described in detail in the following sections, leads in these various cases a), b), c), to the *Lebesgue integral* over J, to the *Lebesgue-Stieltjes integral* over $\langle a, b \rangle$, and to the *Lebesgue integral* over R^n respectively.

It should be emphasized that by proceeding in different ways in our extension of the functional I and its domain of definition $T(J)$ we can arrive, on the one hand, at the *Riemann integral* over J and on the other hand at the *Lebesgue integral.* Whereas in the "*Riemann extension*" in §1.3 the continuity of the functional I was unimportant, it is fully exploited in the "*Lebesgue extension.*" Another important distinction lies in the fact that in the first case integrals are assigned only to bounded functions, but in the second the concept of an integral is so general that certain functions have an integral even though their values include the improper numbers $-\infty, \infty$.[26] Of course, this fact introduces various complications.

2.2. *Seminorms Defined by Abstract Measures*

Let us now assume that the conditions stated at the beginning of the preceding section are satisfied. In order to obtain the "*Lebesgue extension*" of E and μ,[27] we first construct a space F of real-valued functions on A that corresponds to the space B in §1.3. The space $\bar{R}^A$ of all real-valued functions on A with proper or improper values is too inclusive. The construction of F depends on the abstract measure μ.

We first consider functions $g \in \bar{R}^A$ that can be represented as the limits of monotone sequences $t_1 \leqq t_2 \leqq \cdots$ of elementary functions $t_m \geqq 0$ (e.g., $g(x) = \lim_{m \to \infty} t_m(x)$ for every $x \in A$); for such functions we define an upper integral $\mu^*(g)$:

$$\mu^*(g) = \lim_{m \to \infty} \mu(t_m), \quad \text{if} \quad g(x) = \lim_{m \to \infty} t_m(x) \quad t_1, t_2, \ldots \in E. \tag{45}$$
$$\text{for every } x \in A \text{ and } 0 \leqq t_1 \leqq t_2 \leqq \cdots.$$

[26] For computation with $\pm\infty$ we agree, for example, that: $-\infty < \alpha < \infty$, $|-\infty| = |\infty| = \infty$, $\alpha + (\pm\infty) = \pm\infty$, $\alpha - (\pm\infty) = \mp\infty$, $\pm\infty + (\mp\infty) = 0$, $0 \cdot (\pm\infty) = 0$, $\beta \cdot (\pm\infty) = \pm\infty$ or $= \mp\infty$ if $\beta > 0$ or $\beta < 0$ ($\alpha, \beta \in R$). Note that $\bar{R} = R \cup \{-\infty, \infty\}$ is not a vector space over R, since certain of the rules for computation are violated.

[27] Historically this terminology is not justified, since Lebesgue arrived at his notion of an integral in a quite different way. The procedure described here is due to Stone [17]. The reader may also consult the textbooks of Aumann [1], 9.1, . . . , 9.4 and Haupt-Aumann-Pauc [5], Chapter VI.

Then $\mu^*(g)$ is uniquely defined; for if g is defined by two monotone sequences $0 \leqq t_1 \leqq t_2 \leqq \cdots$ and $0 \leqq t'_1 \leqq t'_2 \leqq \cdots$, $t_m, t'_m \in E$, then for $m \to \infty$ the corresponding integrals $\mu(t_m)$, $\mu(t'_m)$ converge to the same number in $\bar{R} = R \cup \{-\infty, \infty\}$. (This fact is essentially a consequence of (44).) We denote the set of such g by M. For arbitrary $f \geqq 0$ in $\bar{R}^A$ we now define

$$(46) \qquad \mu^*(f) = \inf\{\mu^*(g) : g \in M, f \leqq g\} \qquad \text{or} \qquad \mu^*(f) = \infty,$$

according to whether or not functions $g \in M$ with $f \leqq g$ exist for the given f. Then $\mu^*(f)$ is called the *μ-upper integral of f*; it is to be compared with the upper Darboux integral (§1.3). Finally, for arbitrary $f \in \bar{R}^A$ we set

$$(47) \qquad \mu^*(f) = \mu^*(|f|) \qquad f \in \bar{R}^A.$$

Then the abstract measure μ on E has been extended to a function μ^* on $\bar{R}^A$ with values in $\bar{R}$. *The space $F = F(A; \mu^*)$ is now defined as the set of all functions $f \in \bar{R}^A$ for which $\mu^*(f)$ is finite.*

Among the properties of μ^* let us mention the following:

$$(48) \qquad 0 \leqq \mu^*(f) = \mu^*(|f|) \leqq \infty, \qquad \mu^*(0) = 0 \qquad f \in \bar{R}^A,$$

$$(49) \qquad \mu^*(\alpha f) = |\alpha| \mu^*(f) \qquad \alpha \in R, f \in \bar{R}^A,$$

$$(50) \qquad \mu^*(f_1 + f_2) \leqq \mu^*(f_1) + \mu^*(f_2) \qquad f_1, f_2 \in \bar{R}^A,$$

$$(51) \qquad \mu^*(f_1) \leqq \mu^*(f_2), \quad \text{if} \quad |f_1| \leqq |f_2| \qquad f_1, f_2 \in \bar{R}^A.$$

In view of (48), (49), (50) the function μ^* is sometimes called a *norm* or a *seminorm in the wider sense on* $\bar{R}^A$.[28]

Finally, in generalization of (50), (51) we have

$$(52) \qquad \mu^*(f) \leqq \sum_\kappa \mu^*(f_\kappa), \quad \text{if} \quad |f| \leqq \sum_\kappa |f_\kappa| \qquad f, f_\kappa \in \bar{R}^A,$$

where $\sum_\kappa$ denotes a finite or infinite sum.

[28] Let Y be a vector space over R. A function N defined on Y with values in R is called a *seminorm on* Y if:

$$\text{(I)} \qquad N(0) = 0, \quad 0 \leqq N(y) < \infty \quad \text{for} \quad y \in Y;$$
$$\text{(II)} \qquad N(\alpha y) = |\alpha| N(y) \quad \text{for} \quad \alpha \in R, \ y \in Y;$$
$$\text{(III)} \qquad N(y_1 + y_2) \leqq N(y_1) + N(y_2) \quad \text{for} \quad y_1, y_2 \in Y.$$

If the function N has the further property that

$$\text{(I')} \qquad 0 \leqq N(y) < \infty \quad \text{for} \quad y \in Y, \ N(y) = 0 \text{ if and only if } y = 0,$$

then Y is called a *normed space over* R and N is a *norm on* Y. The above function μ^* is a seminorm *in the wider sense*, since μ^* may take the value ∞ and $\bar{R}^A$ is not a vector space over R. In general, the space F also fails to be a vector space over R, so that the restriction of the function μ^* to F can only represent a seminorm in the wider sense on F.

2.3. *Zero Functions and Zero Sets*

In the theory of integration an important role is played by those functions $f \in \bar{R}^A$ for which $\mu^*(f) = 0$. They are called (μ^*-)*zero functions.* A subset B of A is called a (μ^*-)*zero set* if the characteristic function χ_B of B in A (cf. footnote 2) is a μ^*-zero function. The expression "(μ^*-)*almost everywhere*" means "*with the exception of the points of a μ^*-zero set.*"

It is easy to verify the following statements concerning these concepts: *Every subset of a zero set is a zero set. The union of countably many zero sets is a zero set. The function $f \in \bar{R}^A$ is a zero function if and only if $f(x) = 0$ almost everywhere* (*on* A). *Every function $f \in F(A; \mu^*)$ is almost everywhere finite* (*on* A). *If $f, g \in \bar{R}^A$ satisfy the relation $f(x) = g(x)$ almost everywhere, then $\mu^*(f) = \mu^*(g)$.*

Another important concept is that of μ^*-equivalence. Two functions $f, g \in \bar{R}^A$ (or $\in F$) are said to be (μ^*-)*equivalent*, or in symbols $f \sim g$, if $f(x) = g(x)$ almost everywhere on A. Obviously $\sim$ is an equivalence relation.[29] In the theory of measure and integration it is customary to consider this equivalence as an extension of the ordinary concept of equality: two functions f, g are considered as "equal" if their values $f(x), g(x)$ are equal almost everywhere on A.

The space $F = F(A; \mu^*)$, which with respect to the usual concept of equality ($=$) *fails in general to be a vector space over R, is even a vector lattice $\tilde{F} = \tilde{F}(A; \mu^*)$ when considered with respect to the extended notion of equality* ($\sim$)*; the function μ^* is a norm on $\tilde{F}$* (cf. footnote 28).

It is to be noted that the elements of $\tilde{F}$ are not the functions $f \in F$ but the *classes* $[f] = \{f_1 : f_1 \in F, f_1 \sim f\}$ of μ^*-equivalent functions. Definitions which will enable us to convert $\tilde{F}$ into a normed vector lattice are as follows:

(53) $$[f] + [g] = [f_1 + g_1], \quad \text{if} \quad f_1 \in [f], \quad g_1 \in [g];$$

(54) $$\alpha[f] = [\alpha f_1], \quad \text{if} \quad f_1 \in [f], \quad \alpha \in R;$$

(55) $[f] \leqq [g]$ if and only if for suitable $f_1 \in [f]$, $g_1 \in [g]$ we have $f_1 \leqq g_1$ almost everywhere;

(56) $$\mu^*([f]) = \mu^*(f_1), \quad \text{if} \quad f_1 \in [f].$$

2.4. *Integrable Functions and Their Integrals*

Just as for the "*Riemann extension*" in §1.3, we now define a space L^1 of functions, which will include E, as the space of integrable functions

[29] For every two elements $f, g \in \bar{R}^A$ (or $\in F$) either "$f \sim g$" or else "it is not true that $f \sim g$;" for all f, g, h we have: $f \sim f$; from $f \sim g$ follows $g \sim f$, and from $f \sim g$, $g \sim h$ follows $f \sim h$.

and then extend the functional μ from E to L^1. The totality $\mathfrak{R}$ of integrable functions in §1.3 formed a subspace of B; but now L^1 will be defined as a subspace of F: we define $L^1 = L^1(A; \mu)$ as *the set of all functions* $f \in F = F(A; \mu^*)$ *for which there exists an elementary function* $t \in E$ *with* $\mu^*(f-t) < \epsilon$ *for arbitrarily preassigned* $\epsilon > 0$. *The functions* $f \in L^1$ are called the (μ-)*integrable or* (μ-)*summable functions over* A.

On the basis of this definition there exists for every $f \in L^1$ a sequence $t_1, t_2, \ldots$ of elementary functions t_m with $\lim_{m\to\infty} \mu^*(f - t_m) = 0$.[30] For every such sequence the $\lim_{m\to\infty} \mu(t_m)$ exists and the limiting value is independent of the special choice of the sequence $t_1, t_2, \ldots$; in particular $\lim_{m\to\infty} \mu(t_m)$ is identical with $\mu(f)$ for every $f \in E$. Thus $\mu(f)$ is uniquely defined for every $f \in L^1$ if we set

$$(57) \quad \mu(f) = \lim_{m\to\infty} \mu(t_m), \qquad \text{if} \qquad \lim_{m\to\infty} \mu^*(f - t_m) = 0, \qquad t_1, t_2, \ldots \in E.$$

The $\mu(f)$ thus defined is called the (μ-)*integral of* f *over* A and is denoted by

$$(58) \qquad \mu(f) = \int_A f d\mu = \int_A f(x) d\mu(x) \qquad f \in L^1(A; \mu).$$

For nonnegative $f \in L^1(A; \mu)$ we have

$$(58') \qquad \mu(f) = \int_A f d\mu = \mu^*(f) \qquad f \geqq 0, \quad f \in L^1(A; \mu).$$

This extension of the abstract measure and its domain of definition admits an interesting and simple interpretation if we make use of the seminorm μ^* to introduce into F the notion of a distance; by the distance $\rho(f, g)$ of $f, g \in F$ we mean the nonnegative number $\mu^*(f - g)$.[31] *The passage from* E to L^1 *then consists simply of the formation of the closed hull of* E *and* F.

In the sense of the distance ρ *the abstract measure* μ *on* E *is a uniformly continuous function on* E.[32] *The extension* (57) *is then seen to be the* (*uniquely defined*) *extension of the uniform continuous function* μ *to a function continuous on the closed hull* L^1 *of the domain of definition* E.

[30] The sequence $t_1, t_2, \ldots$ is sometimes said to *converge normwise to* f. This concept of convergence is to be distinguished from the usual convergence at every point $x \in A$.

[31] With the exception of the axiom "$\rho(f, g) = 0$ if and only if $f = g$," this distance ρ satisfies all the requirements for a metric. Thus ρ fails to define a metric on F, but defines a uniform structure (cf. III 1, §6). We then have "$\rho(f, g) = 0$ if and only if $f \sim g$." If we make use of the more general concept $\sim$ of equality for functions (in other words if we proceed from F to $\tilde{F}$ and from E, L^1 to the corresponding spaces $\tilde{E}$, $\tilde{L}^1$), then ρ becomes a metric.

[32] I.e., for every $\epsilon > 0$ there exists a $\delta(\epsilon) > 0$ such that $\rho(t, t') < \delta(\epsilon)$ implies $|\mu(t) - \mu(t')| < \epsilon$, for arbitrary $t, t' \in E$.

Among the basic properties of the space $L^1 = L^1(A; \mu)$ and the functional μ on L^1 let us mention the following: *with f the space L^1 also contains every function f_1 that is μ^*-equivalent to f, and the two functions have the same integral $\mu(f) = \mu(f_1)$; in particular every zero function is integrable.* It is sometimes convenient to introduce the extended notion of equality $\sim$, or in other words to proceed from L^1 to the space $\tilde{L}^1 = \tilde{L}^1(A; \mu) \subseteqq \tilde{F}(A; \mu^*)$ of *classes* $[f]$ of μ-integrable functions f. The function μ is then defined on $\tilde{L}^1$ also, by

$$\mu([f]) = \mu(f_1), \quad \text{if} \quad f_1 \in [f] \qquad f, f_1 \in L^1. \tag{59}$$

The space $\tilde{L}^1$ is a vector lattice and μ is a positive linear functional on $\tilde{L}^1$ that is continuous under monotone passage to the limit:

From $[f_1] \geqq [f_2] \geqq \cdots \geqq [0]$ and $\lim_{m \to \infty} f_m(x) = 0$ almost everywhere on A it follows that

$$\lim_{m \to \infty} \mu([f_m]) = 0 \qquad f_1, f_2, \ldots \in L^1. \tag{60}$$

Of course, this theorem can also be expressed, though less strikingly, without the extended notion of equality. We then have statements like: "With f_1, f_2 the function $f_1 + f_2$ is also integrable and

$$\int_A (f_1 + f_2) d\mu = \int_A f_1 d\mu + \int_A f_2 d\mu \qquad f_1, f_2 \in L^1." \tag{61}$$

"If f is integrable and $\geqq 0$ almost everywhere on A, then

$$\int_A f d\mu \geqq 0." \tag{62}$$

The central theorems of the theory refer to the integrability of the limit function of certain sequences of functions. These theorems, which have no analogue in theory of the Riemann integral, run as follows:

Theorem on the integration of a monotonely convergent sequence of functions with uniformly bounded integrals (essentially due to B. Levi, 1906): *let $f_1, f_2, \ldots$ be a sequence of integrable functions for which $f_1 \leqq f_2 \leqq \cdots$ almost everywhere on A. If the sequence of integrals $\int_A f_1 d\mu, \int_A f_2 d\mu, \ldots$ is bounded, then the almost everywhere uniquely defined limit function $f = \lim_{m \to \infty} f_m$*[33] *is integrable and*

$$\int_A f d\mu = \lim_{m \to \infty} \int_A f_m d\mu. \tag{63}$$

Theorem on integration for majorized convergence (essentially due to Lebesgue, 1910): *let $f_1, f_2, \ldots$ be a sequence of integrable functions which*

[33] At points $x \in A$ at which $\lim_{m \to \infty} f_m(x)$ does not exist the function $f(x)$ may be defined arbitrarily.

converges[33] *almost everywhere on A to a limiting value* $f = \lim_{m\to\infty} f_m$. *If there exists an integrable function g with $|f_m| \leqq g$ for all* $m = 1, 2, \ldots,$ *then f is integrable and* (63) *holds.*

2.5. *Examples of Important Methods of Extension*

By suitable specialization of A, E, and μ the above method of extending an abstract measure leads to various important integrals originally defined in other ways.

a) For $A = J$ (bounded closed interval in R^n) and $E = T(J)$ (the space of step functions on J, $\mu = I$ (cf. §2.1,a)) the space $L^1(J; I)$ is the *space of functions that are Lebesgue integrable over J and* $I(f) = \int_J f dI = \int_J f dx = {}^{(L)}\int_J f dx$ *is the (definite) Lebesgue integral of* $f \in L^1(J; I)$ *over J.* The *Lebesgue integrable functions* can also be obtained by extending other underlying functionals. For example, in the textbooks it is customary to begin with $E = C(J)$ (the space of functions continuous on J) and $\mu(t) = I'(t) = \int_J t dx$ (the Riemann integral of $t \in C(J)$ over J).[34]

b) For $A = \langle a, b\rangle$ (a bounded closed interval in R^1), and $E = C$ (the space of functions continuous on $\langle a, b\rangle$, $\mu = I_g$ (cf. §2.1,b)) *the space* $L^1(\langle a, b\rangle; I_g)$ *is the space of functions that are Lebesgue-Stieltjes integrable over $\langle a, b\rangle$ with respect to g, and* $I_g(f) = \int_a^b f dI_g = \int_a^b f dg$ *is the Lebesgue-Stieltjes integral of* $f \in L^1(\langle a, b\rangle; I_g)$ *with respect to g.* These concepts can be extended to the case that g is of bounded variation $\langle a, b\rangle$ (cf. §1.6): if g is written as the difference $p - n$ of the positive and negative variation of g, we set

$$L^1(\langle a, b\rangle; I_g) = L^1(\langle a, b\rangle; I_p) \cap L^1(\langle a, b\rangle; I_n)$$

and

$$I_g(f) = I_p(f) - I_n(f) = \int_a^b f dg \qquad \text{for} \qquad f \in L^1(\langle a, b\rangle; I_g).$$

c) In order to obtain *the space $L^1(R^n; \mu)$ of functions f Lebesgue integrable over R^n, together with their Lebesgue integral*

$$\mu(f) = {}^{(L)}\int_{R^n} f dx = {}^{(L)}\int_{-\infty}^{\infty} \cdots \int_{-\infty}^{\infty} f(x_1, \ldots, x_n) dx_1 \cdots dx_n,$$

by means of extending a measure μ, we may begin with the space $E = K(R^n)$ of functions that are continuous on R^n and have compact support, and with the integral $\mu(t) = \int_{R^n} t dx$ of the $t \in K(R^n)$ defined in §2.1,c). Another possibility is as follows: every step function t_J whose domain of definition

[34] Cf. Bourbaki [2] and MacShane [8]. For a proof that our procedure leads in both cases to the same space $L^1(J; I)$ and to the same concept of an integral see Haupt-Aumann-Pauc [5], 6.1.2 I, II; 6.3.4.

is a closed bounded interval J in R^n is extended to a step function $\tilde{t}$ on R^n by setting $\tilde{t}(x) = 0$ for $x \in (R^n - J)$. The integral $\mu(\tilde{t}) = \int_{R^n} \tilde{t}dx$ can then be defined as

$$\mu(\tilde{t}) = \int_{R^n} \tilde{t}dx = \int_J t_J dx.$$

The $\tilde{t}$ are the elementary functions, and the $\mu(\tilde{t})$ are the elementary integrals.

The terms *Lebesgue integrable, Lebesgue integral*, . . . are often abbreviated to *L-integrable, L-integral*, ...

d) In the *Schwartz theory* of distributions (cf. §3) an important role is played by the concept of *Radon measure on* R^n, a name given to any linear functional μ on the space $K(R^n)$ of functions with compact support that are continuous on R^n and have the following continuity property: if a sequence $t_1, t_2, \ldots$ of functions $t_m \in K(R^n)$, every term of which vanishes outside a fixed bounded closed set in R^n, converges *uniformly* in R^n to 0, then the sequence $\mu(t_1), \mu(t_2), \ldots$ converges to $\mu(0) = 0$. Since *Radon measures* μ are not assumed to be positive, they are not, in general, abstract measures. However, every *Radon measure* μ can be represented as the difference of two positive *Radon measures* and every *positive Radon measure* is an abstract measure, since it satisfies[35] (44) with $E = K(R^n)$.

2.6. *Measurable Functions and Measurable Sets*

The concept of integrability of functions can be generalized to the concept of measurability. For functions $f, g, h \in \bar{R}^A$ we set

$$\text{med}(f, g, h) = \sup(\inf(f, g), \inf(g, h), \inf(h, f)), \tag{64}$$

where med (f, g, h) is called the *medium* of f, g, h. A function $f \in \bar{R}^A$ is said to be (μ-)*measurable* (on A) if for arbitrary μ-integrable functions g, h the function med (f, g, h) is μ-integrable. If A, E, μ are specialized as in §2.5,a) or c), then the term "μ-measurable" is usually replaced by "*Lebesgue-measurable*" and "*L-measurable*."

The space $M = M(A; \mu)$ of μ-measurable functions on A obviously includes the space $L^1 = L^1(A; \mu)$ of functions that are μ-integrable over A. Although by definition every $f \in L^1$ has a finite seminorm $\mu^*(f)$, it may happen for certain $f \in M$ that $\mu^*(f) = \infty$. *A measurable function f is integrable if and only if its seminorm is* $<\infty$ *or if for the given f there exist integrable bounds g, h with* $g \leqq f \leqq h$.

The concept of measurability includes a rather large class of functions. If A, E, μ are specialized as in the preceding section, then practically all the functions that occur in practice for a given A turn out to be measurable.

[35] Cf. Bourbaki [2], Chap. III, §2, Théorème 2; Chap. IV, §1.5, and also Haupt-Aumann-Pauc [5], 6.2.3.III.

Here it is important to note that the space M behaves "even better" than L^1 with respect to limit processes: *the limit f of every convergent (almost everywhere on A) sequence $f_1, f_2, \ldots$ of measurable functions is measurable; and even if $f_1, f_2, \ldots$ does not converge, the functions* $\inf f_m$ *and* $\sup f_m$ *are always measurable.* Moreover, if f_1, f_2 are measurable, then the functions $f_1 + f_2$, $|f_1|$ and αf_1, $0 \neq \alpha \in R$, for example, are also measurable.

For nonnegative $f \in \bar{R}^A$ we have the following criterion of measurability: the function f is measurable if and only if $\inf(f, g)$ is integrable for every nonnegative integrable g. It follows that the constant function 1 is measurable on A if and only if $\inf(1, g)$ is measurable for every nonnegative $g \in L^1$. This condition is satisfied in all cases of practical importance, and in particular for the examples in §2.5.

Other theorems about measurable functions can be obtained by introducing the concept of a measurable set. A subset B of A is said to be (μ-)integrable, or (μ-)measurable, if the characteristic function χ_B of B in A is μ-integrable, or μ-measurable respectively. For measurable subsets B of A the number

$$m(B) = \mu^*(\chi_B) \qquad \chi_B \in M \tag{65}$$

is called the *(μ-)measure of B (in A)*.[36] If B is integrable, then by (58′) we also have

$$m(B) = \int_A \chi_B d\mu \qquad \chi_B \in L^1. \tag{66}$$

If A, E, μ are chosen as in §2.5,a) or §2.5,c), then the expression "μ-measure of B" is usually replaced by one of the expressions "*(n-dimensional) Lebesgue measure of B,*" "*L_n-measure of B,*" *or* "*L-measure of B.*" *A subset B of R^n is of L_n-measure* 0 *if B can be covered by at most countably many open intervals $J^{(1)}, J^{(2)}, \ldots$ in R^n such that the sum of their contents $\sum_\kappa |J^{(\kappa)}|$ is smaller than an arbitrarily preassigned number $\epsilon > 0$.*

If the function 1 is measurable on A (i.e., if the function A is measurable), then there exist certain criteria for measurability which in the classical *Lebesgue theory of integration* serve as the definition of a "measurable function": *a μ^*-almost everywhere finite function $f \in \bar{R}^A$ is measurable if and only if for arbitrary $\alpha, \beta \in R$ with $\alpha < \beta$ the sets $\{x : x \in A, \alpha \leqq f(x) < \beta\}$ are measurable. If f is measurable, then for every $a \in \bar{R}$ the sets*

$$\{x : x \in A, f(x) = a\}, \qquad \{x : x \in A, f(x) \neq a\},$$
$$\{x : x \in A, f(x) > a\}, \qquad \{x : x \in A, f(x) \geqq a\}^{37}$$

are measurable.

[36] More generally, for arbitrary $B \subset A$ we set $m^*(B) = \mu^*(\chi_B)$. Then $m^*(B)$ is called the *outer (μ^*-)measure of B (in A)*.

[37] These sets are usually denoted by the symbols $[f = a]$, $[f \neq a]$, $[f > a]$, $[f \geqq a]$.

In the preceding examples the sets A are measurable with respect to the given measure.

2.7. *Relation to the Theory of Measure of Sets of Points*

By §1.5 there exists a close relation between the *Riemann integral* $I'(f) = \int_J f dx$, or the defining positive linear functional I' on $\mathfrak{R}(J)$, and the *Jordan content* i of measurable point sets. How do matters stand if we start from the extension, as described above, of an abstract measure μ on a set $A \neq \varnothing$? Let $Q(A)$ and $Q_1(A)$ be the set of μ-integrable and μ-measurable subsets of A respectively. On $Q(A)$ let us define a set function m by

$$(67) \qquad B \to m(B) = \int_A \chi_B d\mu \qquad B \in Q(A), \quad \chi_B \in L^1(A;\mu).$$

Then $Q(A)$ and m correspond rather closely to the field of measurable subsets of R^n and the *Jordan content* defined on this field (cf. §1.5, here and below). For $Q(A)$ *is a field of sets*, and m *is a content on* $Q(A)$, i.e., a non-negative finite and additive set function on $Q(A)$. Moreover, m is *complete on* $Q(A)$, i.e., every $B \subseteq A$ for which there exist sets $\underline{A}, \overline{A} \in Q(A)$ with $\underline{A} \subseteq B \subseteq \overline{A}$ and $m(\overline{A} - \underline{A})$ for arbitrary $\epsilon > 0$ also belongs to $Q(A)$: $B \in Q(A)$. The content m is distinguished from an arbitrary content by a property called *σ-additivity* or *total additivity*:

From $B_1, B_2, \ldots \in Q(A)$, $B_\lambda \cap B_\nu = \varnothing$ for $\lambda \neq \nu$

(68) and $\bigcup_{\kappa=1}^{\infty} B_\kappa \in Q(A)$

it follows that $m\left(\bigcup_{\kappa=1}^{\infty} B_\kappa\right) = \sum_{\kappa=1}^{\infty} m(B_\kappa)$.[38]

The set of measurable functions $Q_1(A)$ is also a field of sets, which, with every set of countably many elements $B_1, B_2, \ldots$ also includes their union $\bigcup_{\kappa=1}^{\infty} B_\kappa$ and their intersection $\bigcap_{\kappa=1}^{\infty} B_\kappa$:

(69) From $B_1, B_2, \ldots \in Q_1(A)$ it follows that $\bigcup_{\kappa=1}^{\infty} B_\kappa \in Q_1(A)$ and $\bigcap_{\kappa=1}^{\infty} B_\kappa \in Q_1(A)$.

A field of sets $Q_1(A)$ with the property (69) is called a *σ-field.*

For μ-measurable subsets B of A we now set, in analogy to (67),

$$(70) \qquad B \to m(B) = \mu^*(\chi_B) \qquad B \in Q_1(A), \qquad \chi_B \in M(A;\mu).$$

Then on $Q_1(A)$ we have defined a set function m whose value coincides with the content $m(B)$ for every $B \in Q(A)$. It is to be noted that in general

[38] The *Jordan content* is also totally additive, although this fact was of no importance in the discussion in §1.5.

m is not a content on $Q_1(A)$ in the sense of the word as used up to now, since it may happen that $m(B) = \infty$ for certain measurable $B \subseteqq A$.[39]

In the important special case $A \in Q(A)$[40] the set of μ-integrable functions $Q(A)$ coincides with the set $Q_1(A)$ of μ-measurable functions. Thus $Q(A)$ is a σ-field and the set function m defined by (70) is a *measure in the sense of the theory of measure*,[41] *namely a σ-additive content on a σ-field.*

Even the simplest examples (see footnote 39) show that certain cases in which $Q(A) \neq Q_1(A)$ may still be of practical importance. But these cases all satisfy the *finiteness assumption* that A can be represented as the union of *countably many* μ-integrable subsets of A. It then turns out that every measurable subset B of A admits such a representation. This property is called the *σ-finiteness of m*. Since m has all the other properties of a measure except finiteness, m is called a *σ-finite measure.*

If m satisfies the finite condition, the relation of $Q(A)$ and $Q_1(A)$ to the set functions (67) and (70) defined on these fields of sets can be described as follows.

By a general theorem of extension in the theory of measure, the σ-additive content (67) on $Q(A)$ can be extended in a unique way to a smallest σ-finite complete measure m' whose domain of definition $Q'(A)$ is contained in the σ-field $Q_1(A)$, where by a complete measure we naturally mean a σ-additive complete content on a σ-field. The concept of a "smallest complete measure" is based on the following fact: among all the extensions of a σ-additive content m on $Q(A)$ to complete measures m'' on σ-fields $Q''(A)$ with $Q(A) \subseteqq Q''(A) \subseteqq Q_1(A)$ there exists a smallest, namely the measure on $Q'(A) = \bigcap Q''(A)$. It turns out that under the above assumptions we necessarily have $Q'(A) = Q_1(A)$ and m' coincides with the set function (70).[42]

2.8. *Concluding Remarks on the Theory of Integration and Measure*

The theory of integration and measure can be constructed from widely different points of view. In §§1 and 2 we have given the main outlines of only *one* variant of the theory. In this "functional analytic" variant the theory of integration is subsumed under the theory of positive linear functionals and their extensions. As mentioned above, this procedure

[39] For example, $B = R^n$ is measurable with respect to the n-dimensional *Lebesgue measure* m, and $m(R^n) = \infty$.

[40] In the examples §2.5.a), §2.5.b) this condition is satisfied.

[41] In the theory of measure (theory of additive set functions) a measure is taken to be a *set function* whose domain of definition is a field of sets. On the other hand, in our present discussion an (abstract) measure is taken to be a *functional* on a space of real-valued functions.

[42] Cf., for example, Haupt-Aumann-Pauc [5], 3.2.1.2 extension theorem and 6.3.3 lemma (II).

allows us to arrive very quickly at the central theorems on integration of functions, since the detour through the theory of measure is avoided. It is clear that the construction of a theory of integration such as the one we have called the *Lebesgue theory* requires no assumption on the topology of the domain of definition A of the functions to be integrated. The *Lebesgue integral*, in contradistinction to the situation in the *classical Lebesgue theory*,[43] can thus be introduced without any essential reference to the topology of R^n.

In common with other abstract and general constructions of integration and measure, our procedure in §2 above has the disadvantage that many of the more refined results of the classical *Lebesgue theory* are not immediately obtainable, since they depend on the topological properties of R^n. More generally, lack of space prevents us here from discussing results of the theory of integration and measure that depend on topological assumptions about A.[44]

3. Distributions

3.1. *On the Significance of Distributions*

The distributions of L. Schwartz are a generalization of continuous functions in R^n. In 1945–1948 he published without proof the most important results of his theory of distributions. Then in 1950–1951 there followed a two-volume textbook [13], which we especially recommend to a reader interested in the more detailed features of the theory. The new entities called "distributions" have proved extremely useful in various branches of mathematics (differential equations, Fourier transforms, algebraic topology) and in theoretical physics (quantum mechanics, theory of electricity). From the mathematical point of view it is particularly interesting that a number of difficulties concerning the *concept of differentiability in classical analysis* can be removed to a great extent if we replace functions by distributions.

Let us mention some of the inconvenient properties of classical differentiation:

1. The space $D^1(U)$ of functions that are defined on an open subset $U \neq \varnothing$ in R^n and have partial derivatives there with respect to each of the x_i, $i = 1, \ldots, n$ is surprisingly "small"; it does not even include the space $C(U)$ of functions continuous on U.

[43] For a detailed account of the classical *Lebesgue theory* and some its generalizations consult Kamke [6].

[44] The reader who is interested in general investigations of this sort may consult Bourbaki [2].

2. The space $D^1(U)$ is not "closed" with respect to partial differentiation $f \to D_{x_i}f = \frac{\partial f}{\partial x_i}$, $i = 1, \ldots, n$, i.e., we do not necessarily have $D_{x_i}f \in D^1(U)$.

3. From a certain point of view differentiation is "discontinuous": the limit function f of a uniformly convergent sequence $f_1, f_2, \ldots$ of differentiable functions $f_m \in D^1(U)$ is not necessarily differentiable; and even if $f \in D^1(U)$, it may happen that differentiation D_{x_i} and passage to the lim $m \to \infty$ are not permutable; it is possible that

$$D_{x_i}(\lim_{m \to \infty} f_m) \neq \lim_{m \to \infty} (D_{x_i} f_m).$$

4. Further complications arise, for example, in interchange of the order of partial differentiation: $D_{x_i}(D_{x_j}f) = D_{x_j}(D_{x_i}f)$ is valid only under additional assumptions on f.

In an attempt to avoid these and other difficulties with differentiation we might try to proceed as follows: extend $C(U)$ in some suitable manner, and then make a change in the concepts of differentiation and convergence in $D^1(U)$ in such a way that they can be extended to a larger space including $C(U)$. This basic idea can be realized in numerous ways. The method chosen by L. Schwartz is particularly convenient and leads directly to the space $(D(U))'$ of distributions on U.

The *Schwartz construction* of $(D(U))'$ also solves an important physical problem: the *Dirac delta-function* and its formal properties can be justified, in the setting of the theory of distributions, in a way that is completely satisfactory from both the mathematical and the physical point of view. It is well known how the delta-"function" and its derivatives have been introduced into physical arguments in a manner quite inconsistent with the usual meaning of "point function" and "derivative." For $n = 1$, for example, it is customary to set

$$(71) \quad \delta(x) = 0 \quad \text{for} \quad x \neq 0 \quad \text{and} \quad \int_{-\infty}^{\infty} \delta(x)dx = 1 \qquad x \in R.$$

Then by certain arguments, which are based entirely on plausibility and cannot be discussed here, we arrive at the relations

$$(72) \qquad \int_{-\infty}^{\infty} f(x)\delta(x)dx = f(0), \qquad \int_{-\infty}^{\infty} f(x)\delta(x-a)dx = f(a)$$

$$a \in R, f \in L^1(R^1; I).$$

Moreover, it is convenient to consider δ as an infinitely differentiable "function," so that for the derivatives $\delta', \ldots, \delta^{(r)}$ of δ we set

$$(73) \quad \int_{-\infty}^{\infty} f(x)\delta'(x)dx = [f(x)\delta(x)]_{-\infty}^{\infty} - \int_{-\infty}^{\infty} f'(x)\delta(x)dx = -f'(0)$$

and in general for $r = 1, 2, \ldots$

$$(74)\quad \int_{-\infty}^{\infty} f(x)\delta^{(r)}(x)dx = (-1)^r \int_{-\infty}^{\infty} f^{(r)}(x)\delta(x)dx = (-1)^r f^{(r)}(0).$$

Here the function f is assumed to be sufficiently often differentiable. It is also customary to regard δ as the derivative H' of the *Heaviside function* H,

$$(75)\quad H(x) = 0 \quad \text{for} \quad x < 0, \qquad H(x) = 1 \quad \text{for} \quad x \geqq 0 \qquad x \in R,$$

under the motivation that

$$(76)\quad \int_{-\infty}^{\infty} f(x)dH(x) = f(0) = \int_{-\infty}^{\infty} f(x)\delta(x)dx \qquad f \in C(R^1).$$

Schwartz [13] has shown that these and other equations can be given a rigorous foundation if we consider $H, \delta, \delta', \ldots$ not as functions but as distributions.

3.2. *Axiomatic Treatment of the Theory of Distributions*

The concept "distribution" is to be so defined that it generalizes the concept of "continuous function on an open subset of R^n."[45] Distributions are to be (arbitrarily often) differentiable, and differentiation for distributions must avoid the "disagreeable" properties of classical differentiation, as listed above. These vaguely worded requirements can be made precise as follows:

A 1. *Every function f that is defined and continuous on an open subset U of R^n is a distribution.*

A 2. *Every distribution T*[46] *has an open subset U of R^n as its domain of definition. If T is a continuous function f defined on an open set $U \subseteqq R^n$, then the domain of definition of T, regarded as a distribution, is also U.*

A 3. *The totality of distributions with a common domain of definition U form a vector space over R* (cf. footnote 4); *this vector space includes the vector space $C(U)$ of continuous functions on U.* (By this statement we mean in particular that the concept "sum of two distributions T_1, T_2" agrees with the usual concept of the sum of functions (cf. footnote 3), if T_1, T_2 are continuous functions f_1, f_2 on U.)

A 4. *For every distribution T with domain of definition U, distributions $D_{x_i}T$, $i = 1, \ldots, n$, are defined, with domain of definition U, which are*

[45] In order to avoid unnecessary complications in the process of differentiation it is convenient to allow only open subsets of R^n as domains of definition of distributions.

[46] Distributions are denoted here, in general, by $T, T_1, T_2, \ldots$. These distributions T have nothing to do with the space $T = T(J)$ of step functions in §§1 and 2, and also nothing to do with the "distribution functions" in Chap. III.

called the partial derivatives of T with respect to x_i. *For these distributions* T, T_1, T_2 *defined on U we have*

$$(77)\qquad D_{x_i}(T_1 + T_2) = D_{x_i}T_1 + D_{x_i}T_2 \qquad i = 1, \ldots, n,$$

$$(78)\qquad D_{x_i}(D_{x_j}T) = D_{x_j}(D_{x_i}T) \qquad i, j = 1, \ldots, n.$$

In particular, if T is a continuous function f defined on U with continuous partial derivatives $\dfrac{\partial f}{\partial x_i}$ (in the sense of classical analysis), then

$$D_{x_i}T = \frac{\partial f}{\partial x_i}, \qquad i = 1, \ldots, n.$$

A 5. *For every distribution T with domain of definition U and for every open subset* U_0 *of U the restriction* T_{U_0} *of T on* U_0 *is defined. Here* T_{U_0} *is a distribution with the domain of definition* U_0. *If T is continuous and* $T = f$, *then* T_{U_0} *coincides with the restriction* f_{U_0} *of f on* U_0.[47] *The differentiation* $T \to D_{x_i}T$ *and the restriction* $T \to T_{x_0}$ *are permutable with each other in the sense that*

$$(79)\qquad (D_{x_i}T)_{U_0} = D_{x_i}(T_{U_0}) \qquad i = 1, \ldots, n$$

or arbitrary T and arbitrary $U_0 \subseteqq U$.

A 1, . . ., A 5 are the first five axioms in the axiomatization of the theory of distributions given by Sebastião e Silva [16], §1.14. But the five axioms A 1, . . ., A 5 do not yet uniquely determine the concept "distribution." For this purpose further axioms are necessary. Thus the sixth axiom A 6, to which L. Schwartz [13], Chap. I, Théorème IV has given the very apt name "*principe du recollement des morceaux*," states how *the* "*global*" *behavior of a distribution is determined by its* "*local*" *properties.*

A 6. *Let there be given an open subset U of* R^n. *For every* $x \in U$ *let there be given an open neighborhood* $U_x \subseteqq U$ *of x and a distribution* T_x *with domain of definition* U_x. *If* $U_{x,y} = U_x \cap U_y \neq \varnothing$, *for certain* $x, y \in U$, *then let the restrictions of* T_x, T_y *to* $U_{x,y}$ *coincide. Under these assumptions there exists one and only one distribution T with domain of definition U whose restriction to* U_x *for every* $x \in U$ *is equal to* T_x.

The seventh axiom A 7 determines the local structure of the distributions: *every distribution is* "*locally*" *the derivative of a continuous function*; more precisely:

A 7. *For every T with domain of definition U and for every bounded open interval J of* R^n *whose closed hull* $\bar{J}$ *contains U there exist a function f*

[47] The function f_{U_0} is defined by $f_{U_0}(x) = f(x)$ for $x \in U_0$.

defined and continuous in J and an n-tuple $p = (p_1, \ldots, p_n)$ *of nonnegative integers* p_i *such that* $T_J = D^p f = D_{x_1}^{p_1} \cdots D_{x_n}^{p_n} f$. (Here $D_{x_n}^{p_n} f$ denotes the distribution derived from f, in accordance with A 4, by partial differentiation p_n-times with respect to x_n.)

As in A 7, the requirements in the following axiom A 8 are minimal if the system of axioms A 1,..., A 8 is to admit essentially one and only one realization.

A distribution T is said to be *independent of* x_i if $D_{x_i}T = 0$.

A 8. *Let T be a distribution independent of* x_i *whose domain of definition is an open interval J in* R^n. *Let* $T = D^p f$ *with suitable* $p = (p_1, \ldots, p_n)$ *and* $f \in C(J)$ (cf. A 7). *Then there exists an n-tuple* $q = (q_1, \ldots, q_n)$ *of nonnegative integers* $q_1 \leqq p_1, \ldots, q_n \leqq p_n$ *and a function* $g \in C(J)$, *independent of* x_i *in the usual sense, such that* $T = D^q g$.

Various questions naturally arise: Is the system of axioms A 1, ..., A 8 consistent, i.e., has it at least one realization? Is there more than one nonisomorphic realization? How can realizations be found, assuming they exist? We cannot discuss these questions in detail but must refer to the original literature and in particular to the publications of Sebastião e Silva and Schwartz [13]. Let us merely give the answers: *apart from isomorphisms, there exists exactly one realization of the system of axioms* A 1, ..., A 8; *this realization consists precisely of the distributions that were introduced by* Schwartz *from a somewhat different point of view.*

3.3. *Introduction of Distributions by Schwartz*

Let us now give a brief description[48] of the method adopted by Schwartz for the introduction of distributions. For simplicity we deal only with distributions whose domain of definition is the whole R^n.[49]

The first step in the *Schwartz construction* of distributions is a *partial generalization of the concept of "function (on* R^n*)."* This generalization is to be found in the concept of (*Radon*) *measure* μ (*on* R^n). The μ are in a certain sense "*continuous*" *linear functionals on the space* $K(R^n)$ *of functions continuous on* R^n *with compact support* (cf. §§2.1,c), 2.5,d)).

The connection between functions and measures is set up as follows: if f is a continuous function on R^n, then for every $t \in K(R^n)$ we set

$$\text{(80)} \qquad \mu_f(t) = {}^{(L)}\int_{R^n} ftdx = \int_{-\infty}^{\infty} \cdots \int_{-\infty}^{\infty} f(x_1, \ldots, x_n)t(x_1, \ldots, x_n)dx_1 \cdots dx_n.$$

[48] For details see Schwartz [13], Chap. I, II.

[49] The additional difficulties presented by distributions with an open subset U of R^n for domain of definition are of no theoretical importance.

Here ${}^{(L)}\int_{R^n} ftdx$ is the *Lebesgue integral* of ft over R^n.[50] Thus the continuous function f appears as the *density* of the measure μ_f defined by

$$(81) \qquad t \to \mu_f(t) = {}^{(L)}\int_{R^n} ftdx \qquad t \in K(R^n).$$

But along with the continuous functions it is possible to interpret other functions f as densities of analogous measures μ_f, since in (80), (81) it is only necessary for f to be *locally L-integrable* over R^n, i.e., L-integrable over every bounded closed interval in R^n. Since by (81) locally L-integrable functions f, g define the same measure if and only if f, g are L-almost everywhere equal on R^n, the term "*Radon measure* (*on* R^n)" can be regarded as a generalization of "*the class of L-almost everywhere equal, locally L-integrable functions* (*on* R^n)." Every such class contains at most one continuous function, so that these measures can be regarded as a generalization of continuous functions.

In the *Schwartz theory of distributions* it is customary to identify the measures μ_f with the densities f that define them, $\mu_f = f$. It is to be noted that a function f, say continuous, thereby appears on the one hand as a *point function*, which to every $x \in R^n$ assigns a value $f(x)$, and on the other hand, in view of the identification of f and μ_f, as a *functional* on $K(R^n)$ defined by

$$(82) \qquad t \to f(t) = \mu_f(t) = {}^{(L)}\int_{R^n} ftdx \qquad t \in K(R^n).$$

Not every *Radon measure* is of the form μ_f with locally L-integrable f. The simplest example of such a measure μ is the *Dirac measure* δ defined by

$$(83) \qquad t \to \delta(t) = t(0) \qquad t \in K(R^n).$$

This measure, call it δ, *is the substitute in the theory of distributions for the Dirac delta-"function."*

The second step in the *Schwartz construction* of distributions consists in *generalizing the concept of a* "*Radon measure* (*on* R^n)" to a "*distribution* (*on* R^n)." Schwartz defines a *distribution on* R^n as a "*continuous*" *linear functional T on the space* ($D(R^n)$) *of functions that have compact support*[51] *and have arbitrarily many continuous derivatives on* R^n. *Continuity of the functional T* is to be understood here in the following sense (different from continuity for measures!): if a sequence $t_1, t_2, \ldots$ of functions $t_m \in (D(R^n))$, all of which vanish outside of *one and the same* fixed bounded closed set

[50] In the present case ${}^{(L)}\int_{R^n} ftdx$ is equal to the *Riemann integral* of ft over a bounded closed interval outside of which ft vanishes.

[51] I.e., the $t \in (D(R^n))$ are those $t \in K(R^n)$ for which the partial derivative $\dfrac{\partial^{p_1 + \cdots + p_n} t}{\partial x_1^{p_1} \cdots \partial x_n^{p_n}}$ exists, and is continuous, for all choices of $p_1, \ldots, p_n$.

of R^n, converges *uniformly* in R^n to 0 and if the same situation holds for *every* sequence $\frac{\partial^{p_1+\cdots+p_n}t_1}{\partial t_1^{p_1}\cdots\partial t_n^{p_n}}, \frac{\partial^{p_1+\cdots+p_n}t_2}{\partial x_1^{p_1}\cdots\partial x_n^{p_n}}, \ldots$ of partial derivatives of the t_m, then the sequence $T(t_1), T(t_2), \ldots$ converges to $T(0) = 0$.

Thus the domain of definition of the functional T is a subspace of the space $K(R^n)$. This subspace is "dense" in a certain sense in $K(R^n)$.[52] Every measure μ on R^n defines a distribution T_μ if we restrict the domain of definition of the functional μ from $K(R^n)$ to $(D(R^n))$:

$$T_\mu(t) = \mu(t) \qquad t \in (D(R^n)). \tag{84}$$

It might be thought that different measures μ_1, μ_2 could define the same distribution T_μ in (84). But this is not the case, since $(D(R^n))$ is dense in $K(R^n)$ and the measures are continuous.[53] Every *Radon measure* on R^n can be identified with exactly one distribution T_μ on R^n, $\mu = T_\mu$; *the concept "distribution" generalizes the "Radon measure."* For the distributions just introduced it is now necessary to give suitable definitions of such concepts as "restriction," "sum," "derivative." For example, it must be shown that the totality $(D)'$ of distributions on open sets of R^n satisfies the axioms A 1, . . ., A 8. To do this in detail would far exceed the space at our disposal. It is clear that verification of the individual statements A 1, . . ., A 8 may lead to problems of widely varying degrees of difficulty. Thus A 1 and A 3 offer practically no trouble at all, but the verification of A 7 and A 8 is one of the most important results in the first volume of Schwartz [13].

3.4. *Differentiation of Distributions*

We must give a brief account of the differentiation of distributions in the *Schwartz theory*. By §3.3 every continuous function f on R^n can be identified with a *Radon measure* μ_f and consequently with a distribution $T_f = T_{\mu_f}$; for every $t \in (D(R^n))$ we then have

$$\begin{aligned} f(t) &= T_f(t) = T_{\mu_f}(t) = \mu_f(t) \\ &= \int_{-\infty}^{\infty}\cdots\int_{-\infty}^{\infty} f(x_1, \ldots, x_n)t(x_1, \ldots, x_n)dx_1\cdots dx_n. \end{aligned} \tag{85}$$

If f is assumed to be continuously differentiable with respect to x_n, then by the usual rules for *Riemann integrals* (cf. footnote 50) we obtain

$$\begin{aligned} (D_{x_n}f)(t) &= \int_{-\infty}^{\infty}\cdots\int_{-\infty}^{\infty} (D_{x_n}f(x_1, \ldots, x_n))t(x_1, \ldots, x_n)dx_1\cdots dx_n \\ &= \int_{-\infty}^{\infty}\cdots\int_{-\infty}^{\infty} dx_1\cdots dx_{n-1} \\ &\quad \times \left(\int_{-\infty}^{\infty} (D_{x_n}f(x_1, \ldots, x_n)\right)t(x_1, \ldots, x_n)dx_n) \\ &\qquad t \in (D(R^n)). \end{aligned} \tag{86}$$

[52] Schwartz [13], Chap. I, Théorème I.

[53] Schwartz [13], Chap. I, Théorème III.

Here $D_{x_n}f$ is regarded as the partial derivative of f with respect to x_n in the sense of classical analysis. The inner integral can be evaluated by integration by parts. Since t vanishes outside of a bounded interval of R^n, we first obtain

$$\int_{-\infty}^{\infty} (D_{x_n}f)t\,dx_n = [f(x_1, \ldots, x_n)t(x_1, \ldots, x_n)]_{x_n=-\infty}^{\infty} - \int_{-\infty}^{\infty} f(D_{x_n}t)dx_n$$

$$= -\int_{-\infty}^{\infty} f(D_{x_n}t)dx_n.$$

(For simplicity we have used obvious abbreviations.) Then in view of (86) we further obtain

$$(87)\qquad \begin{aligned}(D_{x_n}f)(t) &= -\int_{-\infty}^{\infty} \cdots \int_{-\infty}^{\infty} dx_1 \cdots dx_{n-1}\left(\int_{-\infty}^{\infty} f(D_{x_n}t)dx_n\right)\\ &= -\int_{-\infty}^{\infty} \cdots \int_{-\infty}^{\infty} f(D_{x_n}t)dx_1 \cdots dx_n.\end{aligned}$$

Finally, from (85) we have

$$(88)\qquad (D_{x_n}f)(t) = -f(D_{x_n}t) \qquad t \in (D(R^n)).$$

The relation (88) can be regarded as an equation which defines partial differentiation with respect to x_n for a special distribution, namely $T = T_{\mu_f} = T_f = f$. However, the relation (88) between $T = f$ and $D_{x_n}T = D_{x_n}f$ remains meaningful if we replace f by an *arbitrary* distribution T:

$$(89)\qquad (D_{x_n}T)(t) = -T(D_{x_n}t) \qquad t \in (D(R^n)).$$

Thus the functional

$$(90)\qquad t \to (D_{x_n}T)(t) = -T(D_{x_n}t) \qquad t \in (D(R^n))$$

is seen to be a distribution $T_1 = D_{x_n}T$ on R^n for every $T \in (D(R^n))'$, which is called the *partial derivative of T with respect to* x_n. The derivatives $D^pT = D_{x_1}^{p_1} \cdots D_{x_n}^{p_n}T$ are defined analogously:

$$(91)\qquad t \to (D_{x_1}^{p_1} \cdots D_{x_n}^{p_n}T)(t) = (-1)^{p_1+\cdots+p_n}T(D_{x_1}^{p_1} \cdots D_{x_n}^{p_n}t) \qquad t \in (D(R^n)).$$

We see at once that *every distribution T is arbitrarily often differentiable and the order of differentiation can be changed at will.* Furthermore the definition of D^pT bears the proper relationship, as postulated by A 4, to the classical process of differentiation.

If in the case of one variable ($n = 1$) we regard the *Heaviside function H* (cf. (75)) as a distribution,

$$(92)\qquad H(t) = \int_{x=-\infty}^{\infty} H(x)t(x)dx \qquad t \in (D(R^1)),$$

its derivative $D_x H = H'$ is seen to be the *Dirac measure* δ:

$$H'(t) = -H(t') = -\int_{-\infty}^{\infty} H(x)t'(x)dx = -\int_{0}^{\infty} H(x)t'(x)dx$$

$$= -\int_{0}^{\infty} t'(x)dx = t(0) = \delta(t). \qquad t \in (D(R^1)). \tag{93}$$

Also, for example,

$$(D_x(D_x H))(t) = H''(t) = \delta'(t) = -\delta(t') = -t'(0) \qquad t \in (D(R^1)). \tag{94}$$

The derivative δ' of the *Dirac measure* δ is the simplest example of a distribution that is not a *Radon measure*.

It is well known that differentiation plays a role in various branches of classical analysis (for example, in the theory of ordinary and partial differential equations). Thus it is natural to make a fresh investigation of these branches of mathematics from the point of view of distributions, by replacing functions with distributions and the classical concept of differentiation of functions by the differentiation defined above for distributions. Investigations of this sort have produced a large number of important results. In particular, distributions may arise as solutions of differential equations. For details see Schwartz [13].

3.5. *Other Methods of Constructing the Theory of Distributions*

After the appearance of the book by Schwartz [13] other methods of constructing the theory of distributions have been published by König [15] and Sebastião e Silva [16]. In both [15] and [16] the purpose was the same as with Schwartz, but the formal methods were different: *the space of distributions* (or some other suitable extension) *is defined by a direct algebraic construction*, whereas the *Schwartz construction*, in view of the conditions of continuity imposed on the *Radon measure* and on the distributions, makes use from the beginning of *topological properties*. Consequently, König has made the comparison that the *Schwartz construction* of $(D(R^n))'$ is related to the construction given by him (König) in the same way as the solution of algebraic equations in the field of complex numbers is related to the symbolic adjunction of algebraic elements.

The direct algebraic constructions of the extensions in [15], [16] have the advantage that the structure of these extensions is determined in advance; for Schwartz, on the other hand, this structure had to be determined subsequently by means of a profound argument. Moreover, the algebraic methods make it clear that *the possibility of constructing the space of distributions depends precisely on the validity of the fundamental theorems of the Lebesgue theory of integration* (cf. in particular [15]). It is obvious that there must be a close relationship between distributions and the *Lebesgue integrability* of functions, since by §3.3 every class of L-almost everywhere equal, locally L-integrable functions f can be identified with a distribution $T = T_f$.

Furthermore, it is clear from the constructions in [15], [16] how matters stand with respect to differentiation of distributions: for every distribution T of the form $T_{\mu_f} = T_f$ *partial differentiation with respect to* x_i is exactly equivalent to *inversion of indefinite Lebesgue integration with respect to* x_i.[54] This fact was clearly recognized by Schwartz, but König and Sebastião e Silva have raised it to a fundamental principle of construction. The constructions of König [15] and Sebastião e Silva [16] cannot be discussed here in detail; we must refer the reader to the original articles.

Bibliography

[1] AUMANN, G.: Reelle Funktionen. Springer, Berlin-Göttingen-Heidelberg 1954.

[2] BOURBAKI, N.: Intégration. Hermann, Paris 1952, 1956.

[3] CARATHÉODORY, C.: Maß und Integral und ihre Algebraisierung. Birkhäuser, Basel 1956.

[4] HALMOS, P. R.: Measure Theory. Van Nostrand, New York 1950.

[5] HAUPT, O., AUMANN, G. and PAUC, C.: Differential- und Integralrechnung III, 2nd ed. de Gruyter, Berlin 1955.

[6] KAMKE, E.: Das Lebesgue-Stieltjes-Integral. Teubner, Leipzig 1956.

[7] MAYRHOFER, K.: Inhalt und Mass. Springer, Vienna 1952.

[8] MACSHANE, E. J.: Integration. Princeton University Press, Princeton 1947.

[9] RIESZ, F. and NAGY, B. SZ.: Vorlesungen über Funktionalanalysis. Deutscher Verlag der Wissenschaften, Berlin 1956.

[10] SAKS, ST.: Théorie de l'intégrale. Warszawa 1933 (out of print).

[54] Let J be a bounded closed interval $\{x : x = (x_\nu), a_\nu \leqq x_\nu \leqq b_\nu, a_\nu < b_\nu\}$ in R^n. For every continuous function f on J it is well known that

$$I_{x_i}f(x) = \int_{a_i}^{x_i} f(x_1, \ldots, x_{i-1}, \xi, x_{i+1}, \ldots, x_n)d\xi$$

exists if $a_i \leq x_i \leqq b_i$ and the x_λ, $\lambda = 1, \ldots, i-1, i+1, \ldots, n$, are chosen in any fixed way with $a_\lambda \leqq x_\lambda \leqq b_\lambda$. Every function F which differs from $I_{x_i}f$ only by an additive *(Riemann)-integrable* function independent of x_i is called an *indefinite integral of f with respect to* x_i. If f is only *Lebesgue integrable* over J, the situation is somewhat more difficult. For then $I_{x_i}f(x)$ exists only L-almost everywhere on J. For those $x' \in J$ for which an L-integral $I_{x_i}f(x')$ does not exist, let $I_{x_i}f(x')$ be defined arbitrarily. If we then form the equivalence classes $I_{x_i}f$ as in §2.3, we again arrive at uniquely defined magnitudes. Every class $[F]$ which differs from $[I_{x_i}f]$ only by an additive class of integrable functions that are "independent" of x_i is to be regarded as the *definite L-integral of* $[f]$ *with respect to* x_i. (Cf. [15].)

[11] SAKS, ST.: Theory of the integral. Stechert, New York 1937.
[12] HALPERIN, I.: Introduction to the theory of distributions. Canad. Math. Congr. Lect. S. No. 1. University of Toronto Press, Toronto 1952.
[13] SCHWARTZ, L.: Théorie des distributions I, II. Hermann, Paris 1950, 1951.
[14] HAUPT, O.: Über die Entwicklung des Integralbegriffes seit Riemann. Schriften-R. Forsch. Inst. Math. Berlin **1**, 303–317 (1957).
[15] KÖNIG, H.: Neue Begründung der Theorie der „Distributionen" von L. Schwartz. Math. Nachr. **9**, 129–148 (1953).
[16] SEBASTIÃO E SILVA, J.: Sur une construction axiomatique de la théorie des distributions. Univ. Lisboa Revista Fac. Si., II. Ser. A **4**, 79–186 (1955).
[17] STONE, M. H.: Notes on integration I, . . ., IV. Proc. Nat. Acad. Sci. U.S.A. **34**, 336–342, 447–455, 483–490 (1948); **35**, 50–58 (1949).
[18] ASPLUND, E. and BUNGART, L.: A first course in integration. Holt, Rinehart and Winston, New York 1966.
[19] BERBERIAN, S. K.: Measure and integration. Macmillan, New York 1965. (A first course).
[20] BURKILL, J. C.: The Lebesgue integral. Cambridge University Press, New York 1951.
[21] PITT, H. R.: Integration, measure and probability. Oliver and Boyd, Edinburgh 1963.
[22] ROGOSINSKI, W. W.: Volume and integral. Oliver and Boyd, Edinburgh; Interscience, New York 1952.
[23] TAYLOR, A. E.: General theory of functions and integration. Blaisdell Pub. Co., New York 1965.

CHAPTER 3a

Fundamental Concepts of the Theory of Probability

1. The Concept of Probability

1.1. *Preliminary Remarks*

The theory of probability originated, as is well known, in games of chance. Its further development in the 18th and 19th centuries is associated with many brilliant names; among them, Jacob Bernoulli, Cauchy, Čebyšev, Gauss, and Poisson. An important impetus was given to the theory by the creation of statistical mechanics (Boltzmann, Maxwell) and of the Fechner theory of collectives, by which v. Mises was much influenced in his treatment of probability as the limiting value of relative frequencies. At the present time the theory of probability is usually based on measure theory and has been put in definitive form by Kolmogorov.

For a long time the theory of probability had an isolated position in the framework of mathematics, but today it has many important applications in mathematics itself. We may mention the theory of potential, boundary-value problems, integral equations, the general theory of differentiation and integration, and finally also number theory. However, the applications in analysis rest almost entirely on the theory of stochastic processes, which cannot be presented here. (See [1] and [3] in the bibliography.)

Before we begin our outline of the measure-theoretic construction of the theory of probability, we wish to raise some questions of an epistemological nature. In daily life, and also in science, the term probability is understood in several senses. Roughly speaking, its various meanings of the term can be divided into two classes: subjective and objective. The first expresses the "degree of knowledge or ignorance of the person expressing an opinion," whereas objective probability refers to events in the external world. For example, let us consider the classical games with dice. If we assume that a given die is completely symmetric and is composed of homogeneous material ("an ideal die"), then the probability of turning up a given number will be estimated in the same way by all observers. The cast of an ideal die is governed by a certain objective regularity, which the observer expresses by saying that for every successive cast the probability of obtaining the given number remains unchanged. This judgment on his part corresponds to the following phenomenon in the external world: if an ideal die is cast N-times, the relative frequencies of the various faces remain practically unchanged when the number N becomes very large. This "stability of the relative frequency" in a long sequence of experiments under constant conditions is by no means restricted to such simple situations as play with dice. In fact, it is of decisive importance for all science. For the situation is always as follows: there exists a set E of events which may be realized in an experiment. To each of these events there is a number assigned, namely its probability. The practical importance of this probability lies in the fact that, for every given event $e \in E$, it is approximated, more and more closely for longer sequences of experiments, by the relative frequency of e in a long sequence of experiments under uniform conditions.

The task of a mathematical theory of probability is to create a method of calculation with probabilities. Just as in geometry the concepts of point and line are no longer directly defined, but only by means of their properties (implicit definition), such concepts as event and probability must be defined implicitly.

1.2. *Events and Probability*

Let E be a set whose elements e will be called events. For every two events e_1 and e_2 let E also contain the event "e_1 or e_2," which we will denote by $e_1 \cup e_2$. Let E also contain the event $e_1 \cap e_2$, which we shall interpret as "both e_1 and e_2." These two connectives satisfy the following conditions:

1a $\quad e_1 \cup e_1 = e_1,$

1b $\quad e_1 \cup e_2 = e_2 \cup e_1,$

1c $\quad (e_1 \cup e_2) \cup e_3 = e_1 \cup (e_2 \cup e_3),$

1d $\quad e_1 \cup (e_1 \cap e_2) = e_1,$

1e $\quad (e_1 \cup e_2) \cap e_3 = (e_1 \cap e_3) \cup (e_2 \cap e_3).$

1a′ $\quad e_1 \cap e_1 = e_1,$

1b′ $\quad e_1 \cap e_2 = e_2 \cap e_1,$

1c′ $\quad (e_1 \cap e_2) \cap e_3 = e_1 \cap (e_2 \cap e_3),$

1d′ $\quad e_1 \cap (e_1 \cup e_2) = e_1.$

From 1d, 1d′, and 1e we can deduce a "dual" distributive law 1e′, where $\cap$ and $\cup$ are interchanged. *A set E, in which two connectives are defined so as to satisfy* 1a to 1e, 1a′ to 1d′ *is called "a distributive lattice."*

We also require the existence of a (uniquely defined) element $o \in E$ *which for all* $e \in E$ *satisfies the condition*

1f $$e \cup o = e.$$

The event o is to be interpreted as an event that is "impossible." From 1d′ and 1f it follows at once that $e \cap o = o$ for all $e \in E$.

We now introduce the concept of an event that is "*certain*" by assuming that there exists in E an element $1 \neq o$ such that

1g $$e \cap 1 = e$$

for all $e \in E$.

The conditions 1a to 1g and 1a′ to 1d do not yet allow us, given an event e_1, to speak of the event "not e_1." Thus we formulate the following *postulate*:

For every $e_1 \in E$ *there exists an* e_2 *such that*

1h $$e_1 \cup e_2 = 1, \qquad e_1 \cap e_2 = o.$$

It is easy to show that e_2, which is called the *complement* of e_1, is uniquely determined; for if there were another e_2' with $e_1 \cup e_2' = 1$, $e_1 \cap e_2' = o$. it would follow that $e_2 \cap (e_1 \cup e_2') = e_2 \cap 1 = e_2$, so that by 1e, 1h, and 1f we would have $e_2 \cap e_2' = e_2$. By interchanging e_2 and e_2' we would obtain $e_2' \cap e_2 = e_2'$ and thus $e_2' = e_2$ in contradiction to the assumption.

It follows that if $e_1 \cup e_3 = e_3$, there exists an e_4 with $e_1 \cup e_4 = e_3$, $e_1 \cap e_4 = o$. The element e_4 is called the *complement* of e_1 with respect to e_3.

To prove this statement we choose e_2 corresponding to e_1 in accordance with 1h, and then show $e_4 = e_2 \cap e_3$. In the first place we have

$$e_1 \cup (e_2 \cap e_3) = (e_1 \cup e_2) \cap (e_1 \cup e_3) = 1 \cap e_3 = e_3.$$

Then from the second condition 1h and from the fact that $o \cap e_3 = o$ it follows at once that $e_1 \cap (e_2 \cap e_3) = o$, so that again e_4 is uniquely determined.

If the conditions 1a to 1h and 1a′ to 1d′ are satisfied, E is called a *Boolean algebra*.

It would be possible to introduce the theory of probability in the following way:[1] *Let E be a Boolean algebra and to every* $e \in E$ *let there be assigned a nonnegative number* $W(e)$, *called the probability of e, such that*

2a $W(1) = 1$, $W(o) = 0$, *and from* $W(e) = 0$ *follows* $e = o$.

2b $W(e_1 \cup e_2) = W(e_1) + W(e_2)$, if $e_1 \cap e_2 = o$.

[1] The reader who is interested in a detailed discussion of probability measures over Boolean algebras may consult [8] in the bibliography. Boolean algebra is also discussed in IA, §9.

The probability function W is thus a positive (finitely) additive measure over a Boolean algebra. The significance of 2a is clear, and 2b is the "law of addition for the theory of probability." However, we prefer to undertake a different construction of the theory of probability, essentially equivalent to the one discussed above. In this connection let us mention the important *theorem of isomorphism* of M. H. Stone: *every Boolean algebra is isomorphic to a field of sets, where the connectives* $\cup$ *and* $\cap$ *of the Boolean algebra correspond to set-theoretic union and intersection.*

1.3. *Probability Measure*

We now proceed to introduce the concept of probability in the following way: *let there be given a set R and a σ-field* (III 3, §2.7)[2] *S of subsets of R, including the whole set R itself. A σ-field of this sort, containing R, is also called a σ-algebra. On S let there be defined a*

(I) *real, nonnegative, totally additive*[3]

set function W (III 3, §2.7) *satisfying the condition*

$$W(R) = 1. \tag{II}$$

Such a function of sets is called a probability distribution over S or a probability measure (III 3, §2.7). Then an event is simply a subset of R that belongs to S. Comparison with the Boolean algebra discussed above shows that we have replaced the field of sets by a σ-field and the condition 2b by the (stronger) condition of complete additivity, the changes being made simply for greater convenience in the mathematical construction of the theory. If $\varnothing$ denotes the empty set, which of course belongs to S, then it follows from the total additivity and finiteness of W that $W(\varnothing) = 0$ But from $M \in S$ and

$$W(M) = 0 \tag{1}$$

it does not necessarily follow that $M = \varnothing$.

From the complete additivity of a probability distribution follows its finite additivity.

Every probability measure (or more generally every finite completely additive measure) has the following *property of continuity*: *let* $\{A_i\}$, $A_i \in S$ *be an increasing or decreasing sequence of sets, i.e.,* $A_1 \subseteqq A_2 \subseteqq \cdots$ or $A_1 \supseteqq A_2 \supseteqq \cdots$. *In the first case we have*

$$\lim_{i \to \infty} W(A_i) = W\left(\bigcup_{i=1}^{\infty} A_i\right)$$

[2] By a σ-field we mean a class of sets which with every collection of countably many sets includes their union and with every two sets includes their difference.

[3] Instead of totally additive it is also customary to say completely additive.

and in the second case

$$\lim_{i \to \infty} W(A_i) = W\left(\bigcap_{i=1}^{\infty} A_i\right).$$

In the most important special case, which is always convenient for purposes of visualization, R is the Euclidean space R^n (III 3, §1.0) and S is the set of Euclidean Borel sets $\subset R^n$.[4] Another important example leads to the classical *definition of probability* according to Laplace: let R be a finite set with 1 elements and let S be the set of all subsets of R. Let W be defined in such a way that $W(x)$ has the same value for every $x \in R$. In view of (II) we then have $W(x) = \frac{1}{l}$. If A is a subset of R with m elements, it follows from the additivity of probability that

$$W(A) = \frac{m}{l} = \frac{\text{number of "favorable cases"}}{\text{number of "possible cases"}}.$$

The pair (R, S) is called a *measurable space* and the triple (R, S, W) is a *probability space*. In what follows we shall always be dealing with a fixed probability space, even when we do not explicitly say so.

Exercises

1. Prove the above "property of continuity" for probability measures.
2. If three cards are drawn at random from a deck of 36 cards (omitting the tens, jacks, queens, and kings), what is the probability of obtaining (*a*) exactly one ace? (*b*) at least one ace?

2. The Distribution Function of a Random Variable

2.1. *Random Variables*

Of basic importance is the concept of a *random variable*. Let there be given a measurable space (R, S), and let k be a mapping of R into the Euclidean space R^1. Then k is said to be *S-measurable*[5] if for every Borel set B^1 in R^1 the set $\{x : k(x) \in B^1\}$ belongs to S. *Every S-measurable mapping k of R into the Euclidean space R^1 is called a (one-dimensional) random variable.* It is easy to show that the sum, difference, product and quotient (with nonvanishing denominator) of two random variables is

[4] The σ-algebra of Borel sets in R^n is the smallest σ-algebra including all intervals of R^n (III 3, §1.0). We denote it by B^n.

[5] Instead of S-measurable we shall, for the most part, merely say measurable. However, if S is a special σ-algebra, we shall always mention that fact. In particular, if S is the σ-algebra of the Borel set of an R^n, then we speak of Borel-measurable functions.

again a random variable. Likewise, the absolute value of a random variable is again a random variable, and more generally for every Borel-measurable real-valued function g whose domain of definition includes the range of k, the function $g \circ k$[6] is again a random variable.

2.2. *Distribution of a Random Variable*

Now let W be a probability distribution over S and let k be a random variable. Over the σ-algebra $\mathfrak{B}^1$ of the one-dimensional Borel sets B^1 we define a *set function* $W^{(k)}$ as follows:

$$W^{(k)}(B^1) = W(\{x : k(x) \in B^1\}).^7 \tag{2}$$

From the definition it follows that $W^{(k)}$ is a completely additive non-negative measure satisfying the condition $W^{(k)}(R^1) = 1$. Thus $W^{(k)}$ is a probability distribution, namely the distribution of the random variable k.

2.3. *One-dimensional Distribution Functions*

In particular, for every interval of the form $(-\infty, y] = I_y$ the probability $W^{(k)}(I_y)$ is defined by (2). The function F of the real variable y which to every y assigns the value $F(y) = W^{(k)}(I_y)$ is known as the *distribution function* (abbreviated d.f., and sometimes called the *cumulative distribution function*) of the random variable k. From the definition (2) it follows that F has the following properties:

E 1 *F is nondecreasing,*[8]

E 2 *F is continuous on the right,*

E 3 $\lim_{y \to -\infty} F(y) = 0, \qquad \lim_{y \to \infty} F(y) = 1.$

In particular, for every probability measure W on $\mathfrak{B}^1$ we have thus defined its distribution function, namely as the distribution function of the random variable $k(x) = x$ for all $x \in R^1$ on the probability space $(R_1, \mathfrak{B}^1, W)$.

Now it is of great importance that conversely, if we start from a function F satisfying the conditions E1 to E3, we can construct a uniquely defined probability measure over the σ-algebra $\mathfrak{B}^1$ in such a way that F is the corresponding d.f. For this purpose we assign to all intervals of the form $(a, b]$ the measure $F(b) - F(a)$ and then apply the general theorem of extension in the theory of measure (cf. III 3, §2.7).

[6] This symbol denotes the composite function which to every $x \in R$ assigns the real number $g(k(x))$.

[7] In place of $\{x : k(x) \in B^1\}$ we may also write $k^{-1}(B^1)$ and call this set the pre-image of B^1 with respect to k.

[8] From E 1 it follows that F has at most countably many points of discontinuity. Let y_0 be a point of discontinuity. Then $W^{(k)}(\{y_0\}) = F(y_0) - F(y_0 - 0) > 0$, where $F(y_0 - 0)$ is the left-sided limit of F at the point y_0.

2.4. *Multidimensional Random Variables*

The concept of a one-dimensional random variable can be extended at once to that of a *multidimensional random variable.* Every measurable mapping $(k_1, \ldots, k_m)$ of R into an R^m will be called an m-dimensional random variable. If for every m-dimensional Borel set B^m we put

$$(3) \qquad W^{k_1, \ldots, k_m}(B^m) = W(\{x : (k_1(x), \ldots, k_m(x)) \in B^m\}),$$

we have thereby defined a probability distribution over the σ-algebra $\mathfrak{B}^m$, namely the probability distribution of the random variable $(k_1, \ldots, k_m)$. For every interval

$$I_{y_1, \ldots, y_m} = (-\infty < x_1 \leqq y_1, \ldots, -\infty < x_m \leqq y_m)$$

we define

$$F(y_1, \ldots, y_m) = W^{(k_1, \ldots, k_m)}(I_{y_1, \ldots, y_m})$$

in accordance with (3). Then F is again called the *distribution function of the random variable* $(k_1, \ldots, k_m)$. The function F has the following properties:

E 1′ $\Delta_{i_1} \cdots \Delta_{i_k} F(y_1, \ldots, y_m) \geqq 0$ *for all* $(y_1, \ldots, y_m) \in R^m$,

where $\{i_1, \ldots, i_k\}$ *is an arbitrary nonempty subset of* $\{1, \ldots, m\}$ *and* $\Delta_i F(y_1, \ldots, y_m)$ *is defined by*

$$F(y_1, \ldots, y_i + h_i, \ldots, y_m) - F(y_1, \ldots, y_i, \ldots, y_m) \quad \textit{with} \quad h_i \geqq 0,$$

and repeated application of Δ *is defined correspondingly.*

E 2′ *F is continuous on the right in each of its arguments.*

E 3′ $\lim_{y_{i_1}, y_{i_2}, \ldots, y_{i_k} \to -\infty} F(y_1, \ldots, y_m) = 0$

uniformly in all the other variables, and

$$\lim_{y_1, \ldots, y_m \to \infty} F(y_1, \ldots, y_m) = 1.$$

In analogy with the one-dimensional case, the general theorem of extension for measures enables us, starting from a given function F that satisfies E 1′ to E 3′, to define a probability distribution over the σ-algebra $\mathfrak{B}^m$ such that F is its d.f.

If $F(y_1, \ldots, y_m)$ can be written for all $y_1, \ldots, y_m$ in the form

$$\int_{-\infty}^{y_1} \cdots \int_{-\infty}^{y_m} f(x_1, \ldots, x_m) dx_1 \cdots dx_m$$

with Lebesgue integrable f, then f is called the *probability density* of $(k_1, \ldots, k_m)$. From E 1′ to E 3′ it follows that

D 1 $f \geqq 0$ *almost everywhere,*

D 2 $\int_{-\infty}^{+\infty} \cdots \int_{-\infty}^{+\infty} f(x_1, \ldots, x_m) dx_1 \cdots dx_m = 1.$

Conversely, every Lebesgue measurable function f that satisfies the condition D 1 to D 2 defines a function F by

$$\int_{-\infty}^{y_1} \cdots \int_{-\infty}^{y_m} f(x_1, \ldots, x_m) dx_1 \cdots dx_m = F(y_1, \ldots, y_m)$$

such that F has the properties E 1′ to E 3′ and is thus a probability distribution.

2.5. *Examples of Probability Distributions on R^1*

1. Let a be an arbitrary real number and let σ be a positive number. Then it is easy to show by calculation that

$$\frac{1}{\sigma\sqrt{2\pi}} e^{-(y-a)^2/2\sigma^2} \qquad -\infty < y < \infty$$

has the properties D 1 to D 2 of a density. The probability distribution defined by this density is called the *normal* or *Gauss distribution.* We shall denote it by $N(a, \sigma^2)$.

2. We now consider the d.f., call it F, defined by $F(y) = 0$ for $y < C$ and $F(y) = 1$ for $y \geqq C$, where C is an arbitrary real number. The corresponding probability distribution $D(C)$, or the corresponding random variable, is said to be *degenerate.* The corresponding probability measure assigns the measure 1 to every Borel set containing the point C.

3. A simple and natural generalization of the degenerate distribution consists of assigning the positive measures p_i to finitely many or countably many points a_i, where $\sum p_i = 1$. The probability distribution thus defined is called a *discrete probability.* Its distribution function is given by $\sum_{a_i \leqq y} p_i$.

4. An important example of a discrete distribution is the *binomial* or *Bernoulli distribution.* Let n be a natural number and p a real number with $0 < p < 1$ and $q = 1 - p$. The d.f. $F^{n,p}$ of the binomial distribution $B_n(p)$ is defined for every real y by $F^{n,p}(y) = \sum_{k \leqq y} \binom{n}{k} p^k q^{n-k}$ where k runs through the integers $0 \leqq k \leqq n$.[9]

5. Let $a > 0$. If we set $F_a(y) = \sum_{k \leqq y} \frac{a^k}{k!} e^{-a}$ for every $y \in R^1$, we thereby define a d.f. F_a, for $k = 0, 1, 2, \ldots$. The probability distribution thus defined is called the *Poisson distribution.*

In the two examples it is easy to verify that the conditions E 1 to E 3 for d.f. are satisfied.

[9] For $y < 0$ or $y > n$ we make the usual conventions.

2.6. *Marginal Distribution*

Let $M_1, \ldots, M_k$ be arbitrary (nonempty) sets. The set of ordered k-tuples $(x_1, \ldots, x_k)$ with $x_i \in M_i$ is called the Cartesian product $M_1 \times \cdots \times M_k$ of the M_i.

Starting from a probability space (R, S, W), we now construct, in accordance with (3), the probability distribution $W^m = W^{(h_1, \ldots, h_m)}$ of an m-dimensional random variable $(h_1, \ldots, h_m)$. Let $(i_1, \ldots, i_l)$ be a subset of $(1, \ldots, m)$. We consider the space R^l formed by the set of all l-tuples of real numbers $(y_{i_1}, \ldots, y_{i_1})$ and define the corresponding R^{m-1}, such that $R^l \times R^{m-1} = R^m$. For $1 \leqq l < m$ we *define* the projection of the probability distribution W^m onto the given R^l: *for every l-dimensional Borel set B^l of the given R^l we set*

$$W^l(B^l) = W^m(B^l \times R^{m-l}). \tag{4}$$

Then W^l is called the *marginal probability distribution* of the l-dimensional random variable $(h_{i_1}, \ldots, h_{i_1})$. It can be seen at once that W^l satisfies the conditions (I) and (II). Moreover, W^l satisfies the following "*compatibility condition of*": if we define W^l by (4) and on the other hand define the probability distribution $W^{(h_{i_1}, \ldots, h_{i_l})}$ by (3), then W^l and $W^{(h_{i_1}, \ldots, h_{i_l})}$ coincide on the σ-algebra $\mathfrak{B}^l$.

Let the distribution function of $(h_1, \ldots, h_m)$ be given for every $(y_1, \ldots, y_m)$ by $F(y_1, \ldots, y_m)$. Then the distribution function of $(h_{i_1}, \ldots, h_{i_l})$ is given for every $(y_{i_1}, \ldots, y_{i_l})$ by

$$\lim_{\substack{y_i \to \infty \\ \text{for all } i \neq i_j}} F(y_1, \ldots, y_m),$$

which is called a *marginal distribution function.*

If we denote by F_i the marginal distribution function of h_i, $i = 1, \ldots, m$, it can be verified at once that $\prod_{i=1}^m F_i$ has the properties E 1′ to E 3′. But in general it is not true that

$$F(y_1, \ldots, y_m) = \prod_{i=1}^{m} F_i(y_i), \tag{5}$$

as can be seen from the so-called multidimensional normal distribution whose density, with $m = 2$, is given for every (y_1, y_2) by

$$\frac{1}{2\pi} \begin{vmatrix} a_{11} & a_{12} \\ a_{21} & a_{22} \end{vmatrix}^{1/2} \exp\left(-\frac{1}{2} \sum_{i,k=1}^{2} a_{ik} y_i y_k\right) \qquad (\text{with } a_{12} = a_{21}),$$

where $\sum_{i,k=1}^{2} a_{ik} y_i y_k$ is a positive definite form. For otherwise we would would have the equation

$$\int_{-\infty}^{y_1}\int_{-\infty}^{y_2} \exp\left(-\frac{1}{2}\sum_{i,k=1}^{2} a_{ik}x_i x_k\right)dx_1 dx_2$$

$$= \frac{\begin{vmatrix} a_{11} & a_{12} \\ a_{21} & a_{22} \end{vmatrix}^{1/2}}{\sqrt{a_{11}a_{22}}} \int_{-\infty}^{y_1} \exp\left[-\frac{1}{2}x_1^2\left(a_{11} - \frac{a_{12}^2}{a_{22}}\right)\right]dx_1$$

$$\times \int_{-\infty}^{y_2} \exp\left[-\frac{1}{2}x_2^2\left(a_{22} - \frac{a_{12}^2}{a_{11}}\right)\right]dx_2,$$

for all real y_1 and y_2, which in general does not hold.

Exercises

1. Let F be the distribution function of a random variable ξ. Show that for $a < b$

$$W(a < \xi \leq b) = F(b) - F(a).$$

2. Prove the properties E 1–E 3 for distribution functions.
3. Prove that the distribution function F of a random variable ξ is continuous at a point x if and only if $W(\xi = x) = 0$.
4. Prove that if ξ is a random variable and g is a Borel-measurable real-valued function whose domain of definition includes all the values of ξ, then $g \circ \xi$ is also a random variable.
5. Let F be the d.f. of $(\xi_1, \ldots, \xi_n)$ and G be the d.f. of $(\xi_1, \ldots, \xi_k)$, with $k < n$. Prove that if F has the density f, then G has a density almost everywhere in R^k and

$$g(x_1, \ldots, x_k) = \int_0^{\infty} \cdots \int_{-\infty}^{\infty} f(x_1, \ldots, x_k; t_{k+1}, \ldots, t_n)dt_{k+1} \cdots dt_n.$$

3. Independence

The relation (5) stands in close connection with the concept of independence. Let I be a (not necessarily countable) index set and let $A_i \in S$ for $i \in I$. *If for every finite subset* $(i_1, \ldots, i_s)$ *of* I *we have*

$$W(A_{i_1} \cdots A_{i_s}) = \prod_{j=1}^{s} W(A_{i_j}),$$

then all the sets A_i *are said to be independent of one another.*

This definition admits an immediate extension to random variables: *let* $k_1, \ldots, k_m$ *be random variables of dimension* $d_1, \ldots, d_m$ *and let* B^{d_i} *be Borel sets of dimensional* d_i, *with* $i = 1, \ldots, m$. *The random variables* $k_1, \ldots, k_m$ *are said to be independent*[10] *if the sets* $k_i^{-1}(B^{d_i}) = \{x : k_i(x) \in B^{d_i}\}$ *are independent for every choice* $B^{d_i}, \ldots, B^{d_m}$ *from* $\mathfrak{B}^{d_1}, \ldots, \mathfrak{B}^{d_m}$. Thus for all subsets $(i_1, \ldots, i_s)$ of $(1, \ldots, m)$ and all Borel sets B^d we have

$$(6)\quad W(\{x : k_{i_1}(x) \in B^{d_{i_1}}\} \cap \cdots \cap \{x : k_{i_s}(x) \in B^{d_{i_s}}\}) = \prod_{r=1}^{s} W(\{x : k_{i_r}(x) \in B^{d_{i_r}}\}).$$

If we denote by F_i the d.f. of k_i for $i = 1, \ldots, m$, it follows from (6) that, in particular,

$$(7)\qquad F^m = \prod_{i=1}^{m} F_i.$$

But this condition (7) is also sufficient for the independence of the random variables $k_1, \ldots, k_m$: for (7) implies (6) by the general theorem on extension. If the corresponding densities f^m and f_i exist, then (7) is equivalent to

$$(7')\qquad f^m = \prod_{i=1}^{m} f_i.$$

From the definition of independence we have the following theorem, which is important in the applications: *if* $k_1, \ldots, k_m$ *are independent random variables and if* g_i, *for* $i = 1, \ldots, m$, *are Borel measurable real-valued functions, defined respectively over the range of the* k_i, *then the functions* $g_i \circ k_i$ *are also independent random variables.*

Exercise

Prove the above theorem.

4. Expectation

4.1. *Measure and Integral*

We now introduce the concept of the *expectation* of a random variable. For this purpose it is necessary to define the integral corresponding to a probability measure W (or more generally to a totally additive finite measure μ defined over S). Here we proceed as follows (cf. III 3, §1.1): let $A_i \in S$ be finitely many pairwise disjoint sets with $\bigcup_i A_i = R$. By c_A we denote the characteristic function of A in R (III 3, §1). *The function* $\sum_{i=1}^{m} \alpha_i c_{A_i}$ *with real* α_i *is called a (generalized) step function.* It is natural to

[10] This definition can be extended at once to a (not necessarily countable) set of random variables.

define $\int_R \sum_{i=1}^m \alpha_i c_{A_i} d\mu$ by $\sum_{i=1}^m \alpha_i \mu(A_i)$. This definition is unique: for if two step functions define the same function, then by the present definition they have the same integral. Moreover, the integral defined over the set of all step functions is a linear functional.[11] Now let f be a measurable function defined over R. It is said to be *integrable* (*more precisely, μ-integrable*) *if there exists a sequence of step functions f_n which converges everywhere to f and has the following property:* $\int_R |f_n - f_m| d\mu \to 0$ for $m, n \to \infty$. Then we *define*: $\int_R f d\mu = \lim_{n\to\infty} \int_R f_n d\mu$. This definition is meaningful, since the limit always exists and, as can easily be shown, is independent of the special choice of the sequence $\{f_n\}$. If f is a step function the present definition corresponds to the one given above. If we wish to refer especially to the elements of R, we write $\int_R f(x) d\mu(x)$. The integral thus defined is a linear functional. Moreover, we have the important inequality:

$$\left|\int_R f d\mu\right| \leqq \int_R |f| d\mu. \tag{8}$$

With f the function $|f|$ is also integrable, and conversely.

Every bounded measurable function is integrable.

For every integrable function f and every set $A \in S$ we define:

$$\int_A f d\mu = \int_R c_A f d\mu.$$

If $A_i \in S$ are at most countably many pairwise disjoint sets with $\bigcup_i A_i = A$, then $\int_A f d\mu = \sum_i \int_{A_i} f d\mu$.

From the definition of the integral we can easily show that if $\mu(A) = 0$,[12] then $\int_A f d\mu = 0$ for every integrable f.

With unimportant changes in the above construction we can also introduce an integral for a σ-finite measure (III 3, §2.7) (example: the Lebesgue integral). The theorem on the integrability of all measurable and bounded functions is then no longer correct, in general. For example, the constant function $\neq 0$ in $-\infty < x < \infty$ is not Lebesgue integrable.

For later use we mention the fact that this definition of the integral can easily be carried over to complex-valued functions f; for we need only consider the real part $\operatorname{Re} f$ and the imaginary part $\operatorname{Im} f$ and define $\int_R f d\mu$ by $\int_R \operatorname{Re} f d\mu + i \int_R \operatorname{Im} f d\mu$.

[11] The set of all step functions can be extended to a vector lattice in an obvious way (III 3, §1.1). Then its connection with the integral introduced in III 3, §1.2 is at once clear.

[12] Such a set A will be called a μ-zero set. But if μ is the Lebesgue measure, this set will simply be called a zero set. A statement is said to be μ-almost everywhere correct if it is correct except for a μ-zero set.

Let us now examine, in particular, the space R^1 and a probability measure W^1 defined over the σ-algebra $\mathfrak{B}^1$. Let the corresponding d.f. be F^1. If k is a continuous function defined over R^1 and if $\int_{R^1} kdW^1$ exists, the Riemann-Stieltjes integral $\int_{-\infty}^{+\infty} kdF^1$ also exists (cf. III 3, §1.6), and conversely. Then

$$\int_{R^1} kdW^1 = \int_{-\infty}^{+\infty} kdF^1. \tag{9}$$

The proof depends on the fact that $\int_a^b kdF^1$ can be approximated for $a < b$ by Riemann sums which are integrals of generalized step functions for the probability measure W^1. If F^1 has a density f^1, we also have $\int_{R_1} kdW^1 = \int_{-\infty}^{+\infty} kf^1dx$, where in general the integral on the right-hand side is a Lebesgue integral.

The right-hand side of (9) is often written, also in the general case, as an equivalent form of the left-hand side.

4.2. *Expectation*

Now let k be a random variable. If $\int_R kdW$ exists, this integral is called the *expectation* of k and is written $E(k)$, or also $E(k; W)$ if we wish to emphasize the underlying probability measure.

Naturally $E(c_1k_1 + c_2k_2) = \sum_{i=1}^{2} c_iE(k_i)$ for two random variables k_1, k_2 and real numbers c_1, c_2, provided the right-hand side is defined. We now have the following important theorem, which shows in particular that the expectation of k can also be obtained by means of the probability distribution $W^{(k)}$ of the random variable k. *Let g be a Borel measurable function defined on R^1. Then*

$$\int_R g \circ kdW = \int_{R^1} gdW^{(k)}, \tag{10}$$

in the sense that the existence of either of the two integrals implies the existence of the other. If g is the identity mapping, then

$$\int_R kdW = \int_{R^1} ydW^{(k)}(y).$$

For the proof of (10) we note that the equality holds if g is a step function. In the general case we approximate g by step functions.

Now let k_1 and k_2 be independent random variables. We have

$$E(k_1k_2) = E(k_1)E(k_2), \tag{11}$$

if the expectations exist.

For step functions this statement follows by direct computation with suitable use of (6), and the general case follows by a passage to the limit.

4.3. *Moments*

If $E(k^n)$ exists (n a natural number), it is called the *n-th (central) moment*. The existence of the n-th moment implies the existence of all m-th moments for $0 \leqq m \leqq n$. The *first moment* is also called the *mean* of k. The quantity $E[(k - E(k))^2]$ is called the *variance* $\sigma^2 = \sigma^2(k)$ of k with $\sigma \geqq 0$. We can show at once by calculation that $\sigma^2(k) = E(k^2) - (E(k))^2$. The variance has the following *minimal property*: $E[(k - c)^2] \geqq \sigma^2(k)$ *for all real c.*

As an example we calculate the mean and the variance of an $N(a, \sigma^2)$. For the first moment we have from (9)

$$\frac{1}{\sigma\sqrt{2\pi}}\int_{-\infty}^{+\infty} y e^{\frac{-(y-a)^2}{2\sigma^2}}\, dy = a$$

and the variance from (10)

$$\frac{1}{\sigma\sqrt{2\pi}}\int_{-\infty}^{+\infty} (y - a)^2 e^{\frac{-(y-a)^2}{2\sigma^2}}\, dy = \sigma^2.$$

These equations make clear the significance of the parameters a and σ^2.

For the mean of a $B_n(p)$ we obtain

$$\int_{-\infty}^{+\infty} y dF^{n,p}(y) = \sum_{0 \leqq k \leqq n} k \binom{n}{k} p^k q^{n-k} = np,$$

and for the variance we get npq.

In the same way we see that the Poisson distribution in §2.5 has the value a for its mean and its variance.

4.4. *Čebyšev Inequality*

From the definition of expectation we obtain the following inequality almost at once. *Let k be a random variable, let f be a nonnegative Borel measurable function over R^1, and let $\inf_{|y| \geqq a} f(y) = b(a)$. If $E(f \circ k)$ exists and $b(a) > 0$, then*

$$W(\{x : |k(x)| \geqq a\}) \leqq \frac{E(f \circ k)}{b(a)}.$$

This inequality follows from

$$\int_R f(k(x))dW(x) \geqq \int_{\{x : k|(x)| \geqq a\}} f(k(x))dW(x)$$
$$\geqq b(a)W\{x : |k(x)| \geqq a\}.$$

In particular, with $f(y) = y^2$ for $-\infty < y < \infty$, we have the *Čebyšev inequality: if $\sigma^2(k)$ exists, then for every real $t > 0$*

$$W(\{x : |k(x) - E(k)| \geqq t\sigma(k)\}) \leqq \frac{1}{t^2}.$$

Exercises

1. Let ξ_1, ξ_2 be independent random variables. Show that $E(\xi_1\xi_2) = E(\xi_1)E(\xi_2)$ if the expectations on the right-hand side exist.
2. Prove the minimal property of the variance: $E[(\xi - c)^2] \geq \sigma^2(\xi)$ for all real c.
3. Calculate the mean and the variance of a Poisson distribution.
4. If a random variable ξ is such that $W(a \leq \xi \leq b) = 1$, then

$$a \leq E(\xi) \leq b \qquad \text{and} \qquad \sigma^2(\xi) \leq \left(\frac{b-a}{2}\right)^2.$$

5. Use the Čebyšev inequality to prove that if $\{\xi_n\}$ is a sequence of pairwise independent random variables and there exists a constant $C > 0$ such that $\sigma^2(\xi_n) \leq C$ for all n, then for arbitrary $\epsilon > 0$

$$\lim_{n\to\infty} W\left\{\left|\frac{1}{n}\sum_{k=1}^{n} \xi_k - \frac{1}{n}\sum_{k=1}^{n} E(\xi_k)\right| < \epsilon\right\} = 1.$$

5. Characteristic Functions

5.1. *Definition of the Characteristic Function*

We now introduce one of the most important concepts of the theory. Let k be a random variable with the probability distribution $W^{(k)}$. Since e^{ity} is bounded and continuous, the integral $\int_{R^1} e^{ity} dW^{(k)}(y)$ exists for every real t and defines a complex-valued function φ of t, called the *characteristic function*[13] of k or of $W^{(k)}$. The function φ is continuous. For we have

$$\left|\int_{R^1} (e^{i(t+h)y} - e^{ity}) dW^{(k)}(y)\right| \leqq \int_{R^1} |e^{i(t+h)y} - e^{ity}| dW^{(k)}(y)$$

$$\leqq \int_I |e^{iyh} - 1| dW^{(k)} + 2\int_{R^1 - I} dW^{(k)},$$

where I is an arbitrary interval. Now $\int_{R^1 - I} dW^{(k)} = W^{(k)}(R^1 - I)$ can be made arbitrarily small by suitable choice of I (continuity of the measure $W^{(k)}$). For a fixed I the integral $\int_I |e^{iyh} - 1| dW^{(k)}(y)$ can be made small by suitable choice of h.

In the same way we can prove the following theorem. *If the n-th moment of a random variable k exists, then the corresponding characteristic function φ is n-times differentiable, and the n-th derivative is continuous.* For under

[13] The characteristic function of a probability distribution is to be distinguished from the characteristic function of a set.

this condition $\frac{d^n\varphi(t)}{dt^n}$ is obtained by n-times differentiating $\int_{R^1} e^{ity}dW^{(k)}(y)$ under the integral sign. In particular,

$$i^n E(k^n) = \varphi^{(n)}(0). \tag{12}$$

5.2. *Theorem of Uniqueness*

The great importance of the characteristic function is made clear by the following theorem. *The d.f. of a random variable is uniquely determined by the characteristic function* (*uniqueness theorem*).

Consequently, knowledge of the characteristic function is equivalent to knowledge of the probability distribution over the σ-algebra of the Borel sets $\mathfrak{B}^1$.

The proof of this theorem rests on the following inversion formula. Let $\varphi(t) = \int_{R^1} e^{ity}dW^{(k)}(y)$. Let F be the d.f. corresponding to $W^{(k)}$. Assume $a < b$ and let F be continuous in a and b. Then

$$F(b) - F(a) = \lim_{N\to\infty} \frac{1}{2\pi}\int_{-N}^{N} \frac{e^{-ita} - e^{-itb}}{it}\varphi(t)dt.$$

For we have

$$\int_{-N}^{N} \frac{e^{-ita} - e^{-itb}}{it}\varphi(t)dt = 2\int_{R_1}\int_{N(y-b)}^{N(y-a)} \frac{\sin t}{t}\,dtdW^{(k)},$$

since interchange of the order of integration can easily be justified. Then for $N \to \infty$ the integral $\int_{N(y-b)}^{N(y-a)} \frac{\sin t}{t}\,dt$ converges boundedly[14] to a limit, namely to π for $a < y < b$, to $\frac{\pi}{2}$ for $y = a$, $y = b$ and to 0 for $y < a$, $y > b$ (Dirichlet discontinuous factor). This bounded convergence justifies passage to the limit for $N \to \infty$ under the integral sign on the right-hand side of (13). Then the assertion follows from the assumed continuity of F in a and b.

5.3. *Examples*

The characteristic function of a $N(a, \sigma^2)$ is seen to be

$$\frac{1}{\sigma\sqrt{2\pi}}\int_{-\infty}^{+\infty} e^{ity}e^{-\frac{(y-a)^2}{2\sigma^2}}\,dy = e^{iat-\frac{t^2\sigma^2}{2}} \qquad \text{for all real } t.$$

[14] A sequence f_n converges boundedly to f if f_n is uniformly bounded and converges pointwise to f.

The characteristic function φ of a degenerate distribution $D(C)$ is given by $\varphi(t) = e^{itC}$ for $-\infty < t < \infty$. The characteristic function of a binomial distribution $B_n(p)$ is

$$\int_{-\infty}^{\infty} e^{ity} dF^{n,p}(y) = \int_{-\infty}^{+\infty} e^{ity} d\left(\sum_{k \leq y} \binom{n}{k} p^k q^{n-k}\right)$$

$$= \sum_{k=0}^{n} \binom{n}{k} e^{itk} p^k q^{n-k} = (pe^{it} + q)^n$$

for $-\infty < t < \infty$.

5.4. *Theorem of Continuity*

Another extremely important property of the characteristic function is expressed by the "*theorem of continuity*." We first make the following definition. *A sequence of d.f.'s F_n is said to be weakly convergent to a nondecreasing function F if $F_n(y)$ converges to $F(y)$ at every point of continuity y of F.*

The limit function F is not necessarily a d.f. For example, if G is the d.f. of a degenerate random variable and $F_n(y) = G(y - n)$ for every real y and natural n, then $F_n(y)$ converges to 0 for every y.

Theorem of continuity. *If a sequence of d.f.'s F_n converges weakly to a d.f. F, then the sequence of corresponding characteristic functions φ_n converges for every t to the characteristic function φ of F. Conversely, if a sequence of characteristic functions φ_n converges pointwise to a continuous function φ then the corresponding sequence of d.f.'s F_n converges weakly to a d.f. F and φ is the characteristic function of F.*

In view of (9) the first part of this theorem is simply a statement about interchange of the Stieltjes integral and the limit, which is seen to be permissible since e^{ity} is continuous and bounded for all real t and y.

The converse lies deeper. To begin with, we require the following theorem:

Theorem of Helly. *Every sequence $\mathfrak{F}$ of d.f.'s contains a subsequence that is weakly convergent (to a nondecreasing function).*

The basic idea of the proof is as follows. In $-\infty < y < \infty$ we choose a countable dense set $r_1, r_2, \ldots$; for example, the rational numbers. The set of $F(r_1)$ with $F \in \mathfrak{F}$ is bounded. Thus there exists a convergent sequence $(F_{1n}(r_1))$ with $F_{1n} \in \mathfrak{F}$. The set of $\{F_{1n}(r_2)\}$ is bounded. Thus we choose a subset (F_{2n}) of the (F_{1n}) such that $(F_{2n}(r_2))$ converges. Now if (F_{kn}) is already defined, an infinite subsequence $(F_{(k+1)n})$ is so chosen that $F_{(k+1n)}(r_{k+1})$ converges. The diagonal sequence (F_{ii}) converges for all r_j. If we set $\lim_{i\to\infty} F_{ii}(r_j) = G(r_j)$, a limit function G is defined on the set $\{r_j\}$ and is nondecreasing there. For every $x \neq r_j$ we define $G(x) = \inf_{rj>x} G(r_j)$. Then G is nondecreasing on the real line and it is easy to show that F_{ii} converges weakly to G.

By means of the Helly theorem we now choose from the set of F_n a subsequence (F_{n_k}) converging weakly to a nondecreasing function F, for which we naturally have $0 \leqq F \leqq 1$.

We now assume that $\lim_{x\to\infty} F(x) - \lim_{x\to-\infty} F(x) = d < 1$. We choose an ϵ with $0 < 4\epsilon < 1 - d$. Since φ is continuous and $\varphi(0) = 1$, there exists a $u > 0$ with

$$\frac{1}{u}\left|\int_0^u \varphi dt\right| > d + 4\epsilon. \tag{14}$$

We now choose a $y_1 \geqq \dfrac{1}{u\epsilon}$ such that F is continuous at y_1 and at $-y_1$. By assumption there exists an N with

$$|F_{n_k}(y_1) - F_{n_k}(-y_1)| \leqq d + \epsilon$$

for all $n_k \geqq N$. Since $\left|\int_0^u e^{ity}dt\right|$ is bounded on the one hand by u, and on the other hand by $2/y_1$ (for $y \geqq y_1$, as can be seen by carrying out the integration) we obtain for $n_k \geqq N$

$$\frac{1}{u}\left|\int_0^u \varphi_{n_k}dt\right| = \frac{1}{u}\left|\int_{R_1}\int_0^u e^{ity}dtdF_{n_k}\right| \leqq \frac{1}{u}\left|\int_{|y|\leqq y_1}\int_0^u e^{ity}dtdF_{n_k}\right|$$
$$+ \frac{1}{u}\left|\int_{|y|\geqq y_1}\int_0^u e^{ity}dtdF_{n_k}\right| \leqq (d+\epsilon) + 2\epsilon \leqq d + 3\epsilon.$$

The change in the order of integration can easily be justified. But since φ_{n_k} converges boundedly to φ, it follows that $\int_0^u \varphi_{n_k}dt \to \int_0^u \varphi dt$ and thus $\frac{1}{u}\left|\int_0^u \varphi dt\right| \leqq d + 3\epsilon$, in contradiction to (14).

Consequently, $\lim_{x\to\infty} F(x) - \lim_{x\to-\infty} F(x) = 1$, and the first part of the continuity theorem shows that φ is the characteristic function of F.

But now, if the sequence F_n were itself not weakly convergent to F, there would exist a subsequence (F_{n_j}) weakly converging to a d.f. F^* distinct from F but with the same characteristic function, in contradiction to the uniqueness theorem. Thus the theorem of continuity is proved.

5.5. *Multidimensional Case*

There is no difficulty in extending the concept of characteristic function to multidimensional random variables. If W^m is the probability distribution of an m-dimensional random variable, the characteristic function is defined by $\int_{R^m} e^{i(t_1y_1+\cdots+t_my_m)}dW^m$ for all $(t_1, \ldots, t_m)$. The theorems of uniqueness and continuity are easily proved.

Exercises

1. Calculate the characteristic function of a random variable with Poisson distribution.
2. Formulate (and prove) the theorems of the uniqueness and continuity for multidimensional random variables.
3. Prove that the functions

$$\varphi_1(t) = \sum_{k=0}^{\infty} a_k \cos kt, \qquad \varphi_2(t) = \sum_{k=0}^{\infty} a_k e^{i\lambda_k t}, \qquad -\infty < t < \infty,$$

with $a_k \geq 0$ and $\sum_{k=0}^{\infty} a_k = 1$ are characteristic functions; determine the corresponding probability distributions.

6. Sums of Independent Random Variables

6.1. *Variance and Characteristic Function; Convolution*

Let k_1, k_2 be two independent random variables with the probability distributions $W_1^{(k_1)}, W_2^{(k_2)}$ and the d.f.'s F_1, F_2. Let $k_3 = k_1 + k_2$. Then for the variances we have $\sigma^2(k_3) = \sigma^2(k_1) + \sigma^2(k_2)$. For without loss of generality we may assume that $E(k_1) = E(k_2) = 0$. Since $E(k_1k_2) = E(k_1)E(k_2)$, it follows from (11) that $E(k_3^2) = E(k_1 + k_2)^2 = E(k_1^2) + E(k_2^2)$.

For the characteristic functions φ_i of k_i $(i = 1, 2, 3)$ we have:

$$\varphi_3(t) = \varphi_1(t)\varphi_2(t) \tag{15}$$

for every t, since it follows from (11) that

$$\varphi_3(t) = E(e^{ik_3t}) = E(J^{i(k_1+k_2)t}) = E(e^{ik_1t})E(e^{ik_2t}).$$

Thus, somewhat more generally: *the characteristic function of a sum of independent random variables is equal to the product of the characteristic functions of the summands.*

We show that the d.f. of k_3 for every real y is given by

$$\int_{-\infty}^{+\infty} F_1(y - z)dF_2(z)$$

or also by

$$\int_{-\infty}^{+\infty} F_2(y - z)dF_1(z).$$

These two expressions coincide, as can be seen by integration by parts in case F_1 and F_2 are continuous. Moreover it is easy to see that the function F_3 defined by either of these two expressions is actually a d.f. But now

$$\int_{-\infty}^{+\infty} e^{ity}dF_3 = \int_{-\infty}^{+\infty} e^{ity}d\left(\int_{-\infty}^{+\infty} F_1(y - z)dF_2(z)\right),$$

and this latter expression is identical with $\varphi_1(t)\varphi_2(t)$, as can be shown formally by merely interchanging the order of integration; an interchange which can easily be justified. By comparison with (15) the uniqueness theorem shows that F_3 is the d.f. of k_3. The function F_3 is usually called the *convolution* of F_1 and F_2. If k_1 and k_2 have densities f_1 and f_2 respectively, then for the density f_3 of k_3 we obtain

$$f_3(y) = \int_{-\infty}^{+\infty} f_1(y - z)f_2(z)dz, \qquad -\infty < y < \infty.$$

6.2. *Examples*

Let k_1, k_2 be independent and let k_i be distributed according to an $N(a_i, \sigma_i^2)$, $i = 1, 2$. The characteristic function of $k_3 = k_1 + k_2$ is given for every t by

$$e^{ia_1t-(\sigma_1^2t^2/2)}e^{ia_2t-(\sigma_2^2t^2/2)} = e^{i(a_1+a_2)t-[(\sigma_1^2+\sigma_2^2)/2]t^2}.$$

Thus k_3 is also normally distributed.

Let h_1 be distributed according to $B_1(p)$. The characteristic function is given for every t by $pe^{it} + q$. Thus if we consider n independent random variables with this distribution, the characteristic function of their sum is given for every t by $(pe^{it} + q)^n$, which is the characteristic function of a $B_n(p)$.

Exercises

1. Prove that the sum of two Poisson-distributed random variables is also Poisson-distributed.
2. Let $\xi_1, \ldots, \xi_n$ be independent random variables with the $N(0, 1)$-distribution. Show by induction that the random variable $\eta_n = \xi_1^2 + \cdots + \xi_n^2$ has the following density:

$$g_n(y) = \frac{1}{2^{n/2}\Gamma\left(\frac{n}{2}\right)} y^{n/2-1}e^{-y/2} \qquad \text{for } y > 0,$$

$$g_n(y) = 0 \qquad \text{for } y \leq 0$$

(the so-called χ^2-distribution of Helmert-Pearson).

7. Conditional Probability and Conditional Expectation

7.1. *Definition of Conditional Probability*

Let $A, B \in S$ and $W(B) \neq 0$. As suggested by certain classical definitions (Bayes) we define the *conditional probability* $W(A|B)$ of A under the hypothesis B by

$$W(A|B) = \frac{W(A \cap B)}{W(B)}. \tag{16}$$

We see at once that $W(A|B)$ with fixed B satisfies the conditions (I) and (II) as a measure over S.

Let $A_1 \in S$ for $i = 1, 2, \ldots$; let the A_i be pairwise disjoint and let $\bigcup_{i=1}^{\infty} A_i = R$. Also let $W(A_i) \neq 0$. For every $B \in S$ we then have

$$W(B) = W\left(B \cap \bigcup_{i=1}^{\infty} A_i\right) = \sum_{i=1}^{\infty} W(B \cap A_i).$$

If also $W(B) \neq 0$, it follows from the definition of conditional probability that

$$W(A_i|B) = \frac{W(B|A_i)W(A_i)}{\sum_{j=1}^{\infty} W(B|A_j)W(A_j)}$$

for $i = 1, 2, \ldots$. This is the *Bayes theorem*, which formerly played an important role in mathematical statistics.

In order to penetrate more deeply into the theory we now change our point of view and investigate $W(A|B)$ for fixed A in "dependence on B." For this purpose we consider the σ-algebra $\mathfrak{S}$[15] generated by B. This algebra consists of the sets $\varnothing$, B, $R - B$, R where $\varnothing$ denotes the empty set. We leave the two sets $\varnothing$ and R aside and define

$$(17) \qquad W(A|\mathfrak{S}) = W(A|B)c_B + W(A|R - B)c_{R-B}),$$ [16]

where we assume that $W(B)$ and $W(R - B)$ are not equal to 0. Thus $W(A|\mathfrak{S})$ is defined as a function over R (not, as might perhaps be thought, as a function of $\mathfrak{S}$) and is trivially $\mathfrak{S}$-measurable. We note that from (17) it follows that

$$(18) \qquad W(A \cap B) = \int_B W(A|\mathfrak{S})dW$$

and also

$$(19) \qquad W(A \cap R - B) = \int_{R-B} W(A|\mathfrak{S})dW.$$

These remarks can be extended at once to the case of countably many pairwise disjoint sets $B_i \in S$ with $\bigcup_{i=1}^{\infty} B_i = R$. We again let $\mathfrak{S}$ denote the σ-algebra generated by the B_i and then define a mapping $W(A|\mathfrak{S})$ of R into the real numbers by

$$(20) \qquad W(A|\mathfrak{S}) = \sum_{i=1}^{\infty} W(A|B_i)c_{B_i},$$

where for the time being we must assume that $W(B_i) \neq 0$ for all i. But if this condition is discarded, we can still use (20) as a definition of $W(A|\mathfrak{S})$,

[15] That is, the intersection of all the σ-algebras containing B; this algebra is again a σ-algebra.

[16] For the meaning of c_B see §4.1.

provided we give up the definition on a W-zero set. For $W(A|\mathfrak{S})$ is always defined on the complement of the union M of B_i with $W(B_i) = 0$, and M is itself a W-zero set since it is the union of at most countably many W-zero sets. In any case $W(A|\mathfrak{S})$ is again $\mathfrak{S}$-measurable.[17] For every set $B \in \mathfrak{S}$, since it is the union of suitably chosen B_i, we again have

$$W(A \cap B) = \int_B W(A|\mathfrak{S})dW. \tag{21}$$

Thus it is clear that $W(A|B)$ can be defined in terms of $W(A|\mathfrak{S})$ for all $B \in \mathfrak{S}$ with $W(B) \neq 0$. Then $W(A|\mathfrak{S})$ is called the *conditional probability* of A under the hypothesis that $\mathfrak{S}$ is given.

For an arbitrary σ-algebra $\mathfrak{S}$ that is a subalgebra of S we cannot proceed at once from (20) to define $W(A|\mathfrak{S})$ for every $A \in S$ as an $\mathfrak{S}$-measurable function over R. On the other hand, by means of the Radon-Nikodym theorem (cf. [12]) we can prove the existence of an $\mathfrak{S}$-measurable function $W(A|\mathfrak{S})$ satisfying (21) for all $B \in \mathfrak{S}$. Of course, this latter function is uniquely defined only up to zero sets which in general depend upon A. Thus, generally speaking, we must give a negative answer to the natural question whether $W(A|\mathfrak{S})$ at a fixed point $x \in R$ is a probability over S when considered as a function of A.[18] However, we can prove from (21) that (I) and (II) are valid W-almost everywhere. So for pairwise disjoint $A_i \in S$ we have the equation $W(\bigcup_{i=1}^{\infty} A_i|\mathfrak{S}) = \sum_{i=1}^{\infty} W(A_i|\mathfrak{S})$ up to W-zero sets, although the latter cannot, in general, be chosen once for all but depend on the A_i.

For the applications it is of great importance that σ-subalgebras of S can be obtained by means of arbitrary mappings of R into an arbitrary set. We restrict our attention here to random variables. Let k be a one-dimensional random variable. By definition $\mathfrak{B}^1$ has measurable preimages under k, i.e., for every $B^1 \in \mathfrak{B}^1$ we have $k^{-1}(B^1) \in S$. From the fact that

$$k^{-1}\left(\bigcup_{i=1}^{\infty} B_i^1\right) = \bigcup_{i=1}^{\infty} k^{-1}(B_i^1),$$

$$k^{-1}\left(\bigcap_{i=1}^{\infty} B_i^1\right) = \bigcap_{i=1}^{\infty} k^{-1}(B_i^1),$$

and

$$k^{-1}(B_0^1 - B_{00}^1) = k^{-1}(B_0^1) - k^{-1}(B_{00}^1)$$

for arbitrary Borel sets $B_i^1, B_0^1, B_{00}^1 \in \mathfrak{B}^1$ it follows that the set $k^{-1}(\mathfrak{B}^1)$ of all $k^{-1}(B^1)$ with $B^1 \in \mathfrak{B}^1$ is a σ-algebra. Of course, $k^{-1}(\mathfrak{B}^1)$ is a subalgebra

[17] We here make the following remark once for all: if f is a mapping of R into the space R^1 that is defined only up to W-zero sets, then f is said to be $\mathfrak{S}$-measurable if for all real α the preimage of $(-\infty, \alpha)$ under f belongs to $\mathfrak{S}$ up to W-zero sets.

[18] For this and related questions see [15] and the literature cited there.

of S. For the mapping defined by $W(A|k^{-1}(\mathfrak{B}^1))$ we also write $W_R(A|k)$. *Then* $W_R(A|k)$ *is also called the probability of* A *under the hypothesis that* k *is given.*

The above discussion suggests that for given k the *conditional probability* should be defined in such a way that it appears not as a function over R but as a function over R^1. We shall not carry out this idea in detail but merely mention the result: *for every random variable* k *and every* $A \in S$ *a Borel measurable mapping, which we shall denote by* $W_{R^1}(A|k)$, *can be defined over* R^1 *in such a way that it has the following property: for all* $B^1 \in \mathfrak{B}^1$

$$W(A \cap k^{-1}(B^1)) = \int_{B^1} W_{R^1}(A|k)dW^{(k)}. \tag{22}$$

Here $W_{R^1}(A|k)$ *is uniquely determined only up to Borel-zero sets.*

From (10) it is easy to see the connection between $W_R(A|k)$ and $W_{R^1}(A|k)$: the mappings $W_R(A|k)$ and $W_{R^1}(A|k) \circ k$ are (up to zero sets) identical. $W_{R^1}(A|k)$ is a $\mathfrak{B}^1$-measurable function defined on R^1, and $W_R(A|k)$ is a $k^{-1}(\mathfrak{B}^1)$-measurable function defined on R. The analogous remark holds for the conditional expectation defined below.

7.2. *Conditional Distribution Functions*

Let h be a random variable and let $A_y = h^{-1}((-\infty, y])$ for every real y. We see from (16) that $W(A_y|B)$ is defined for every y and every $B \in S$ with $W(B) \neq 0$. The function F_B^h defined for every real y by

$$F_B^h(y) = W(A_y|B) \tag{23}$$

is called the *conditional* d.f. *of* h *under the hypothesis* B. In fact, F_B^h satisfies the conditions E 1 to E 3. The preceding discussion leads, on the basis of (23), to the following question: given any two random variables h, k, can we define a real-valued function F_k^h on R^2 mapping (y, z) say into $F_k^h(y|z)$ in such a way that $y \to F_k^h(y|z)$[19] is a probability distribution for fixed z and $z \to F_k^h(y|z)$ is a Borel measurable function for fixed y, and the relation

$$W(A_y \cap k^{-1}(B^1)) = \int_{B^1} F_k^h dW^k(z) \tag{24}$$

holds for all y and all $B^1 \in \mathfrak{B}^1$? If the answer is affirmative, the mapping $y \to F_k^h(y|z)$ is called the *conditional* d.f. *of* h *under the hypothesis* $k(x) = z$. Of course, it is natural here to make use of (22) and for every fixed y to define $z \to F_k^h(y|z)$ as the mapping $W_{R^1}(A_y|k)$, so that (24) would then follow directly from the definition of F_k^h. But the crucial difficulty lies in the fact that then $z \to F_k^h(y|z)$ for every y is defined only up to W^k-zero sets and there is no guarantee, at least at first sight, that the definition of the W^k-zero sets (which depend on y) can in each case be extended in such

[19] It is customary to denote a mapping g of a set R into any other set by $x \to g(x)$.

a way that $y \to F_k^h(y|z)$ always defines a d.f. However, this can actually be shown to be the case, and in fact we even have the following theorem, which is of great importance for the applications: *if the probability distribution of the random variable* (h, k) *has a density* f, *then for all real* y *and almost all* z

$$F_k^h(y|z) = \frac{\int_{-\infty}^{y} f(v, z)dv}{\int_{-\infty}^{+\infty} f(v, z)dv}.$$

In this case we can also introduce a conditional probability density. Let

$$f_k^h(y|z) = \frac{f(y, z)}{\int_{-\infty}^{+\infty} f(v, z)dv}.$$

The mapping $y \to f_k^h(y|z)$ is called the conditional density under the hypothesis $k(x) = z$.

Example. Let us assume that the two-dimensional random variable (h, k) has the density of a normal distribution as defined in §2.6. A simple calculation shows that the conditional d.f. $F_k^h(y|z)$ is given by

$$\sqrt{\frac{a_{11}}{2\pi}} \int_{-\infty}^{y} e^{-(1/2)a_{11}[v + (a_{12}/a_{11})z]^2} dv$$

for all real y and z.

7.3. *Definition of Conditional Expectation*

Let h be a random variable for which the expectation exists. By

$$(25) \qquad E(h|B) = \frac{1}{W(B)} \int_B hdW$$

we define the *conditional expectation of* h *under the hypothesis* B (with $W(B) \neq 0$). The definition (25) is equivalent to $\int_B E(h|B)dW = \int_B hdW$. More generally, let us consider pairwise disjoint $B_i \in S$ with $\bigcup_{i=1}^{\infty} B_i = R$; if $\mathfrak{S}$ is again the smallest σ-algebra generated by the B_i, we define $E(h|\mathfrak{S})$ as a real-valued function over R (where again the definition is unique only up to W-zero sets) by setting

$$(26) \qquad E(h|\mathfrak{S}) = \sum_{i=1}^{\infty} c_{B_i} \frac{1}{W(B_i)} \int_{B_i} hdW.$$

The number $E(h|\mathfrak{S})$, which is obviously an $\mathfrak{S}$-measurable function over R, is called the conditional expectation[20] of h under the hypothesis $\mathfrak{S}$. For every $B \in \mathfrak{S}$ it follows from (26) that

$$(27) \qquad \int_B E(h|\mathfrak{S})dW = \int_B hdW.$$

[20] We point out that the use of the definite article here is really not correct, since $E(h|\mathfrak{S})$ is determined only up to W-zero sets.

The relation (27) can again be taken as the starting point for the definition of $E(h|\mathfrak{S})$ for more general subalgebras $\mathfrak{S}$ of S, when we again obtain an $\mathfrak{S}$-measurable function.

If we choose $h = c_A$ with $A \in S$, we obtain $E(c_A|\mathfrak{S}) = W(A|\mathfrak{S})$.[21]

With $B = R$ we have from (27) the important equation:

$$E(E(h|\mathfrak{S})) = E(h).$$

Just as the conditional probability at every point x behaves "essentially" like a probability for every x, here also the conditional expectation behaves like an expectation up to W-zero sets.

More precisely, we have the following typical theorems:

If h and k are two random variables whose expectations exist, then $E(h + k|\mathfrak{S}) = E(h|\mathfrak{S}) + E(k|\mathfrak{S})$ holds W-almost everywhere.

For every real number c the equation $E(ch|\mathfrak{S}) = cE(h|\mathfrak{S})$ holds W-almost everywhere.

From $h \leqq k$ it follows that $E(h|\mathfrak{S}) \leqq E(k|\mathfrak{S})$ holds W-almost everywhere, and thus $|E(h|\mathfrak{S})| \leqq E(|h|\ |\mathfrak{S})$ also holds W-almost everywhere.

Other important properties are expressed by the following theorems:

Let h be defined as above and let k be an arbitrary $\mathfrak{S}$-measurable bounded random variable. Then

$$E(kh|\mathfrak{S}) = kE(h|\mathfrak{S}) \qquad W\text{-almost everywhere.}$$

It should be pointed out that every $\mathfrak{S}$-measurable function is, of course, also S-measurable and is thus a random variable in the sense of our definition.

Furthermore: *let $h_1 \leqq h_2 \leqq \cdots$ be an increasing sequence of random variables for each of which the expectation exists. Let h_n converge to h and let $E(h)$ exist. Then it follows that $E(h_n|\mathfrak{S}) \to E(h|\mathfrak{S})$.*

If to each W-integrable function h we assign its conditional expectation $E(h|\mathfrak{S})$ (for fixed $\mathfrak{S}$), we thus obtain a linear mapping. This mapping has other important properties, as we have seen above, and conversely we might ask whether the conditional expectation of an operator with such properties could be characterized independently of the definition given here. This can in fact be done (see [16]).

Exercises

1. Compute the example in 7.2.
2. Prove the theorems in 7.3.

[21] Of course, this equation holds only up to zero sets. Thus it would have been sufficient to define the conditional expectation, since the conditional probability is obtained as a special case.

8. Some Limit Theorems

8.1. *Inequality of Kolmogorov*

Let $h_1, \ldots, h_m$ be independent random variables, with $E(h_i) = 0$, for which the variances $\sigma^2(h_i)$, $i = 1, \ldots, m$ exist. Let $k_1 = \sum_{i=1}^{l} h_i$ and $A_l = \{x : |k_1(x)| < \epsilon\} \cap \{x : |k_2(x)| < \epsilon\} \cap \cdots \cap \{x : |k_\iota(x)| < \epsilon\}$ for an $\epsilon > 0$ and $1 \leqq 1 \leqq m$. *Then*

$$W(A_m) \geqq 1 - \frac{1}{\epsilon^2} \sum_{i=1}^{m} \sigma^2(h_i)$$

or equivalently

$$W(\{x : \max_{1 \leqq i \leqq m} |k_i(x)| \geqq \epsilon\}) \leqq \frac{1}{\epsilon^2} \sum_{i=1}^{m} \sigma^2(h_i). \tag{28}$$

This inequality is obviously a generalization of the Čebyšev inequality. For the proof we let $B_l = A_{l-1} \cap \{x : |k_l(x)| \geqq \epsilon\}$, $1 \leqq l \leqq m$, with $A_0 = R$. The B_1 are pairwise disjoint and

$$\{x : \max_{1 \leqq i \leqq m} |k_i(x)| \geqq \epsilon\} = \bigcup_{i=1}^{m} B_i,$$

so that

$$W(\{x : \max_{1 \leqq i \leqq m} |k_i(x)| \geqq \epsilon\}) = \sum_{l=1}^{m} W(B_l).$$

Now

$$E(k_m^2) = \int_R k_m^2 dW \geqq \sum_{l=1}^{m} \int_{B_l} k_m^2 dW \tag{29}$$

and

$$\begin{aligned} \int_{B_i} k_m^2 dW &= \int_{B_i} (k_i + h_{i+1} + \cdots + h_m)^2 dW \\ &= \int_{B_i} k_i^2 dW + 2 \sum_{j>i} \int_{B_i} k_\iota h_j dW + \int_{B_i} (k_m - k_i)^2 dW. \end{aligned}$$

Further

$$\int_{B_i} k_i h_j dW = \int_R c_{B_i} k_i h_j dW = \int_R c_{B_i} k_i dW \int_R h_j dW,$$

since the h_i are independent. But since $E(h_j) = 0$, the last expression vanishes. Thus

$$\int_{B_i} k_m^2 dW \geqq \epsilon^2 W(B_i). \tag{30}$$

In view of $E(k_m^2) = \sum_{i=1}^{m} \sigma^2(h_i)$, the desired inequality (28) follows from (29) and (30).

8.2. *Convergence in Probability*

A sequence h_i, $i = 1, 2, \ldots$, of random variables h_i is said to be *convergent in probability* to a real number c if

$$\lim_{i\to\infty} W(\{x : |h_i(x) - c| \geqq \epsilon\}) = 0 \tag{31}$$

for every $\epsilon > 0$.[22]

It can be seen at once that c is unique.

Let F_i be the d.f. of h_i for $i = 1, 2, \ldots$. The sequence h_i converges in probability to c if and only if the sequence F_i converges weakly to the d.f. of $D(c)$. For if y is an arbitrary real number with $y < c$, then for every sufficiently small $\epsilon > 0$ we also have $y + \epsilon \leqq c$. Since $\{x : h_i(x) \leqq y\} \subset \{x : |h_i(x) - c| \geqq \epsilon\}$, it follows at once from (31) that $F_i(y) \to 0$ for $y < c$. But if $y > c$, then it also follows that $F_i(y) \to 1$.

The proof of the converse is equally simple.

8.3. *Laws of Large Numbers*

In the introduction we discussed the fact that the stability of the relative frequency in a long series of experiments—a phenomenon called the (empirical) law of large numbers—is a decisive factor in the application of probability methods. Thus, in the mathematical theory of probability, it is a very important fact that certain theorems correspond, under a suitable interpretation, to the empirical situation.

Until further notice the symbol h_i will always denote a sequence of independent random variables such that for all i the corresponding expectation exists and

$$E(h_i) = 0, \qquad i = 1, 2, \ldots . \tag{32}$$[23]

Moreover, we shall write k_n for the sum $h_1 + \cdots + h_n$.

8.4. *The Weak Law of Large Numbers*

Assume that all the h_i have the same d.f., denoted by F. Then k_n/n converges in probability to 0 for $n \to \infty$.[24]

It is sufficient to show that the sequence of the d.f.'s of the k_n/n converges weakly to the d.f. of a $D(0)$. But if φ is the characteristic function of h_1, then φ^n is the characteristic function of k_n. Thus we must show that for

[22] Convergence in probability of a sequence of random variables h_i to a random variable h is defined in exactly the same way. Here h is uniquely determined only W-almost everywhere.

[23] The condition (32) is not a restriction, since we could always write $h_i - E(h_i)$ instead of h_i.

[24] If $\sigma^2(h_i)$ exists, this theorem is an immediate consequence of the Čebyšev inequality.

every real t the sequence $\psi_n(t) = \varphi^n(t/n)$ converges to 1 (cf. §5.3). Now we have

$$\varphi\left(\frac{t}{n}\right) = 1 + \epsilon\left(\frac{t}{n}\right)\frac{t}{n},$$

from which it follows by continuity from (12) and from (32) that $\epsilon\left(\frac{t}{n}\right) \to 0$ for every real t. Thus $\psi_n(t) = \left(1 + \epsilon\left(\frac{t}{n}\right)\frac{t}{n}\right)^n \to 1$, as was to be proved.

A particular case is the *Bernoulli law of large numbers*: if k_n is distributed according to a $B_n(p)$, then $\frac{k_n}{n} - p$ converges in probability to 0.

8.5. *The Strong Law of Large Numbers*

The weak law of large numbers can be sharpened to the *strong law of large numbers which states: under the assumptions of the weak law of large numbers, $k_n/n \to 0$ is convergent W-almost everywhere.*[25]

For the proof we employ the important method of truncation. Let the random variables h_n^* be defined as follows:

$$h_n^*(x) = \begin{cases} h_n(x) & \text{for} \quad \{x : |h_n(x)| \leqq n\}, \\ 0 & \text{for} \quad \{x : |h_n(x)| > n\}. \end{cases}$$

Then let h be a random variable with the same d.f., namely F, as each of the h_i, and let $A_l = \{x : l - 1 < |h(x)| \leqq l\}$, $l = 1, 2, \ldots$. Since h_n^* is bounded, the integral $\int_R h_n^{*2} dW$ exists and

$$\sigma^2(h_n^*) \leqq \int_R h_n^{*2} dW = \sum_{l=1}^{n} \int_{A_l} h_n^{*2} dW \leqq \sum_{l=1}^{n} l^2 W(A_l). \tag{33}$$

Furthermore

$$\sum_{l=1}^{\infty} l W(A_l) \leqq 1 + 2\sum_{l=1}^{\infty} (l-1) W(A_l) \leqq 1 + 2\sum_{l=1}^{\infty} \int_{A_l} |h| dW,$$

and therefore

$$\sum_{l=1}^{\infty} l W(A_l) \leqq 1 + 2\int_{R^1} |y| dF. \tag{34}$$

[25] We also say that k_n/n converges to 0 with probability 1. It is easy to show that convergence W-almost everywhere implies convergence in probability.

By (33) and (34) we thus have

$$\sum_{n=1}^{\infty} \frac{\sigma^2(h_n^*)}{n^2} \leqq \sum_{n=1}^{\infty} \frac{1}{n^2} \sum_{l=1}^{n} l^2 W(A_l)$$

$$= \sum_{l=1}^{\infty} l^2 W(A_l) \sum_{n=l}^{\infty} \frac{1}{n^2} < K \sum_{l=1}^{\infty} l W(A_l) \leqq K + 2K \int_{R_1} |y| dF,$$

where $K > 0$ is a suitable constant. Consequently, $\sum_{n=1}^{\infty} \frac{\sigma^2(h_n^*)}{n^2}$ converges.

We first show:

$$(35) \qquad \frac{1}{n} \sum_{i=1}^{n} (h_i - h_i^*) \to 0,$$

W-almost everywhere.

For let $B_n = \{x : h_n^*(x) = h_n(x)\}$ and $\bigcap_{n \geqq m} B_n = C_m$. Since

$$\left| \sum_{i=1}^{s} (h_i - h_i^*) \right|$$

is finite for every s, it follows that (35) holds for $x \in \bigcup_{m=1}^{\infty} C_m$.

Now we have $R - \bigcup_{m=1}^{\infty} C_m = \bigcap_m \bigcup_{n \geqq m} (R - B_n) \subset \bigcup_{n \geqq r} (R - B_n)$ for every $r \geqq 1$. Thus $W(R - \bigcup_{m=1}^{\infty} C_m) \leqq W(\bigcup_{n \geq r} (B - B_n))$, and the right-hand side is $\leqq \sum_{n=r}^{\infty} W(R - B_n)$.[26] Thus if we can show that $\sum_{n=1}^{\infty} W(R - B_n)$ converges, we can make $W(R - \bigcup_{m=1}^{\infty} C_m)$ arbitrarily small, i.e., $W(R - \bigcup_{m=1}^{\infty} C_m)$ vanishes. But a simple calculation shows that $\sum_{n=1}^{\infty} W(R - B_n) = \sum_{j=1}^{\infty} j W(A_{j+1})$ and this series is convergent by (34).

But now $\lim_{n \to \infty} E(h_n^*) = 0$, so that also

$$\lim_{n \to \infty} \frac{\sum_{i=1}^{n} E(h_i^*)}{n} = 0.$$

In fact

$$|E(h_n) - E(h_n^*)| = |E(h_n^*)| \leqq \int_{\{x:\ |h_n(x)| > n\}} |h_n| dW \leqq \sum_{l=n+1}^{\infty} l W(A_l) \to 0$$

by (34).

Thus it only remains to prove the following lemma. Let h_i' be a sequence of independent random variables, not necessarily with the same probability distribution. Let $\sigma^2(h_i')$ exist for every i and let $\sum_{j=1}^{\infty} \frac{\sigma^2(h_j')}{j^2}$ converge. Then $\frac{k_n'}{n} = \frac{1}{n} \sum_{i=1}^{\infty} h_i' \to 0$ converges W-almost everywhere.

[26] From the fact that a probability measure W is nonnegative and totally additive it follows at once that $W(\bigcup_{i=1}^{\infty} A_i) \leqq \sum_{i=1}^{\infty} W(A_i)$ for every sequence (A_i), $A_i \in S$.

For the proof we choose a sequence $\epsilon_i > 0$, $1 \leqq i < \infty$ converging monotonely to 0. The assertion of the lemma is correct for all x belonging to the set

$$\bigcap_{i=1}^{\infty} \bigcup_{m=1}^{\infty} \bigcap_{n \geqq m} \{x : |k_n'(x)/n| < \epsilon_i\}.$$

The complement of this set, to be denoted by M, is

$$\bigcup_{i=1}^{\infty} \bigcap_{m=1}^{\infty} \bigcup_{n \geqq m} \{x : |k_n'(x)/n| \geqq \epsilon_i\}.$$

The proof of our theorem is complete if we show $W(M) = 0$. For this purpose we prove that for every sufficiently small $\eta > 0$ we can choose m in such a way that $W(E_{mi}) < \eta^i$, where

$$E_{mi} = \bigcup_{n \geqq m} \{x : |k_n'(x)/n| \geqq \epsilon_i\}.$$

It then follows that $W(M) < \sum_{i=1}^{\infty} \eta^i = \dfrac{\eta}{1 - \eta}$, i.e., $W(M) = 0$.

Now let $n_1 < n_2 < \cdots$ be a sequence of positive numbers with

$$n_{s+1}/n_s \to c > 1 \qquad \text{and} \qquad n_r < m \leqq n_{r+1}.$$

Let

$$\{x : \max_{n_s < l \leqq n_{s+1}} |k_l'(x)/l| \geqq \epsilon_i\} = F_{si}.$$

Then $E_{mi} \subset \bigcup_{s=r}^{\infty} F_{si}$. But since

$$F_{si} \subset \{x : \max_{n_s < l \leqq n_{s+1}} |k_l'(x)| \geqq n_s \epsilon_i\},$$

it follows from the Kolmogorov inequality that

$$W(F_{si}) \leqq \frac{1}{n_s^2 \epsilon_i^2} \sum_{l=1}^{n_{s+1}} \sigma^2(h_l'),$$

and thus

$$W(E_{mi}) \leqq \sum_{s=r}^{\infty} W(F_{si}) \leqq \sum_{s=r}^{\infty} \frac{1}{n_s^2 \epsilon_i^2} \sum_{l=1}^{n_{s+1}} \sigma^2(h_l')$$

$$< \frac{1}{\epsilon_i^2} \left[C_1 \frac{1}{n_{r+1}^2} \sum_{l=1}^{n_{r+1}} \sigma^2(h_l') + C_2 \sum_{l=n_{r+1}+1}^{\infty} \frac{\sigma^2(h_l')}{l^2} \right],$$

where C_1 and C_2 depend only on c. If we choose m (i.e., r) sufficiently large, we have $W(E_{mi}) < \eta^i$, since $\sum_{l=1}^{\infty} \dfrac{\sigma^2(h_l')}{l^2}$ converges. Consequently, the proof of the strong law of large numbers is complete.

The strong law of large numbers is the simplest of the so-called *ergodic theorems: if the* h_i *are independent and have the same d.f., namely F, and if* $\int_{-\infty}^{+\infty} ydF$ *exists, then the sequence* $\frac{h_1 + \cdots + h_n}{n} \to \int_{-\infty}^{+\infty} ydF$ *converges W-almost everywhere.* In the ergodic theory, which cannot be discussed here, this theorem is proved for a much more general class of random variables h_i, not necessarily independent. Cf., e.g. [3], in the bibliography.

By specializing the strong law of large numbers to the $B_n(p)$ we obtain the *theorem of Borel-Cantelli*, which is the strong counterpart of the *theorem of Bernoulli*.

Up to now we have assumed that all the h_i have the same distribution. But if we take a closer look at the lemma on p. 126, we see at once that it contains the following theorem: let the h_i be an arbitrary sequence of independent random variables with first and second moment. For simplicity we let $E(h_i) = 0$, $i \geqq 1$. From the convergence of $\sum_{j=1}^{\infty} \frac{\sigma^2(h_j)}{j^2}$ it follows that $\frac{1}{n} \sum_{i=1}^{n} h_i$ converges to 0 with probability 1.

8.6. *Limit Theorems in the Theory of Probability*

In all the applications of the theory of probability an important role is played by the notion that the distribution of the sum of independent variables is "approximately" normal provided that the number of summands is "sufficiently large." The limit theorems in the mathematical theory of probability reflect this situation to a certain extent. In recent years these problems have been under active investigation and the results are now in more or less definitive form. Only a few of the basic ones can be mentioned here, the interested reader being referred to [7] in the bibliography.

We first state a theorem which, in spite of its simple form, is particularly important for the applications to mathematical statistics. *Let the* h_i *be independent and let them all have the same distribution. Let the variance* $E(h_i^2) = \sigma^2$ *exist and be* > 0, *and let the d.f.'s of*

$$k_n/\sigma\sqrt{n} = (h_1 + \cdots + h_n)/\sigma\sqrt{n}$$

be denoted by F_n. *Then the sequence* F_n *converges weakly to the d.f. of an* $N(0, 1)$.

The proof is analogous to that of the weak law of large numbers. We need only note that the assumed existence of the variance of the h_i permits us to expand the characteristic function of the h_i in a Taylor series up to the term of second order.

The theorem can be concisely stated as follows: $k_n/\sigma\sqrt{n}$ is asymptotically distributed according to an $N(0, 1)$.

In particular we have the theorem of de Moivre-Laplace: *if* k_n *is distributed according to a* $B_n(p)$, *then* $\frac{k_n - np}{\sqrt{npq}}$ *is asymptotically distributed according to an* $N(0, 1)$.

If all the h_i do not necessarily have the same distribution, we can still prove a similar limit theorem.

Theorem of Ljapunov: *let* $\int_R |h_i|^3 dW$ *exist and let*

$$\int_R |h_i|^3 dW \neq 0 \tag{36}$$

for every i.

We denote $\sum_{i=1}^n \sigma^2(h_i)$ *by* S_n^2, *and assume that*

$$\frac{1}{S_n^3} \sum_{i=1}^n \int_R |h_i|^3 dW \to 0. \tag{37}$$

Let the d.f. of k_n/S_n *be denoted by* F_n. *Then the sequence* F_n *converges weakly to a d.f. of an* $N(0, 1)$.

We give an outline of the proof. Let φ_i be the characteristic function of h_i.

For every t we have

$$\varphi_i\left(\frac{t}{S_n}\right) = 1 - \frac{1}{2}\frac{\sigma^2(h_i)}{S_n^2} t^2 + \frac{t^3}{\sigma S_n^3} \eta_i\left(\frac{t}{S_n}\right)$$

with

$$\left|\eta_i\left(\frac{t}{S_n}\right)\right| \leqq 2 \int_R |h_i^3| dW.$$

From (37) and (36) it follows that $S_n \to \infty$ and it is easily shown that $\max_{1 \leqq i \leqq n} \frac{\sigma(h_i)}{S_n} \to 0$. Thus the condition (37) implies that $\sum_{i=1}^n \log \varphi_i\left(\frac{t}{S_n}\right)$ converges to $\frac{-t^2}{2}$ as $n \to \infty$, for every real t. Application of the continuity theorem completes the proof.

This theorem is a special case of the theorem of Lindeberg-Feller: *let the d.f. of* h_i *be denoted* F_i. *Let the variance* $\sigma^2(h_i)$ *exist and again set* $\sum_{i=1}^n \sigma^2(h_i) = S_n^2$. *Then* k_n/S_n *is asymptotically distributed according to* $N(0, 1)$ *and* $\max_{1 \leqq i \leqq n} \frac{\sigma_i(h)}{S_n} \to 0$ *if and only if*

$$\frac{1}{S_n^2} \sum_{j=1}^n \int_{|y| > \epsilon S_n} y^2 dF_j \to 0 \qquad \textit{for every} \qquad \epsilon > 0. \tag{38}$$

We have no space for the rest of the proof and merely remark that

$$\sum_{j=1}^{n} \int_{|y|>\epsilon S_n} y^2 dF_j \leqq \frac{1}{\epsilon S_n} \sum_{j=1}^{n} \int_{-\infty}^{+\infty} |y|^3 dF_j,$$

so that (38) follows from (37).

8.7. *Infinitely Divisible Distributions*

We shall now mention some results due to Hinčin, Feller, Gnedenko, Kolmogorov, Lévy, and others. The limit theorems considered up to now lead to the following generalization. Let there be given an infinite matrix of random variables:

$$\begin{array}{c} h_{11}, \ldots, h_{1l_1} \\ \cdot \quad \cdot \quad \cdot \quad \cdot \\ h_{n1}, \ldots, h_{nl_n} \\ \cdot \quad \cdot \quad \cdot \quad \cdot \end{array}$$

with $1_n \to \infty$. In each row let the random variables be independent. The problem is to find the asymptotic distribution of the random variables $s_n = \sum_{i=1}^{l_n} h_{ni}$, provided it exists, which without some restriction is certainly not the case.[27] Thus we set up the following condition:

$$(39) \qquad \max_{1 \leqq j \leqq l_n} W(\{x : |h_{nj}(x)| \geqq \epsilon\}) \to 0, \qquad n \to \infty,$$

for every $\epsilon > 0$.

The limit theorems proved up to now might suggest the conjecture that under the condition (39) only a normal distribution could arise as an asymptotic distribution of the s_n. But this conjecture is not correct, as can be seen from the following example: Let $a > 0$ and let the h_{nj} for $n = 1, 2, \ldots$ and $1 \leqq j \leqq n$ be distributed according to a $B_1(a/n)$. Then s_n is distributed according to a $B_n(a/n)$. But we can calculate at once that for the corresponding sequence of d.f.'s we have

$$\sum_{k \leqq x} \binom{n}{k} \left(\frac{a}{n}\right)^k \left(1 - \frac{a}{n}\right)^{n-k} \to \sum_{k \leqq x} e^{-a} \frac{a^k}{k!}$$

for every x. However, it is possible to give a complete answer to the question of the conditions under which the distribution of the s_n is asymptotically normal (if the asymptotic distribution exists at all). *If the*

[27] For example, let all the $h_{2k+1,1}$ be distributed according to the same Poisson distribution and let the $h_{2k,1}$ be distributed according to an $N(0, 1)$, $1 \leqq k < \infty$. Let all the other h_{ij} be distributed according to a $D(0)$. Then it is obvious that no asymptotic distribution exists for the s_n. Note that the condition (39) is not satisfied by this example.

sequence of the d.f.'s of the s_n *is (weakly) convergent, then the limit distribution function is normal and* (39) *is satisfied if and only if* $\max_{1 \leq j \leq l_n} |h_{nj}|$ *converges in probability to* 0.

Such questions stand in close relation to the concept of an infinitely divisible distribution. A d.f. F, or the corresponding probability measure, is said to be infinitely divisible if for every natural n there exists a d.f. F_n such that F is the n-fold convolution of F_n. Alternatively, we may say that the n-th root of the characteristic function φ of F (with $\sqrt[n]{\varphi(0)} = 1$) is again a characteristic function.

Then we have the important theorem of Lévy-Hinčin. *An infinitely divisible probability distribution is characterized by the fact that the logarithm of its characteristic function* φ *can be represented in the form*

$$\log \varphi(t) = ita + \int_{-\infty}^{+\infty} \left(e^{ity} - 1 - \frac{ity}{1 + y^2}\right) \frac{1 + y^2}{y^2} \, dG(y).$$

Here a *is an arbitrary real number and* G *is a monotone nondecreasing function such that* $\lim_{y \to -\infty} G(y) - \lim_{y \to -\infty} G(y)$ *is finite.*

It is easy to see that the logarithm of the characteristic function of a normal or Poisson distribution can be represented in this form.

We now have the following fundamental theorem:

The set of all possible limit distributions of the sequence s_n *(under the condition* (39)) *is identical with the set of all infinitely divisible distributions.*

Exercises

1. Prove that for $n \to \infty$

$$\frac{\sqrt{n^n}}{\sqrt{2^n}\,\Gamma\left(\frac{n}{2}\right)} \int_0^{1+\sqrt{z/n}} z^{n/2-1} e^{-nz/2} dz \to \frac{1}{\sqrt{2\pi}} \int_{-\infty}^{t} e^{-\frac{z^2}{2}} \, dz.$$

(Hint. Apply the Ljapunov theorem to the χ^2-distribution).

2. By applying the Ljapunov theorem to the Poisson distribution prove that

$$\lim_{n \to \infty} e^{-n} \sum_{k=1}^{n} \frac{n^k}{k!} = \frac{1}{2}.$$

Bibliography

[1] Blanc-Lapierre, A. and Fortet, R.: Théorie des fonctions aléatories. Masson et Cie, Paris 1953.

[2] Cramér, H.: Mathematical Methods of Statistics. Princeton University Press 1958.

[3] Doob, J. L.: Stochastic Processes. John Wiley, New York 1953.
[4] Feller, W.: An Introduction to Probability Theory and Its Applications, 2nd ed., John Wiley, New York 1957.
[5] Fisz, M.: Wahrscheinlichkeitsrechnung und Mathematische Statistik. Deutscher Verlag der Wissenschaften, Berlin 1958.
[6] Gnedenko, B. W.: Lehrbuch der Wahrscheinlichkeitsrechnung. Akademie-Verlag, Berlin 1957.
[7] Gnedenko, B. W. and Kolmogorov, A. N.: Grenzverteilungen von Summen unabhängiger Zufallsgrößen. Akademie-Verlag, Berlin 1959.
[8] Kappos, D. A.: Strukturtheorie der Wahrscheinlichkeitsfelder und -räume. Springer, Berlin-Göttingen-Heidelberg 1960.
[9] Kolmogorov, A. N.: Grundbegriffe der Wahrscheinlichkeitsrechnung. Springer, Berlin 1933.
[10] Lehmann, E. L.: Testing Statistical Hypotheses. John Wiley, New York 1959.
[11] Loève, M.: Probability Theory. Van Nostrand, Toronto-New York-London 1955.
[12] Richter, H.: Wahrscheinlichkeitstheorie. Springer, Berlin-Göttingen-Heidelberg 1956.
[13] Schmetterer, L.: Einführung in die mathematische Statistik. Springer, Wien 1956.
[14] Van der Waerden, B. L.: Mathematische Statistik. Springer, Berlin-Göttingen-Heidelberg 1957.
[15] Blackwell, D.: Proc. of the Third Berkeley Symposium on Math. Statistics and Probability. Vol. II, 1956, Berkeley and Los Angeles, 1–6.
[16] Shu-Teh Chen Moy: Pacific Journ. of Math. IV (1954), 47–63.
[17] Athen, H.: Einführung in die Statistik, Erg.-Heft 2. Hannover 1955.
[18] Bangen, G. and Stender, R.: Wahrscheinlichkeitsrechnung und mathematische Statistik, 2nd rev. ed., Otto Salle Verlag, Frankfurt 1961.
[19] Gnedenko, B. W.: Lehrbuch der Wahrscheinlichkeitsrechnung. Akademie-Verlag, Berlin 1962. (Revised and enlarged edition of [6].)
[20] Hodges, J. L. Jr.; Lehmann, E. L.: Basic concepts of probability and statistics. Holden-Day, Inc., San Francisco-London-Amsterdam 1964.
[21] Kolmogorov, A. N.: Foundations of the theory of probability. Chelsea Publishing Co., New York 1950. (Translation by N. Morrison of [9].)
[22] Parzen, E.: Modern probability theory and its applications. John Wiley and Sons, Inc., New York-London 1960.
[23] Pfeiffer, P. E.: Concepts of probability theory. McGraw-Hill Book Co., San Francisco-Toronto-London 1965.

CHAPTER 4

Alternating Differential Forms

Introduction

The theory of alternating differential forms plays an important role in many branches of modern mathematics. Its significance is due to the simple laws for transformation of the so-called alternating differentials, which are the expressions that occur under the integral sign in multiple integrals. Elie Cartan (1869–1951), who is the true creator of the theory, although the algebraic part of it goes back to H. Grassmann (1809–1877), has developed it, together with the theory of infinitesimal transformations, into one of the most powerful instruments of modern mathematics. Its power depends essentially on its simple algebraic structure, which enables us to see clearly into very complicated relationships without being required at every turn to make calculations with the correspondingly complicated components of functions, vectors and differential representations. Where such calculations are unavoidable, the clear-cut nature of the calculus of alternating differential forms allows us to make the necessary computations with a minimum of labor.

In its applications this calculus extends far into the theory of differential equations, into the theory of functions of n real variables, into vector analysis, into differential geometry, especially the theory of *Riemannian manifolds*, into the theory of functions of several complex variables, into the theory of complex differential geometry, and into modern topology. Some knowledge of it is part of the fundamental equipment of every mathematician, and it is studied in the earliest courses in the university. Consequently, the present work must contain some account of its basic facts. The applications given here are taken from vector analysis, where we show that the most important theorems can be obtained at once as special cases of its general results.

If the only purpose of the calculus of differential forms were to represent well-known facts in a simpler way, it would scarcely have acquired such far-ranging wide importance. Its essential feature is that it makes it possible to solve problems whose technical complications represent insurmountable obstacles to the use of classical methods. But in spite of its great importance we must admit that it cannot dispose of all the problems in a given field of application as though it were some complicated, powerful machine. It remain only a calculus and must almost always be supplemented by the specific methods of the branch of mathematics in which it is being applied. For example, differential geometry must continue to deal with the calculation of covariant and contravariant vector and tensor fields (in particular, with the theory of infinitesimal transformations, i.e., of contravariant vector fields) in contrast to the skew-symmetric covariant vector and tensor fields of the theory of alternating differential forms. In the theory of functions of complex variables we must still take advantage of the specific function-theoretic properties of complex structures; and in all these fields of application an important role is played by the methods of topology, whenever there arises a question of differential geometry or function theory in the large. For then it is necessary to make use of homology and cohomology, which again, however, are closely connected with differential forms.

The calculus of differential forms is in the first place a theory in the small. We consider these forms at the individual points and on small segments of curves and surfaces, where the coordinates of a *Euclidean space* can be chosen as the local parameters. For the most part, the applications are also confined to these small segments, from which it is then possible to piece together larger segments of curves and surfaces. At the end of the chapter we will indicate briefly how differential forms can be defined on differentiable and *Riemannian manifolds*, without any need for embedding such manifolds in a surrounding space.

A. The Laws for Multiple Integrals

1. Integrals over Curves in the Plane and in Space

In the plane of two real variables x^1 and x^2 let there be given an *arc C* of a curve, namely a one-to-one continuous image of a bounded closed interval I of the t-axis. Let the interval I be defined by the restriction $\alpha \leqq t \leqq \beta$ with $\alpha < \beta$. The values α and β are called the *boundary points of I*, the points t with $\alpha < t < \beta$ are *interior points of I*. Let the mapping of the interval I be given by the two functions

(1) $$x^\kappa = \varphi^\kappa(t), \qquad \kappa = 1, 2; \qquad \alpha \leqq t \leqq \beta.$$

If $x = (x^1, x^2)$ denotes the point with coordinates x^1 and x^2, the mapping of I onto C can also be written in the form

$$x = \varphi(t), \qquad \alpha \leqq t \leqq \beta, \tag{2}$$

where $\varphi(t) = (\varphi^1(t), \varphi^2(t))$ is defined by the pair of functions $\varphi^1(t)$, $\varphi^2(t)$. The equations (1) or (2) give a *parametric representation of the arc C* by means of the *parameter t*. The points $a = \varphi(\alpha)$ and $b = \varphi(\beta)$ are called the *boundary points of C*, and the images of the interior points of I are the *interior points of C*.

If t traverses the interval I from α to β, then x traverses the arc C from a to b. The interval I, and with it the arc C also, is thereby provided with a direction or, as we shall say, with an *orientation*. The point a is called the *initial point* and b the *end point* of the now *oriented* arc C. If the parameter t traverses the interval I in the opposite sense, namely from β to α, then x traverses the arc C from b to a, in the sense opposite to the original sense. The interval provided with this *opposite orientation* will be denoted by $-I$ and the corresponding arc by $-C$ (see Figure 1).

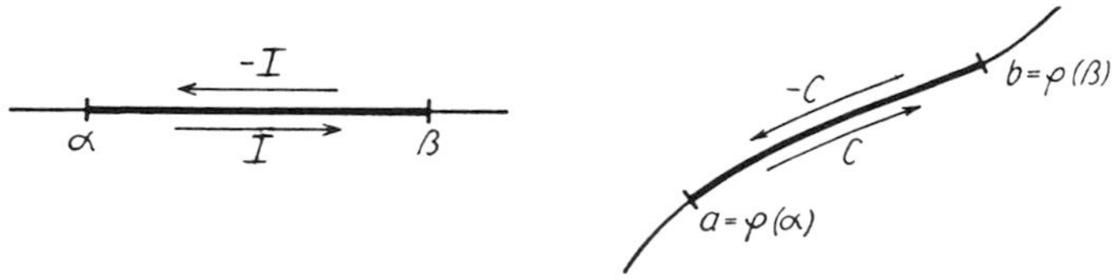

Fig. 1

The order (a, b) of the end points of an arc can also be given by attaching signs to the points, a plus sign (+) to the end point and a minus sign (−) to the initial point. A point to which a sign has been attached is called an *oriented point*. The boundary points of the arc C oriented in this way are called the *oriented boundary of C* and we write:

$$\text{Bd } C = +b - a = b - a$$

or also

$$\partial C = +b - a = b - a.$$

It is clear that *an orientation of C induces an orientation of its boundary* Bd C *and conversely*. Thus the process of forming the boundary in the sense of topology has been algebrized and, just as in topology, we can now compute with multiples of arcs and boundary points. For example, we may define $3C$ as the oriented arc $+C$ traversed three times in the positive sense, $-2C$ as the oriented arc $-C$ traversed twice, $2a$ as the positively oriented point a taken twice, $-3a$ as the negatively oriented boundary $-a$

taken three times, and $0 \cdot C$ and $0 \cdot a$ as the empty set, for which we then simply write 0. For $1 \cdot a = +a$ we also write simply a.

If the functions $\varphi^\kappa(t)$, $\kappa = 1, 2$ are continuously differentiable in the closed interval $\alpha \leqq t \leqq \beta$ (by which we mean that they are defined and continuously differentiable for some distance beyond the boundary points) and if the two derivatives $\dfrac{d\varphi^\kappa(t)}{dt}$, $\kappa = 1, 2$, do not vanish simultaneously at any point of the interval, then we say that the arc is *smooth*.

Now let there be given a smooth curve C and on C two continuous functions $f_1(x^1, x^2)$ and $f_2(x^1, x^2)$. Then the Riemann integral

$$\int_I \Phi(t)dt = \int_\alpha^\beta \Phi(t)dt, \tag{3}$$

taken over the interval I from α to β for the function

$$\Phi(t) = f_1(\varphi^1(t), \varphi^2(t))\frac{d\varphi^1(t)}{dt} + f_2(\varphi^1(t), \varphi^2(t))\frac{d\varphi^2(t)}{dt}, \tag{4}$$

is called a *curvilinear integral over the oriented arc* $+C$ and for it we write more concisely:

$$\int_{+C} (f_1(x^1, x^2)dx^1 + f_2(x^1, x^2)dx^2). \tag{5}$$

From the fact that $-C$ is the image of $-I$ we have the rule:

$$\begin{aligned}\int_{-C} &(f_1(x^1, x^2)dx^1 + f_2(x^1, x^2)dx^2)\\ &= \int_{-I} \Phi(t)dt = \int_\beta^\alpha \Phi(t)dt = -\int_\alpha^\beta \Phi(t)dt = -\int_I \Phi(t)dt\\ &= -\int_C (f_1(x^1, x^2)dx^1 + f_2(x^1, x^2)dx^2).\end{aligned} \tag{6}$$

Exactly as in the plane, we can define *curvilinear integrals* for smooth, oriented arcs in the three-dimensional space R^3, or more generally *in the n-dimensional space* R^n, by simply letting the index κ run from 1 to n, or in other words by considering n functions $\varphi^\kappa(t)$ instead of two and, instead of two functions $f_\nu(x^1, x^2)$ of two variables, n functions $f_\nu(x^1, \ldots, x^n)$ of n variables on C. Instead of (4) we then have

$$\Phi(t) = \sum_{\nu=1}^{n} f_\nu(\varphi^1(t), \ldots, \varphi^n(t))\frac{d\varphi^\nu(t)}{dt}, \tag{7}$$

and instead of (5):

$$\int_{+C} \sum_{\nu=1}^{n} f_\nu(x^1, \ldots, x^n)dx^\nu. \tag{8}$$

The assumption that the n derivatives $d\varphi^\nu/dt$ do not vanish simultaneously at any point in the interval I means geometrically that at every point x_0 the curve has a tangent. The half-line of the tangent in the direction of the orientation of the curve consists of all points

$$x = x_0 + \lambda \frac{dx}{dt} \tag{9}$$

with positive factor λ. Here

$$\frac{dx}{dt} = \left(\frac{d\varphi^1}{dt}, \frac{d\varphi^2}{dt}, \ldots, \frac{d\varphi^n}{dt}\right)$$

is a nonvanishing vector, called the *tangent vector* to the curve C at the point x_0. In an orthonormal[1] coordinate system $(x^1, \ldots, x^n)$ in R^n the tangent vector forms with the positive x^μ-axis an angle α_μ whose cosine can be calculated from the derivatives $[d\varphi^\mu(t)/dt]$ by the formulas

$$\cos \alpha_\mu = \frac{\dfrac{d\varphi^\mu}{dt}}{\sqrt{\displaystyle\sum_{\nu=1}^{n} \left(\frac{d\varphi^\nu}{dt}\right)^2}}, \qquad \mu = 1, 2, \ldots, n.$$

If we take these functions $\cos \alpha_\mu$ as the functions f_μ on C, then by (3), (7), and (8) we have

$$\int_{+C} \sum_{\mu=1}^{n} \cos \alpha_\mu dx^\mu = \int_\alpha^\beta \sum_{\mu=1}^{n} \frac{\dfrac{d\varphi^\mu}{dt}}{\sqrt{\displaystyle\sum_{\nu=1}^{n} \left(\frac{d\varphi^\nu}{dt}\right)^2}} \frac{d\varphi^\mu}{dt}\, dt = \int_\alpha^\beta \sqrt{\sum_{\nu=1}^{n} \left(\frac{d\varphi^\nu}{dt}\right)^2}\, dt.$$

But the last of these integrals is precisely the *arc-length $L(C)$ of the arc C*, for which we thus have the formula

$$L(C) = \int_{+C} \sum_{\mu=1}^{n} \cos \alpha_\mu dx^\mu.$$

This result is also written formally as

$$L(C) = \int_{+C} v\, dx,$$

where $v = (\cos \alpha_1, \cos \alpha_2, \ldots, \cos \alpha_n)$ is the *tangent vector of length* 1 to the curve $+C$ in the direction of its orientation and $dx = (dx^1, dx^2, \ldots, dx^n)$ is called the *vector line-element*.

The functions f_κ are often defined not only on C, but also in a domain G containing C. For the integral (8) it is obvious that only the values of f_κ

[1] The coordinate vectors are mutually orthogonal and of unit length.

on C affect the result; but now we can calculate this integral for all the arcs in the region G, and in the applications it is often of interest to compare its values over various arcs, e.g., joining two fixed points.

If the values of the functions f_κ at each point of the domain G are regarded as components of a vector, these functions determine a *vector field* f in G:

$$f = f(x^1, \ldots, x^n) = (f_1(x^1, \ldots, x^n), \ldots, f_n(x^1, \ldots, x^n)).$$

In this notation we have

$$\Phi(t) = f\frac{dx}{dt},$$

where the right-hand side is the scalar product of the "vectors" f and $\frac{dx}{dt}$. For the integral (8) we can then write concisely:

$$\int_{+C} f dx.$$

In the applications the integrand f is often a vector field that has been obtained from other functions or vector fields. For example, if $\psi(x^1, x^2, x^3)$ is a continuously differentiable function in a domain of R^3, the vector field $\left(\frac{\partial\psi}{\partial x^1}, \frac{\partial\psi}{\partial x^2}, \frac{\partial\psi}{\partial x^3}\right)$ is called the *gradient* of the function ψ and is written

$$\operatorname{grad} \psi = \left(\frac{\partial\psi}{\partial x^1}, \frac{\partial\psi}{\partial x^2}, \frac{\partial\psi}{\partial x^3}\right).$$

Substituting in (10), we obtain

$$\int_{+C} \operatorname{grad} \psi dx = \int_{+C} \left(\frac{\partial\psi}{\partial x^1} dx^1 + \frac{\partial\psi}{\partial x^2} dx^2 + \frac{\partial\psi}{\partial x^3} dx^3\right).$$

The expression under the integral sign on the right-hand side is called the *total differential* of the function ψ and is written $d\psi$. Carrying out the integration by (3) and (4) we obtain the well-known theorem of vector analysis

$$\int_{+C} \operatorname{grad} \psi dx = \int_{+C} d\psi = \psi(b) - \psi(a).$$

The above discussion holds not only for smooth arcs but also for piecewise smooth curves C, i.e., for curves composed of finitely many smooth arcs $C_1, \ldots, C_k$ in such a way that C_κ and $C_{\kappa+1}$ for $\kappa = 1, 2, \ldots, k - 1$ have exactly one boundary point in common. If we orient one of the arcs C_λ, then for each arc C_κ it is obvious that there exists exactly one orientation, or in other words one order of its boundary points (a_κ, b_κ), such that

(11) $$b_\kappa = a_{\kappa+1}, \qquad \kappa = 1, 2, \ldots, k - 1.$$

We say that the arcs C_κ are *oriented in the same sense*, and speak of an *orientation of the curve* C. The points a_1 and b_k are called the boundary points, a_1 the *initial point* and b_k the *end point* of C. Again C can be oriented in exactly two different ways, which we shall call $+C$ and $-C$. Algebraically the situation is described as follows

$$C = \sum_{\kappa=1}^{k} C_\kappa$$

and

$$\text{Bd } C = \sum_{\kappa=1}^{k} \text{Bd } C_\kappa;$$

for by (11) we have

$$\sum_{\kappa=1}^{k} (b_\kappa - a_\kappa) = b_\kappa - a_1.$$

If the end points of C coincide, or in other words if

(12) $$b_k = a_1,$$

then C is said to be *closed*, and we have

(13) $$\text{Bd } C = 0.$$

Conversely, it follows from (13) that C is closed. It is possible that the arcs C_κ have other points in common in addition to the points (11) and (12), in which case it is said to have *double points*. If the curve C has no double points, it is said to be *simple*.

If on C we are given the functions $f_\nu(x^1, \ldots, x^n)$, $\nu = 1, 2, \ldots, n$, where n is the dimension of the space in which C lies, then by the *integral of the form* (8) *over* C we mean the value

$$\int_C = \sum_{\kappa=1}^{k} \int_{C_\kappa}.$$

2. Integrals over Surfaces in Space

We have seen that piecewise smooth curves in the plane or in space can be constructed from arcs of one-to-one and continuously differentiable mappings of intervals I; in the same way we can construct *piecewise smooth surfaces in space* from segments of one-to-one and continuously differentiable mappings of domains in the plane. For each of these segments of surfaces in R^3 we require three continuously differentiable functions of two parameters t^1 and t^2:

(14) $$x^\kappa = \varphi^\kappa(t^1, t^2), \qquad \kappa = 1, 2, 3,$$

where the parameters traverse a domain G of the (t^1, t^2)-plane which in the present context will always be taken to be *simply connected* (G contains the whole interior of any closed curve lying in G) and *bounded by a simple closed piecewise smooth curve*. Let such a system (14) be defined for a short distance beyond the boundary curve of the bounded and closed domain G and assume that in the closed domain G the rank of the functional matrix is everywhere equal to 2:

$$\text{Rank}\left(\frac{\partial\varphi^\kappa}{\partial t^\lambda}\right) = 2. \tag{15}$$

Then the image of G is called a *smooth segment* F of a surface in the space R^3. The images of the inner points of G are called *inner points of* F, and the images of boundary points of G are called *boundary points of* F.

On F we can now define *surface integrals* in the same way as on arcs of a curve. For this purpose we require, just as for arcs, a concept of the orientation of a segment of a surface.

In the case of arcs of curves, we determined the orientation by the direction in which we traversed the parametric interval I, and the calculation of curvilinear integrals was reduced to this parametric interval and its orientation. Analogously, we define the *orientation of the* (t^1, t^2)*-plane as the order* t^1, t^2 *of the coordinates*, which we call *positive* in contrast to the order t^2, t^1, which we call *negative*. These two different orders define two directions of rotation in the (t^1, t^2)-plane, which are obtained in the first case by rotating the positive half-line of the t^1-axis into the positive half-line of the t^2-axis through the shortest possible angle and in the second by rotating the positive t^2-axis into the positive t^1-axis (see Figure 2).

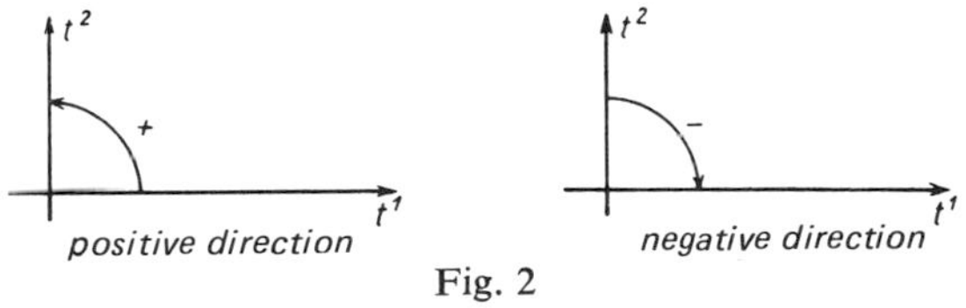

Fig. 2

It has become customary, in diagrams of the coordinate-plane, to place the t^1- and t^2-axes in such a way that the positive direction of rotation is counterclockwise. The domains G in the (t^1, t^2)-plane are now provided with an *orientation* by assigning to them a sign + or − corresponding to the two orders (t^1, t^2) and (t^2, t^1) of the axes. Thus $+G$ is the domain corresponding to the order (t^1, t^2) and $-G$ is the domain corresponding to the order (t^2, t^1). For $+G$ we often write simply G. These orientations can be visualized as assigning to all the points of the domain either a positive or a negative sense of rotation (see Figure 3).

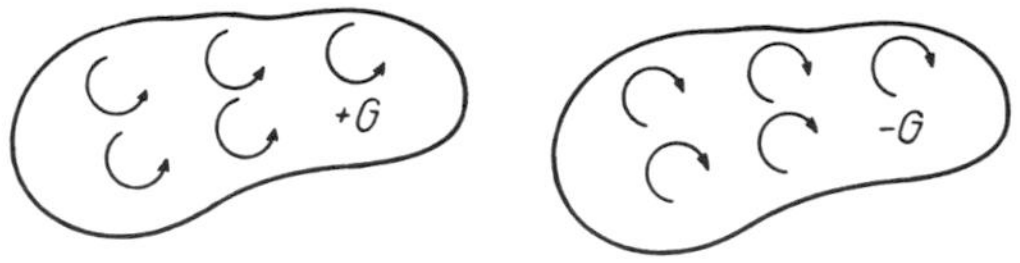

Fig. 3

By this last remark we mean more precisely the following: through a point $t_0 = (t_0^1, t_0^2)$ of the domain we consider two oriented, continuously differentiable arcs C_1 and C_2:

$$\left.\begin{array}{l} C_1 : t^\kappa = g_1^\kappa(s^1) \\ C_2 : t^\kappa = g_2^\kappa(s^2) \end{array}\right\}, \qquad \kappa = 1, 2, \tag{16}$$

whose orientations are given by the increasing parameters s^1 and s^2, respectively. We now consider the vectors v_1 and v_2 with components

$$\left.\begin{array}{l} v_1^\kappa = \dfrac{dg_1^\kappa(s^1)}{ds^1} \\[2ex] v_2^\kappa = \dfrac{dg_2^\kappa(s^2)}{ds^2} \end{array}\right\}, \qquad \kappa = 1, 2,$$

tangent to these arcs at the point t_0. We say that the "*bilateral*" (v_1, v_2) *consisting of the two linearly independent vectors v_1 and v_2 is positively or negatively oriented* according to whether the determinant $\begin{vmatrix} v_1^1 & v_2^1 \\ v_1^2 & v_2^2 \end{vmatrix}$ is positive or negative. Consequently, the bilateral (v_1, v_2) is oppositely oriented to the bilateral (v_2, v_1). Geometrically speaking, for positive orientation the vector v_1 can be rotated in the positive direction through an angle less than π so as to coincide in direction with the vector v_2, and for negative orientation it will be rotated in the negative direction. Thus an orientation of G can be regarded as the class of all positively (or of all negatively) oriented bilaterals of tangent vectors to any two curves through a point in G.

If we now consider an orientation of the simple closed boundary curve of G, we define it as the *orientation induced* by the oriented domain G, provided that the following situation holds: if through a boundary point that is an inner point of a smooth boundary arc we choose a continuously differentiable oriented arc C^* cutting a C_κ exactly once under an angle between 0 and π and oriented from the interior to the exterior of G, then *the bilateral (v^*, v_κ) consisting of the outwardly directed tangent vector v^* to C^* and the vector v_κ tangent to C_κ* oriented in the direction of the orientation of C_κ *has the same orientation as G*. This definition is independent of the choice of point and curve. When the boundary of an oriented domain G is oriented in this way we denote it by Bd G or also by ∂G. Thus

$$\text{Bd}\,(-G) = -\text{Bd}\,G.$$

Since the boundary of the domain G is a simple closed curve, we have

$$\text{Bd Bd } G = 0.$$

If the positive orientation of the domain G is counterclockwise, then the domain lies to our left when we traverse the boundary of G in the sense of the induced orientation.

The orientation of the region G in the (t^1, t^2)-plane is closely related to the process of integration over this domain. If G is a closed and bounded domain, as we have assumed above, and if the function $\Phi(t^1, t^2)$ is continuous in G, then by

$$\int_{+G} \Phi(t^1, t^2)dt^1dt^2 \tag{17}$$

we denote the Riemann integral

$$\int_G \Phi(t^1, t^2)dt^1dt^2,$$

where dt^1dt^2 is the positive volume element of the domain G. If $\Phi(t^1, t^2)$ is positive, the integral (17) is also positive. Correspondingly, by

$$\int_{-G} \Phi(t^1, t^2)dt^1dt^2$$

we denote the value

$$-\int_G \Phi(t^1, t^2)dt^1dt^2.$$

Given a surface-segment F in R^3, as described above, we can provide F with a positive or negative orientation by assigning to it the orientation of the parametric domain G. By this we mean the following: The two tangent vectors $v_1 = \begin{pmatrix} v_1^1 \\ v_1^2 \end{pmatrix}$ and $v_2 = \begin{pmatrix} v_2^1 \\ v_2^2 \end{pmatrix}$ to two curves C_1 and C_2 through a point t_0 in G are transformed by §§1, (1) and 2, (14) and (15) into the linearly independent tangent vectors

$$w_1 = \begin{pmatrix} w_1^1 \\ w_1^2 \\ w_1^3 \end{pmatrix} = \begin{pmatrix} \dfrac{\partial \varphi^1}{\partial t^1} v_1^1 + \dfrac{\partial \varphi^1}{\partial t^2} v_1^2 \\[2ex] \dfrac{\partial \varphi^2}{\partial t^1} v_1^1 + \dfrac{\partial \varphi^2}{\partial t^2} v_1^2 \\[2ex] \dfrac{\partial \varphi^3}{\partial t^1} v_1^1 + \dfrac{\partial \varphi^3}{\partial t^2} v_1^2 \end{pmatrix}$$

and

$$w_2 = \begin{pmatrix} w_2^1 \\ w_2^2 \\ w_2^3 \end{pmatrix} = \begin{pmatrix} \dfrac{\partial \varphi^1}{\partial t^1} v_2^1 + \dfrac{\partial \varphi^1}{\partial t^2} v_2^2 \\[2ex] \dfrac{\partial \varphi^2}{\partial t^1} v_2^1 + \dfrac{\partial \varphi^2}{\partial t^2} v_2^2 \\[2ex] \dfrac{\partial \varphi^3}{\partial t^2} v_2^1 + \dfrac{\partial \varphi^3}{\partial t^2} v_2^2 \end{pmatrix}.$$

Thus we have determined an order for the sides of the bilateral (w_1, w_2) consisting of the tangent vectors w_1 and w_2. Consequently, an orientation for F means simply an order for two linearly independent tangent vectors at each point x_0 in F. The image of the oriented boundary of G under the mapping (14) is an oriented simple closed curve, which we call the *boundary of the oriented surface-segment* F and denote by Bd F. Then

$$\mathrm{Bd}\,(-F) = -\mathrm{Bd}\,F.$$

On F let there now be given three continuous functions $g_{12}(x^1, x^2, x^3)$, $g_{13}(x^1, x^2, x^3)$, and $g_{23}(x^1, x^2, x^3)$. Then we form the integral

$$\text{(18)} \qquad \int_G \sum_{i<j} g_{ij}(\varphi^1(t^1, t^2), \varphi^2(t^1, t^2), \varphi^3(t^1, t^2)) \frac{\partial(\varphi^i, \varphi^j)}{\partial(t^1, t^2)}\, dt^1 dt^2,$$

where

$$\frac{\partial(\varphi^i, \varphi^j)}{\partial(t^1, t^2)} = \begin{vmatrix} \dfrac{\partial\varphi^i}{\partial t^1} & \dfrac{\partial\varphi^i}{\partial t^2} \\ \dfrac{\partial\varphi^j}{\partial t^1} & \dfrac{\partial\varphi^j}{\partial t^2} \end{vmatrix},$$

and call it the *surface-integral over the positively oriented segment* $+F$. We also write this integral in the concise form

$$\text{(19)} \qquad \int_F (g_{12}(x^1, x^2, x^3)dx^1 \wedge dx^2 + g_{13}(x^1, x^2, x^3)dx^1 \wedge dx^3 + g_{23}(x^1, x^2, x^3)dx^2 \wedge dx^3).$$

By the symbol "$\wedge$" we express the fact that by $dx^i \wedge dx^j$ we do not mean a product in the usual sense of the volume elements but a product in the sense of formula (18). The vectors

$$\text{(20)} \qquad \left(\frac{\partial\varphi^1}{\partial t^1}, \frac{\partial\varphi^2}{\partial t^1}, \frac{\partial\varphi^3}{\partial t^1}\right) \quad \text{and} \quad \left(\frac{\partial\varphi^1}{\partial t^2}, \frac{\partial\varphi^2}{\partial t^2}, \frac{\partial\varphi^3}{\partial t^2}\right)$$

at a point x_0 with the coordinates $x^\kappa = \varphi^\kappa(t_0^1, t_0^2)$, $\kappa = 1, 2, 3$ are linearly independent vectors tangent to the curves through x_0 which are the images of the two lines $t^2 = t_0^2$ and $t^1 = t_0^1$ in the (t^1, t^2)-plane. If we now expand the determinants

$$\begin{vmatrix} \dfrac{\partial\varphi^1}{\partial t^\lambda} & \dfrac{\partial\varphi^2}{\partial t^\lambda} & \dfrac{\partial\varphi^3}{\partial t^\lambda} \\ \dfrac{\partial\varphi^1}{\partial t^1} & \dfrac{\partial\varphi^2}{\partial t^1} & \dfrac{\partial\varphi^3}{\partial t^1} \\ \dfrac{\partial\varphi^1}{\partial t^2} & \dfrac{\partial\varphi^2}{\partial t^2} & \dfrac{\partial\varphi^3}{\partial t^2} \end{vmatrix} = 0, \qquad \lambda = 1, 2,$$

by the first row, we see that the "vector"

$$n = \left(\frac{\partial(\varphi^2, \varphi^3)}{\partial(t^1, t^2)}, -\frac{\partial(\varphi^1, \varphi^3)}{\partial(t^1, t^2)}, \frac{\partial(\varphi^1, \varphi^2)}{\partial(t^1, t^2)}\right)$$

is perpendicular to the vectors (20) and thus is perpendicular to the plane through x_0 tangent to F. In a linear mapping the length of n is equal to the ratio of the area of a parallelogram to the area of its preimage in the (t^1, t^2)-plane. Consequently, in vector analysis the expression

$$df = (dx^2 \wedge dx^3, -dx^1 \wedge dx^3, dx^1 \wedge dx^2)$$

$$(21) \qquad = \left(\frac{\partial(\varphi^2, \varphi^3)}{\partial(t^1, t^2)}, -\frac{\partial(\varphi^1, \varphi^3)}{\partial(t^1, t^2)}, \frac{\partial(\varphi^1, \varphi^2)}{\partial(t^1, t^2)}\right) dt^1 dt^2$$

is called a "*vector surface-element*" of F at the point in question; for if we take dt^1 and dt^2 to be small finite coordinate-differences, then the length of df is approximately equal to the surface area of the image of the rectangle with sides dt^1 and dt^2.

If the functions g_{23}, $-g_{12}$ are also regarded as a "vector-function" on F:

$$(22) \qquad g = (g_{23}, -g_{13}, g_{12}),$$

then the integrals (18) and (19) can be written in the form

$$(23) \qquad \int_F g df.$$

In the neighborhood of the surface-segment F let there now be given a continuously differentiable vector field $v = (f_1, f_2, f_3)$. From v we form a new "vector field," called the *rotation* (or *curl*) of v:

$$\operatorname{rot} v = \left(\frac{\partial f_3}{\partial x^2} - \frac{\partial f_3}{\partial x^3}, -\frac{\partial f_3}{\partial x^1} + \frac{\partial f_1}{\partial x^3}, \frac{\partial f_2}{\partial x^1} - \frac{\partial f_1}{\partial x^2}\right).$$

An important theorem in vector analysis is the (special) *Stokes theorem*:

$$\int_F \operatorname{rot} v df = \int_{\mathrm{Bd} F} v dx,$$

which in our present terminology is written as follows:

$$\int_F \left[\left(\frac{\partial f_3}{\partial x^2} - \frac{\partial f_2}{\partial x^3}\right) dx^2 \wedge dx^3 - \left(-\frac{\partial f_3}{\partial x^1} + \frac{\partial f_1}{\partial x^3}\right) dx^1 \wedge dx^3\right.$$

$$(24) \qquad \left. + \left(\frac{\partial f_2}{\partial x^1} - \frac{\partial f_1}{\partial x^2}\right) dx^1 \wedge dx^2\right] = \int_{\mathrm{Bd} F} (f_1 dx^1 + f_2 dx^2 + f_3 dx^3).$$

This theorem and many other integral theorems will be seen below to be simple specializations of the so-called (general) *Stokes theorem*.

From the *simple* surface-segments with piecewise smooth boundary considered above we now wish to construct *general* surface-segments. In topology it is shown that every simple surface-segment F is triangulable, i.e., can be dissected into finitely many triangles. By a *triangle D on F* we mean a compact point set (every infinite subset has at least one limit point) which is the one-to-one and continuous image of a Euclidean triangle $\triangle$ of a (τ^1, τ^2)-plane. The images of the vertices of $\triangle$ are the vertices of D, and the images of the edges of $\triangle$ are the edges of D. We now define: F is said to be *triangulable* if F can be dissected into finitely many triangles in such a way that 1. every point of F belongs to at least one triangle, 2. two triangles have in common either no point at all or one vertex or one edge (see Figure 4), 3. every edge that contains an inner point of F belongs to exactly two triangles, 4. at every inner point of F that coincides with a vertex the triangles form a cycle in which every triangle has an edge in common with the succeeding triangle (e.g., the point a in Figure 4), 5. every boundary point of F lies on exactly one edge of a triangle (e.g., the point b in Figure 4) or is a vertex of triangles $D_{i_1}, D_{i_2}, \ldots, D_{i_k}$, where D_{i_ν} has exactly one edge in common with $D_{i_{\nu+1}}$ for $\nu = 1, 2, \ldots, k - 1$ (e.g., the point c in Figure 4).

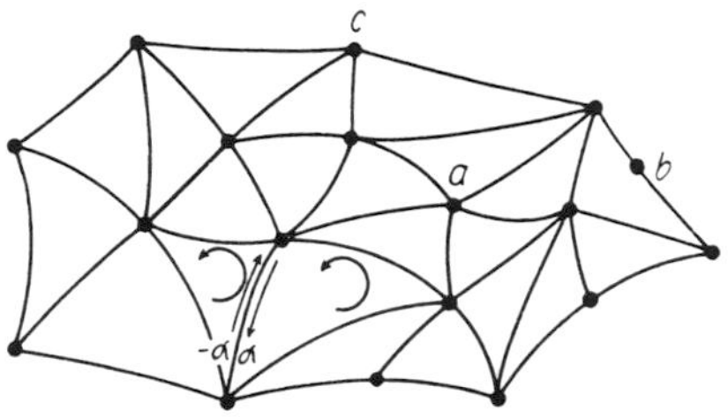

Fig. 4

It can be shown that the simply connected parameter domain G in the (t^1, t^2)-plane described above can be triangulated by triangles D^* in such a way that the mapping functions $t^1 = \psi^1(\tau^1, \tau^2)$, $t^2 = \psi^2(\tau^1, \tau^2)$ of every triangle D^* have continuous partial derivatives in the closed triangle $\triangle$ of the (τ^1, τ^2)-plane. The triangles and the mapping can be so chosen that the functional determinant $\dfrac{\partial(t^1, t^2)}{\partial(\tau^1, \tau^2)}$ vanishes at most at the vertices. If we then map G by the system (14) onto the simple surface-segment F, we obtain on F a triangulation in which the partial derivatives of the mapping functions of the triangles have the same continuity properties as the corresponding partial derivatives of the mapping functions in G.

If an integral (19) is defined on F and if F is triangulated by the triangles $D_1, D_2, \ldots, D_k$, then:

$$\int_F = \sum_{\kappa=1}^{k} \int_{D_\kappa},$$

since the theorems of the infinitesimal calculus as applied to the integral (18) show that in the (t^1, t^2)-plane

$$\int_G = \sum_{\kappa=1}^{k} \int_{D^*_\kappa}$$

where the D^*_κ are the preimages of the triangles D_κ in the (t^1, t^2)-plane.

If G and therefore F are oriented, then the individual triangles in G, and therefore the triangles on F, are also oriented Each of these triangles has a boundary which is the sum of three oriented edges. If we form the sum of all these edges, then every edge containing inner points of F occurs exactly twice, with opposite orientations, so that their sum is zero (e.g., the edge α in Figure 4). Thus there remain only those edges which form the oriented boundary of F, and we have

$$\text{Bd } F = \sum_{\kappa=1}^{k} \text{Bd } D. \tag{25}$$

By a *general* piecewise smooth surface-segment F in R^3 we mean a compact point set in R^3 which can be dissected into finitely many *simple* surface-segments F_κ, $\kappa = 1, 2, \ldots, k$ in such a way that each pair of these segments can have only boundary points in common and the surface-segments F_κ in each case have a triangulation whose triangles form a triangulation of F in the sense of the above definition. Then a point is called a *boundary point of* F if it lies on exactly one edge of exactly one triangle or if it is a vertex of one or more triangles that do not form part of a cycle. The other points are called *inner points of* F. Thus they also include points on the edges and the vertices of two or more segments F_κ.

If the triangles of F can be oriented in such a way that every edge belonging to two triangles occurs with opposite orientations, then F is said to be *orientable*. The orientability of a surface-segment is independent of the special choice of the triangulation The familiar Möbius strip is not orientable. The surface of a torus is orientable, and every segment of an orientable surface is orientable.

By the boundary of an oriented segment F we mean the sum of the boundaries of the simple segments F_κ, and by (25) this boundary is equal to the sum of all the boundaries of the triangles of a triangulation:

$$\text{Bd } F = \sum_{\kappa=1}^{k} \text{Bd } F_\kappa = \sum_{\lambda=1}^{l} \text{Bd } D_\lambda.$$

The integrals discussed below will always be taken over oriented surface-segments.

On F we can now form *surface-integrals* of the form (19), if the functions g_{12}, g_{13}, and g_{23} are defined on F; to do this, we form the sum of all the surface-integrals over the surface-segments F_κ:

$$\int_F = \sum_{\kappa=1}^{k} \int_{F_\kappa}.$$

As was shown above, the integrals over the F_κ are composed of the integrals over the D_λ; consequently, a surface-integral over F may also be written as the sum of the integrals over all the triangles D_λ, $\lambda = 1, 2, \ldots, l$:

$$\int_F = \sum_{\lambda=1}^{l} \int_{D_\lambda}.$$

This integral is independent of the special choice of triangulation of F.

3. Integrals over Manifolds M^r in the Space R^n

In the preceding section we introduced a positive and a negative orientation in the (t^1, t^2)-plane by our choice for the order of the positive coordinate axes; in the same way we now introduce into R^3 a positive and a negative orientation by choice of the order of the positive coordinate axes; to the order (t^1, t^2, t^3) and to the orders (t^2, t^3, t^1) and (t^3, t^1, t^2) arising from it by an even permutation we assign the positive orientation. To the orders (t^1, t^3, t^2), (t^2, t^1, t^3), and (t^3, t^2, t^1) arising from (t^1, t^2, t^3) by an odd permutation we assign the negative orientation. An orientation of this kind can be defined for every space R^r; the order $(t^1, t^2, \ldots, t^r)$ and all orders $(t^{i_1}, t^{i_2}, \ldots, t^{i_r})$ arising from it by an even permutation are positive, and the orders arising by an odd permutation are negative (cf. IB 2, §15.3.2). These orientations are also assigned to every domain that is part of the space R^3 or of R^r.

In all the spaces R^r the domains G with piecewise smooth boundaries can be defined recursively from dimension to dimension, and then we can consider their images M^r in spaces of higher dimension, and these images can again be used for the definition of piecewise smooth boundaries in spaces of still higher dimension. The configurations arising here can be divided up into simplexes and, wherever necessary, they can be oriented, since the triangles on surfaces now correspond to simplexes of dimension r. By a *simplex* of dimension r we mean the image under a one-to-one bicontinuous mapping (i.e., the topological image) of a rectilinear simplex in R^r, obtained as the convex hull of $r + 1$ points, where the r vectors leading from one of the points to r other points are linearly independent. In R^1 a rectilinear simplex is a line-segment, in R^2 a triangle and in R^3 a tetrahedron.

On the orientable manifolds M^r of dimension r constructed in this way in R^n we can now define integrals of the form

$$(26)\qquad \int_{M^r} \sum_{i_1 < i_2 < \cdots < i_r} f_{i_1 i_2 \cdots i_r}(x^1, \ldots, x^n) dx^{i_1} \wedge dx^{i_2} \wedge \cdots \wedge dx^{i_r}$$

as follows: let M^r be subdivided into simple oriented smooth segments M^r_κ,

$\kappa = 1, 2, \ldots, k$ with piecewise smooth boundaries (e.g., into simplexes), which by means of continuously differentiable functions

$$x^\nu = \varphi^\nu(t^1, \ldots, t^r), \qquad \nu = 1, 2, \ldots, n,$$

with

$$\operatorname{rank}\left(\frac{\partial \varphi^\nu}{\partial t^\rho}\right) = r,$$

are defined in each case as the image of an oriented domain G^r_κ in a parameter space R^r. Then the integral (26), taken over M^r_κ, is defined as the Riemann integral

$$(27) \quad \int_{M^r_\kappa} = \int_{G^r_\kappa} \sum_{i_1 < \cdots < i_r} f_{i_1 \cdots i_r}(\varphi^1(t^1, \ldots, t^r), \ldots, \varphi^n(t^1, \ldots, t^r)) \times \frac{\partial(\varphi^{i_1}, \ldots, \varphi^{i_r})}{\partial(t^1, \ldots, t^r)}\, dt^1 \cdots dt^r,$$

and the integral (26) is

$$\int_{M^r} = \sum_{\kappa=1}^{k} \int_{M^r_\kappa}.$$

Since we shall be dealing below with integral formulas over oriented manifolds M^r and over their $(r - 1)$-dimensional boundaries, we must now define what we mean by the *induced orientations of the boundaries.* This may be done in the same way as for the two-dimensional surfaces in R^3. The boundary of the simple segments M^r of r-dimensional manifolds as considered above consists of simple smooth segments M^{r-1}_κ, $\kappa = 1, 2, \ldots, k$, of $(r - 1)$-dimensional manifolds. Let us assume that M^r is parametrically represented and has an orientation. Let us also assume that each of the simple boundary segments M^{r-1}_κ has an orientation, so that at the same time an orientation is given for each of its preimages $M^{r-1}_{0,\kappa}$ in the parameter space of the domain G^r, where $M^{r-1}_{0,\kappa}$ is a segment of the boundary of G^r. These orientations induce orientations on the linearly independent r-hedrals of tangent vectors in the order say $(v_1, \ldots, v_r)$ in G^r and also on the linearly independent $(r - 1)$-hedrals of vectors in the order say $(w_2, \ldots, w_r)$ tangent to the segments $M^{r-1}_{0,\kappa}$. At every interior point of $M^{r-1}_{0,\kappa}$ let there also be given a vector w pointing toward the exterior of G^r, and let w be linearly independent of the vectors $w_2, \ldots, w_r$. Then we say that the orientation of the boundary of G^r is induced by the orientation of the domain G^r if the orientations of the r-hedrals $(v_1, \ldots, v_r)$ and $(w, w_2, \ldots, w_r)$ in this order coincide with each other for all boundary points of G^r that are inner points of boundary segments $M^{r-1}_{0,\kappa}$. It can be shown that an orientation of this kind provides an orientation for all the boundary segments $M_{0,\kappa}^{r-1}$, which is called the *induced orientation*. For the oriented boundary we write

$$\text{Bd } M^r = \sum_{\kappa=1}^{k} M_\kappa^{r-1}.$$

Looking more closely at the boundary of Bd M^r we note that every $(r-2)$-dimensional boundary segment of the segments M_κ^{r-1} occurs exactly twice, in opposite orientations, so that again we have:

$$\text{Bd Bd } M^r = 0.$$

If we are given a general segment of a manifold M^r composed of several simple segments M_λ^r, $\lambda = 1, 2, \ldots, l$, in the manner described above for surfaces, then M^r is said to be *orientable* if all the segments M_λ^r can be oriented in such a way that the two induced orientations of a boundary segment common to two segments $M_{\lambda_1}^r$ and $M_{\lambda_2}^r$ are opposite to each other. For surface-segments of this kind we can define integrals of the form (26):

$$\int_{M^r} = \sum_{\lambda=1}^{l} \int_{M_\lambda^r}.$$

4. Transformations of the Parameters

In the preceding sections, in order to define curvilinear and surface-integrals, and also integrals over the manifolds M^r, we have made use (without motivation, for the time being) of the integrals §§1, (3), (4), 2, (18), and 3, (27) in the parameter spaces, and then we have gone on to write these integrals in the form §§1, (5), 2, (19), and 3, (26), where these parameter spaces are not indicated in any way. Now these strange formations §§1, (3), (4), 2, (18), and 3, (27), and also the possibility of writing the integrals without indication of the parameters, depend on the fundamental fact that the integrals must be defined in exactly this way if they are to be independent of the arbitrary choice of parameters. If we look at some of the unproved statements above, e.g., the integral theorem of Stokes, we see that they can make sense only if the choice of parameters has no influence on the value of the integrals. We must discuss this matter in greater detail.

Let us first consider curvilinear integrals in the plane. Again let the arc C be the image of an oriented integral I under the functions §1, (1), and let C be also the image of an oriented interval I^* under the functions $\varphi_* = (\varphi_*^1, \varphi_*^2)$:

$$(28) \qquad x^\kappa = \varphi_*^\kappa(t_*), \qquad \kappa = 1, 2, \quad \alpha_* \leqq t_* \leqq \beta_*,$$

where the derivatives $\dfrac{d\varphi_*^\kappa(t_*)}{dt_*}$ of these $\varphi_*^\kappa(t_*)$ do not vanish simultaneously at any point in the interval I^* and we have $\varphi_*(\alpha_*) = a$ and $\varphi_*(\beta_*) = b$. Then the arc C is given the same orientation by the system (28) as by the functions §1, (1). On account of the one-to-one relationship between points of C and points of the intervals I and I^* the latter intervals can also be put into one-to-one correspondence:

$$t_* = f(t).$$

We now subdivide the arc C into finitely many subarcs $C_1, \ldots, C_m$ (the fact that this is possible is the content of the familiar Heine-Borel theorem) such that on every subarc C_μ one of the derivatives $\dfrac{d\varphi_*^\kappa(t_*)}{dt_*}$, say $\dfrac{d\varphi_*^1(t_*)}{dt_*}$, does not vanish. Then on C_μ we can solve the equation $x^1 = \varphi_*^1(t_*)$ for $t_* : t_* = \psi(x^1)$, where t_* is continuously differentiable in x^1. Substituting the first equation §1, (1), we obtain the parameter t_* as a continuously differentiable function of t everywhere on C_μ; but this function is precisely $t_* = f(t)$:

$$t_* = \psi(\varphi^1(t)) = f(t).$$

Since this equation holds for every arc C_μ, the function $f(t)$ is continuously differentiable everywhere in I, and naturally its inverse $t = g(t_*)$ is also continuously differentiable. Thus from $t \equiv g(f(t))$ it follows that

$$g'(t_*)f'(t) = 1$$

and consequently that $f'(t)$ does not vanish in I nor does $g'(t_*)$ vanish in I^*. From $\varphi^\kappa(t) = \varphi_*^\kappa(f(t))$ it follows that

$$\text{(29)} \qquad \frac{d\varphi^\kappa(t)}{dt} = \frac{d\varphi_*^\kappa(f(t))}{dt_*}\frac{df(t)}{dt}, \qquad \kappa = 1, 2.$$

If we now calculate the integral §1, (5) in terms of the parameter t_* in accordance with §1, (3) and (4), and then calculate the same integral by means of the substitution $t_* = f(t)$, we obtain from (29):

$$\text{(30)} \qquad \int_{I^*} \Phi_*(t_*)dt_* = \int_I \Phi_*(f(t))\frac{df(t)}{dt}\,dt = \int_I \Phi(t)dt$$

with

$$\begin{aligned}\Phi(t) &= f_1(\varphi_*^1(f(t)), \varphi_*^2(f(t)))\frac{d\varphi_*^1(f(t))}{dt_*}\frac{df(t)}{dt} \\ &\quad + f_2(\varphi_*^1(f(t)), \varphi_*^2(f(t)))\frac{d\varphi_*^2(f(t))}{dt_*}\frac{df(t)}{dt} \\ &= f_1(\varphi^1(t), \varphi^2(t))\frac{d\varphi^1(t)}{dt} + f_2(\varphi^1(t), \varphi^2(t))\frac{d\varphi^2(t)}{dt}.\end{aligned}$$

Thus the integral §1, (5) has the same value with respect to the parameter t_* as with respect to the parameter t. So it is meaningful to write this integral in the form §1, (5), since the choice of a parameter makes no difference in the calculation of the integral.

Since $\alpha < \beta$, $\alpha_* < \beta_*$, $\alpha_* = f(\alpha)$, and $\beta_* = f(\beta)$, the nowhere-vanishing derivative $f'(t)$ is everywhere positive in I, i.e., I^* is positively oriented with I. If we choose I^* so that $\alpha_* > \beta_*$, then the interval I^* is negatively oriented by the mapping $t^* = f(t)$. But this choice of I^* does not change the value of the integral §1, (5), since now the derivative $f'(t)$ in (30) is negative and therefore compensates for the negative orientation of I^*.

The same argument holds for surface-segments and segments of manifolds M^r. If an oriented simple smooth surface-segment is referred on the one hand to an oriented parameter domain G of the (t^1, t^2)-plane by functions §2, (14), and on the other hand is referred to an oriented parameter domain G^* of the (t^1_*, t^2_*)-plane by functions

$$x^\kappa = \varphi^\kappa_*(t^1_*, t^2_*), \qquad \kappa = 1, 2, 3,$$

with

$$\operatorname{rank}\left(\frac{\partial \varphi^\kappa_*}{\partial t^\lambda_*}\right) = 2$$

in G^*, then the parameter domains G and G^* are mapped onto each other by bicontinuously differentiable functions

$$t^\lambda_* = \psi^\lambda(t^1, t^2), \qquad \lambda = 1, 2, \tag{31}$$

with

$$\operatorname{rank}\left(\frac{\partial \psi^\lambda}{\partial t^\mu}\right) = 2. \tag{32}$$

In view of the identities

$$x^\kappa = \varphi^\kappa_*(\psi^1(t^1, t^2), \psi^2(t^1, t^2)) = \varphi^\kappa(t^1, t^2), \qquad \kappa = 1, 2, 3,$$

the integral §2, (19), computed in terms of the parameters t^1_*, t^2_*, then takes the following form, where these parameters are replaced by the t^μ according to (31) and account is taken of §2, (18) and the rule for transformation of integrals:

$$\int_G \sum_{i<j} g_{ij}(\varphi^1_*(\psi^1(t^1, t^2), \psi^2(t^1, t^2)), \varphi^2_*(\psi^1(t^1, t^2), \psi^2(t^1, t^2)),$$

$$\varphi^3_*(\psi^1(t^1, t^2), \psi^2(t^1, t^2)))\frac{\partial(\varphi^i_*, \varphi^j_*)}{\partial(t^1_*, t^2_*)}\frac{\partial(\psi^1, \psi^2)}{\partial(t^1, t^2)}\,dt^1dt^2$$

$$= \int_G \sum_{i<j} g_{ij}(\varphi^1(t^1, t^2), \varphi^2(t^1, t^2), \varphi^3(t^1, t^2))\frac{\partial(\varphi^i, \varphi^j)}{\partial(t^1, t^2)}\,dt^1dt^2. \tag{33}$$

Thus the integral §2, (18) has exactly the same value for the parameters (t_*^1, t_*^2) as for the parameter (t^1, t^2).

Since the rank of the functional matrix (32) is equal to 2, the functional determinant $\frac{\partial(\psi^1, \psi^2)}{\partial(t^1, t^2)}$ is everywhere different from zero, and thus it is either everywhere positive or everywhere negative, in view of its continuity and of the fact that G is connected. Correspondingly, since G is positively oriented, G^* will be positively or negatively oriented by the mapping. From (33) we then see that the definition of the surface integral §2, (19) is independent not only of the choice of parameters but also of the orientation of the parameter domain: for if G^* is negatively oriented, then its negative orientation is compensated for by the sign -1 of the determinant

$$\frac{\partial(\psi^1, \psi^2)}{\partial(t^1, t^2)}.$$

In a completely analogous way we can show that the integrals §3, (26) are independent of the special choice of parameters for the segments of manifolds M^r.

5. Transformation of Coordinates. The Concept of an Alternating Differential Form

In the previous discussion we have made use of the fact that if for a Riemann integral over an oriented domain in R^n we make a one-to-one and bicontinuously differentiable transformation of the domain G to a domain G^*:

$$t^\nu = \varphi^\nu(t_*^1, \ldots, t_*^n), \qquad \nu = 1, 2, \ldots, n,$$

the integrand is multiplied by the functional determinant

$$\int_G f(t^1, \ldots, t^n)dt^1 \cdots dt^n$$

$$= \int_{G^*} f(\varphi^1(t_*^1, \ldots, t_*^n), \ldots, \varphi^n(t_*^1, \ldots, t_*^n)) \frac{\partial(\varphi^1, \ldots, \varphi^n)}{\partial(t_*^1, \ldots, t_*^n)} dt_*^1 \cdots dt_*^n.$$

It is this fact that explains why the integrals defined over curves, surfaces and manifolds are independent of the choice of paramcters. We now raise the question how the integrals in §§1, (5), 2, (19), and 3, (26) change when we map the $(x^1, \ldots, x^n)$-space in which the manifolds are embedded into a $(y^1, \ldots, y^n)$-space in a bicontinuously differentiable way. A mapping of this sort can be regarded as the *introduction of new coordinates* in the $(x^1, \ldots, x^n)$-space: every n-tuple $(y^1, \ldots, y^n)$ determines a point $(x^1, \ldots, x^n)$.

Let us now consider an oriented arc C_x determined on a plane by the functions

$$x^\kappa = \varphi^\kappa(t), \qquad \kappa = 1, 2,$$

with

$$\left(\frac{d\varphi^1}{dt}, \frac{d\varphi^2}{dt}\right) \neq (0, 0)$$

in the interval I: $\alpha \leqq t \leqq \beta$. Let it be required to calculate the curvilinear integral

$$\int_{C_x} (f_1(x^1, x^2)dx^1 + f_2(x^1, x^2)dx^2)$$

over C_x. Now let us suppose that a neighborhood U_x of the arc C_x is mapped by the continuously differentiable functions

$$(34) \qquad x^\kappa = \psi^\kappa(y^1, y^2) \qquad \text{with} \qquad \operatorname{rank}\left(\frac{\partial\psi^\kappa}{\partial y^\lambda}\right) = 2$$

in a one-to-one way onto a region U_y that is a neighborhood of the image C_y of C_x. Then the arc C_y has a parametric representation

$$y^\lambda = \varphi_*^\lambda(t), \qquad \lambda = 1, 2, \qquad \text{with} \qquad \left(\frac{d\varphi_*^1}{dt}, \frac{d\varphi_*^2}{dt}\right) \neq (0, 0)$$

in the interval I, where

$$\varphi^\kappa(t) = \psi^\kappa(\varphi_*^1(t), \varphi_*^2(t)), \qquad \kappa = 1, 2.$$

Then

$$\int_{C_x} (f_1 dx^1 + f_2 dx^2) = \int_I \left(f_1 \frac{d\varphi^1}{dt} + f_2 \frac{d\varphi^2}{dt}\right)dt$$

$$= \int_I \left[\left(f_1 \frac{\partial\psi^1}{\partial y^1} + f_2 \frac{\partial\psi^2}{\partial y^1}\right)\frac{d\varphi_*^1}{dt} + \left(f_1 \frac{\partial\psi^1}{\partial y^2} + f_2 \frac{\partial\psi^2}{\partial y^2}\right)\frac{d\varphi_*^2}{dt}\right]dt$$

$$= \int_{C_y} (f_1^* dy^1 + f_2^* dy^2).$$

Thus the integral over the vector field $(f_1(x^1, x^2), f_2(x^1, x^2))$ on C_x has the same value as the integral over the vector field $(f_1^*(y^1, y^2), f_2^*(y^1, y^2))$ on C_y with

$$f_\lambda^* = \sum_{\kappa=1}^{2} f_\kappa \frac{\partial\psi^\kappa}{\partial y^\lambda}, \qquad \lambda = 1, 2.$$

Vectors which transform in this way under the transformation (34) are said to be *covariant*.

Analogously, we see that in R^3, or generally in R^n, under a transformation of coordinates

(35) $x^\kappa = \psi^\kappa(y^1, \ldots, y^n), \qquad \kappa = 1, 2, \ldots, n, \quad \text{with} \quad \text{rank}\left(\frac{\partial\psi^\kappa}{\partial y^\lambda}\right) = n,$

a curvilinear integral satisfies the law of transformation

$$\int_{C_x} \sum_{\kappa=1}^{n} f_\kappa dx^\kappa = \int_{C_y} \sum_{\lambda=1}^{n} f_\lambda^* dy^\lambda,$$

where the vector fields $(f_1, \ldots, f_n)$ on C_x and $(f_1^*, \ldots, f_n^*)$ on C_y are transformed into each other covariantly:

(36) $$f_\lambda^* = \sum_{\kappa=1}^{n} f_\kappa \frac{\partial\psi^\kappa}{\partial y^\lambda}, \qquad \lambda = 1, 2, \ldots, n.$$

The expression $\sum_{\kappa=1}^{n} f_\kappa(x^1, \ldots, x^n)dx^\kappa$, defined on a curve C_x in R^n over which we can construct an integral in accordance with §1, (4) to (8) and which obeys the rule (36) for transformation of coordinates, is said to be a *differential form of first degree* or a *Pfaffian form.*

If in a domain of R^n we are given a continuously differentiable *function*

(37) $$u = \varphi(x^1, \ldots, x^n),$$

we can construct the differential form

$$\sum_{\kappa=1}^{n} \frac{\partial\varphi}{\partial x^\kappa} dx^\kappa,$$

which is called the *total differential* of the function $\varphi(x_1, \ldots, x_n)$, and for which we also write du or $d\varphi$:

(38) $$du = d\varphi = \sum_{\kappa=1}^{n} \frac{\partial\varphi}{\partial x^\kappa} dx^\kappa.$$

If in (37) we introduce new coordinates by means of (35), then for the function $u = \varphi^*(y^1, \ldots, y^n) = \varphi(\psi^1(y^1, \ldots, y^n), \ldots, \psi^n(y^1, \ldots, y^n))$ we obtain:

(39) $$du = \sum_{\lambda=1}^{n} \frac{\partial\varphi^*}{\partial y^\lambda} dy^\lambda = \sum_{\lambda=1}^{n} \left(\sum_{\kappa=1}^{n} \frac{\partial\varphi}{\partial x^\kappa} \frac{\partial\psi^\kappa}{\partial y^\lambda}\right) dy^\lambda,$$

so that the law of transformation (36) holds for the total differential.

A special set of total differentials is obtained if we consider the functions $u = x^\kappa$. For them we have by (38)

$$du = 1 \cdot dx^\kappa, \qquad \kappa = 1, 2, \ldots, n,$$

for which we also write simply dx^κ.

We can now regard the differential forms $\sum_{\kappa=1}^{n} f_\kappa(x^1, \ldots, x^n)dx^\kappa$ as linear combinations of the total differentials dx^κ with coefficients from the domain of continuous functions $f(x^1, \ldots, x^n)$ on C_x. Then these differential forms constitute a *module of dimension n over the ring of continuous functions*[2] on C_x, if equality and addition of two such differential forms, and multiplication by a continuous function, are defined as follows:

$$\sum_{\kappa=1}^{n} f_\kappa dx^\kappa = \sum_{\kappa=1}^{n} g_\kappa dx^\kappa$$

if and only if $f_\kappa = g_\kappa$ for $\kappa = 1, 2, \ldots, n$,

$$\sum_{\kappa=1}^{n} f_\kappa dx^\kappa + \sum_{\kappa=1}^{n} g_\kappa dx^\kappa = \sum_{\kappa=1}^{n} (f_\kappa + g_\kappa)dx^\kappa, \tag{40}$$

$$f\sum_{\kappa=1}^{n} f_\kappa dx^\kappa = \sum_{\kappa=1}^{n} ff_\kappa dx^\kappa. \tag{41}$$

If we introduce a new coordinate system by (35), we see from (39) that the total differentials dx^κ are transformed as follows:

$$dx^\kappa = \sum_{\lambda=1}^{n} \frac{\partial\psi^\kappa}{\partial y^\lambda} dy^\lambda. \tag{42}$$

The law of transformation (36) for the differential forms $\sum_{\kappa=1}^{n} f_\kappa dx^\kappa$ can now be very simply obtained by subjecting the total differentials in them to the law of transformation (42) and then forming the linear combinations of these total differentials with the functions f_κ, expressed in the new variables $(y^1, \ldots, y^n)$:

$$\sum_{\kappa=1}^{n} f_\kappa dx^\kappa = \sum_{\kappa=1}^{n} f_\kappa \sum_{\lambda=1}^{n} \frac{\partial\psi^\kappa}{\partial y^\lambda} dy^\lambda = \sum_{\lambda=1}^{n} \left(\sum_{\kappa=1}^{n} f_\kappa \frac{\partial\psi^\kappa}{\partial y^\lambda}\right) dy^\lambda.$$

It follows at once that the operations (40) and (41) of the module are invariant under transformation of coordinates.

[2] By a (left) *module M over a ring R* we mean an additively written Abelian group for which there is defined a multiplication αu for $\alpha \in R$ and $u \in M$ such that for $\alpha, \beta \in R$ and $u, v \in M$

1. $\alpha u \in M$,
2. $\alpha(u + v) = \alpha u + \alpha v$,
3. $(\alpha + \beta)u = \alpha u + \beta u$,
4. $\alpha(\beta u) = (\alpha\beta)u$.

The module is said to be of dimension n if it contains n linearly independent elements $u_1, \ldots, u_n$ but does not contain $n + 1$ linearly independent elements.

Let us now consider the laws for transformation of surface-integrals. Here we start from a simple oriented surface-segment F_x in the (x^1, x^2, x^3)-space, which is the image of an oriented domain G in the (t^1, t^2)-plane:

(43) $F_x: x^\kappa = \varphi^\kappa(t^1, t^2)$, $\kappa = 1, 2, 3$,   with   rank $\left(\dfrac{\partial \varphi^\kappa}{\partial t^\lambda}\right) = 2$,

and a bicontinuously differentiable transformation:

(44) $x^\kappa = \psi^\kappa(y^1, y^2, y^3)$, $\kappa = 1, 2, 3$,   with   rank $\left(\dfrac{\partial \psi^\kappa}{\partial y^\lambda}\right) = 3$,

of a neighborhood of F_x, where F_x is transformed into a simple oriented surface-segment F_y in the (y^1, y^2, y^3)-space:

(45) $F_y: y^\lambda = \varphi_*^\lambda(t^1, t^2)$, $\lambda = 1, 2, 3$,   with   rank $\left(\dfrac{\partial \varphi_*^\lambda}{\partial t^\sigma}\right) = 2$.

By (43), (44), and (45) we have

(46) $\varphi^\kappa(t^1, t^2) = \psi^\kappa(\varphi_*^1(t^1, t^2), \varphi_*^2(t^1, t^2), \varphi_*^3(t^1, t^2))$, $\kappa = 1, 2, 3$.

On F_x we now consider an integral

(47) $$\int_{F_x} \sum_{i<j} f_{ij}\, dx^i \wedge dx^j = \int_G \sum_{i<j} f_{ij} \frac{\partial(\varphi^i, \varphi^j)}{\partial(t^1, t^2)}\, dt^1 dt^2.$$

Since

(48) $$\frac{\partial(\varphi^i, \varphi^j)}{\partial(t^1, t^2)} = -\frac{\partial(\varphi^j, \varphi^i)}{\partial(t^1, t^2)}, \qquad i, j = 1, 2, 3,$$

we have

$$\int_G \sum_{i<j} f_{ij} \frac{\partial(\varphi^i, \varphi^j)}{\partial(t^1, t^2)}\, dt^1 dt^2 = \int_G \sum_{i<j} (-f_{ij}) \frac{\partial(\psi^j, \varphi^i)}{\partial(t^1, t^2)}\, dt^1 dt^2$$

(49) $$= \int_G \sum_{i>j} (-f_{ji}) \frac{\partial(\varphi^i, \varphi^j)}{\partial(t^1, t^2)}\, dt^1 dt^2.$$

If we now set

(50) $$g_{ij} = \left\{ \begin{array}{lll} f_{ij} & \text{for} & i < j \\ -f_{ij} & \text{for} & i > j \\ 0 & \text{for} & i = j \end{array} \right\} \quad i, j = 1, 2, 3,$$

then

(51) $$g_{ij} = -g_{ji}, \qquad i, j = 1, 2, 3.$$

If now for $i \geqq j$ we also define the integrals $\int f_{ij} dx^i \wedge dx^j$ by setting

$$\int_{F_x} f_{ij} dx^i \wedge dx^j = \int_G f_{ij} \frac{\partial(\varphi^i, \varphi^j)}{\partial(t^1, t^2)} dt^1 dt^2, \tag{52}$$

then from (47) to (52) we have

$$\begin{cases} \int_{F_x} \sum_{i<j} f_{ij} dx^i \wedge dx^j = \frac{1}{2} \int_G \sum_{i,j=1}^{3} g_{ij} \frac{\partial(\varphi^i, \varphi^j)}{\partial(t^1, t^2)} dt^1 dt^2 \\ = \frac{1}{2} \int_G \sum_{i,j=1}^{3} g_{ij} \frac{\partial \varphi^i}{\partial t^1} \frac{\partial \varphi^j}{\partial t^2} dt^1 dt^2 - \frac{1}{2} \int_G \sum_{i,j=1}^{3} g_{ij} \frac{\partial \varphi^j}{\partial t^1} \frac{\partial \varphi^i}{\partial t^2} dt^1 dt^2 \\ = \int_G \sum_{i,j=1}^{3} g_{ij} \frac{\partial \varphi^i}{\partial t^1} \frac{\partial \varphi^j}{\partial t^2} dt^1 dt^2. \end{cases} \tag{53}$$

Consequently, we can easily state the law of transformation: in view of (46) we have

$$\begin{aligned} \int_G \sum_{i,j=1}^{3} g_{ij} \frac{\partial \varphi^i}{\partial t^1} \frac{\partial \varphi^j}{\partial t^2} dt^1 dt^2 &= \int_G \sum_{i,j=1}^{3} g_{ij} \sum_{k=1}^{3} \frac{\partial \psi^i}{\partial y^k} \frac{\partial \varphi_*^k}{\partial t^1} \sum_{l=1}^{3} \frac{\partial \psi^j}{\partial y^l} \frac{\partial \varphi_*^l}{\partial t^2} dt^1 dt^2 \\ &= \int_G \sum_{k,l=1}^{3} \left(\sum_{i,j=1}^{3} g_{ij} \frac{\partial \psi^i}{\partial y^k} \frac{\partial \psi^j}{\partial y^l} \right) \frac{\partial \varphi_*^k}{\partial t^1} \frac{\partial \varphi_*^l}{\partial t^2} dt^1 dt^2 \\ &= \int_G \sum_{k,l=1}^{3} g_{kl}^* \frac{\partial \varphi_*^k}{\partial t^1} \frac{\partial \varphi_*^l}{\partial t^2} dt^1 dt^2 \end{aligned} \tag{54}$$

with

$$g_{kl}^* = \sum_{i,j=1}^{3} g_{ij} \frac{\partial \psi^i}{\partial y^k} \frac{\partial \psi^j}{\partial y^l}, \qquad k, l = 1, 2, 3. \tag{55}$$

Here it follows from (51) that

$$g_{kl}^* = -g_{lk}^*. \tag{56}$$

A system of 3×3 functions g_{ij} which under a transformation of coordinates are transformed according to the rule (55) is called a *covariant tensor of rank* 2. If this system also satisfies the conditions (51), it is clear from (56) that the same conditions are satisfied by the transformed system (55). We then say that the tensor is *skew-symmetric*. If we now set

$$g_{kl}^* = f_{kl}^* \qquad \text{for} \qquad k < 1,$$

then it follows finally from (50) to (56) that

$$\int_{F_x} \sum_{i<j} f_{ij} dx^i \wedge dx^j = \int_{F_y} \sum_{k<l} f^*_{kl} dy^k \wedge dy^l, \tag{57}$$

where

$$f^*_{kl} = \sum_{i<j} f_{ij} \frac{\partial(\psi^i, \psi^j)}{\partial(y^k, y^l)}. \tag{58}$$

It is to be noted here that the functions f_{ij} are the *determining components of a skew-symmetric covariant tensor of rank* 2 (cf. IB3, §2.6). *Thus they are not transformed like the components of a vector but like the components of a skew-symmetric tensor.* We shall pay particular attention to this matter below in connection with the fact that the rotation of a vector field is again called a "vector field," whereas the formula §2, (24) shows that in fact we are dealing here with a tensor field. The present discussion for 2-dimensional surface-segments in R^3 applies without restriction to 2-dimensional surface-segments in R^n. It is only necessary to allow the indices i, j, k, l to run from 1 through n, instead of 1 through 3.

Let us now consider some special integrals in these spaces. By our definition we have, for arbitrary i and j:

$$\int_{F_x} dx^i \wedge dx^j = \int_G \frac{\partial(\varphi^i, \varphi^j)}{\partial(t^1, t^2)} dt^1 dt^2$$

and

$$\int_{F_x} dx^j \wedge dx^i = \int_G \frac{\partial(\varphi^j, \varphi^i)}{\partial(t^1, t^2)} dt^1 dt^2 = -\int_G \frac{\partial(\varphi^i, \varphi^j)}{\partial(t^1, t^2)} dt^1 dt^2.$$

Consequently

$$\int_{F_x} dx^i \wedge dx^j = -\int_{F_x} dx^j \wedge dx^i, \qquad i, j = 1, 2, \ldots, n,$$

and in particular

$$\int_{F_x} dx^i \wedge dx^i = 0, \qquad i = 1, 2, \ldots, n.$$

Also, by transformation of coordinates

$$\int_{F_x} dx^i \wedge dx^j = \int_{F_y} \sum_{k<l} \frac{\partial(\psi^i, \psi^j)}{\partial(y^k, y^l)} dy^k \wedge dy^l,$$

and by (57) and (58)

$$\int_{F_x} \sum_{i<j} f_{ij} dx^i \wedge dx^j = \int_{F_y} \sum_{k<l} \left(\sum_{i<j} f_{ij} \frac{\partial(\psi^i, \psi^j)}{\partial(y^k, y^l)} \right) dy^k \wedge dy^l.$$

An expression of the form

$$\sum_{i<j} f_{ij} dx^i \wedge dx^j,$$

whose coefficients f_{ij} obey the law of transformation (58) and for which it is possible to construct integrals over surface-segments in accordance with §2, (18) is called a *differential form of second order*. If we regard such differential forms as linear combinations of the elementary differential forms $dx^i \wedge dx^j$, they constitute a *module of dimension* $\binom{n}{2}$ *over the ring of continuous functions on* F_x, provided that equality and addition of two forms, and multiplication by a continuous function on F_x, are defined in the following way:

$$\sum_{i<j} f_{ij} dx^i \wedge dx^j = \sum_{i<j} g_{ij} dx^i \wedge dx^j$$

if and only if $f_{ij} = g_{ij}$ for all pairs $i < j$, and

$$\sum_{i<j} f_{ij} dx^i \wedge dx^j + \sum_{i<j} g_{ij} dx^i \wedge dx^j = \sum_{i<j} (f_{ij} + g_{ij}) dx^i \wedge dx^j, \tag{59}$$

$$f \sum_{i<j} f_{ij} dx^i \wedge dx^j = \sum_{i<j} f f_{ij} dx^i \wedge dx^j. \tag{60}$$

Together with the integrals over the differential forms of this module we have also discussed integrals containing sums over expressions $f_{ij} dx^i \wedge dx^j$, $i \geqq j$. But since we always have

$$\int_{F_x} f_{ij} dx^i \wedge dx^j = \int_{F_x} (-f_{ij}) dx^j \wedge dx^i$$

and thus

$$dx^i \wedge dx^i = 0,$$

for the theory of integration we can identify a summand $f_{ij} dx^i \wedge dx^j$, $i > j$, with the expression $-f_{ij} dx^j \wedge dx^i$ and an expression $f_{ii} dx^i \wedge dx^i$ with the zero element of the module:

$$f_{ij} dx^i \wedge dx^j = -f_{ij} dx^j \wedge dx^i, \tag{61a}$$

$$f_{ii} dx^i \wedge dx^i = 0. \tag{61b}$$

The module of differential forms of the second order is related in a remarkable way to the module of differential forms of the first order. Under a transformation of coordinates the differential form

$$\sum_{i<j} f_{ij} dx^i \wedge dx^j$$

is transformed into

$$\sum_{k<l}\left(\sum_{i<j} f_{ij}\frac{\partial(\psi^i, \psi^j)}{\partial(y^k, y^l)}\right)dy^k \wedge dy^l,$$

which shows at once that the operations (59) and (60) of the module are invariant under transformation of coordinates. In particular, the form $dx^i \wedge dx^j$ is transformed into

$$\sum_{k<l}\frac{\partial(\psi^i, \psi^j)}{\partial(y^k, y^l)}\,dy^k \wedge dy^l.$$

But this latter transformation can be obtained from the total differentials

(62) $$dx^i = \sum_{k=1}^{n}\frac{\partial\psi^i}{\partial y^k}\,dy^k \quad\text{and}\quad dx^j = \sum_{l=1}^{n}\frac{\partial\psi^j}{\partial y^l}\,dy^l$$

by constructing the product $dx^i \wedge dx^j$ termwise, i.e., by multiplying the coefficients together, constructing the differential form $dy^k \wedge dy^l$ from the two factors dy^k and dy^l in that order, and then finally taking the sum of these forms under the conditions (61a) and (61b):

$$dx^i \wedge dx^j = \sum_{k,l=1}^{n}\frac{\partial\psi^i}{\partial y^k}\frac{\partial\psi^j}{\partial y^l}\,dy^k \wedge dy^l = \sum_{k<l}\left(\frac{\partial\psi^i}{dy^k}\frac{\partial\psi^j}{\partial y^l} - \frac{\partial\psi^j}{\partial y^k}\frac{\partial\psi^i}{\partial y^l}\right)dy^k \wedge dy^l$$

$$= \sum_{k<l}\frac{\partial(\psi^i, \psi^j)}{\partial(y^k, y^l)}\,dy^k \wedge dy^l.$$

This method of constructing a product is called *exterior multiplication* of the differential forms (62). It obviously satisfies the rules

(63) $$dx^i \wedge dx^j = -dx^j \wedge dx^i, \qquad i, j = 1, 2, \ldots, n,$$

and thus

$$dx^i \wedge dx^i = 0, \qquad i = 1, 2, \ldots, n.$$

In view of the rule (63) this multiplication is also called *alternating*. We can now extend the concept of exterior multiplication to arbitrary differential forms of first order.[3] As exterior product $\omega_1 \wedge \omega_2$ of the differential forms of first order

$$\omega_1 = \sum_{i=1}^{n} f_{1i}dx^i \quad\text{and}\quad \omega_2 = \sum_{j=1}^{n} f_{2j}dx^j,$$

[3] It is convenient to use the *d*-free notation ω, Θ^2, Θ^r and so forth for differential forms of first, second, *r*-th order instead of $d\omega$, $d^2\Theta$ and the like. Expressions like df, $d\Theta$, $dd\Theta$ are reserved exclusively for the total derivatives of differential forms. The reader should keep this convention in mind when he encounters integrals in the possibly unfamiliar notation $\int\omega$, $\int\Theta^2$, $\int\Theta^r$, etc.

we obtain the differential form of second order:

$$\Theta^2 = \omega_1 \wedge \omega_2 = \sum_{i=1}^{n} \sum_{j=1}^{n} f_{1i} f_{2j} dx^i \wedge dx^j \tag{64}$$
$$= \sum_{i<j} (f_{1i} f_{2j} - f_{1j} f_{2i}) dx^i \wedge dx^j.$$

Here also we recognize from (64) the validity of the rules

$$\omega_1 \wedge \omega_2 = -\omega_2 \wedge \omega_1 \qquad \text{and} \qquad \omega_m \wedge \omega_m = 0, \qquad m = 1, 2.$$

Moreover, this multiplication is independent of transformation of coordinates. Since

$$\omega_m = \sum_{k=1}^{n} \sum_{i=1}^{n} f_{mi} \frac{\partial \psi^i}{\partial y^k} dy^k, \qquad m = 1, 2,$$

we see that in the y-coordinates:

$$\omega_1 \wedge \omega_2 = \sum_{k<l} \left(\sum_{i,j=1}^{n} f_{1i} f_{2j} \frac{\partial \psi^i}{\partial y^k} \frac{\partial \psi^j}{\partial y^l} - \sum_{i,j=1}^{n} f_{1j} f_{2i} \frac{\partial \psi^i}{\partial y^l} \frac{\partial \psi^j}{\partial y^k} \right) dy^k \wedge dy^l$$
$$= \sum_{k<l} \left(\sum_{i<j} (f_{1i} f_{2j} - f_{1j} f_{2i}) \frac{\partial(\psi^i, \psi^j)}{\partial(y^k, y^l)} \right) dy^k \wedge dy^l.$$

But this is exactly the rule (58) for transformation of the differential form Θ^2 in (64).

There exists a second possibility for obtaining differential forms of second order from differential forms of first order. Let

$$\omega = \sum_{j=1}^{n} f_j(x^1, \ldots, x^n) dx^j$$

be a form of first degree whose coefficients are defined and continuously differentiable in a neighborhood of F_x. Then by total differentiation of its coefficients and exterior multiplication with the total differentials dx^j we can obtain a new form, called a *total differential of second order*:

$$d\omega = \sum_{j=1}^{n} \sum_{i=1}^{n} \frac{\partial f_j}{\partial x^i} dx^i \wedge dx^j = \sum_{i<j} \left(\frac{\partial f_j}{\partial x^i} - \frac{\partial f_i}{\partial x^j} \right) dx^i \wedge dx^j. \tag{65}$$

This total differentiation is also independent of the coordinates, since $d\omega$ is a linear combination of the exterior products of the differential forms df_j and dx^j. If ω is already a total differential df of a twice continuously differentiable function $f(x^1, \ldots, x^n)$, then

$$d\omega = ddf = \sum_{i<j} \left(\frac{\partial^2 f}{\partial x^i \partial x^j} - \frac{\partial^2 f}{\partial x^j \partial x^i} \right) dx^i \wedge dx^j = 0.$$

A form like (65) has already occurred, namely in the formula §2, (24). In our present terminology the *Stokes law* formulated there is now written simply:

$$\int_F d\omega = \int_{\mathrm{Bd}F} \omega.$$

This theorem will be proved in a still more general setting in R^n below. There we will consider integrals over differential forms of r-th order

$$\Theta^r = \sum_{i_1 < \cdots < i_r} f_{i_1 \cdots i_r}(x^1, \ldots, x^n) dx^{i_1} \wedge \cdots \wedge dx^{i_r}, \tag{66}$$

which by the formula §3, (26) are to be taken over r-dimensional oriented manifolds in the sense of §3, (27). Under a transformation to new coordinates $y^1, \ldots, y^n$:

$$x^\kappa = \psi^\kappa(y^1, \ldots, y^n), \qquad \kappa = 1, 2, \ldots, n, \qquad \operatorname{rank}\left(\frac{\partial \psi^\kappa}{\partial y^\lambda}\right) = n,$$

we obtain, exactly as for surfaces in R^3, the form

$$\Theta^r = \sum_{j_1 < \cdots < j_r} \left(\sum_{i_1 < \cdots < i_r} f_{i_1 \cdots i_r} \frac{\partial(\psi^{i_1}, \ldots, \psi^{i_r})}{\partial(y^{j_1}, \ldots, y^{j_r})} \right) dy^{j_1} \wedge \cdots \wedge dy^{j_r}.$$

Thus we can arrive at this law of transformation by forming the exterior products of the differentials dx^i:

$$\begin{aligned} dx^{i_1} \wedge \cdots \wedge dx^{i_r} &= \sum_{j_1=1}^{n} \frac{\partial \psi^{i_1}}{\partial y^{j_1}} dy^{j_1} \wedge \cdots \wedge \sum_{j_r=1}^{n} \frac{\partial \psi^{i_r}}{\partial y^{j_r}} dy^{j_r} \\ &= \sum_{j_1, \ldots, j_r = 1}^{n} \frac{\partial \psi^{i_1}}{\partial y^{j_1}} \cdots \frac{\partial \psi^{i_r}}{\partial y^{j_r}} dy^{j_1} \wedge \cdots \wedge dy^{j_r} \\ &= \sum_{j_1 < \cdots < j_r} \frac{\partial(\psi^{i_1}, \ldots, \psi^{i_r})}{\partial(y^{j_1}, \ldots, y^{j_r})} dy^{j_1} \wedge \cdots \wedge dy^{j_r}, \end{aligned}$$

and substituting them into (66); for in analogy with the products of two differential forms of first order we here obtain the *alternating rule of multiplication for r differential forms of first order:*

$$\omega_1 \wedge \omega_2 \wedge \cdots \wedge \omega_r = \operatorname{sgn}(i_1, i_2, \ldots, i_r) \omega_{i_1} \wedge \omega_{i_2} \wedge \cdots \wedge \omega_{i_r},$$

where $\operatorname{sgn}(i_1, i_2, \ldots, i_r) = +1$ or -1 according to whether the permutation $(i_1, i_2, \ldots, i_r)$ of the indices $1, 2, \ldots, r$ is even or odd. The differential forms Θ^r constitute a *module of dimension* $\binom{n}{r}$ *over the ring of continuous functions on* M^r (cf. footnote 2). Thus for every dimension r, $0 \leqq r \leqq n$, we have defined differential forms of order r (where as differential forms of order zero we take the functions themselves). Then by exterior multiplication of two differential forms Θ^r and Θ^s of r-th and s-th order we can construct differential forms $\Theta^r \wedge \Theta^s$ of $(r + s)$-th order and from a form Θ^r of r-th order with continuously differentiable coefficients we can obtain a form of $(r + 1)$-th order, denoted by $d\Theta^r$, by differentiation of the coefficients. The properties of this new form will be discussed in the next section. So we will be confident that the operations undertaken there are permissible in the domain of skew-symmetric covariant tensor fields, over whose components we can construct integrals on the corresponding manifolds in the sense of §3, (26) and (27).

This discussion of integrals and their integrands has necessarily led us to the concept of *alternating differential forms*, to the laws of the algebra and analysis of these forms, to the rules for their transformation, and to their connection with integrals. Thus we have created a firm foundation for the following discussion, in which we will set up these rules at the very beginning and then develop from them the algebra, analysis, integration, and applications of differential forms.

B. The Calculus of Alternating Differentials

6. The Grassmann Algebra of Alternating Differential Forms

On a point set M of the Euclidean space R^n of the variables $(x^1, \ldots, x^n)$ let real functions f and $f_{i_1 \ldots i_p}$ be defined for $p = 1, 2, 3, \ldots$ and $1 \leqq i_\nu \leqq n$. Then by an *alternating differential form Θ^p of order p on M* we mean an expression

$$(67) \qquad \begin{cases} \Theta^0 = f & \text{for } p = 0, \\ \Theta^p = \displaystyle\sum_{i_1, \ldots, i_p} f_{i_1 \cdots i_p} dx^{i_1} \wedge \cdots \wedge dx^{i_p}, & p = 1, 2, 3, \ldots, \end{cases}$$

the following rules are satisfied:

0. For the expressions $1 \cdot dx^{i_1} \wedge \cdots \wedge dx^{i_p}$ we write simply $dx^{i_1} \wedge \cdots \wedge dx^{i_p}$ and call them *elementary differentials of p-th degree.*

1. The differential forms of p-th order constitute a module A^p over the ring of functions defined on M, if addition and multiplication by a function f on M are defined as follows:
for

$$\Theta^p = \Theta_1^p = \sum_{i_1, \ldots, i_p} f_{i_1 \cdots i_p} dx^{i_1} \wedge \cdots \wedge dx^{i_p}$$

and

$$\Theta_2^p = \sum_{i_1, \ldots, i_p} g_{i_1 \cdots i_p} dx^{i_1} \wedge \cdots \wedge dx^{i_p}$$

let

$$\Theta_1^p + \Theta_2^p = \sum_{i_1,\ldots,i_p} (f_{i_1\cdots i_p} + g_{i_1\cdots i_p}) dx^{i_1} \wedge \cdots \wedge dx^{i_p} \tag{68}$$

and

$$f\Theta^p = \sum_{i_1,\ldots,i_p} f f_{i_1\cdots i_p} dx^{i_1} \wedge \cdots \wedge dx^{i_p}. \tag{69}$$

We call A^p a module of rank p.

2. The elementary differentials are skew-symmetric in the indices:

$$dx^{i_1} \wedge \cdots \wedge dx^{i_p} = \operatorname{sgn}(\nu_1, \ldots, \nu_p) dx^{i_{\nu_1}} \wedge \cdots \wedge dx^{i_{\nu_p}}, \tag{70}$$

where $\operatorname{sgn}(\nu_1, \ldots, \nu_p) = +1$ or -1 according to whether the indices $(\nu_1, \ldots, \nu_p)$ form an even or an odd permutation of the indices $(1, \ldots, p)$.

3. The $\binom{n}{p}$ elementary differentials

$$dx^{i_1} \wedge \cdots \wedge dx^{i_p}, \qquad 1 \leqq i_1 < i_2 < \cdots < i_p \leqq n$$

are linearly independent.

4. For two differential forms

$$\Theta_1^p = \sum_{i_1,\ldots,i_p} f_{i_1\cdots i_p} dx^{i_1} \wedge \cdots \wedge dx^{i_p}$$

and

$$\Theta_2^q = \sum_{j_1,\ldots,j_q} g_{j_1\cdots j_q} dx^{j_1} \wedge \cdots \wedge dx^{j_q}$$

an *exterior multiplication* $\Theta_1^p \wedge \Theta_2^q$ is defined as follows:

$$\Theta_1^p \wedge \Theta_2^q = \sum_{i_1,\ldots,i_p} \sum_{j_1,\ldots,j_q} f_{i_1\cdots i_p} g_{j_1\cdots j_q} dx^{i_1} \wedge \cdots \wedge dx^{i_p} \wedge dx^{j_1} \wedge \cdots \wedge dx^{j_q}. \tag{71}$$

This exterior multiplication is obviously associative and distributive, i.e.,

$$(\Theta_1^p \wedge \Theta_2^q) \wedge \Theta_3^r = \Theta_1^p \wedge (\Theta_2^q \wedge \Theta_3^r) \tag{72}$$

and

$$\begin{cases} \Theta_1^p \wedge (\Theta_2^q + \Theta_3^q) = \Theta_1^p \wedge \Theta_2^p + \Theta_1^q \wedge \Theta_3^q, \\ (\Theta_1^p + \Theta_2^p) \wedge \Theta_3^q = \Theta_1^p \wedge \Theta_3^q + \Theta_1^p \wedge \Theta_3^q. \end{cases} \tag{73}$$

Furthermore

$$dx^{i_1} \wedge \cdots \wedge dx^{i_\mu} \wedge \cdots \wedge dx^{i_\nu} \wedge \cdots \wedge dx^{i_p} = 0, \quad \text{if} \quad i_\nu = i_\mu \quad \text{for} \quad \nu \neq \mu; \tag{74}$$

in particular, $dx^i \wedge dx^j = -dx^j \wedge dx^i$ and $dx^i \wedge dx^i = 0$. Also, it follows from (74) that

$$dx^{i_1} \wedge \cdots \wedge dx^{i_p} = 0 \qquad \text{for every } p > n, \tag{75}$$

since two of the elementary differentials dx^{i_ν} in this form are equal to each other. Consequently the modules of rank higher than the n-th consist only of the zero element.

On the basis of the alternating rule (70) and the rule (68) every differential form Θ^p can be brought into the uniquely determined *normal form*

$$\Theta^p = \sum_{i_1 < \cdots < i_p} f_{i_1 \cdots i_p} dx^{i_1} \wedge \cdots \wedge dx^{i_p},$$

from which it follows by rule (3) that the module A^p of these forms is of dimension $\binom{n}{p}$.

A differential form of first order $\omega = \sum_{i=1}^n f_i dx^i$ is called a *Pfaffian form* (see p. 145). From the definition of exterior multiplication it follows at once that for k such forms

$$\omega_1 \wedge \omega_2 \wedge \cdots \wedge \omega_k = \operatorname{sgn}(i_1, i_2, \ldots, i_k)\omega_{i_1} \wedge \omega_{i_2} \wedge \cdots \wedge \omega_{i_k}, \tag{76}$$

if $(i_1, i_2, \ldots, i_k)$ is some permutation of the numbers $1, 2, \ldots, k$. Furthermore, from (71) and (70) we have the rule for change of order

$$\Theta_1^p \wedge \Theta_2^q = (-1)^{pq} \Theta_2^q \wedge \Theta_1^p. \tag{77}$$

Thus *a form of even order p is permutable with every other form.*

If

$$\omega_\kappa = \sum_{i=1}^n f_{\kappa i} dx^i, \qquad \kappa = 1, 2, \ldots, k; \quad k \leqq n, \tag{78}$$

then

$$\omega_1 \wedge \omega_2 \wedge \cdots \wedge \omega_k = \sum_{i_1 < i_2 < \cdots < i_k} \begin{vmatrix} f_{1i_1} f_{1i_2} \cdots f_{1i_k} \\ f_{2i_1} f_{2i_2} \cdots f_{2i_k} \\ \cdot \quad \cdot \quad \cdot \quad \cdot \\ f_{ki_1} f_{ki_2} \cdots f_{ki_k} \end{vmatrix} dx^{i_1} \wedge dx^{i_2} \wedge \cdots \wedge dx^{i_k}. \tag{79}$$

This result is easily proved if in the expression

$$\omega_1 \wedge \omega_2 \wedge \cdots \wedge \omega_k = \sum_{i_1, i_2, \ldots, i_k} f_{1i_1} f_{2i_2} \cdots f_{ki_k} dx^{i_1} \wedge dx^{i_2} \wedge \cdots \wedge dx^{i_k} \tag{80}$$

we set all those terms equal to zero which, as in (74), have two equal indices i_ν, and if we then make use of (70) to arrange the other terms in the order of increasing indices

$$\begin{aligned}(81)\qquad &\sum_{i_1,i_2,\ldots,i_k} f_{1i_1}f_{2i_2}\cdots f_{ki_k}dx^{i_1}\wedge dx^{i_2}\wedge\cdots\wedge dx^{i_k}\\ &= \sum_{i_1<i_2<\cdots<i_k}\left(\sum_{(\nu_1,\nu_2,\ldots,\nu_k)} \operatorname{sgn}(\nu_1,\nu_2,\ldots,\nu_k)f_{1i\nu_1}f_{2i\nu_2}\cdots f_{ki\nu_k}\right)\\ &\qquad\qquad dx^{i_1}\wedge dx^{i_2}\wedge\cdots\wedge dx^{i_k}.\end{aligned}$$

The expression in parentheses is precisely the determinant in (79).

From (79) we now have the theorem that k Pfaffian forms $\kappa = 1, 2, \ldots, k$, are linearly independent at a point of M if and only if their exterior product vanishes at that point; in other words, we can then find n functions $f_1, f_2, \ldots, f_k$, defined on M, which do not vanish simultaneously at any point of M but are such that

$$(82)\qquad f_1\omega_1 + f_2\omega_2 + \cdots + f_k\omega_k = 0.$$

To every differential form Θ^p, $0 \leqq p \leqq n$ we now assign *an adjoint form of order* $n - p$, denoted by $*\,\Theta^p$, in the following way. Let

$$\begin{aligned}(83)\qquad *\,\Theta^p &= *\sum_{i_1,\ldots,i_p} f_{i_1\cdots i_p}dx^{i_1}\wedge\cdots\wedge dx^{i_p}\\ &= \sum_{i_1,\ldots,i_p} f_{i_1\cdots i_p}\operatorname{sgn}(i_1,\ldots,i_p,i_{p+1},\ldots,i_n)dx^{i_{p+1}}\wedge\cdots\wedge dx^{i_n},\end{aligned}$$

where $i_{p+1}, \ldots, i_n$ is any fixed permutation of those numbers among $1, 2, \ldots, n$ that do not occur among the $i_1, \ldots, i_p$. In particular,

$$\begin{aligned} *\,f dx^1\wedge\cdots\wedge dx^n &= f,\\ *\,f &= f dx^1\wedge\cdots\wedge dx^n.\end{aligned}$$

For the normal forms we thus have

$$\begin{aligned} *\,\Theta^p &= *\sum_{i_1<\cdots<i_p} f_{i_1\cdots i_p}dx^{i_1}\wedge\cdots\wedge dx^{i_p}\\ &= \sum_{i_{p+1}<\cdots<i_n}\operatorname{sgn}(i_1,\ldots,i_p,i_{p+1},\ldots,i_n)f_{i_1\cdots i_p}dx^{i_{p+1}}\wedge\cdots\wedge dx^{i_n},\end{aligned}$$

where $(i_1, \ldots, i_p, i_{p+1}, \ldots, i_n)$ is the uniquely determined permutation of the numbers $(1, 2, \ldots, n)$ for which $i_{p+1} < \cdots < i_n$. For example, in R^5:

$$\begin{aligned} &*\,(f_{13}dx^1\wedge dx^3 + f_{24}dx^2\wedge dx^4 + f_{45}dx^4\wedge dx^5)\\ &\quad= \operatorname{sgn}(1,3,2,4,5)f_{13}dx^2\wedge dx^4\wedge dx^5\\ &\qquad + \operatorname{sgn}(2,4,1,3,5)f_{24}dx^1\wedge dx^3\wedge dx^5\\ &\qquad + \operatorname{sgn}(4,5,1,2,3)f_{45}dx^1\wedge dx^2\wedge dx^3\\ &\quad= -f_{13}dx^2\wedge dx^4\wedge dx^5 - f_{24}dx^1\wedge dx^3\wedge dx^5 + f_{45}dx^1\wedge dx^2\wedge dx^3.\end{aligned}$$

For this star operator $*$ we have:

(84) $$** \Theta^p = (-1)^{p(n-p)} \Theta^p,$$

as follows from the fact that

$$\operatorname{sgn}(i_{p+1} \cdots i_n i_1 \cdots i_p) = (-1)^{p(n-p)} \operatorname{sgn}(i_1 \cdots i_p i_{p+1} \cdots i_n).$$

In R^3 and in every space of odd dimension we thus have

$$** \Theta^p = \Theta^p \qquad \text{for all } p,$$

but in R^2 and in every space of even dimension,

$$** \Theta^p = (-1)^p \Theta^p.$$

Then

$$\Theta_1^p \wedge * \Theta_2^p = \sum_{i_1 < \cdots < i_p} f_{i_1 \cdots i_p} g_{i_1 \cdots i_p} dx^1 \wedge dx^2 \wedge \cdots \wedge dx^n,$$

and thus

$$* (\Theta_1^p \wedge * \Theta_2^p) = \sum_{i_1 < \cdots < i_p} f_{i_1 \cdots i_p} g_{i_1 \cdots i_p},$$

so that in particular:

(85) $$* (\Theta^p \wedge * \Theta^p) = \sum_{i_1 < \cdots < i_p} (f_{i_1 \cdots i_p})^2.$$

The following relation is also of importance

(86) $$* (* (\omega_1 \wedge \omega_2) \wedge \omega_3) = \omega_2 * (\omega_1 \wedge * \omega_3) - \omega_1 * (\omega_2 \wedge * \omega_3).$$

For we have

$$\omega_1 \wedge \omega_2 = \sum_{i_1, i_2} f_{1i_1} f_{2i_2} dx^{i_1} \wedge dx^{i_2},$$

$$* (\omega_1 \wedge \omega_2) = \sum_{i_1, i_2} f_{1i_1} f_{2i_2} \operatorname{sgn}(i_1, i_2, i_3, \ldots, i_n) dx^{i_3} \wedge \cdots \wedge dx^{i_n},$$

$$\begin{aligned} * (\omega_1 \wedge \omega_2) \wedge \omega_3 &= \sum_{i = i_1, i_2} \sum_{i_1, i_2} f_{1i_1} f_{2i_2} f_{3i} \operatorname{sgn}(i_1, \ldots, i_n) dx^{i_3} \wedge \cdots \wedge dx^{i_n} \wedge dx_i \\ &= \sum_{i_1, i_2} f_{1i_1} f_{2i_2} f_{3i_1} \operatorname{sgn}(i_1, \ldots, i_n) dx^{i_3} \wedge \cdots \wedge dx^{i_n} \wedge dx^{i_1} \\ &\quad + \sum_{i_1, i_2} f_{1i_1} f_{2i_2} f_{3i_2} \operatorname{sgn}(i_1, \ldots, i_n) dx^{i_3} \wedge \cdots \wedge dx^{i_n} \wedge dx^{i_2}, \end{aligned}$$

$$
\begin{aligned}
((\omega_1 \wedge \omega_2) \wedge \omega_3) \\
&= \sum_{i_1} f_{1i_1} f_{3i_1} \sum_{i_2} \operatorname{sgn}(i_1, \ldots, i_n) \operatorname{sgn}(i_3, \ldots, i_n, i_1, i_2) f_{2i_2} dx^{i_2} \\
&\quad + \sum_{i_2} f_{2i_2} f_{3i_2} \sum_{i_1} \operatorname{sgn}(i_1, \ldots, i_n) \operatorname{sgn}(i_3, \ldots, i_n, i_2, i_1) f_{1i_1} dx^{i_1} \\
&= \omega_2 \sum_i f_{1i} f_{3i} - \omega_1 \sum_i f_{2i} f_{3i},
\end{aligned}
$$

from which (86) follows.

The star operator provides a one-to-one linear mapping of the module of differential forms of order p onto the module of order $n - p$; i.e.,

$$*(\Theta_1^p + \Theta_2^p) = *\,\Theta_1^p + *\,\Theta_2^p \qquad \text{and} \qquad *f\Theta^p = f *\,\Theta^p.$$

It is often desirable to make the differential forms into a closed set with respect to the operations of the ring. Then we must consider all formal sums

$$\Omega = \Theta^0 + \Theta^1 + \cdots + \Theta^n,$$

which form a module of dimension 2^n and which, under formal addition and exterior multiplication of two such expressions, lead again to expressions of the same kind. Thus these forms Ω constitute a (noncommutative) ring. This ring is the direct sum (cf. IB6, §6) of the modules $A^0, A^1, \ldots, A^n$, which is called a *graded ring* in the sense of algebra, namely the *Grassmann ring of differential forms* on M.

In this section we have not made any assumptions about the functions $f_{i_1 \cdots i_p}$. Consequently our discussion refers purely to points, i.e., our differential forms are considered essentially in isolation at each point of M. There they are equivalent to the *Grassmann algebra* of an n-dimensional real linear vector space, namely the vector space of Pfaffian forms at the point in question. Thus the notation we have chosen for the basis elements dx^i and their products is for the time purely formal. We now wish to relate these forms to analysis, and in particular to the differential calculus, and thereby to motivate our choice of notation.

7. The Differential Operations for the Alternating Differential Forms

We now assume that the coefficients of our alternating differential forms Θ^p have continuous partial derivatives in an n-dimensional domain G

of R^n. Then we define the *total differential* $d\Theta^p$ of such a form in G as follows:

$$d\Theta^p = d \sum_{i_1,\ldots,i_p} f_{i_1\cdots i_p} dx^{i_1} \wedge \cdots \wedge dx^{i_p}$$

$$(87) \qquad = \sum_{i_1,\ldots,i_p} \sum_{i} \frac{\partial f_{i_1\cdots i_p}}{\partial x^i} dx^i \wedge dx^{i_1} \wedge \cdots \wedge dx^{i_p}.$$

This derivation provides a mapping of the module of continuously differentiable differential forms of order p into the module of continuous differential forms of order $p + 1$. If Θ^p is a continuously differentiable function: $\Theta^0 = f$, then

$$df = \sum_{i=1}^{n} \frac{\partial f}{\partial x^i} dx^i$$

is a Pfaffian form, called the *total differential of the function f*.

For the total differential we have by (87) the rules

$$d(\Theta_1^p + \Theta_2^p) = d\Theta_1^p + d\Theta_2^p$$

and

$$(88) \qquad d(\Theta_1^p \wedge \Theta_2^q) = d\Theta_1^p \wedge \Theta_2^q + (-1)^p \Theta_1^p \wedge \Theta_2^q.$$

In particular

$$d(f\Theta^p) = df \wedge \Theta^p + fd\Theta^p.$$

In view of this last relation the mapping d of the continuously differentiable differential forms from the module A^p into the module A^{p+1} is not linear with respect to the operations of the module.

If the differential form Θ^p is twice continuously differentiable, the total differential $d\Theta^p$ can be differentiated once more. Then we have the following lemma.

First Poincaré lemma:

$$(89) \qquad dd\Theta^p = 0.$$

The proof is very simple:

$$dd \sum_{i_1,\ldots,i_p} f_{i_1\cdots i_p} dx^{i_1} \wedge \cdots \wedge dx^{i_p}$$

$$= d \sum_{i_1,\ldots,i_p} \sum_{i} \frac{\partial f_{i_1\cdots i_p}}{\partial x^i} dx^i \wedge dx^{i_1} \wedge \cdots \wedge dx^{i_p}$$

$$= \sum_{i_1,\ldots,i_p} \sum_{i} \sum_{j} \frac{\partial^2 f_{i_1\cdots i_p}}{\partial x^j \partial x^i} dx^j \wedge dx^i \wedge dx^{i_1} \wedge \cdots \wedge dx^{i_p}$$

$$= \sum_{i_1,\ldots,i_p} \sum_{j<i} \left(\frac{\partial^2 f_{i_1\cdots i_p}}{\partial x^i \partial x^j} - \frac{\partial^2 f_{i_1\cdots i_p}}{\partial x^j \partial x^i} \right) dx^i \wedge dx^j \wedge dx^{i_1} \wedge \cdots \wedge dx^{i_p} = 0.$$

Thus the form $d\Theta^p = \Theta^{p+1}$ is such that $d\Theta^{p+1} = 0$. Now a form Θ^p for which $d\Theta^p = 0$ is said to be *closed* or *integrable*, so that the first Poincaré lemma states that the total differential of a twice continuously differentiable differential form Θ^p is closed.

From (88) it follows that $\Theta_1^p \wedge \Theta_2^q$ is closed if Θ_1^p and Θ_2^q are continuously differentiable and closed.

On the other hand, the question whether every closed continuously differentiable differential form Θ^q is the total differential of a twice continuously differentiable form Θ^{q-1} must be answered in general in the negative, since the answer depends on the topological structure of the domain G. But if this domain can be shrunk to an interior point by means of twice differentiably continuous transformations, then the answer is affirmative. In the following lemma we prove a slightly weaker statement.

Second Poincaré lemma:

If the once continuously differentiable differential form Θ^p is closed, i.e., $d\Theta^p = 0$ in G, then for every point $x_0 = (x_0^1, \ldots, x_0^n)$ in G there exists a neighborhood U_{x_0} and in U_{x_0} a differential form Θ^{p-1} such that in U_{x_0}

$$\Theta^p = d\Theta^{p-1}. \tag{90}$$

For the proof we consider at the point x_0 an open neighborhood U_{x_0} which with every point x contains all the points $x_0 + t(x - x_0)$, $0 \leqq t \leqq 1$. Such a domain U_{x_0} is said to be *star-shaped about the center* x_0. Now let

$$\Theta^p = \sum_{i_1,\ldots,i_p} f_{i_1\cdots i_p}(x)dx^{i_1} \wedge \cdots \wedge dx^{i_p}$$

with $x = (x^1, \ldots, x^n)$. Then the condition $d\Theta^p = 0$ can be written

$$\sum_{i_0<i_1<\cdots<i_p} \left(\sum_{(\nu_0,\nu_1,\ldots,\nu_p)} \operatorname{sgn}(\nu_0, \nu_1, \ldots, \nu_p) \frac{\partial f_{i_{\nu_1}\ldots i_{\nu_p}}(x)}{\partial x^{i_{\nu_0}}} \right) \times dx^{i_0} \wedge dx^{i_1} \wedge \cdots \wedge dx^{i_p} = 0.$$

Consequently, in view of the linear independence of the elementary differentials we have:

$$\sum_{(\nu_0,\nu_1,\ldots,\nu_p)} \operatorname{sgn}(\nu_0, \nu_1, \ldots, \nu_p) \frac{\partial f_{i_{\nu_1}\ldots i_{\nu_p}}(x)}{\partial x^{i_{\nu_0}}} \equiv 0 \tag{91}$$

in G, where the summation is taken over all $(p + 1)!$ permutations of the numbers $(0, 1, \ldots, p)$ and $\operatorname{sgn}(\nu_0, \nu_1, \ldots, \nu_p)$ denotes the sign of such a permutation.

Here we may assume that $x_0 = (0, \ldots, 0)$, since otherwise we could everywhere replace $tx = (tx^1, \ldots, tx^n)$ by

$$x_0 + t(x - x_0) = (x_0^1 + t(x^1 - x_0^1), \ldots, x_0^n + t(x^n - x_0^n))$$

and x^i by $x^i - x_0^i$, $i = 1, 2, \ldots, n$. Now let us explicitly write down a differential form Θ^{p-1} of the desired kind. We set

$$\Theta^{p-1} = \sum_{i_1, \ldots, i_p} \int_0^1 f_{i_1 \cdots i_p}(tx) t^{p-1} dt \sum_{\nu=1}^{p} (-1)^{\nu-1} x^{i_\nu} dx^{i_1} \wedge \cdots \wedge \widehat{dx^{i_\nu}} \wedge \cdots \wedge dx^{i_p}.$$

Here the symbol $\wedge$ over the differential dx^{i_ν} means that this differential is to be omitted in the sequence $dx^{i_1} \wedge \cdots \wedge dx^{i_p}$.

Then for the partial derivatives of the coefficients in $d\Theta^{p-1}$ we have

$$\frac{\partial}{\partial x^{i_0}} \left\{ \int_0^1 f_{i_1 \cdots i_p}(tx) t^{p-1} dt (-1)^{\nu-1} x^{i_\nu} \right\}$$

$$= \begin{cases} \displaystyle\int_0^1 \left. \frac{\partial f_{i_1 \cdots i_p}}{\partial x^{i_0}} \right|_{tx} t^p dt (-1)^{\nu-1} x^{i_\nu} & \text{for} \quad i_0 \neq i_\nu, \\ \displaystyle\int_0^1 \left. \frac{\partial f_{i_1 \cdots i_p}}{\partial x^{i_0}} \right|_{tx} t^p dt (-1)^{\nu-1} x^{i_0} + \int_0^1 f_{i_1 \cdots i_p}(tx) t^{p-1} dt (-1)^{\nu-1} & \text{for} \quad i_0 = i_\nu. \end{cases}$$

Setting this expression in $d\Theta^{p-1}$ we obtain

$$\begin{aligned} d\Theta^{p-1} = {} & \sum_{i_0, i_1, \ldots, i_p} \int_0^1 \left. \frac{\partial f_{i_1 \cdots i_p}}{\partial x^{i_0}} \right|_{tx} t^p dt \sum_{\nu=1}^{p} (-1)^{\nu-1} x^{i_\nu} dx^{i_0} \wedge dx^{i_1} \wedge \cdots \wedge \widehat{dx^{i_\nu}} \wedge \cdots \wedge dx^{i_p} \\ & + \sum_{i_1, \ldots, i_p} \int_0^1 f_{i_1 \cdots i_p}(tx) t^{p-1} dt \sum_{\nu=1}^{p} (-1)^{\nu-1} dx^{i_\nu} \wedge dx^{i_1} \wedge \cdots \wedge \widehat{dx^{i_\nu}} \wedge \cdots \wedge dx^{i_p}. \end{aligned} \tag{92}$$

The last sum leads to $p dx^{i_1} \wedge \cdots \wedge dx^{i_\nu} \wedge \cdots \wedge dx^{i_p}$. Taking the factor p under the sign of integration and integrating by parts, and then writing $f_{i_1 \cdots i_p}(tx) = g(t)$, $pt^{p-1} = h'(t)$, and thus $t^p = h(t)$, we obtain

$$\begin{aligned} \int_0^1 f_{i_1 \cdots i_p}(tx) p t^{p-1} dt &= \int_0^1 g(t) h'(t) dt = g(t) h(t) |_0^1 - \int_0^1 g'(t) h(t) dt \\ &= f_{i_1 \cdots i_p}(tx) t^p |_0^1 - \int_0^1 \sum_{i_0} \left. \frac{\partial f_{i_1 \cdots i_p}}{\partial x^{i_0}} \right|_{tx} x^{i_0} t^p dt \\ &= f_{i_1 \cdots i_p}(x) - \sum_{i_0} \int_0^1 \left. \frac{\partial f_{i_1 \cdots i_p}}{\partial x^{i_0}} \right|_{tx} t^p dt x^{i_0}. \end{aligned}$$

Setting this result in (92), we get

$$
\begin{aligned}
d\Theta^{p-1} = &\sum_{i_1,\ldots,i_p} f_{i_1\cdots i_p}(x)dx^{i_1} \wedge \cdots \wedge dx^{i_p} \\
(93) \qquad &+ \sum_{i_0,i_1,\ldots,i_p} \int_0^1 \frac{\partial f_{i_1\cdots i_p}}{\partial x^{i_0}}\bigg|_{tx} t^p dt \sum_{\nu=0}^{p} (-1)^{\nu-1}x^{i_\nu}dx^{i_0} \wedge dx^{i_1} \wedge \cdots \\
&\wedge \widehat{dx^{i_\nu}} \wedge \cdots \wedge dx^{i_p}.
\end{aligned}
$$

The expression

$$\Theta^p_{i_0 i_1\cdots i_p} = \sum_{\nu=0}^{p} (-1)^{\nu-1}x^{i_\nu}dx^{i_0} \wedge dx^{i_1} \wedge \cdots \wedge \widehat{dx^{i_\nu}} \wedge \cdots \wedge dx^{i_p}$$

s skew-symmetric, i.e., under a permutation of the indices $(i_0, i_1, \ldots, i_p)$ it is merely multiplied by sgn $(i_0, i_1, \ldots, i_p)$:

$$\Theta^p_{i_{\nu_0} i_{\nu_1}\cdots i_{\nu_p}} = \operatorname{sgn}(\nu_0, \nu_1, \ldots, \nu_p)\,\Theta^p_{i_0 i_1\cdots i_p}.$$

For if we interchange two neighboring indices i_ρ and $i_{\rho+1}$, then a summand in which $\nu \neq \rho$, $\rho + 1$ changes sign, and the two monomials for which $\nu = \rho$ and $\nu = \rho + 1$ change places with a consequent change of sign, so that the entire expression is multiplied by the factor -1. Since every permutation can be generated by transposition of successive indices, the desired result follows. Thus the expression vanishes if two of the indices i and i_μ are equal, and the last sum in (93) can be written

$$
\begin{aligned}
&\sum_{i_0<i_1<\cdots<i_p} \int_0^1 \sum_{(\nu_0,\nu_1,\ldots,\nu_p)} \operatorname{sgn}(\nu_0, \nu_1, \ldots, \nu_p) \frac{\partial f_{i_{\nu_1}\cdots i_{\nu_p}}}{\partial x^{i_{\nu_0}}}\bigg|_{tx} \\
&\qquad \times t^p dt \sum_{\nu=0}^{p} (-1)^{\nu-1}x^{i_\nu}dx^{i_0} \wedge \cdots \wedge dx^{i_\nu} \wedge \cdots \wedge dx^{i_p}.
\end{aligned}
$$

Since the integrands vanish because of (91), our statement is proved.

Of course, the form Θ^{p-1} for which (90) is satisfied is not uniquely determined. Two such forms Θ^{p-1}_1 and Θ^{p-1}_2 differ in a common neighborhood of the point x_0 by a closed $(p-1)$-form and every closed $(p-1)$-form added to a form Θ^{p-1} that satisfies (90) produces again a form of the same kind.

Along with the differential operator d, which to every continuously differentiable form Θ^p assigns the total differential $\Theta^{p-1} = d\Theta^p$, there is a second operator that plays an important role in applications of the calculus of alternating differential forms, namely the operator δ defined as follows:

$$(94) \qquad \delta\Theta^p = (-1)^{np+n+1} * d * \Theta^p.$$

The process of forming a differential form Θ^{p-1} in this way is called *codifferentiation* and $\delta\Theta^p$ is the *codifferential of* Θ^p. If

$$\Theta^p = \sum_{i_1<\cdots<i_p} f_{i_1\cdots i_p} dx^{i_1} \wedge \cdots \wedge dx^{i_p}, \tag{95}$$

then by §§6, (83) and 7, (87) a simple calculation gives:

$$\delta\Theta^p = \sum_{i_1<\cdots<i_p} \sum_{\nu=1}^{p} (-1)^{\nu} \frac{\partial f_{i_1\cdots i_p}}{\partial x^{i_\nu}} dx^{i_1} \wedge \cdots \wedge \widehat{dx^{i_\nu}} \wedge \cdots \wedge dx^{i_p}, \tag{96}$$

where the symbol $\widehat{\ }$ over dx^{i_ν} means that this differential is to be omitted.

For the codifferential operator δ we have theorems analogous to the first and second Poincaré lemmas:

1. *If Θ^p is twice continuously differentiable in the domain G, then*

$$\delta\delta\Theta^p = 0. \tag{97}$$

2. *If Θ^p is continuously differentiable in G and if*

$$\delta\Theta^p = 0,$$

then for every point in G there exists a neighborhood and in it a twice continuously differentiable form Θ^{p+1} such that

$$\Theta^p = \delta\Theta^{p+1}.$$

Formula (97) follows immediately from (94) together with (89) and §6, (84). The second statement follows from the second Poincaré lemma: $\delta\Theta^p = 0$ is equivalent to $d * \Theta^p = 0$. From the lemma it then follows that the form $* \Theta^p$ can be represented locally as the total differential of a form Θ^{n-p-1}. Consequently, the form $\Theta^{p+1} = (-1)^{np+1} * \Theta^{n-p-1}$ has the desired property.

The operators d and δ are closely related to the *Laplace operator*

$$\Delta = \sum_{i=1}^{n} \frac{\partial^2}{(\partial x^i)^2}.$$

For we have

$$d\delta + \delta d = -\Delta$$

for every twice continuously differentiable differential form Θ^p, i.e.,

$$d\delta\Theta^p + \delta d\Theta^p = -\Delta\Theta^p. \tag{98}$$

This equality follows from (87) and (96) by a simple computation: for if Θ^p is given by (95), then

$$\begin{aligned} d\delta\,\Theta^p = {} & \sum_{i_1<\cdots<i_p} \sum_{\mu=1}^{p} (-1) \frac{\partial^2 f_{i_1\cdots i_p}}{(\partial x^{i_\mu})^2}\, dx^{i_1} \wedge \cdots \wedge dx^{i_p} \\ & + \sum_{i_1<\cdots<i_p} \sum_{\nu=1}^{p} \sum_{\mu=p+1}^{n} (-1)^{\nu} \frac{\partial^2 f_{i_1\cdots i_p}}{\partial x^{i_\mu}\partial x^{i_\nu}}\, dx^{i_\mu} \wedge dx^{i_1} \wedge \cdots \\ & \wedge \widehat{dx^{i_\nu}} \wedge \cdots \wedge dx^{i_p} \end{aligned} \tag{99}$$

and

$$\begin{aligned} \delta d\Theta^p = {} & \sum_{i_1<\cdots<i_p} \sum_{\mu=p+1}^{n} \sum_{\nu=1}^{p} (-1)^{\nu+1} \frac{\partial^2 f_{i_1\cdots i_p}}{\partial x^{i_\nu}\partial x^{i_\mu}}\, dx^{i_\mu} \wedge dx^{i_1} \wedge \cdots \\ & \wedge \widehat{dx^{i_\nu}} \wedge \cdots \wedge dx^{i_p} \\ & + \sum_{i_1<\cdots<i_p} \sum_{\mu=p+1}^{n} (-1) \frac{\partial^2 f_{i_1\cdots i_p}}{(\partial x^{i_\mu})^2}\, dx^{i_1} \wedge \cdots \wedge dx^{i_p}, \end{aligned} \tag{100}$$

from which follows the desired equality (98).

In the literature the operator $d\delta + \delta d$ is often denoted by Δ, so that the Laplace operator is $-\Delta$.

8. Transformation of Coordinates

In §5 we have discussed in detail how the alternating differential forms are to be transformed under a transformation of coordinates if the corresponding integrals are to remain invariant under the transformations. We now set up these rules at the very beginning of the discussion and use them to define a mapping of the differential forms.

Let there be given a continuously differentiable mapping $G_y \xrightarrow{\varphi} G_x$ of a domain G_y in $(y^1, \ldots, y^m)$-space into a domain G_x of $(x^1, \ldots, x^n)$-space by means of n continuously differentiable functions

$$x^i = \varphi^i(y^1, \ldots, y^m), \qquad i = 1, 2, \ldots, n.$$

To the total differentials dx^i in the x-space we assign the total differentials

$$\sum_{j=1}^{m} \frac{\partial \varphi^i}{\partial y^j}\, dy^j, \qquad i = 1, 2, \ldots, n$$

in the y-space and to every differential form

$$\Theta_x^p = \sum_{i_1,\ldots,i_p} f_{i_1\cdots i_p} dx^{i_1} \wedge \cdots \wedge dx^{i_p}$$

in the x-space we assign the differential form

$$\text{(101)}\quad \Theta_y^p = \sum_{i_1,\ldots,i_p} f_{i_1\cdots i_p} \sum_{j_1=1}^{m} \frac{\partial\varphi^{i_1}}{\partial y^{j_1}}\, dy^{j_1} \wedge \cdots \wedge \sum_{j_p=1}^{m} \frac{\partial\varphi^{i_p}}{\partial y^{j_p}}\, dy^{j_p}$$

$$= \sum_{j_1,\ldots,j_p} \left(\sum_{i_1,\ldots,i_p} f_{i_1\cdots i_p} \frac{\partial\varphi^{i_1}}{\partial y^{j_1}} \cdots \frac{\partial\varphi^{i_p}}{\partial y^{j_p}} \right) dy^{j_1} \wedge \cdots \wedge dy^{j_p}$$

$$= \sum_{j_1<\cdots<j_p} \left(\sum_{i_1,\ldots,i_p} f_{i_1\cdots i_p} \sum_{(\nu_1,\ldots,\nu_p)} \operatorname{sgn}(\nu_1,\ldots,\nu_p) \frac{\partial\varphi^{i_1}}{\partial y^{j_{\nu_1}}} \cdots \frac{\partial\varphi^{i_p}}{\partial y^{j_{\nu_p}}} \right)$$
$$\times\, dy^{j_1} \wedge \cdots \wedge dy^{j_p}$$

$$= \sum_{j_1<\cdots<j_p} \left(\sum_{i_1,\ldots,i_p} f_{i_1\cdots i_p} \frac{\partial(\varphi^{i_1},\ldots,\varphi^{i_p})}{\partial(y^{j_1},\ldots,y^{j_p})} \right) dy^{j_1} \wedge \cdots \wedge dy^{j_p}$$

in the y-space.

To a mapping of G_y into G_x we have thereby assigned a mapping $\check{\varphi}$ of all differential forms in G_x into the differential forms in G_y:

$$\text{(102)}\qquad \Theta_x^p \xrightarrow{\check{\varphi}} \Theta_y^p.$$

We see at once that this mapping obeys the following rules

$$\Theta_{x,1}^p + \Theta_{x,2}^p \xrightarrow{\check{\varphi}} \Theta_{y,1}^p + \Theta_{y,2}^p,$$
$$f\Theta_x^p \xrightarrow{\check{\varphi}} f\Theta_y^p,$$
$$\Theta_{x,1}^p \wedge \Theta_{x,2}^q \xrightarrow{\check{\varphi}} \Theta_{y,1}^p \wedge \Theta_{y,2}^q.$$

If Θ_x^p is continuously differentiable and if the mapping φ is twice continuously differentiable, then Θ_y^p is continuously differentiable and we have

$$\text{(103)}\qquad d\Theta_x^p \xrightarrow{\check{\varphi}} d\Theta_y^p.$$

This rule follows from the fact that in the x-space

$$df_{i_1\ldots i_p} dx^{i_1} \wedge \cdots \wedge dx^{i_p} = \sum_{i=1}^{n} \frac{\partial f_{i_1\cdots i_p}}{\partial x^i}\, dx^i \wedge dx^{i_1} \wedge \cdots \wedge dx^{i_p}$$

but in the y-space

$$d\left(f_{i_1\cdots i_p} \sum_{j_1=1}^{m} \frac{\partial\varphi^{i_1}}{\partial y^{j_1}}\, dy^{j_1} \wedge \cdots \wedge \sum_{j_p=1}^{m} \frac{\partial\varphi^{i_p}}{\partial y^{j_p}}\, dy^{j_p} \right)$$

$$= df_{i_1\cdots i_p} \wedge \sum_{j_1=1}^{m} \frac{\partial\varphi^{i_1}}{\partial y^{j_1}}\, dy^{j_1} \wedge \cdots \wedge \sum_{j_p=1}^{m} \frac{\partial\varphi^{i_p}}{\partial y^{j_p}}\, dy^{j_p}$$
$$+ f_{i_1\cdots i_p} d\left(\sum_{j_1=1}^{m} \frac{\partial\varphi^{i_1}}{\partial y^{j_1}}\, dy^{j_1} \wedge \cdots \wedge \sum_{j_p=1}^{m} \frac{\partial\varphi^{i_p}}{\partial y^{j_p}}\, dy^{j_p} \right)$$
$$= df_{i_1\cdots i_p} \wedge \sum_{j_1=1}^{m} \frac{\partial\varphi^{i_1}}{\partial y^{j_1}}\, dy^{j_1} \wedge \cdots \wedge \sum_{j_p=1}^{m} \frac{\partial\varphi^{i_p}}{\partial y^{j_p}}\, dy^{j_p},$$

since the total differentials

$$\sum_{j_\nu=1}^{m} \frac{\partial \varphi^{i_\nu}}{\partial y^{j_\nu}} dy^{j_\nu}$$

are closed and thus their product is also closed. But the total differential $df_{i_1 \ldots i_p}$ in the y-space is precisely

$$df_{i_1 \cdots i_p} = \sum_{j=1}^{m} \frac{\partial f_{i_1 \cdots i_p}}{\partial y^j} dy^j = \sum_{j=1}^{m} \sum_{i=1}^{n} \frac{\partial f_{i_1 \cdots i_p}}{\partial x^i} \frac{\partial \varphi^i}{\partial y^j} dy^j,$$

and by (101) this is the image of the differential

$$\sum_{i=1}^{n} \frac{\partial f_{i_1 \cdots i_p}}{\partial x^i} dx^i.$$

From (103) it also follows that the image of a closed form is closed: for from (102) and

$$d\Theta_x^p = 0 \qquad \text{it follows that} \qquad d\Theta_y^p = 0.$$

Further, from $d\Theta_x^p = 0$ and $\Theta_x^p = d\Theta_x^{p-1}$ we have the same relations for the images:

$$d\Theta_y^p = 0 \qquad \text{and} \qquad \Theta_y^p = d\Theta_y^{p-1}.$$

If $n = m$ and if φ is a one-to-one continuously differentiable mapping between the domain G_y of the $(y_1, \ldots, y_n)$-space and the domain G_x of the $(x_1, \ldots, x_n)$-space, we may regard the variables $(y_1, \ldots, y_n)$ as new coordinates in the $(x_1, \ldots, x_n)$-space. The law of transformation (101) then enables us to express a form Θ^p in the new coordinates $(y_1, \ldots, y_n)$. In this case we identify the two forms. If we write them as

$$\Theta^p = \sum_{i_1 < \cdots < i_p} f_{i_1 \cdots i_p} dx^{i_1} \wedge \cdots \wedge dx^{i_p} = \sum_{j_1 < \cdots < j_p} f^*_{j_1 \ldots j_p} dy^{j_1} \wedge \cdots \wedge dy^{j_p},$$

their coefficients are transformed like the coefficients of a skew-symmetric covariant tensor field of rank p.

Let us now see how the *adjoint differential forms* $* \ \Theta^p$ *are transformed under a change of coordinates*. For this purpose we first consider a *basis transformation* of the differentials $dx^1, \ldots, dx^n$. Let $\omega^1, \ldots, \omega^n$ be n Pfaffian forms in G_x:

$$\omega^j = \sum_{i=1}^{n} h_i^j dx^i, \qquad j = 1, 2, \ldots, n,$$

where the determinant $|h^i_j|$ does not vanish in G_x. Then the dx^i are linear combinations of the ω^j:

$$dx^i = \sum_{j=1}^{n} \check{h}^i_j \omega^j, \qquad j = 1, 2, \ldots, n,$$

where

$$\sum_{i=1}^{n} h^j_i \check{h}^i_k = \delta^i_k = \begin{cases} 1 & \text{for } j = k, \\ 0 & \text{for } j \neq k. \end{cases} \tag{104}$$

Thus we may use the Pfaffian forms ω^j as a basis for all the differential forms in G_x, which can therefore be expressed in terms of the ω^j:

$$\Theta^p = \sum_{i_1, \ldots, i_p} f_{i_1 \cdots i_p} dx^{i_1} \wedge \cdots \wedge dx^{i_p} = \sum_{j_1, \ldots, j_p} f^*_{j_1 \cdots j_p} \omega^{j_1} \wedge \cdots \wedge \omega^{j_p},$$

where

$$f^*_{j_1 \ldots j_p} = \sum_{i_1, \ldots, i_p} f_{i_1 \cdots i_p} \check{h}^{i_1}_{j_1} \cdots \check{h}^{i_p}_{j_p}.$$

If we write Θ^p in the normal form

$$\Theta^p = \sum_{i_1 < \cdots < i_p} f_{i_1 \cdots i_p} dx^{i_1} \wedge \cdots \wedge dx^{i_p} = \sum_{j_1 < \cdots < j_p} f^*_{j_1 \ldots j_p} \omega^{j_1} \wedge \cdots \omega^{j_p},$$

we have

$$f^*_{j_1 \cdots j_p} = \sum_{i_1 < \cdots < i_p} f_{i_1 \cdots i_p} \sum_{(\nu_1, \ldots, \nu_p)} \operatorname{sgn}(\nu_1, \ldots, \nu_p) \check{h}^{i_1}_{j_{\nu_1}} \cdots \check{h}^{i_p}_{j_{\nu_p}}.$$

For the algebra of differential forms with respect to this new basis, the formulas §6, (67) to (82) remain valid if everywhere the functions $f_{i_1 \cdots i_p}$ are replaced by $f^*_{j_1 \cdots j_p}$ and the differentials $dx^1, \ldots, dx^n$ by $\omega^1, \ldots, \omega^n$.

The rules for transformation of the adjoint form $* \, \Theta^p$ are somewhat more complicated. These rules become simple only when the basis differentials ω^j are *orthonormal* (i.e., orthogonal and normal) or, in other words, only when the coefficients h^j_i form an orthonormal matrix, i.e., when

$$\sum_{i=1}^{n} h^j_i h^k_i = \delta^{jk} = \begin{cases} 1 & \text{for } j = k \\ 0 & \text{for } j \neq k \end{cases} \tag{105}$$

(cf. also IB3, §3.2). Here the determinant $|h^j_i| = \pm 1$. From the equations (104) and (105) we have in this case

$$\check{h}^i_k = h^k_i, \qquad i, k = 1, 2, \ldots, n. \tag{106}$$

If for the basis differentials ω^j we calculate the expressions $* \, (\omega^j \wedge * \, \omega^k)$, we see that

$$* \, (\omega^j \wedge * \, \omega^k) = \sum_{i=1}^{n} h^j_i h^k_i.$$

Thus we have the following theorem.

The basis $\omega^1, \ldots, \omega^n$ is orthonormal if and only if it satisfies the relations

$$* (\omega^j \wedge * \omega^k) = \delta^{jk} = \begin{cases} 1 & \text{for } j = k \\ 0 & \text{for } j \neq k. \end{cases}$$

For an orthonormal basis we have

$$\begin{aligned} * \Theta^p &= * \sum_{j_1, \ldots, j_p} f^*_{j_1 \cdots j_p} \omega^{j_1} \wedge \cdots \wedge \omega^{j_p} \\ &= * \sum_{i_1, \ldots, i_p} \left(\sum_{j_1, \ldots, j_p} f^*_{j_1 \cdots j_p} h^{j_1}_{i_1} \cdots h^{j_p}_{i_p} \right) dx^{i_1} \wedge \cdots \wedge dx^{i_p} \\ &= \sum_{i_1, \ldots, i_p} \left(\sum_{j_1, \ldots, j_p} f^*_{j_1 \cdots j_p} h^{j_1}_{i_1} \cdots h^{j_p}_{i_p} \right) \operatorname{sgn}(i_1, \ldots, i_p, i_{p+1}, \ldots, i_n) dx^{i_{p+1}} \\ &\qquad \wedge \cdots \wedge dx^{i_n}. \end{aligned}$$

If we sum this equation over all possible combinations of $i_{p+1}, \ldots, i_n$, every term appears $(n-p)!$ times, so that we may also write

$$\begin{aligned} * \Theta^p &= \frac{1}{(n-p)!} \sum_{i_1, \ldots, i_n} \sum_{j_1, \ldots, j_p} f^*_{j_1 \cdots j_p} h^{j_1}_{i_1} \cdots h^{j_p}_{i_p} \operatorname{sgn}(i_1, \ldots, i_n) dx^{i_{p+1}} \wedge \\ &\qquad \cdots \wedge dx^{i_n} \\ &= \frac{1}{(n-p)!} \sum_{j_1, \ldots, j_p} f^*_{j_1 \cdots j_p} \sum_{j_{p+1}, \ldots, j_p} \sum_{i_1, \ldots, i_n} \operatorname{sgn}(i_1, \ldots, i_n) \\ &\qquad \times h^{j_1}_{i_1} \cdots h^{j_p}_{i_p} \check{h}^{i_{p+1}}_{j_{p+1}} \cdots \check{h}^{i_n}_{j_n} \omega^{j_{p+1}} \wedge \cdots \wedge \omega^{j_n}. \end{aligned}$$

From (106) and $|h^j_i| = \pm 1$ we have

$$\begin{aligned} \sum_{i_1, \ldots, i_n} \operatorname{sgn}(i_1, \ldots, i_n) h^{j_1}_{i_1} \cdots h^{j_p}_{i_p} \check{h}^{i_{p+1}}_{j_{p+1}} \cdots \check{h}^{i_n}_{j_n} &= \sum_{i_1, \ldots, i_n} \operatorname{sgn}(i_1, \ldots, i_n) h^{j_1}_{i_1} \cdots h^{j_n}_{i_n} \\ &= \operatorname{sgn}(j_1, \ldots, j_n) |h^j_i|. \end{aligned}$$

Thus we obtain:

$$* \Theta^p = |h^j_i| \frac{1}{(n-p)!} \sum_{j_1, \ldots, j_p} f^*_{j_1 \cdots j_p} \sum_{j_{p+1}, \ldots, j_n} \operatorname{sgn}(j_1, \ldots, j_n) \omega^{j_{p+1}} \wedge \cdots \wedge \omega^{j_n}.$$

The last sum again provides $(n-p)!$ equal summands, so that finally:

$$(107) \quad * \Theta^p = |h^j_i| \sum_{j_1, \ldots, j_p} f^*_{j_1 \cdots j_p} \operatorname{sgn}(j_1, \ldots, j_p, j_{p+1}, \ldots, j_n) \times \omega^{j_{p+1}} \wedge \cdots \wedge \omega^{j_n},$$

if $j_{p+1}, \ldots, j_n$ is a fixed permutation of those numbers in $1, 2, \ldots, n$ that do not occur in $j_1, \ldots, j_p$. Consequently, we have the following important result.

If $\omega^1, \ldots, \omega^n$ is an orthonormal basis of Pfaffian forms, then for a differential form

$$\Theta^p = \sum_{j_1 < \cdots < j_p} f^*_{j_1 \cdots j_p} \omega^{j_1} \wedge \cdots \wedge \omega^{j_p}$$

the adjoint differential form is

$$* \Theta^p = \pm \sum_{j_{p+1} < \cdots < j_n} f^*_{j_1 \cdots j_p} \operatorname{sgn}(j_1, \ldots, j_p, j_{p+1}, \ldots, j_n) \omega^{j_{p+1}} \wedge \cdots \wedge \omega^{j_n}.$$

Here the sign + or − is taken according to whether the orientation of the basis is the same as that of $dx^1, \ldots, dx^n$ or opposite to it.

In a study of total differentiation as related to basis transformations of the differentials $dx^1, \ldots, dx^n$ the formulas are naturally somewhat more complicated since then the Pfaffian forms ω^j must themselves be differentiated. If we assume that the coefficients h^j_i are continuously differentiable, we obtain

$$d\omega^j = \sum_{\alpha=1}^{n} \sum_{\beta=1}^{n} \frac{\partial h^j_\alpha}{\partial x^\beta} dx^\beta \wedge dx^\alpha = \sum_{i,k} \left(\sum_{\alpha,\beta} \frac{\partial h^j_\alpha}{\partial x^\beta} \check{h}^\beta_i \check{h}^\alpha_k \right) \omega^i \wedge \omega^k,$$

and thus

$$d\omega^j = \sum_{i,k} \Phi^j_{ik} \omega^i \wedge \omega^k \quad \text{with} \quad \Phi^j_{ik} = \sum_{\alpha,\beta} \frac{\partial h^j_\alpha}{\partial x^\beta} \check{h}^\beta_i \check{h}^\alpha_k, \quad i, j, k = 1, 2, \ldots, n.$$

Furthermore

$$df = \sum_{i=1}^{n} \frac{\partial f}{\partial x^i} dx^i = \sum_{j=1}^{n} \left(\sum_{i=1}^{n} \frac{\partial f}{\partial x^i} \check{h}^i_j \right) \omega^j.$$

In this way the total differentials of arbitrary forms

$$\Theta^p = \sum_{j_1, \ldots, j_p} f^*_{j_1 \cdots j_p} \omega^{j_1} \wedge \cdots \wedge \omega^{j_p}$$

can be calculated and expressed in terms of the basis differentials. For example, for a Pfaffian form

$$\omega = \sum_{i=1}^{n} f_i \omega^i$$

the total differential is

$$d\omega = \sum_{j,k} \left(\sum_{i=1}^{n} \frac{\partial f_k}{\partial x^i} \check{h}^i_j + \sum_{i=1}^{n} f_i \Phi^i_{jk} \right) \omega^j \wedge \omega^k.$$

We have now developed the subject up to the gateway of modern differential geometry, in particular of Riemannian geometry, a subject which cannot be discussed here for lack of space. Thus we shall content ourselves with applications of interest in vector analysis, for which we must first derive the centrally important Stokes theorem.

9. The Stokes Theorem

In R^n, the $(x^1, \ldots, x^n)$-space, let there be given a segment of an r-dimensional oriented manifold M^r with the boundary Bd M^r, as described in §3. Let this segment consist of finitely many simple segments M^r_κ, $\kappa = 1, 2, \ldots, k$ with the boundaries Bd M^r_κ:

$$M^r = \sum_{\kappa=1}^{k} M^r_\kappa \quad \text{and} \quad \text{Bd } M^r = \sum_{\kappa=1}^{k} \text{Bd } M^r_\kappa.$$

Let each segment M^r be the one-to-one continuously differentiable image of an oriented domain G^r in the $(t^1, \ldots, t^r)$-space:

$$G^r_\kappa \xrightarrow{\varphi} M^r_\kappa \colon x^i = \varphi^i(t^1, \ldots, t^r), \qquad i = 1, 2, \ldots, n.$$

Then the mapping φ maps the oriented boundary of G^r_κ onto the oriented boundary of M^r_κ:

$$\text{Bd } G^r_\kappa \xrightarrow{\varphi} \text{Bd } M^r_\kappa.$$

Now let there be given in the neighborhood of M^r a continuous differential form

$$\Theta^r_x = \sum_{i_1 < \cdots < i_r} f_{i_1 \ldots i_r} dx^{i_1} \wedge \cdots \wedge dx^{i_r}.$$

Then by §3, (26)

$$\int_{M^r} \Theta^r_x = \sum_{\kappa=1}^{k} \int_{M^r_\kappa} \Theta^r_x, \tag{108}$$

and by §3, (27)

$$\int_{M^r_\kappa} \Theta^r_x = \int_{G^r_\kappa} \sum_{i_1 < \cdots < i_r} f_{i_1 \ldots i_r} \frac{\partial(x^{i_1}, \ldots, x^{i_r})}{\partial(t^1, \ldots, t^r)} \, dt^1 \cdots dt^r.$$

On the right-hand side we have precisely the integral over the image Θ^r_t of the differential Θ^r_x under the mapping $\check{\varphi}$: $\Theta^r_x \xrightarrow{\check{\varphi}} \Theta^r_t$, so that

$$\int_{M^r_\kappa} \Theta^r_x = \int_{G^r_\kappa} \Theta^r_t, \tag{109}$$

in view of the fact that for the identical mapping $t^\rho = t^\rho$, $\rho = 1, 2, \ldots, r$, of the oriented domain G^r_κ onto itself we have by §3, (27)

$$\int_{G^r_\kappa} f dt^1 \wedge \cdots \wedge dt^r = \int_{G^r_\kappa} f dt^1 \cdots dt^r.$$

If Θ_x^{r-1} is a differential form defined in a neighborhood of the boundaries of the M_κ^r, then it is also defined in a neighborhood of the boundary of M^r and we have the relation:

$$\int_{\mathrm{Bd}M^r} \Theta_x^{r-1} = \sum_{\kappa=1}^{k} \int_{\mathrm{Bd}M_\kappa^r} \Theta_x^{r-1} = \sum_{\kappa=1}^{k} \int_{\mathrm{Bd}G_\kappa^r} \Theta_t^{r-1}, \tag{110}$$

where Θ_t^{r-1} is the image of the differential form Θ_x^{r-1}.

Now from the theorems of topology we can construct a simplicial decomposition of the domains G_κ^r, i.e., we can divide them into finitely many r-dimensional oriented simplexes S_λ^r in such a way that

$$G_\kappa^r = \sum_{\lambda=1}^{l} S_\lambda^r \qquad \text{and} \qquad \mathrm{Bd}\, G_\lambda^r = \sum_{\lambda=1}^{l} \mathrm{Bd}\, S_\lambda^r,$$

and the S_λ^r are defined in the $(\tau^1, \ldots, \tau^r)$-space by one-to-one continuously differentiable mappings of the oriented Euclidean simplexes E_λ^r:

$$E_\lambda^r \xrightarrow{\psi} S_\lambda^r \colon t^i = \psi^i(\tau^1, \ldots, \tau^r), \qquad i = 1, 2, \ldots, r.$$

Then the relations (109) hold for these mappings and we have

$$\int_{G_\kappa^r} \Theta_t^r = \sum_{\lambda=1}^{l} \int_{S_\lambda^r} \Theta_t^r = \sum_{\lambda=1}^{l} \int_{E_\lambda^r} \Theta_\tau^r,$$

where the differential forms Θ_τ^r are defined in each case by the mappings $\Theta_t^r \xrightarrow{\check{\psi}} \Theta_\tau^r$. If the differential form Θ_t^{r-1} is also defined in a neighborhood of the boundaries $\mathrm{Bd}\, S_\lambda^r$ of the simplexes S_λ^r, then we also have

$$\int_{\mathrm{Bd}\, G_\kappa^r} \Theta_t^{r-1} = \sum_{\lambda=1}^{l} \int_{\mathrm{Bd}\, S_\lambda^r} \Theta_t^{r-1} = \sum_{\lambda=1}^{l} \int_{\mathrm{Bd}\, E_\lambda^r} \Theta_\tau^{r-1}, \tag{112}$$

where the Θ_τ^{r-1} are defined in each case by $\Theta_t^{r-1} \xrightarrow{\check{\psi}} \Theta_\tau^{r-1}$. Now the following result holds:

Let the function f be continuously differentiable on the Euclidean, positively oriented simplex

$$E^r \colon 0 \leqq \tau^i \leqq 1, \qquad i = 1, 2, \ldots, r, \qquad \sum_{i=1}^{r} \tau^i \leqq 1 \tag{113}$$

of the $(\tau^1, \ldots, \tau^r)$*-space. Then for the differential form*

$$\Theta^{r-1} = f d\tau^2 \wedge \cdots \wedge d\tau^r$$

we have the integral theorem

$$\int_{E^r} d\Theta^{r-1} = \int_{\mathrm{Bd}\, E^r} \Theta^{r-1}. \tag{114}$$

Written out explicitly this theorem reads:

$$\int_{E^r} \frac{\partial f}{\partial \tau^1} d\tau^1 \wedge \cdots \wedge d\tau^r = \int_{\mathrm{Bd}\, E^r} f d\tau^2 \wedge \cdots \wedge d\tau^n. \tag{115}$$

In order to calculate the integral on the left-hand side we make use of the parameters $\tau^1, \ldots, \tau^r$, where $\tau^2, \ldots, \tau^r$ traverse the simplex

$$\bar{E}_1^{r-1} : 0 \leqq \tau^\rho \leqq 1, \qquad \tau^2 + \cdots + \tau^r \leqq 1,$$

and for fixed values of $\tau^2, \ldots, \tau^r$ the parameter τ^1 traverses the interval $0 \leqq \tau^1 \leqq 1 - \tau^2 - \cdots - \tau^r$. Then we have

$$\int_{E^r} \frac{\partial f}{\partial \tau^1} d\tau^1 \wedge \cdots \wedge d\tau^r = \int_{E^r} \frac{\partial f}{\partial \tau^1} d\tau^1 d\tau^2 \cdots d\tau^r$$

$$= \int_{\bar{E}_1^{r-1}} \left(\int_0^{1-\tau^2-\cdots-\tau^r} \frac{\partial f}{\partial \tau^1} d\tau^1 \right) d\tau^2 \cdots d\tau^r \tag{116}$$

$$= \int_{\bar{E}_1^{r-1}} [f(1 - \tau^2 - \cdots - \tau^r, \tau^2, \ldots, \tau^r) - f(0, \tau^2, \ldots, \tau^r)] d\tau^2 \cdots d\tau^r.$$

The boundary of E^r now consists of the r simplexes ($\nu = 1, 2, \ldots, r$)

$$E_\nu^{r-1}: 0 \leqq \tau^\rho \leqq 1, \quad \rho = 1, 2, \ldots, r, \quad \tau^1 + \cdots + \tau^r \leqq 1, \quad \tau^\nu = 0$$

and the simplex

$$E_0^{r-1}: 0 \leqq \tau^\rho \leqq 1, \qquad \rho = 1, 2, \ldots, r, \qquad \tau^1 + \cdots + \tau^r = 1,$$

where these simplexes are to be given the induced orientation. For a simplex E_ν^{r-1}, $\nu = 2, \ldots, r$, we have $\tau^\nu = 0$. As parameters here we may choose the variables $\tau^1, \ldots, \tau^{\nu-1}, \tau^{\nu+1}, \ldots, \tau^r$. Then the mapping functions are:

$$\tau^\rho = \varphi^\rho(\tau^1, \ldots, \tau^{\nu-1}, \tau^{\nu+1}, \ldots, \tau^r) \equiv \tau^\rho \qquad \text{for} \qquad \rho \neq \nu,$$

$$\tau^\nu = \varphi^\nu(\tau^1, \ldots, \tau^{\nu-1}, \tau^{\nu+1}, \ldots, \tau^r) \equiv 0,$$

so that in this case we have

$$\frac{\partial(\varphi^2, \ldots, \varphi^r)}{\partial(\tau^1, \ldots, \tau^{\nu-1}, \tau^{\nu+1}, \ldots, \tau^\nu)} = 0$$

for these surfaces, and consequently

(117)

$$\int_{E_\nu^{r-1}} f d\tau^2 \wedge \cdots \wedge d\tau^r = \pm \int_{\bar{E}_\nu^{r-1}} f \frac{\partial(\varphi^2, \ldots, \varphi^r)}{\partial(\tau^1, \ldots, \tau^{\nu-1}, \tau^{\nu+1}, \ldots, \tau^r)} d\tau^1 \cdots d\tau^{\nu-1} d\tau^{\nu+1} \cdots d\tau^r = 0,$$

where $\bar{E}_\nu^{r-1}$ is the simplex in the $(\tau^1, \ldots, \tau^{\nu-1}, \tau^{\nu+1}, \ldots, \tau^r)$-space with the orientation $(\tau^1, \ldots, \tau^{\nu-1}, \tau^{\nu+1}, \ldots, \tau^r)$.

For the surface E_1^{r-1} we choose the mapping functions

$$\tau^\rho = \varphi^\rho(\tau^2, \dots, \tau^r) \equiv \tau^\rho, \qquad \rho = 2, 3, \dots, r,$$

$$\tau^1 = \varphi^1(\tau^2, \dots, \tau^r) \equiv 0.$$

Since E^r is positively oriented, and is thus determined by the order of the vectors $(v_1, \dots, v_r)$ in the direction of the positive coordinate axes, and since these vectors have the components $v_i = (0, \dots, t_i, \dots, 0)$, $i = 1, 2, \dots, r$, $t_i > 0$, whereas the outward-directed vector $v = (t, 0, \dots, 0)$ with $t < 0$ leads, when combined with $v_2, \dots, v_r$, to a negative orientation for the r-hedral $(v, v_2, \dots, v_r)$, it follows that E_1^{r-1} is negatively oriented with respect to the order $\tau^2, \dots, \tau^r$ of the simplex $\bar{E}_1^{r-1}$ in the $(\tau^2, \dots, \tau^r)$-space. Thus

$$(118) \quad \int_{E_1^{r-1}} f d\tau^2 \wedge \cdots \wedge d\tau^r = -\int_{\bar{E}_1^{r-1}} f(0, \tau^2, \dots, \tau^r) d\tau^2 \cdots d\tau^r.$$

Finally, for E_0^{r-1} we also choose the parameters $\tau^2, \dots, \tau^r$. Then the simplex E_0^{r-1} has the parametric representation

$$\tau^1 = \varphi^1(\tau^2, \dots, \tau^r) \equiv 1 - \tau^2 - \cdots - \tau^r,$$

$$\tau^\rho = \varphi^\rho(\tau^2, \dots, \tau^r) \equiv \tau^\rho, \qquad \rho = 2, 3, \dots, r.$$

Here an outward-directed vector has the form $v = (t, 0, \dots, 0)$ with $t > 0$, which together with the vectors $v_2, \dots, v_r$ determines a positive orientation, so that E_0^{r-1} has the same orientation as $\bar{E}_1^{r-1}$. Thus

(119)

$$\int_{E_0^{r-1}} f d\tau^2 \wedge \cdots \wedge d\tau^r = \int_{\bar{E}_1^{r-1}} f(1 - \tau^2 - \cdots - \tau^r, \tau^2, \dots, \tau^r) d\tau^2 \cdots d\tau^r.$$

From (117), (118), and (119) it therefore follows that

$$\begin{aligned} \int_{\mathrm{Bd}\, E^r} f d\tau^2 \wedge \cdots \wedge d\tau^r &= \sum_{\nu=0}^{r} \int_{E_\nu^{r-1}} f d\tau^2 \wedge \cdots \wedge d\tau^r \\ &= \int_{E_0^{r-1}} f d\tau^2 \wedge \cdots \wedge d\tau^r + \int_{E_1^{r-1}} f d\tau^2 \wedge \cdots \wedge d\tau^r \\ &= \int_{\bar{E}_1^{r-1}} [f(1 - \tau^2 - \cdots - \tau^r, \tau^2, \dots, \tau^r) \\ &\qquad - f(0, \tau^2, \dots, \tau^r)] d\tau^2 \cdots d\tau^r, \end{aligned}$$

which together with (115) and (116) proves the assertion (114).

Now the general Stokes integral theorem can be reduced to this simple special case. For to begin with, the integral theorem (114) holds for every form

$$\Theta_{\tau}^{r-1} = f_i d\tau^1 \wedge \cdots \wedge \widehat{d\tau^i} \wedge \cdots \wedge d\tau^r,$$

since such a form can be reduced to the first case by a transformation

$$\begin{aligned} \tau^j &= \varphi^j(\tau_*^1, \ldots, \tau_*^r) \equiv \tau_*^{j+1}, \qquad j = 1, \ldots, i-1, \\ \tau^i &= \varphi^i(\tau_*^1, \ldots, \tau_*^r) \equiv \tau_*^1, \\ \tau^j &= \varphi^j(\tau_*^1, \ldots, \tau_*^r) \equiv \tau_*^j, \qquad j = i+1, \ldots, r, \end{aligned}$$

where Θ_{τ}^{r-1} is taken into $\Theta_{\tau_*}^{r-1}$, $d\Theta_{\tau}^{r-1}$ into $d\Theta_{\tau_*}^{r-1}$, E^r into $(-1)^{i-1}E_*^r$ and Bd E^r into $(-1)^{i-1}$ Bd E_*^r. Thus (114) holds for every form

$$\Theta^{r-1} = \sum_{i=1}^{r} f_i d\tau^1 \wedge \cdots \wedge \widehat{d\tau^i} \wedge \cdots \wedge d\tau^r.$$

Then the theorem holds for an arbitrary oriented simplex E_λ^r in the $(\tau^1, \ldots, \tau^r)$-space, since this simplex can be mapped by an affine transformation onto the special simplex (113). Furthermore, the theorem holds, by (111) and (112), for every simplex S_λ^r and thus for G_κ^r, and finally, by (108), (109), and (110) it holds for every segment of an oriented manifold M^r in R^n. Here we have made constant use of the fact that under a mapping φ of one space into another the derivatives of the differentials are transformed into the differentials of the derivatives in accordance with §8, (102) and (103). Thus we have finally proved the following important theorem:

Stokes integral theorem. In R^n let there be given an oriented segment of an r-dimensional manifold M^r, composed of finitely many simple segments. Let Bd M^r *denote the oriented boundary of M^r. In the neighborhood of M^r let there be given a continuously differentiable differential form Θ^{r-1}, and let $d\Theta^{r-1}$ be its total differential. Then we have*

$$\int_{M^r} d\Theta^{r-1} = \int_{\mathrm{Bd}\, M^r} \Theta^{r-1}. \tag{120}$$

In order to use the theorem in this formulation for curves $C = M^1$, we must define the integral of a function $\Theta^0 = f$ taken over an oriented point p as the value

$$\int_p f = \pm f(p)$$

with positive or negative sign according to whether p is positively or negatively oriented. The formula (120) then has the well-known appearance,

$$\int_C \sum_{i=1}^{n} \frac{\partial f}{\partial x^i}\, dx^i = f(b) - f(a),$$

where a is the initial point and b is the final point of the arc C from a to b.

C. Applications of the Calculus of Alternating Differential Forms

10. Differential Forms in the Euclidean Plane

In the *Euclidean* (x, y)*-plane* the calculus of alternating differential forms reduces to the study of the functions $f(x, y)$, differentials $\omega = f(x, y)dx + g(x, y)dy$ and differentials $\Theta^2 = f(x, y)dx \wedge dy$. The *Stokes integral theorem* applied to a curve C with initial point $a = (x_1, y_1)$ and final point $b = (x_2, y_2)$ and to a function f on C then becomes

$$\int_C df = \int_C \frac{\partial f}{\partial x}\,dx + \frac{\partial f}{\partial y}\,dy = f(b) - f(a).$$

If we apply this theorem to a differential form $\omega = fdx + gdy$ in a domain G of the plane, we have

$$\int_G \left(\frac{\partial g}{\partial x} - \frac{\partial f}{\partial y}\right)dxdy = \int_{\mathrm{Bd}\,G} fdx + gdy, \tag{121}$$

which is the well-known *Gauss integral theorem in the plane.* If $d\omega = 0$ in G, i.e., if $\dfrac{\partial g}{\partial x} = \dfrac{\partial f}{\partial y}$, then

$$\int_{\mathrm{Bd}\,G} fdx + gdy = 0.$$

The second Poincaré lemma states if $d\omega = 0$, then ω is locally the derivative of a function f:

$$\omega = df. \tag{122}$$

Two such functions differ by a function $g(x, y)$ with vanishing derivative $\dfrac{\partial g}{\partial x}\,dx + \dfrac{\partial g}{\partial y}\,dy$, so that the function is necessarily constant. Thus, in a domain G in which $d\omega = 0$, a function f defined in a subdomain in which $df = \omega$ can be extended continuously along a curve in only one way, if we are to have $df = \omega$ in the neighborhood of the curve. If the region is simply connected, the function f can then be extended uniquely into the whole region. But if the region G is not simply connected, this result does not necessarily follow. For then a differential ω with $d\omega = 0$ does not necessarily correspond to a function f for which the relation (122) holds everywhere in G. For example, let us consider the differential

$$\omega = -\frac{y}{x^2 + y^2}\,dx + \frac{x}{x^2 + y^2}\,dy$$

in the annulus $0 < x^2 + y^2 < 1$. If we introduce polar coordinates (r, φ) with $x = r\cos\varphi$, $y = r\sin\varphi$, we have

$$\begin{cases} dx = \cos\varphi dr - r\sin\varphi d\varphi, \\ dy = \sin\varphi dr + r\cos\varphi d\varphi, \end{cases} \tag{123}$$

and thus $\omega = d\varphi$, so that $d\omega = 0$.

Now for the corresponding function f we may take any $f = \varphi - \varphi_0$. But such a function changes under passage around the origin along a circle with radius r, $0 < r < 1$, precisely by the value $\pm 2\pi$, i.e., it cannot be extended in a one-valued way throughout the whole of G.

If we apply (121) to the differential $\omega = xdy$, we obtain, for a positively oriented domain G:

$$\int_{\mathrm{Bd}\, G} xdy = \int_G dx \wedge dy = \int_G dxdy = F,$$

where F is the area of G. Under passage to polar coordinates we have $dx \wedge dy = rdr \wedge d\varphi$, and thus for the differential form $\omega = \frac{1}{2}r^2 d\varphi$ we see that

$$\int_{\mathrm{Bd}\, G} \frac{r^2}{2}\, d\varphi = \int_G rdr \wedge d\varphi = F.$$

We now consider the adjoint forms

$$\begin{aligned} *f &= fdx \wedge dy, \\ *(fdx + gdy) &= -gdx + fdy, \\ *fdx \wedge dy &= f, \end{aligned}$$

and also the codifferential operator $\delta = -*d*$, from which we construct the Laplace operator Δ:

$$\Delta = d*d* + *d*d.$$

In particular, since $d*f = 0$ and $d(fdx \wedge dy) = 0$, we have:

$$\Delta f = *d*df \qquad \text{and} \qquad \Delta(fdx \wedge dy) = d*d*(fdx \wedge dy).$$

These relations can be used to calculate the operator Δ in other orthogonal coordinates. For example, by (123) the differentials $\omega^1 = dr$ and $\omega^2 = rd\varphi$ form an orthonormal basis with positive orientation. Thus from (107) we readily obtain the expression for the Laplace operator Δ in polar coordinates:

$$\begin{aligned} \Delta f &= *d*df = *d*\left(\frac{\partial f}{\partial r}\, dr + \frac{1}{r}\frac{\partial f}{\partial \varphi}\, rd\varphi\right) \\ &= *d\left(-\frac{1}{r}\frac{\partial f}{\partial \varphi}\, dr + \frac{\partial f}{\partial r}\, rd\varphi\right) = *\left(\frac{1}{r}\frac{\partial^2 f}{\partial \varphi^2} + \frac{\partial^2 f}{\partial r^2}\, r + \frac{\partial f}{\partial r}\right) dr \wedge d\varphi \\ &= *\left(\frac{\partial^2 f}{\partial r^2} + \frac{1}{r}\frac{\partial f}{\partial r} + \frac{1}{r^2}\frac{\partial^2 f}{\partial \varphi^2}\right) dr \wedge rd\varphi = \frac{\partial^2 f}{\partial r^2} + \frac{1}{r}\frac{\partial f}{\partial r} + \frac{1}{r^2}\frac{\partial^2 f}{\partial \varphi^2}. \end{aligned}$$

11. Differential Forms in Euclidean Three-Dimensional Space. Vector Analysis

From the theory of alternating differential forms in the Euclidean (x^1, x^2, x^2)-space we can obtain many of the theorems of vector calculus.

Let us first consider some of the theorems of vector algebra. At a point of the space let the vector $\mathfrak{v} = (v_1, v_2, v_3)$ of a given vector field be put in correspondence with the Pfaffian form $v_1 dx^1 + v_2 dx^2 + v_3 dx^3$ at the same point, namely

$$\mathfrak{v} = (v_1, v_2, v_3) \leftrightarrow \omega = v_1 dx^1 + v_2 dx^2 + v_3 dx^3.$$

This correspondence is an isomorphic mapping of the linear vector space of vectors $\mathfrak{v}$ at the point onto the linear vector space of Pfaffian forms ω at this point, so that we can now proceed to translate the usual notation of vector algebra into the language of differential forms.

Let

$$\mathfrak{v}_i = (v_{i1}, v_{i2}, v_{i3}) \leftrightarrow \omega_i = v_{i1} dx^1 + v_{i2} dx^2 + v_{i3} dx^3, \qquad i = 1, 2, 3, \ldots,$$

from which by §6, (85) we have the following relation for the *scalar product* $\mathfrak{v}_1 \cdot \mathfrak{v}_2 = v_{11}v_{21} + v_{12}v_{22} + v_{13}v_{23}$:

$$\mathfrak{v}_1 \cdot \mathfrak{v}_2 = *(\omega_1 \wedge *\omega_2) = *(\omega_2 \wedge *\omega_1). \tag{124}$$

Then for the *vector product*

$$\mathfrak{v}_1 \times \mathfrak{v}_2 = ((v_{12}v_{23} - v_{13}v_{22}), (v_{13}v_{21} - v_{11}v_{23}), (v_{11}v_{22} - v_{12}v_{21}))$$

we have the correspondence

$$\mathfrak{v}_1 \times \mathfrak{v}_2 \leftrightarrow *(\omega_1 \wedge \omega_2). \tag{125}$$

Since $** \Theta^p = \Theta^p$ in R^3 it follows from (124), (125), and §6, (79) that

$$\mathfrak{v}_1(\mathfrak{v}_2 \times \mathfrak{v}_3) = *(\omega_1 \wedge **(\omega_2 \wedge \omega_3)) = *(\omega_1 \wedge \omega_2 \wedge \omega_3)$$
$$= \begin{vmatrix} v_{11} & v_{12} & v_{13} \\ v_{21} & v_{22} & v_{23} \\ v_{31} & v_{32} & v_{33} \end{vmatrix}.$$

Furthermore, by (125)

$$(\mathfrak{v}_1 \times \mathfrak{v}_2) \times \mathfrak{v}_3 \leftrightarrow *(*(\omega_1 \wedge \omega_2) \wedge \omega_3).$$

But by §6, (86) this last expression is equal to

$$\omega_2 *(\omega_1 \wedge *\omega_3) - \omega_1 *(\omega_2 \wedge *\omega_3).$$

Thus by (124)

$$(\mathfrak{v}_1 \times \mathfrak{v}_2) \times \mathfrak{v}_3 = (\mathfrak{v}_1 \cdot \mathfrak{v}_3)\mathfrak{v}_2 - (\mathfrak{v}_2 \cdot \mathfrak{v}_3)\mathfrak{v}_1.$$

Let us now turn to the rules for *vector analysis*. A *curvilinear integral* over a vector field $\mathfrak{v}$ on a curve which is the image of a parametric interval I is written in the form

$$\int_C \mathfrak{v} \cdot d\mathfrak{s} = \int_I \left(v_1 \frac{\partial \varphi^1}{\partial t} + v_2 \frac{\partial \varphi^2}{\partial t} + v_3 \frac{\partial \varphi^3}{\partial t} \right) dt.$$

Here the notation $\mathfrak{v}\cdot d\mathfrak{s}$ indicates the scalar product of the vector $\mathfrak{v}$ with the tangential line-element $d\mathfrak{s} = \left(\frac{\partial\varphi^1}{\partial t}, \frac{\partial\varphi^2}{\partial t}, \frac{\partial\varphi^3}{\partial t}\right)dt$. But then we have precisely the definition given above for the curvilinear integral over the differential form ω. Thus we may identify the scalar product $\mathfrak{v}\cdot d\mathfrak{s}$ with the form ω, provided that the correspondence $\mathfrak{v} \leftrightarrow \omega$ holds:

$$\mathfrak{v}\cdot d\mathfrak{s} = \omega.$$

Now in vector analysis a *surface integral* over an oriented segment F lying in a vector field $\mathfrak{v}$ is written, in accordance with §2, (21) to (23), in the form

$$\int_F \mathfrak{v}\cdot d\mathfrak{f} = \int_G \left(v_1 \frac{\partial(\varphi^2, \varphi^3)}{\partial(t^1, t^2)} + v_2 \frac{\partial(\varphi^3, \varphi^1)}{\partial(t^1, t^2)} + v_3 \frac{\partial(\varphi^1, \varphi^2)}{\partial(t^1, t^2)}\right) dt^1 dt^2.$$

But this is precisely the definition given above in §2, (18) and (19) for the surface-integral over the differential form $*\,\omega$. Consequently, we identify the scalar product $\mathfrak{v}\cdot d\mathfrak{f}$ with $*\,\omega$:

$$\mathfrak{v}\cdot d\mathfrak{f} = *\,\omega.$$

Finally, the *volume-integral*, taken over a positively oriented domain G in R^3, of a function f in G is defined by

$$\int_G f dV = \int_G f dx^1 dx^2 dx^3.$$

On the right-hand side we have an integral over the differential form $*\,f$, and so we may identify:

$$f dV = *\,f.$$

We must now go on to translate the differential operators of vector analysis into the language of differential forms. In the first place, in R^3:

$$** \,\Theta^p = \Theta^p, \qquad p = 0, 1, 2, 3,$$
$$\delta\,\Theta^p = (-1)^p * d * \Theta^p, \qquad p = 0, 1, 2, 3,$$

and in particular

$$\delta f = 0.$$

Furthermore,

$$\Delta\,\Theta^p = (-d\delta - \delta d)\,\Theta^p = (-1)^p(* d * d - d * d *)\,\Theta^p,$$

and in particular

$$\Delta f = * d * df,$$
$$\Delta\omega = (d * d * - * d * d)\omega,$$
$$\Delta\,\Theta^2 = (* d * d - d * d *)\,\Theta^2,$$
$$\Delta\,\Theta^3 = d * d * \Theta^3.$$

We now have the following correspondence:

$$\mathfrak{v} = (v_1, v_2, v_3) \leftrightarrow \omega = v_1 dx^1 + v_2 dx^2 + v_3 dx^3,$$

$$\text{grad } u = \left(\frac{\partial u}{\partial x^1}, \frac{\partial u}{\partial x^2}, \frac{\partial u}{\partial x^3}\right) \leftrightarrow du,$$

$$\text{div } \mathfrak{v} = \text{div } (v_1, v_2, v_3) = \frac{\partial v_1}{\partial x^1} + \frac{\partial v_2}{\partial x^2} + \frac{\partial v_3}{\partial x^3} \leftrightarrow -\delta\omega = * d * \omega,$$

$$\text{rot } \mathfrak{v} = \text{rot } (v_1, v_2, v_3) = \left(\frac{\partial v_3}{\partial x^2} - \frac{\partial v_2}{\partial x^3}, \frac{\partial v_1}{\partial x^3} - \frac{\partial v_3}{\partial x^1}, \frac{\partial v_2}{\partial x^1} - \frac{\partial v_1}{\partial x^2}\right) \leftrightarrow * d\omega.$$

Finally, the *nabla operator* $\nabla = \left(\frac{\partial}{\partial x^1}, \frac{\partial}{\partial x^2}, \frac{\partial}{\partial x^3}\right)$ is assigned to the "vector operator"

$$d = dx^1 \frac{\partial}{\partial x^1} + dx^2 \frac{\partial}{\partial x^2} + dx^3 \frac{\partial}{\partial x^3}$$

for which the operation $d \wedge \Theta^p$ is defined by $d\Theta^p$:

$$\nabla \leftrightarrow d.$$

For the ∇-operator it is customary to define the following operations, for which we can now also give their translation into the language of differential forms:

$$\begin{aligned} \nabla u &= \text{grad } u \leftrightarrow du, \\ \nabla \cdot \mathfrak{v} &= \text{div } \mathfrak{v} \leftrightarrow * d * \omega, \\ \nabla \times \mathfrak{v} &= \text{rot } \mathfrak{v} \leftrightarrow * d\omega, \\ \nabla \cdot \nabla &= \Delta \leftrightarrow \Delta. \end{aligned}$$

By means of these correspondences a great part of vector analysis is subsumed under the calculus of alternating differential forms. We now give a number of such formulas and theorems in each of the two languages, without any attempt at completeness:

$$\begin{aligned} \text{grad } (uv) &= (\text{grad } u)v + u \text{ grad } v \leftrightarrow d(uv) = (du)v + udv, \\ \text{div } (u\mathfrak{v}) &= (\text{grad } u) \cdot \mathfrak{v} + u \text{ div } \mathfrak{v} \leftrightarrow * d * (u\omega) \\ &= * d(u * \omega) = * (du \wedge * \omega) + u * d * \omega, \\ \text{rot } (u\mathfrak{v}) &= (\text{grad } u) \times \mathfrak{v} + u \text{ rot } \mathfrak{v} \leftrightarrow * d(u\omega) \\ &= * du \wedge \omega + u * d\omega, \\ \text{div } (\mathfrak{v}_1 \times \mathfrak{v}_2) &= \mathfrak{v}_2 \cdot \text{rot } \mathfrak{v}_1 - \mathfrak{v}_1 \cdot \text{rot } \mathfrak{v}_2 \leftrightarrow * d * (* (\omega_1 \wedge \omega_2)) \\ &= * d(\omega_1 \wedge \omega_2) = * (d\omega_1 \wedge \omega_2) - * (\omega_1 \wedge d\omega_2) \\ &= * (\omega_2 * (* d\omega_1)) - * (\omega_1 \wedge * (* d\omega_2)), \\ \text{rot grad } u &= 0 \leftrightarrow * ddu = 0, \\ \text{div rot } \mathfrak{v} &= 0 \leftrightarrow * d ** d\omega = * dd\omega = 0, \\ \text{rot rot } \mathfrak{v} &= \text{grad div } \mathfrak{v} - \Delta\mathfrak{v} \leftrightarrow * d * d\omega = d * d * \omega - \Delta\omega, \\ \text{div grad } u &= \Delta u \leftrightarrow * d * du = \Delta u. \end{aligned}$$

In domains that are topologically equivalent to a sphere we have the following rules:

From rot $\mathfrak{v} = 0$ follows $\mathfrak{v} = \text{grad } u$
$\leftrightarrow$ from $* d\omega = 0$ follows $\omega = du$.
From div $\mathfrak{v}_1 = 0$ follows $\mathfrak{v}_1 = \text{rot } \mathfrak{v}_2$
$\leftrightarrow$ from $* d * \omega = 0$ follows $* \omega_1 = d\omega_2$, so that $\omega_1 = * d\omega_2$.
From rot $\mathfrak{v} = 0$ and div $\mathfrak{v} = 0$ follows $\mathfrak{v} = \text{grad } u$ with $\Delta u = 0$.
$\leftrightarrow$ From $* d\omega = 0$ and $* d * \omega = 0$ follows $\omega = du$ with $* d * du = \Delta u = 0$.

Then we have the following integral theorems, each of which is a special case of the (general) Stokes integral theorem:

Stokes integral theorem:

$$\int_F \text{rot } \mathfrak{v} \cdot d\mathfrak{f} = \int_{\text{Bd } F} \mathfrak{v} \cdot d\mathfrak{s} \leftrightarrow \int_F * (* \, d\omega) = \int_F d\omega = \int_{\text{Bd } F} \omega.$$

$$\int_{\text{Bd } G} \text{rot } \mathfrak{v} \cdot d\mathfrak{f} = 0 \leftrightarrow \int_{\text{Bd } G} d\omega = \int_G dd\omega = 0.$$

Gauss integral theorem:

$$\int_G \text{div } \mathfrak{v} dV = \int_{\text{Bd } G} \mathfrak{v} \cdot d\mathfrak{f} \leftrightarrow \int_G * (* \, d * \omega) = \int_G d * \omega = \int_{\text{Bd } G} * \, \omega.$$

Also, the integral theorems

$$\int_F d\mathfrak{f} \times \text{grad } u = \int_{\text{Bd } F} u \cdot d\mathfrak{s}$$

and

$$\int_G \text{rot } \mathfrak{v} dV = \int_{\text{Bd } F} d\mathfrak{f} \times \mathfrak{v},$$

where

$$d\mathfrak{f} \times \mathfrak{v} = \left(v_3 \frac{\partial(\varphi^3, \varphi^1)}{\partial(t^1, t^2)} - v_2 \frac{\partial(\varphi^1, \varphi^2)}{\partial(t^1, t^2)}, v_1 \frac{\partial(\varphi^1, \varphi^2)}{\partial(t^1, t^2)} - v_3 \frac{\partial(\varphi^2, \varphi^3)}{\partial(t^1, t^2)}, v_2 \frac{\partial(\varphi^2, \varphi^3)}{\partial(t^1, t^2)} - v_1 \frac{\partial(\varphi^3, \varphi^1)}{\partial(t^1, t^2)} \right) dt^1 dt^2$$

$$= (v_3 dx^3 \wedge dx^1 - v_2 dx^1 \wedge dx^2, v_1 dx^1 \wedge dx^2 - v_3 dx^2 \wedge dx^3, v_2 dx^2 \wedge dx^3 - v_1 dx^3 \wedge dx^1)$$

are obtained by application of the (general) *Stokes theorem* to the individual components of the integrals.

Green formulas:

$$
\begin{aligned}
\int_G (u\Delta v + \operatorname{grad} u \cdot \operatorname{grad} v)dV &= \int_{\mathrm{Bd}\,G} u \operatorname{grad} v \cdot d\mathfrak{f} \leftrightarrow \int_{\mathrm{Bd}\,G} * udv \\
&= \int_G d(* udv) = \int_G *[* d * (udv)] \\
&= \int_G *[* d(u * dv)] \\
&= \int_G * [u * d * dv + * (du \wedge * dv)],
\end{aligned} \tag{126}
$$

$$
\begin{aligned}
\int_G (\Delta(uv) - v\Delta u)dV &= \int_{\mathrm{Bd}\,G} u \operatorname{grad} v \cdot d\mathfrak{f} \leftrightarrow \int_{\mathrm{Bd}\,G} * u\, dv \\
&= \int_G d * (udv) = \int_G * [* d * (udv)] \\
&= \int_G * [* d * d(uv) - (* d * du)v].
\end{aligned} \tag{127}
$$

If we interchange u and v in (126) or (127) and subtract, we obtain

$$\int_G (u\Delta v - v\Delta u)dV = \int_{\mathrm{Bd}\,G} (u \operatorname{grad} v - v \operatorname{grad} u)d\mathfrak{f}.$$

The calculus of differentials enables us, without much difficulty, to transform all these expressions to an arbitrary orthogonal coordinate system. For polar coordinates we have

$$
\begin{aligned}
x^1 &= r \sin \vartheta \cos \varphi, \\
x^2 &= r \sin \vartheta \sin \varphi, \\
x^3 &= r \cos \vartheta:
\end{aligned}
$$

$$
\begin{aligned}
dx^1 &= \sin \vartheta \cos \varphi dr + r \cos \vartheta \cos \varphi d\vartheta - r \sin \vartheta \sin \varphi d\varphi, \\
dx^2 &= \sin \vartheta \sin \varphi dr + r \cos \vartheta \sin \varphi d\vartheta + r \sin \vartheta \cos \varphi d\varphi, \\
dx^3 &= \cos \vartheta dr - r \sin \vartheta d\vartheta,
\end{aligned}
$$

or in matrix notation:

$$
\begin{pmatrix} dx^1 \\ dx^2 \\ dx^3 \end{pmatrix} = \begin{pmatrix} \sin \vartheta \cos \varphi, & \cos \vartheta \cos \varphi, & -\sin \varphi \\ \sin \vartheta \sin \varphi, & \cos \vartheta \sin \varphi, & \cos \varphi \\ \cos \vartheta, & -\sin \vartheta, & 0 \end{pmatrix} \begin{pmatrix} dr \\ rd\vartheta \\ r \sin \vartheta d\varphi \end{pmatrix}.
$$

Consequently

$$\omega^1 = dr, \qquad \omega^2 = rd\vartheta, \qquad \omega^3 = r \sin \vartheta d\varphi$$

is an orthonormal basis with positive orientation. Now if (v_1, v_2, v_3) are the components of a vector $\mathfrak{v}$ with respect to the basis ω^1, ω^2, ω^3, the corresponding form is

$$\mathfrak{v} \leftrightarrow \omega = v_1\omega^1 + v_2\omega^2 + v_3\omega^3 = v_1 dr + rv_2 d\vartheta + r \sin\vartheta v_3 d\varphi$$

and thus for the function u

$$\text{(128)} \qquad \begin{aligned} \operatorname{grad} u \leftrightarrow du &= \frac{\partial u}{\partial r} dr + \frac{\partial u}{\partial \vartheta} d\vartheta + \frac{\partial u}{\partial \varphi} d\varphi \\ &= \frac{\partial u}{\partial r}\omega^1 + \frac{1}{r}\frac{\partial u}{\partial \vartheta}\omega^2 + \frac{1}{r\sin\vartheta}\frac{\partial u}{\partial \varphi}\omega^3, \end{aligned}$$

so that grad u has the components

$$\operatorname{grad} u = \left(\frac{\partial u}{\partial r}, \frac{1}{r}\frac{\partial u}{\partial \vartheta}, \frac{1}{r\sin\vartheta}\frac{\partial u}{\partial \varphi}\right).$$

Furthermore

$$\begin{aligned} d\omega = &\left(\frac{\partial(r\sin\vartheta v_3)}{\partial \vartheta} - \frac{\partial(rv_2)}{\partial \varphi}\right) d\vartheta \wedge d\varphi \\ &+ \left(\frac{\partial v_1}{\partial \varphi} - \frac{\partial(r\sin\vartheta v_3)}{\partial r}\right) d\varphi \wedge dr + \left(\frac{\partial(rv_2)}{\partial r} - \frac{\partial v_1}{\partial \vartheta}\right) dr \wedge d\vartheta \\ = &\frac{1}{r\sin\vartheta}\left(\frac{\partial(\sin\vartheta v_3)}{\partial \vartheta} - \frac{\partial v_2}{\partial \varphi}\right)\omega^2 \wedge \omega^3 + \left(\frac{1}{r\sin\vartheta}\frac{\partial v_1}{\partial \varphi} - \frac{1}{r}\frac{\partial(rv_3)}{\partial r}\right)\omega^3 \wedge \omega^1 \\ &+ \frac{1}{r}\left(\frac{\partial(rv_2)}{\partial r} - \frac{\partial v_1}{\partial \vartheta}\right)\omega^1 \wedge \omega^2, \end{aligned}$$

and thus

$$\begin{aligned} \operatorname{rot} \mathfrak{v} = \Bigg[&\frac{1}{r\sin\vartheta}\left(\frac{\partial(\sin\vartheta v_3)}{\partial \vartheta} - \frac{\partial v_2}{\partial \varphi}\right), \left(\frac{1}{r\sin\vartheta}\frac{\partial v_1}{\partial \varphi} - \frac{1}{r}\frac{\partial(rv_3)}{\partial r}\right), \\ &\frac{1}{r}\left(\frac{\partial(rv_2)}{\partial r} - \frac{\partial v_1}{\partial \vartheta}\right)\Bigg]. \end{aligned}$$

Also

$$\begin{aligned} *\,\omega &= v_1\omega^2 \wedge \omega^3 + v_2\omega^3 \wedge \omega^1 + v_3\omega^1 \wedge \omega^2 \\ &= r^2 \sin\vartheta v_1 d\vartheta \wedge d\varphi + r\sin\vartheta v_2 d\varphi \wedge dr + rv_3 dr \wedge d\vartheta, \end{aligned}$$

and thus

$$\begin{aligned} d *\omega &= \left(\sin\vartheta \frac{\partial(r^2 v_1)}{\partial r} + r\frac{\partial(\sin\vartheta v_2)}{\partial \vartheta} + r\frac{\partial v_3}{\partial \varphi}\right) dr \wedge d\vartheta \wedge d\varphi \\ &= \left(\frac{1}{r^2}\frac{\partial(r^2 v_1)}{\partial r} + \frac{1}{r\sin\vartheta}\frac{\partial(\sin\vartheta v_2)}{\partial \vartheta} + \frac{1}{r\sin\vartheta}\frac{\partial v_3}{\partial \varphi}\right)\omega^1 \wedge \omega^2 \wedge \omega^3 \end{aligned}$$

and consequently

$$\operatorname{div} \mathfrak{v} = \frac{1}{r^2}\frac{\partial(r^2 v_1)}{\partial r} + \frac{1}{r \sin\vartheta}\frac{\partial(\sin\vartheta\, v_2)}{\partial\vartheta} + \frac{1}{r\sin\vartheta}\frac{\partial v_3}{\partial\varphi}.$$

From (128) it also follows that

$$* \, du = r^2 \sin\vartheta \frac{\partial u}{\partial r}\, d\vartheta \wedge d\varphi + \sin\vartheta \frac{\partial u}{\partial\vartheta}\, d\varphi \wedge dr + \frac{1}{\sin\vartheta}\frac{\partial u}{\partial\varphi}\, dr \wedge d\vartheta,$$

and thus

$$d * du = \left[\sin\vartheta \frac{\partial}{\partial r}\left(r^2 \frac{\partial u}{\partial r}\right) + \frac{\partial}{\partial\vartheta}\left(\sin\vartheta \frac{\partial u}{\partial\vartheta}\right) + \frac{1}{\sin\vartheta}\frac{\partial^2 u}{\partial\varphi^2}\right] dr \wedge d\vartheta \wedge d\varphi,$$

so that

$$\Delta u = \frac{1}{r^2}\frac{\partial}{\partial r}\left(r^2\frac{\partial u}{\partial r}\right) + \frac{1}{r^2\sin\vartheta}\frac{\partial}{\partial\vartheta}\left(\sin\vartheta\frac{\partial u}{\partial\vartheta}\right) + \frac{1}{r^2\sin^2\vartheta}\frac{\partial^2 u}{\partial\varphi^2}.$$

With these results we must bring to an end our account of the applications of the calculus of differential forms to vector analysis.

12. Differential Forms on Differentiable and Riemannian Manifolds

In the preceding discussion we have examined differential forms on r-dimensional manifolds M^r embedded in Euclidean space. Consequently these differential forms underwent very simple transformations when the Euclidean space was subjected to a transformation of coordinates. Moreover, the differential forms were easily calculated when one space was mapped into another, and in particular we could consider them in the parameter spaces. Moreover, the rules for transformation of these forms were so simple that we could calculate straightforwardly how they would change from one space to another. But these facts indicate that the differential forms can also be defined on arbitrary differentiable manifolds, which do not need to be embedded in a Euclidean space.

By an *r-dimensional ρ-times continuously differentiable manifold* M^r we mean a topological space which locally has the structure of an r-dimensional Euclidean space with certain properties; i.e., for each point of the manifold there exists a neighborhood U that is mapped topologically, i.e., one-to-one and bicontinuously, onto a domain V (we call V a *coordinate neighborhood*) of an r-dimensional Euclidean space: $U \xrightarrow{\varphi} V$. If a point in M^r belongs to two neighborhoods U_1 and U_2, the first of which is mapped by a mapping φ_1 onto a domain V_1 of the $(t^1, \ldots, t^r)$-space and the other by a mapping φ_2 onto a domain V_2 of the $(\tau^1, \ldots, \tau^r)$-space, then in the intersection $U_1 \cap U_2$, there exists a neighborhood U_3 of the point which is mapped by φ_1 onto a neighborhood $V_3 \subset V_1$ and by φ_2 onto a neighborhood $V_3^* \subset V_2$. Consequently, under the mapping $V_3^* \xrightarrow{\varphi_1 \check{\varphi}_2} V_3$, these neighborhoods V_3^* and V_3 are mapped onto each other in a one-to-one and bicontinuous way by the functions

$$t^i = \psi^i(\tau^1, \ldots, \tau^r), \qquad i = 1, 2, \ldots, r. \tag{129}$$

Here $\check{\varphi}_2$ denotes the inverse mapping of φ_2 and $\varphi_1\check{\varphi}_2$ denotes the $\check{\varphi}_2$ followed by φ_1. Then M^r is said to be a ρ-times continuously differentiable manifold if its system of neighborhoods $\{U\}$, together with the corresponding mappings $U \xrightarrow{\varphi} V$ onto the neighborhoods $\{V\}$, has the property that the images of the intersection of two neighborhoods U_1 and U_2 are mapped onto each other, in the mapping $\varphi_1\check{\varphi}_2$, by means of ρ-times continuously differentiable functions (129).

Now in each coordinate neighborhood V_ι let there be given an alternating differential form

$$\Theta_\iota^s = \sum_{i_1, \ldots, i_s} a_{i_1 \cdots i_s} dt^{i_1} \wedge \cdots \wedge dt^{i_s}.$$

If in the images of an intersection of two neighborhoods the differential forms Θ_i^s are related to each other according to the rules for the transformation of coordinates, we identify them there and say that the totality of all the differential forms Θ_ι^s thus defined in the V_ι constitute *an alternating differential form* Θ^s *on* M^r.

The essential feature in this concept is as follows: Only local coordinates have been assigned on M^r, so that a differential form can be defined in the first place only in terms of the local coordinates; nevertheless, the same expression is always obtained for it, whether we consider its original definition at a given point or calculate it from some other form defined primarily in another neighborhood. Since all the above results are essentially of a local character, we can now study the differential forms in each case in a coordinate neighborhood. Here the majority of the rules and theorems stated above will remain valid, and under passage to another coordinate neighborhood all the foregoing statements will be transformed into corresponding statements in the other system.

A manifold M^r is said to be *orientable* if the system of neighborhoods can be so chosen that in the passage (129) from one coordinate system to another the functional determinant is always positive:

$$\frac{\partial(\psi^1, \ldots, \psi^r)}{\partial(t^1, \ldots, t^r)} > 0.$$

An orientable manifold always has two orientations, which can be transformed into each other by simultaneously reorienting the coordinate neighborhoods V.

In order to construct the adjoint form $*\,\Theta^s$ of a given form Θ^s it is necessary to have a *Riemannian metric* on M^r, whereby M^r is made into a *Riemannian manifold.* Such a metric exists if in every neighborhood V_ι there is given a positive definite symmetric form for the line-element ds:

$$ds^2 = \sum_{i,j} g_{ij} dt^i dt^j,$$

where the g_{ij} transform, under passage to another system of coordinates in the neighborhood of a point, like a covariant tensor of second rank; i.e., under such a transformation of coordinates:

$$ds^2 = \sum_{i,j} g_{ij} dt^i dt^j = \sum_{k,l} \sum_{i,j} g_{ij} \frac{\partial \psi^i}{\partial \tau^k} \frac{\partial \psi^j}{\partial \tau^l} d\tau^k d\tau^l = \sum_{k,l} g^*_{kl} d\tau^k d\tau^l.$$

With respect to this metric the orthonormality of two differential forms can be defined as follows.

A system of *Pfaffian forms*

$$\omega^k = \sum_{i=1}^{r} b_i^k dt^i, \qquad k = 1, 2, \ldots, r,$$

is said to be an orthonormal basis if

$$\sum_{i,j} g^{ij} b_i^k b_j^l = \delta^{kl} = \begin{cases} 1 & \text{for } k = 1, \\ 0 & \text{for } k \neq 1, \end{cases}$$

where the tensors g^{ij} are defined by

$$\sum_{j=1}^{m} g_{ij} g^{jm} = \delta_i^m = \begin{cases} 1 & \text{for } i = m, \\ 0 & \text{for } i \neq m. \end{cases}$$

If a differential form Θ^s is expressed in terms of the basis $\omega^1, \ldots, \omega^r$, the adjoint form $* \Theta^s$ is defined by the relation §8, (107). Then the theory of adjoint forms can be developed on a *Riemann manifold* in the same way as above for the special case R^n.

Bibliography

[1] Behnke, H.: Infinitesimalrechnung II. Aschendorff, Münster 1961.

[2] Blaschke, W. and Reichardt, H.: Einführung in die Differentialgeometrie. Springer, Berlin-Göttingen-Heidelberg 1960.

[3] Kähler, E.: Einführung in die Theorie der Systeme von Differentialgleichungen. Teubner, Leipzig-Berlin 1934.

[4] Lagally, M. and Franz, W.: Vorlesungen über Vektorrechnung. Akademische Verlagsgesellschaft, Leipzig 1959.

[5] Lichnerowicz, A.: Lineare Algebra und lineare Analysis. VEB Deutscher Verlag der Wissenschaften, Berlin 1956.

[6] MAAK, W.: Differential- und Integralrechnung. Vandenhoeck & Ruprecht, Göttingen 1960.
[7] REICHARDT, H.: Vorlesungen über Vektor- und Tensorrechnung. VEB Deutscher Verlag der Wissenschaften, Berlin 1957.
[8] TIETZ, H.: Geometrie. Handbuch der Physik. Bd. II. Springer, Berlin-Göttingen-Heidelberg 1955.

Bibliography added in Translation

LICHNEROWICZ, A.: Elements of tensor calculus (translated from the French by J. W. Leech and D. J. Newman). Methuen and Co. Ltd.; London; John Wiley and Sons, Inc., New York 1962.
SPRINGER, C. E.: Tensor and vector analysis. With applications to differential geometry. The Ronald Press Co., New York 1962.

CHAPTER 5

Complex Numbers
The Foundations of Analysis in the Complex Plane

1. The Complex Numbers

1.1. *Definition and Basic Concepts*

We start from the *field K of the real numbers*, whose basic properties are here assumed (IB1, §4). Let us briefly recall (IB8, §1) that the *complex numbers* can be introduced (1) as ordered pairs of real numbers, (2) as classes of polynomials mod $(x^2 + 1)$ in $K[x]$ and finally (3) as two-rowed matrices over K of the special form $\begin{pmatrix} a & -b \\ b & a \end{pmatrix}$ occurring in the rotation-dilatations of the plane.

1. For an ordered *pair of numbers* (a, b) *over* K we define equality "componentwise":

$$(a_1, a_2) = (b_1, b_2) \text{ if and only if } a_1 = b_1 \quad \text{and} \quad a_2 = b_2;$$

addition is also defined "componentwise":

$$(a_1, a_2) + (b_1, b_2) = (a_1 + b_1, a_2 + b_2);$$

and multiplication is defined by the special law:

$$(a_1, a_2)\cdot(b_1, b_2) = (a_1b_1 - a_2b_2, a_1b_2 + a_2b_1),$$

from which it follows that $(0, 1)\cdot(0, 1) = (-1, 0)$. The complex numbers (a, b) form a field $K(i)$ with zero element $(0, 0)$ and unit element $(1, 0)$. The correspondence $(a, 0) \rightarrow a$ is a one-to-one mapping of the subset of all pairs of the particular form $(a, 0)$ onto the set K of real numbers, with preservation of the four fundamental rules of arithmetic, so that we may identify corresponding elements: $(a, 0) = a$. If we also set $(0, 1) = i$, we have $(a, b) = a + bi$. Since $i^2 + 1 = 0$, the set of complex numbers is an extension of the field K of degree 2 (IB7, §2). This set is called $K(i)$.

2. Consider the *set of polynomials* $p(x)$ (of arbitrary degree) in one indeterminate x with arbitrary real numbers as coefficients (the "polynomial ring" $K[x]$ over K). In this set we may define a *congruence* as follows: $p_1(x) \equiv p_2(x) \bmod (x^2 + 1)$ if $p_1(x) - p_2(x)$ is exactly divisible by the polynomial $x^2 + 1$, i.e., if there exists a polynomial $q(x) \in K[x]$ with $p_1(x) = p_2(x) + (x^2 + 1)\cdot q(x)$. The set of all polynomials in $K[x]$ congruent in this sense to a given $p_0(x)$ constitutes a *congruence class*. In every congruence class there is exactly one polynomial of the form $a + bx$ (a, b real). The operations of addition and multiplication, applied to any representatives $p(x)$ and $q(x)$ of two congruence classes, each produce a polynomial $r(x)$ whose congruence class is uniquely determined by the given congruence classes and is therefore independent of the special choice of representatives p, q. The correspondence $a + bx \to a + bi$ maps the congruence classes one-to-one onto the complex numbers with preservation of addition and multiplication.

3. In Cartesian coordinates a *rotation-dilatation* of the (Euclidean) plane leaving the origin fixed is written in the following way:

$$x^* = ax - by,$$

$$y^* = bx + ay;$$

this is a pure rotation if $a^2 + b^2 = 1$, and the angle of rotation is φ if $a = \cos\varphi$, $b = \sin\varphi$. Every rotation-dilatation immediately determines a matrix $\begin{pmatrix} a & -b \\ b & a \end{pmatrix}$, and the addition and multiplication of matrices (in the sense of the calculus of matrices) again leads to matrices of the same structure, so that addition and multiplication are exactly the same as for the complex numbers $a + bi$. This method of introducing complex numbers leads in a very natural way to the law of multiplication; and at the same time, multiplication of corresponding determinants produces the theorem for the product of norms (cf. §1.2).

1.2. Let us briefly recall the *fundamental laws* for computation with complex numbers.

Addition is commutative and associative. The complex numbers form a *vector space of dimension* 2 over K (e.g., with the basis $1, i$) (cf. IB3, §1.3). The equation $\alpha + \xi = \beta$ always has exactly one solution $\xi = \beta + (-1)\alpha$, which we also denote by $\beta - \alpha$. *Multiplication* is commutative, associative and without divisors of zero (cf. IB5, §1.7). The usual distributive law is also valid.

The complex numbers form a *group* with respect to addition with 0 as the neutral element and, if 0 is omitted, they form a group with respect to multiplication, with 1 as the neutral element

For $\alpha = a_1 + ia_2$ with a_1, a_2 real, we have the following notation

$$a_1 = \operatorname{Re}\alpha = \text{real part of } \alpha,$$

$$a_2 = \operatorname{Im}\alpha = \text{imaginary part of } \alpha.$$

The *conjugate* of $\alpha = a_1 + ia_2$ is the number $\bar{\alpha} = a_1 - ia_2 = \alpha - 2i\,\mathrm{Im}\,\alpha$, for which

$$a_1 = \frac{1}{2}(\alpha + \bar{\alpha}), \qquad a_2 = -\frac{i}{2}(\alpha - \bar{\alpha}).$$

The operation of passing from α to $\bar{\alpha}$ is permutable with addition and multiplication.

The *norm of* α is the number $N\alpha = \alpha\bar{\alpha} = a_1^2 + a_2^2$, for which $N\alpha \geqq 0$ for $\alpha \gtreqless 0$, $N\bar{\alpha} = N\alpha$, $N(\alpha\beta) = N\alpha N\beta$ ("theorem for the product of norms").

The *reciprocal of* α (for $\alpha \neq 0$) is defined as the unique solution ξ of $\alpha\xi = 1$ and is denoted by $\frac{1}{\alpha}\cdot$ We see that

$$\xi = \frac{1}{N\alpha}\bar{\alpha}.$$

Division. For $\alpha \neq 0$ the equation $\alpha\xi = \beta$ has exactly one solution, which we denote by $\frac{\beta}{\alpha}\cdot$ Then

$$\xi = \frac{1}{N\alpha}\beta\bar{\alpha}.$$

The *absolute value of* α is $|\alpha| = \sqrt{N\alpha} \geqq 0$ for $\alpha \gtreqless 0$ (for the rules governing the absolute value, see IB1, §3.4). This absolute value is identical with the Euclidean distance of the point (a, b) from the point $(0, 0)$ and thus satisfies the *triangle inequality* (see IB8, §1.1 and II 7, §2.5).

The *argument* φ of α (for $\alpha \neq 0$), $\varphi = \arg\alpha$, is defined by $\alpha = \rho e(\varphi)$, $\rho = |\alpha|$, $e(\varphi) = \cos\varphi + i\cdot\sin\varphi$, mod 2π. Thus $|e(\varphi)| = 1$, $e(\varphi + \psi) = e(\varphi)e(\psi)$, $e(n\varphi) = [e(\varphi)]^n$ and consequently

$$\arg(\alpha\beta) = \arg\alpha + \arg\beta, \qquad \arg(\alpha^n) = n\arg\alpha \ (\mathrm{mod}\ 2\pi).$$

Here ρ, φ are the polar coordinates of the point (a, b) in the Euclidean plane. For $e(\varphi)$ it is better at this stage not to use the customary notation $e^{i\varphi}$, since in the present method of introducing complex analysis the exponential function e^z is not yet at our disposal for nonreal z.

1.3. *Solution of the Equation* $\alpha^n = \beta$

We first investigate the case $n = 2$, since here we can show that the solution is always expressible in terms of square roots in the real field. We have

$$\alpha^2 = (a_1^2 - a_2^2) + 2a_1a_2i = b_1 + ib_2 = \beta.$$

Thus in view of the equations

$$a_1^2 - a_2^2 = b_1, \qquad 2a_1a_2 = b_2, \qquad a_1^2 + a_2^2 = |\beta|,$$

we can at once write down the solutions:

$$\text{for} \quad b_2 \neq 0 \quad a_2 = \pm\sqrt{\frac{1}{2}(|\beta| - b_1)} \quad \text{and} \quad a_1 = \frac{b_2}{2a_2}$$

$$\text{for} \quad b_2 = 0, \; b_1 \geqq 0 \quad a_2 = 0, \; a_1 = \pm\sqrt{b_1},$$

$$\text{for} \quad b_2 = 0, \; b_1 < 0 \quad a_1 = 0, \; a_2 = \pm\sqrt{-b_1}.$$

In this way the problem of solving an equation of the second degree in complex numbers or, in particular, of extracting a square root in the complex field, has been reduced to the operation of extracting a square root in the real field.

The equation $\alpha^n = \beta$ with $\beta = \delta e(\psi)$, $\delta = |\beta|$, $\psi = \arg\beta$ can be rewritten, in view of the *de Moivre equation* $[\rho e(\varphi)]^n = \rho^n e(n\varphi)$, in the form $\rho^n e(n\varphi) = \delta e(\psi)$, which can then be split into the two equations: $\rho^n = \delta$, $e(n\varphi) = e(\psi)$. Thus we have precisely the n solutions $\alpha = \rho e(\varphi)$, with

$$\rho = \delta^{1/n}, \qquad \varphi = \frac{1}{n}(\psi + k\cdot 2\pi), \qquad k = 0, 1, \ldots, n-1.$$

For $n = 3$ this formula provides the so-called trigonometric solution of the *cubic equation in the casus irreducibilis*: from the relation

$$\alpha^3 = a_1(a_1^2 - 3a_2^2) + ia_2(3a_1^2 - a_2^2) = b_1 + ib_2 = \beta$$

we have $a_1^2 + a_2^2 = (\beta\bar{\beta})^{1/3}$ or $a_2^2 = (\beta\bar{\beta}) - a_1^2$; inserting this result in $a_1(a_1^2 - 3a_2^2) = b_1$, we obtain the cubic equation:

$$4a_1^3 - 3Pa_1 - Q = 0 \quad \text{with} \quad P = (\beta\bar{\beta})^{1/3} > 0, \qquad Q = b_1,$$
$$P^3 - Q^2 = \beta\bar{\beta} - b_1^2 = b_2^2 > 0,$$

which consequently has the real zeros:

$$|\beta|^{1/3}\cos\frac{\psi + k\cdot 2\pi}{3}, \qquad k = 0, 1, 2,$$

where $|\beta|^{1/3}$ and ψ are to be calculated from

$$|\beta|^{1/3} = P^{1/2}, \qquad Q = P^{3/2}\cos\psi.$$

2. The Relation of Complex Numbers to Elementary Geometry

2.1. *Number, Point, Vector, Line*

If to every complex number $\alpha = a_1 + ia_2$ we assign in standard fashion the *point* with coordinates (a_1, a_2) in a Cartesian coordinate system of the

Euclidean plane, we obtain the *Gauss representation of the complex numbers in the Euclidean plane.* This correspondence is one-to-one, so that the two expressions "complex number α" and "point of the complex plane"[1] are often used in a completely synonymous way. This practice is quite legitimate and, for deep-lying reasons, is extremely useful. The *Euclidean distance* of two "points" α, β is given by $|\beta - \alpha|$. We can also say: the totality of the complex numbers is made into the *Euclidean plane* (in Cartesian representation) by the definition of distance $E(\alpha, \beta) = |\beta - \alpha|$. Many concepts in the Euclidean plane are useful for studying the field $K(i)$, and conversely the complex numbers are extremely well adapted to the description of Euclidean geometry, and indeed of non-Euclidean geometry in two dimensions, or more generally the geometry of any subgroup of the group of *linear-fractional transformations* (Möbius circle-preserving transformations) of the form $w = \dfrac{az + b}{cz + d}$.

Since the complex numbers form a vector space of dimension 2 over K, they can themselves be considered as *vectors* in the plane (IB8, §1.1), with the advantage over the vector notation that now the *rotation* of a vector through the angle φ (with the origin as center) corresponds to multiplication of the complex number α by $e(\varphi)$. The *scalar product* of the vectors α, β is given by $\operatorname{Re} \bar{\alpha}\beta$, and $\operatorname{Im} \bar{\alpha}\beta$ is the *area of the parallelogram* spanned by the two vectors α, β (in that order). The *area F of the triangle with vertices* $\alpha_1, \alpha_2, \alpha_3$ is expressed by:

$$F = \tfrac{1}{2} \operatorname{Im} P \qquad \text{with} \qquad P = \bar{\alpha}_1\alpha_2 + \bar{\alpha}_2\alpha_3 + \bar{\alpha}_3\alpha_1.$$

For two distinct points z_0, z_1 the equation

$$z - z_0 = \gamma t \qquad \text{with} \qquad \gamma = \frac{z_2 - z_0}{|z_2 - z_0|}$$

gives a parametric representation of the (directed) segment z_0, z_2 (for $0 \leqq t \leqq |z_2 - z_0|$) or of the entire *line* directed in the same sense (for t arbitrarily real; cf. Figure 1). Here γ is the unit *direction* vector. If we now introduce the corresponding unit *normal* vector $\alpha = i\gamma$, then from $\operatorname{Re}(-i\gamma\bar{\gamma}t) = 0$ we obtain the parameter-free equation

$$\operatorname{Re} \alpha(z - z_0) = 0$$

of the line through z_0, where the unit vector α gives the direction of the normal. Alternatively, (setting $p = \operatorname{Re} \bar{\alpha}z_0$) we may write this equation in the *Hesse normal form*

$$\operatorname{Re} \bar{\alpha}\, z = p,$$

or also

$$\bar{\alpha}z + \alpha\bar{z} - 2p = 0.$$

[1] Linguistically it would be more correct to speak of the "plane of the complex numbers" rather than the "complex plane," as has become customary on account of its brevity.

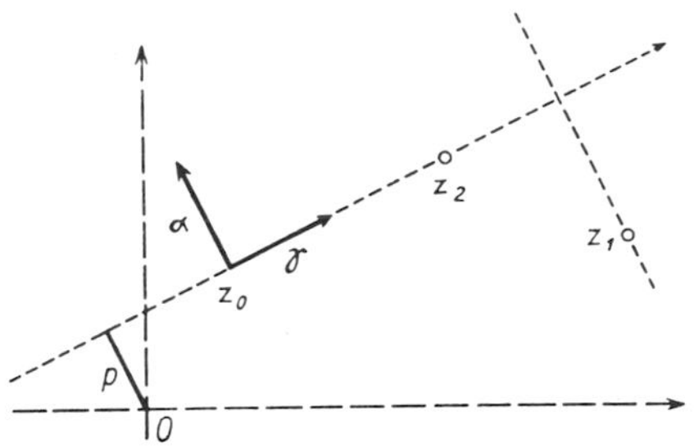

Fig. 1

The equation of the perpendicular from z_1 to this line is given by $\operatorname{Im} \bar{\alpha}(z - z_1) = 0$. Since $|\alpha| = 1$, the quantity $\operatorname{Re} \bar{\alpha}(z_1 - z_0)$ represents the (signed) distance of the point z_1 from the line. Here the sign factor is ± 1, according to whether or not the directed normal points to the side of the line on which z_1 lies. Thus, of the two half-planes bounded by the line, $\operatorname{Re} \bar{\alpha}(z - z_0) > 0$ represents that one into which the directed normal points.

2.2. *Circles and Lines, Involutions*

The equation (1) $f \equiv az\bar{z} + \bar{\alpha}z + \alpha\bar{z} + b = 0$, a, b real, $D = \alpha\bar{\alpha} - ab > 0$, represents an arbitrary *line* if $a = 0$, and if $a \neq 0$ an arbitrary *circle* (with center at $-\alpha/a$, and radius $r = \sqrt{D}|a|$).

If we introduce the expression (2) $F \equiv a\zeta\bar{z} + \bar{\alpha}\zeta + \alpha\bar{z} + b$, then (cf. Figure 2) (3) $\operatorname{Re} F = 0$ represents the *polar* of z with respect to $f = 0$, (4)

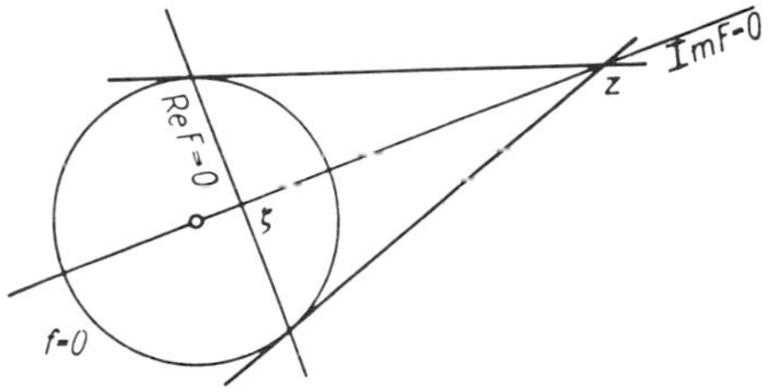

Fig. 2

$\operatorname{Im} F = 0$ represents the *perpendicular* from z to $f = 0$; and the *mirror image* ζ of z with respect to the circle $f = 0$ is the intersection of the two lines (3) and (4), i.e., the solution of the equation (5) $F = 0$:

$$\zeta = \frac{-\alpha\bar{z} - b}{a\bar{z} + \bar{\alpha}}. \tag{6}$$

It is easy to show that an arbitrary involutory mapping of the form

$$\zeta = \frac{A\bar{z} + B}{C\bar{z} + D} \tag{7}$$

with complex coefficients, i.e., a sense-reversing linear-fractional transformation T with the involutory property T^2 = identity, must always be of the form (6) with $\alpha\bar{\alpha} - ab \neq 0$. If $\alpha\bar{\alpha} - ab > 0$, we have a *reflection* in line or circle $f = 0$, whose points are therefore fixed points, whereas in the case $\alpha\bar{\alpha} - ab < 0$ there are no fixed points. A typical example of the second case is the mapping $\zeta = \frac{1}{z}$ (called a "*diametrical point relation*"), since when represented by the stereographic projection (from the point (0, 0, 1)) of the plane $x_3 = 0$, $z = x_1 + ix_2$ onto the sphere $(\sum_{\nu=1}^{3} x_\nu^2 = 1)$ it associates diametrically opposed points with each other (see Carathéodory [2], p. 36–44).

2.3. *The Group of Linear-fractional (Circle-Preserving) Transformations and Its Subgroups*

Even for the above mapping (§2.2, (6)) it is convenient to introduce the "*closure*" (see III 7, §§1 and 3) of the "complex plane" by *adjoining the point* "∞" in the special way suggested by stereographic projection, and this closure is indispensable in a study of the *group of linear-fractional (circle-preserving) transformations* $w = \frac{\alpha z + \beta}{\gamma z + \delta}$, supplemented if necessary by the sense-reversing transformation $w = \frac{\alpha\bar{z} + \beta}{\gamma\bar{z} + \delta}$, with $\alpha\delta - \beta\gamma \neq 0$. Among the important subgroups are the following.

1. The group of *equiform* mappings $w = \alpha z + \beta$, $\alpha \neq 0$. For $\alpha = 1$ this subgroup contains the translations, for $\alpha \neq 1$ the general rotation-dilatations with $\frac{\beta}{1 - \alpha}$ as fixed point, and for $|\alpha| = 1$ the *Euclidean motions* $w = \alpha z + \beta$ with $\alpha = e(\varphi)$, where (for $\alpha \neq 1$) the point $\frac{\beta}{1 - \alpha}$ is the center and φ is the angle of rotation.

2. An arbitrary *non-Euclidean motion* is written in the form

$$w = \frac{\alpha z + \beta}{-\epsilon\bar{\beta}z + \bar{\alpha}} \qquad \text{with} \qquad \alpha\bar{\alpha} + \epsilon\beta\bar{\beta} \neq 1,$$

where in the *hyperbolic* case $\epsilon = -1$ and $z\bar{z} < 1$, and in the *spherical* case $\epsilon = +1$ and z is arbitrary (including ∞). We are dealing here with *angle-preserving* realizations of these geometries in which the geodesics are represented by certain easily characterized circular arcs, and the above form of the equation for arbitrary motions constitutes the starting point of an extremely elegant calculus for the whole of non-Euclidean geometry. The sense-reversing non-Euclidean mappings are obtained by substituting $\bar{z}$ for z (see Carathéodory [2], pp. 54–88, Peschl [10], p. 42).

3. An arbitrary real-linear mapping of the plane is given by

$$w = \alpha z + \beta\bar{z} + \gamma \qquad \text{with} \qquad \Delta = \alpha\bar{\alpha} - \beta\bar{\beta} \neq 0,$$

which for $\Delta > 0$ is sense-preserving and for $\Delta < 0$ is sense-reversing.

3. Fundamental Theorem of Algebra

3.1. The so-called *fundamental theorem of algebra* states that *the field $K(i)$ is algebraically closed.* The theorem reads: *every polynomial $p(x) = \sum_{\nu=0}^{n} a_\nu x^\nu$ of degree $n \geqq 1$* (i.e., $a_n \neq 0$) *with coefficients $a_\nu \in K(i)$ has at least one zero in $K(i)$.* There are many proofs of this theorem: see Pickert IB8, §2.3, Gauss [3], Behnke-Sommer [1], p. 146, Knopp [6], pp. 115 and 140, and Mangoldt-Knopp [8] II, p. 546. If α is a zero of $p(z)$ and if after the substitution $z = \alpha + h$ the polynomial $p(z)$ is arranged in powers of $h = z - \alpha$, so that $p(z) = \sum_{\nu=0}^{n} b_\nu(z - \alpha)^\nu$, then there exists a well-defined natural number k, $1 \leqq k \leqq n$, with the property:

$$b_\nu = 0 \qquad \text{for all} \qquad \nu = 0, 1, \ldots, k - 1 \qquad \text{and} \qquad b_k \neq 0;$$

this number k is called the *multiplicity of the zero* α. We have

$$p(z) = (z - \alpha)^k p_1(z) \qquad \text{with} \qquad p_1(\alpha) = b_k \neq 0.$$

In other words: if α is a zero of $p(z)$ with multiplicity k, then the polynomial is divisible by $(z - \alpha)^k$ but not by $(z - \alpha)^{k+1}$. If $k < n$, the polynomial $p_1(z)$ is of degree $n - k \geqq 1$ and can be discussed in exactly the same way. Proceeding thus, we obtain the theorem:

Every polynomial $p(z)$ of degree $n \geqq 1$ with coefficients in $K(i)$ can be written as the product of n polynomials of first degree:

$$p(z) = a_n \prod_{\nu=1}^{l} (z - \alpha_\nu)^{n_\nu} \qquad \text{with} \qquad \sum_{\nu=1}^{l} n_\nu = n \qquad \text{(IB8, §2.3).}$$

Here the complex numbers α_ν run through all the *distinct* zeros of $p(z)$, and n_ν is the multiplicity of the zero α_ν. This decomposition into n linear factors is unique up to the order in which the α_ν factors are written. It is customary to say that a polynomial of n-th degree with coefficients in $K(i)$ has *exactly n zeros* in $K(i)$, where each zero is counted according to its *multiplicity.*

3.2. If *all the coefficients a_ν in $p(z) = \sum_{\nu=0}^{n} a_\nu z^\nu$ are real,* then on the basis of the rules for computation with complex numbers we have $p(\bar{z}) = \overline{p(z)}$, from which it follows that if $p(\alpha) = 0$, then $p(\bar{\alpha}) = 0$; thus

$$p_0(z) = z - \alpha, \qquad \text{for real } \alpha$$

$$p_0(z) = (z - \alpha)(z - \bar{\alpha}) \qquad \text{for nonreal } \alpha$$

is a polynomial with real coefficients that divides $p(z)$: $p(z) = p_0(z)p_1(z)$. Since $p_1(z)$ also has real coefficients it readily follows that if all the coefficients of $p(z)$ are real and if α is a nonreal zero with multiplicity k, then $\bar{\alpha}$ is also a zero with the same multiplicity k. We thus obtain the following representation of $p(z)$ as a product

$$p(z) = a_n \prod_{\nu=1}^{h} [(z - \alpha_\nu)(z - \bar{\alpha}_\nu)]^{n_\nu} \prod_{\lambda=1}^{l} (z - \beta_\lambda)^{m_\lambda}$$

with

$$2 \sum_{\nu=1}^{h} n_\nu + \sum_{\lambda=1}^{l} m_\lambda = n.$$

Here the α_ν run through all the *distinct nonreal zeros with positive imaginary part* and the β_λ run through all *distinct real zeros*; since for odd n the second product is certainly not empty, it follows that a polynomial of odd degree with real coefficients has at least one real root.

4. Sets and Sequences of Complex Numbers, Basic Topological Concepts

The basic concepts for sets and sequences of complex numbers and for the corresponding sets of points can be derived from the concept of *neighborhood.*

4.1. By an *elementary neighborhood* $U_\epsilon(z_0)$ of the complex number z_0 (or of the point z_0) we mean the set of all numbers (points) z for which $|z - z_0| < \epsilon$, with $\epsilon > 0$. Such a neighborhood consists of all the points in the interior of a circle with center z_0 and radius ϵ. By a *neighborhood of* z_0 in general we mean any set $\mathfrak{M}$ which contains an elementary neighborhood $U_\epsilon(z_0)$.

The *Euclidean distance* $|z - z_0|$ thus establishes in the z-plane a topology, derived from the ordinary two-dimensional Euclidean metric, which forms the basis for analysis in the domain of complex numbers. The set of complex numbers z with this definition of distance and the corresponding topology is usually called C^1. If we write $z = x + iy$, then C^1 is also a Euclidean space R^2 of the points (x, y), which is called the "corresponding" R^2.

4.2. Of primary importance in the theory of sets of points in the plane are the two concepts *limit point* and *convergence of a sequence.*

If by an infinite sequence of complex numbers (points of the plane) we agree to mean a correspondence $n \to z_n$, $n = 1, 2, \ldots$ which to every natural number n assigns a complex number z_n, then the complex number z_0 is called a *limit number* or *limit point* (*or point of accumulation*) *of the sequence* if for every neighborhood of z_0 there exists an infinite subsequence of the z_n whose terms all belong to the neighborhood. A sequence $\{z_n\}$ is said to be convergent if there exists a complex number α such that for arbitrarily preassigned $\epsilon > 0$ there exists an $n_0(\epsilon)$ with the property that for all $n > n_0(\epsilon)$ the z_n are elements of $U_\epsilon(\alpha)$: $|z_n - \alpha| < \epsilon$; the number α is called the *limit of the numerical sequence* $\{z_n\}$ or the *limit of the sequence of points* $\{z_n\}$; in symbols: $\alpha = \lim_{n\to\infty} z_n$, or $z_n \to \alpha$ for $n \to \infty$; and it is clear that the limit of a convergent sequence is its only limit point. Of great importance is the Cauchy criterion for the convergence of a sequence $\{z_n\}$ (see IB8, §2, footnote 2 and III 1, §3.3, theorem 10).

4.3. *Sets of Points in the Complex Plane*

The basic properties of neighborhoods of sets of points can be derived exclusively from the two concepts of neighborhood and open set, and the second of these two concepts can be defined in terms of the first.

A *set of points* $\mathfrak{M}$ is said to be *open* if it is a neighborhood for every point $z_0 \in \mathfrak{M}$. Thus for every point z_0 of an open set of points $\mathfrak{M}$ there exists an elementary neighborhood $U_\epsilon(z_0)$ which is a subset of $\mathfrak{M}$. For neighborhoods of complex numbers we have the following theorem (see H. Kneser [5], pp. 18–29):

1. *If a neighborhood of z_0 is a subset of a set $\mathfrak{M}$, then $\mathfrak{M}$ is itself a neighborhood of z_0.*

2. *The intersection $\mathfrak{U}$ of finitely many neighborhoods of z_0 is itself a neighborhood of z_0.*

For if the n neighborhoods of z_0 contain the n elementary neighborhoods $U_{\epsilon_\nu}(z_0)$, $\nu = 1, 2, \ldots, n$, from the ϵ_ν we select the smallest ϵ_k. Then $U_{\epsilon k}(z_0)$ belongs to the intersection and therefore, by (1), is itself a neighborhood.

3. *Every neighborhood of z_0 contains an open neighborhood of z_0 as a subset.*

4. *For two distinct points z_{01} and z_{02} there exist neighborhoods $U_{\epsilon_1}(z_{01})$ and $U_{\epsilon_2}(z_{02})$ without common points.*

For if $|z_{01} - z_{02}| = 2\epsilon$, we need only take $\epsilon_1, \epsilon_2 < \epsilon$.

The concepts of open set and interior point stand in close relation to each other. Every point of a set $\mathfrak{M}$ for which $\mathfrak{M}$ is a neighborhood is called an interior point. Thus every open set consists exclusively of interior points. For an arbitrary set $\mathfrak{M}$ the set $\mathfrak{M}_0$ of its interior points is open. (This result follows from (3) and (1).)

5. *The union $\mathfrak{M}$ of arbitrarily many open sets of points $\mathfrak{M}_i$ is itself open,* since for every point z_0 in $\mathfrak{M}$ the set $\mathfrak{M}$ is a neighborhood of z_0, in view of the fact that z_0 belongs to one of the $\mathfrak{M}_i$ (1).

6. *The intersection $\mathfrak{M}$ of finitely many open sets of points is itself open;* since every point $z_0 \in \mathfrak{M}$ has every $\mathfrak{M}_\nu$ as a neighborhood, it follows from (2) that $\mathfrak{M}$ is also a neighborhood of z_0.

The concept of an interior point of a set leads us to define limit points for sets, although up to now we have defined them only for sequences.

A point z_0 is called a *limit point of set* $\mathfrak{M}$ if every neighborhood of z_0 contains at least one point of $\mathfrak{M}$ distinct from z_0. From the definition of neighborhood it then follows that every neighborhood of z_0 contains infinitely many points of $\mathfrak{M}$. (For a sequence these infinitely many points could coincide, either partly or altogether; e.g., for $1, \frac{1}{2}, 1, \frac{1}{3}, 1, \frac{1}{4}, \ldots$ the two points 0 and 1 are both limit points of the sequence, but 1 is not a limit point of the set $\mathfrak{M}$ of elements of the sequence.) A point of a set which is not a limit point of the set is called an *isolated* point. Not every limit point of a set necessarily belongs to the set itself, and a set which contains all its limit points is said to be *closed*. The properties of closed sets of points are easily obtained from the concept of a complementary set. The set of all points of C^1 that do not belong to $\mathfrak{M}$ is called the *complementary set* of $\mathfrak{M}$ and is usually denoted by $\mathfrak{M}'$.

7. *A set $\mathfrak{M}$ is closed if and only if its complementary set $\mathfrak{M}'$ is open or empty.*

Remark. The latter case (M' is empty) occurs only if $\mathfrak{M}$ is identical with C^1. This is the only case in which $\mathfrak{M}$ is both open and closed.

For if $\mathfrak{M}'$ is open, then every point $z_0 \in \mathfrak{M}'$ has a neighborhood belonging entirely to $\mathfrak{M}'$, so that z_0 cannot be a limit point of $\mathfrak{M}$. Conversely, if $\mathfrak{M}$ is closed, then no point z_0 of $\mathfrak{M}'$ can be a limit point of $\mathfrak{M}$, since z_0 must have a neighborhood not belonging to $\mathfrak{M}$.

8. From (5) it follows that *the intersection of arbitrarily many closed sets $\mathfrak{M}_i$ is again closed.*

The union of the open complementary sets $\mathfrak{M}_i'$ is again open and forms the complementary set of the intersection.

Correspondingly we have:

9. *The union of finitely many closed sets $\mathfrak{M}_\nu$ is again closed.*

The intersection of the open complementary sets $\mathfrak{M}_\nu'$ is the complementary set of the union of the $\mathfrak{M}_\nu$.

If to every point $z_0 \in \mathfrak{M}$ there is assigned a neighborhood containing z_0, the set of neighborhoods is called a *neighborhood system* for $\mathfrak{M}$. The set $\mathfrak{M}$ is said to be *compact* if from every neighborhood system of $\mathfrak{M}$ it is possible to select finitely many neighborhoods covering $\mathfrak{M}$. The answer to the question: when is a point set compact? is given by the following two theorems.

10. The theorem of Heine-Borel. *Every bounded closed point set is compact.*

11. *Every compact set of points is bounded and closed.*

Proof of Theorem 10. The fact that $\mathfrak{M}$ is bounded means that for the (x, y)-coordinates of all the points $z \in \mathfrak{M}$ there exist real numbers a, b, c, d with $a \leqq x \leqq b$ and $c \leqq y \leqq d$ such that $b - a = d - c$ is the side of a square Q_0

containing all the points of $\mathfrak{M}$. We show that the assumption that $\mathfrak{M}$ is not coverable, i.e., that there exists a system of neighborhoods from which it is impossible to select finitely many neighborhoods covering $\mathfrak{M}$, leads to a contradiction. For then there would exist a sequence of squares Q_n, nested inside one another, containing noncoverable subsets of $\mathfrak{M}$, where the Q_n are constructed as follows: Q_0 is divided into four closed subsquares by the parallels through its midpoint; then each of these subsquares contains a subset of $\mathfrak{M}$, or perhaps no point at all of $\mathfrak{M}$. Thus $\mathfrak{M}$ can be regarded as the union of at most four closed subsets formed in this way. At least one of these subsets must then be noncoverable. Let one of the squares containing a noncoverable subset be called Q_1; then let Q_1 be again divided into four parts and find a square Q_2 that contains a noncoverable subset of $\mathfrak{M}$, and so forth. In this way there arises a sequence of squares Q_n, the coordinates of whose vertices are determined by the monotone sequences:

$$a_n \leqq a_{n+1} < b_{n+1} \leqq b_n \qquad \text{and} \qquad c_n \leqq c_{n+1} < d_{n+1} \leqq d_n,$$

where the length of the sides

$$b_n - a_n = d_n - c_n = \frac{1}{2^n}(b - a) \qquad \text{for} \qquad n \to \infty$$

form a null sequence. The nondecreasing a_n and c_n, and the nonincreasing b_n and d_n, form bounded sequences with the common limit x_0 and y_0, respectively. The point $z_0 = x_0 + iy_0$ belongs to all the squares of the sequence $\{Q_n\}$ and is thus a limit point for $\mathfrak{M}$; and therefore, since $\mathfrak{M}$ is closed, z_0 is itself a point of $\mathfrak{M}$. If $U(z_0)$ is the neighborhood assigned to z_0 in the given system of neighborhoods $\mathfrak{M}$ and if $U_\epsilon(z_0)$ is an elementary neighborhood contained in $U(z_0)$, then in the sequence of the Q_n there exists a Q_k that lies entirely in $U_\epsilon(z_0)$. But this is a contradiction, since Q_k is covered by the single neighborhood $U_\epsilon(z_0)$.

From the covering theorem it follows at once that every infinite subset $\mathfrak{N}$ of a compact set $\mathfrak{M}$ has at least one limit point in $\mathfrak{M}$.

For if $\mathfrak{M}$ contained no limit point of the subset $\mathfrak{N}$, it would be possible to assign to every point $z \in \mathfrak{M}$ a neighborhood $U(z)$ containing from $\mathfrak{N}$ at most the point z itself but no other point. Finitely many of these neighborhoods would cover $\mathfrak{M}$. But these can contain only finitely many points of $\mathfrak{N}$. Thus in this case $\mathfrak{N}$ can consist of only finitely many points.

Thus theorem 11 is proved; for if the compact set $\mathfrak{M}$ were not closed, there must exist at least one infinite subset whose limit points do not belong to $\mathfrak{M}$.

4.4. *Domains, Boundary*

An open point set $\mathfrak{M}$ is called a *domain* if it is connected, i.e., if for every pair of points z_1, z_2 there exists a polygonal line (formed from finitely many line segments) connecting z_1 with z_2 and belonging entirely to the set $\mathfrak{M}$. The importance of domains is largely due to the role played by them as sets of points on which functions are defined in the theory of functions of a complex variable.

A domain G is said to be *simply connected* if the interior of every simple closed polygonal line contained in G also belongs to G. The fact that every simple closed polygonal line S has an "interior," i.e., a bounded domain whose boundary is the polygonal line S, is the statement of the *Jordan curve theorem*, applied to the special case of polygonal lines as curves.

By the *boundary* of a domain G we mean the set of all limit points of G that do not belong to G.

In addition to the domains in the plane, the set of points on which a function is defined often consists of a two-dimensional manifold with complex structure (a so-called *Riemann surface*), but these surfaces would require a special discussion.

5. Functions, Real and Complex Differentiability, and Differentials

5.1. *Functions of a Complex Variable*

Let $\mathfrak{M} \in C^1$ be a set of complex numbers and to each number $z \in \mathfrak{M}$ let there be assigned a complex number $w \in C^1$; then a *complex-valued function* $w = f(z)$ of one variable is thereby defined on $\mathfrak{M}$. The set $\mathfrak{M}$ is called the *domain* (or set of arguments) of the function $f(z)$; thus the domain of a function is not necessarily a domain in the above sense, although in fact this is usually the case; the set of all w for which there exists (at least) one $z \in \mathfrak{M}$ with $w = f(z)$ is called the *range* (or set of values) $\mathfrak{W}$ of $f(z)$ over $\mathfrak{M}$. Thus a function $f(z)$ is simply a mapping of the set $\mathfrak{M} \subset C^1$ into C^1, where the "range" $\mathfrak{W}$ is the (f)-image of $\mathfrak{M}$ under this mapping. (If $\mathfrak{M}$ is a domain in the above sense it is sometimes necessary to be more careful and to call $\mathfrak{W}$ the "carrier set of the image" of $\mathfrak{M}$, namely when the "image" is itself defined as a "covering domain.")

If we write $z = x + iy$, $w = u + iv$, then to such a function $w = f(z)$ there corresponds a one-valued mapping

$$A^*: (x, y) \to (u, v) = A^*(x, y),$$
$$(x, y) \in \mathfrak{M}^* \subset R^2, \qquad (u, v) \in \mathfrak{W}^* \subset R^2$$

(the "real interpretation" of $A: z \to f(z)$ and conversely).

Let the complex-valued function $f(z)$ be defined in $\mathfrak{M}$ and let z_0 be a limit point of $\mathfrak{M}$. Also let there exist a complex number α with the following property:

For every $\epsilon > 0$ there exists a $\delta > 0$ such that $|f(z) - \alpha| < \epsilon$ for all z with $0 < |z - z_0| < \delta$ *and* $z \in \mathfrak{M}$.

Then α is called the limit of $f(z)$ as $z \to z_0$, $z \in \mathfrak{M}$, or in symbols:

$$\alpha = \lim_{\substack{z \to z_0 \\ z \in \mathfrak{M}}} f(z).$$

A necessary and sufficient condition for the existence of this limit is that $\lim_{n \to \infty} f(z_n) = \alpha$ for every sequence $z_n \to z_0$ $(z_n \neq z_0, z_n \in \mathfrak{M})$.

A complex-valued function $f(z)$ is said to be *continuous* at a point z_0 (where $z_0 \in \mathfrak{M}$ and z_0 is a limit point of $\mathfrak{M}$) if

$$\lim_{\substack{z \to z_0 \\ z \in \mathfrak{M}}} f(z) = f(z_0).$$

The function $f(z)$ is said to be continuous in $\mathfrak{M}$ if it is continuous at every limit point of $\mathfrak{M}$.

The most important case occurs when z_0 is an interior point of $\mathfrak{M}$. In this case we denote the limit by $\lim_{z \to z_0} f(z)$.

If $\mathfrak{M}$ is the unit interval $0 \leqq x \leqq 1$, $y = 0$ and if $f(z)$ is continuous on $\mathfrak{M}$, then $w = f(z)$ represents a continuous curve K. If this curve is simple (free of double points), i.e., if $f(x_1) \neq f(x_2)$ for arbitrary $x_1 \neq x_2$ ($\in \mathfrak{M}$), then K is called a *Jordan arc*. A continuous curve is said to be closed if $f(1) = f(0)$, and is called a *closed Jordan curve* if it is free of double points, i.e., if $f(x_1) \neq f(x_2)$ for arbitrary $x_1 < x_2$, except for the single case $x_2 = 1$, $x_1 = 0$.

A complex-valued function $f(z)$ defined over a domain G is said to be *domain-continuous* in G if the image under $w = f(z)$ of each open subset of G is an open set.

5.2. *Real Differentiability, Real Differential (expressed in the real variables x, y or also in the variables z, $\bar{z}$), Partial Derivatives f_x, f_y, f_z, $f_{\bar{z}}$*

Since a complex-valued function represents in general a mapping from R^2 into R^2, we shall use a concept of real differentiability which proves to be especially convenient for functions of two real variables. The central feature of the concept we are about to introduce consists of approximability at a point by a real-linear mapping (see §2.3, 3). A (homogeneous) mapping of this sort can be written as

(1) $$U = a_{11}X + a_{12}Y, \qquad V = a_{21}X + a_{22}Y$$

or, bringing the terms together, as $(W = U + iV)$:

(2) $$W = a_1 X + a_2 Y$$

with

(3) $$a_1 = a_{11} + ia_{21}, \qquad a_2 = a_{12} + ia_{22}.$$

Expressing the real variables in terms of Z, $\bar{Z}$:

(4) $$X = \frac{1}{2}(Z + \bar{Z}), \qquad Y = \frac{i}{2}(-Z + \bar{Z}),$$

we obtain

(5) $$W = \alpha Z + \beta \bar{Z}$$

with

(6)
$$\alpha = \tfrac{1}{2}(a_1 - ia_2) = \tfrac{1}{2}((a_{11} + a_{22}) + i(a_{21} - a_{12})),$$
$$\beta = \tfrac{1}{2}(a_1 + ia_2) = \tfrac{1}{2}((a_{11} - a_{22}) + i(a_{21} + a_{12})),$$

and correspondingly

(7)
$$\overline{W} = \bar{\beta}Z + \bar{\alpha}\overline{Z}.$$

A (complex-valued) function $f(z)$ is said to be *real-differentiable* at an interior point $z_0 = x_0 + iy_0$ of $\mathfrak{M}$ *if there exists a linear mapping* (2) *with* $a_1, a_2 \in K(i)$ (where $W = w - w_0$, $Z = X + iY = z - z_0$), *such that:*

(8)
$$\frac{1}{|z - z_0|}(f(z) - w_0 - a_1(x - x_0) - a_2(y - y_0)) \to 0$$
$$\text{for} \quad |z - z_0| \to 0, \quad z \in \mathfrak{M}, \quad z \neq z_0.$$

Then

(9)
$$a_1 dx + a_2 dy = df$$

is called the (uniquely determined) (*total or complete*) *differential of* $f(z)$ *at* z_0. The existence of this real differential at z_0 and the real differentiability of z_0 are thus equivalent concepts. From the real differentiability of $f(z)$ at z_0 it follows that the partial derivatives f_x, f_y exist at z_0, and

(10)
$$f_x = a_1, \qquad f_y = a_2.$$

However, the converse is not true. From the existence of f_x, f_y at z_0 it does not follow in general that $f(z)$ is real-differentiable at z_0.

If we rewrite the linear mapping (2) in the complex form (5) in terms of the variables $Z, \overline{Z}$, we obtain a *second equivalent form of the definition of the real differentiation of* $f(z)$ *at* z_0: $f(z)$ *is said to be real-differentiable at an interior point* z_0 *of* $\mathfrak{M}$ *if there exists a pair of complex numbers* α, β *such that*

(11)
$$\frac{1}{|z - z_0|}(f(z) - w_0 - \alpha(z - z_0) - \beta(\bar{z} - \bar{z}_0)) \to 0$$
$$\text{for} \quad |z - z_0| \to 0, \quad z \in \mathfrak{M}, \quad z \neq z_0.$$

Thus the expression

(12)
$$\alpha dz + \beta d\bar{z} = df$$

is also called the real (*total or complete*) *differential of* $f(z)$ *at* z_0 *and in this case the partial complex derivatives* f_z, $f_{\bar{z}}$ *at* z_0 *are defined by*

(13)
$$\frac{\partial}{\partial z} f = f_z = \alpha, \qquad \frac{\partial}{\partial \bar{z}} f = f_{\bar{z}} = \beta \qquad \text{(with (6))}.$$

Then if $f(z)$ is real-differentiable at z_0, the partial derivatives $f_x, f_y, f_z, f_{\bar{z}}$ exist at this point and

$$f_z = \tfrac{1}{2}(f_x - if_y), \qquad f_{\bar{z}} = \tfrac{1}{2}(f_x + if_y). \tag{14}$$

This concept of real differentiability, which is due to O. Stolz, has proved to be particularly convenient in topology and analysis, where it is widely used (see Ostrowski [9], p. 226). It has very important *properties*:

D(1) If $f(z)$ is real-differentiable at z_0, then $f(z)$ is continuous at z_0.

This is a very remarkable fact, since the continuity of $f(z)$ at z_0 cannot be deduced if we assume only the existence of f_x and f_y at z_0 (see Mangoldt-Knopp [8], II, p. 335).

D(2a) The functions $f \equiv c$ (c a complex constant), $z, \bar{z}$ have the differentials

$$0(= 0\,dz + 0 d\bar{z}),\ dz,\ d\bar{z}, \text{ respectively.}$$

D(2b) If f is real-differentiable at z_0, then so is $\bar{f}$.

D(3) If f and g are real-differentiable at z_0, then so are $f + g$ and fg, and we have the usual rules

$$d(f+g) = df + dg, \qquad d(fg) = gdf + fdg,$$
$$(f+g)_z = f_z + g_z, \qquad (f+g)_{\bar{z}} = f_{\bar{z}} + g_{\bar{z}},$$
$$(fg)_z = f_z g + f g_z, \qquad (fg)_{\bar{z}} = f_{\bar{z}} g + f g_{\bar{z}}.$$

In other words, $d(\cdots)$, $\dfrac{\partial}{\partial z}, \dfrac{\partial}{\partial \bar{z}}$ are *linear operators.*

D(4) If $f(z_0) \neq 0$ and f is real-differentiable at z_0, then so is $1/f$ and we have the usual rule

$$d\left(\frac{1}{f}\right) = -\frac{1}{f^2}\,df, \qquad \left(\frac{1}{f}\right)_z = -\frac{1}{f^2} f_z, \qquad \left(\frac{1}{f}\right)_{\bar{z}} = -\frac{1}{f^2} f_{\bar{z}}.$$

D(5) If $f(z)$ is real-differentiable at z_0 and $g(w)$ is real-differentiable at $w_0 = f(z_0)$, and if the process of substitution $[g(w)]_{w=f(z)}$ has been defined, then $h(z) = g(f(z))$ is also real-differentiable at z_0, and we have the *chain rule*:

$$h_z = g_w(f) f_z + g_{\bar{w}}(f) \bar{f}_z, \qquad h_{\bar{z}} = g_w(f) f_{\bar{z}} + g_{\bar{w}}(f) \bar{f}_{\bar{z}}.$$

It is well known that the existence of the partial derivatives f_x, f_y at z_0 is not sufficient, in general, to guarantee the validity of the chain rule at z_0.

D(6) We have $\dfrac{\partial}{\partial z}\bar{f} = \overline{\left(\dfrac{\partial}{\partial \bar{z}} f\right)}$, and thus (with $\bar{g} = f$):

$$\frac{\partial}{\partial \bar{z}}\bar{g} = \overline{\left(\frac{\partial}{\partial z} g\right)}.$$

D(7) For a real-valued function $f\ (\bar{f} = f)$ we have:

$$f_{\bar{z}} = \overline{(f_z)}.$$

D(8) For $f = u + iv$, u, v real, we have:

$$2f_z = (u_x + v_y) + i(v_x - u_y), \qquad 2f_{\bar{z}} = (u_x - v_y) + i(v_x + u_y),$$

from which it follows that

D(9) The system of *Cauchy-Riemann differential equations* $u_x - v_y = 0$, $v_x + u_y = 0$ *is equivalent*, in the case of real differentiability, *to* $f_{\bar{z}} = 0$.

5.3. *Complex Differentiability*

Let $f(z)$ be defined in an elementary neighborhood $U_\epsilon(z_0)$ of z_0 and be real-differentiable at z_0 (in the sense of 5.2); then by 5.2 (11) there exist two complex numbers α, β (with $\alpha = f_z(z_0)$, $\beta = f_{\bar{z}}(z_0)$) such that for arbitrary z in $U_\epsilon(z_0)$

$$f(z) = f(z_0) + \alpha(z - z_0) + \beta(\bar{z} - \bar{z}_0) + \epsilon_1|z - z_0| \quad \text{with} \quad \epsilon_1 \to 0 \quad \text{for} \quad z \to z_0, \tag{1}$$

from which it follows that

$$\frac{f(z) - f(z_0)}{z - z_0} = \alpha + \eta\cdot\beta + \epsilon_2,$$

$$\eta = \frac{\bar{z} - \bar{z}_0}{z - z_0}, \qquad \epsilon_2 = \epsilon_1 \frac{|z - z_0|}{z - z_0} \to 0 \quad \text{for} \quad z \to z_0, \quad z \neq z_0, \tag{2}$$

where $|\eta| = 1$ and the argument of η can actually assume every value.

We say that $f(z)$ is *complex-differentiable at* z_0 *if*

$$\lim_{\substack{z \to z_0 \\ z \neq z_0}} \frac{f(z) - f(z_0)}{z - z_0} = \alpha \tag{3}$$

exists and we denote this limiting value by

$$\alpha = f'(z_0) = \left(\frac{df}{dz}\right)_{z = z_0}.$$

Then α is called the *complex derivative* of $f(z)$ at z_0 and $\alpha dz = f'(z_0)dz$ is the *complex (total) differential of* $f(z)$ *at* z_0. Since (3) is equivalent to

$$\frac{1}{z - z_0}(f(z) - f(z_0) - \alpha(z - z_0)) = \epsilon_2 \to 0 \quad \text{for} \quad z \to z_0, \quad z \neq z_0, \tag{4}$$

we have the following theorem:

A complex-valued function $f(z)$, *defined in the neighborhood* $U_\epsilon(z_0)$, *is complex-differentiable at* z_0 *if and only if* 1. *it is real-differentiable there and* 2. *at that point*

$$f_{\bar{z}} = 0, \tag{5}$$

or (equivalently by §5.2, D(9)) *if the Cauchy-Riemann differential equations are satisfied at that point.*

We have the following elementary rules:

D*(1) If f and g are complex-differentiable at z_0, then so are $f \pm g, fg$, and f/g, where in the last case we must assume $g \neq 0$ at that point.

D*(2) If $f(z)$ is complex-differentiable at z_0 and $g(w)$ is complex-differentiable at $w_0 = f(z_0)$, then $h(z) = g(f(z))$ is also complex-differentiable at z_0, and we have the *chain rule*:

$$\frac{dh}{dz} = \left[\frac{dg}{dw}\right]_{w=f(z)} \cdot \frac{df}{dz}.$$

D*(3) If f is complex-differentiable at z_0, then at that point we have the equations ($u = \operatorname{Re} f$, $v = \operatorname{Im} f$):

a) $f' = \dfrac{df}{dz} = f_z = f_x = -if_y = 2u_z = 2iv_z,$

b) $u_z = (\operatorname{Re} f)_z = \dfrac{1}{2}f', \qquad v_z = (\operatorname{Im} f)_z = -\dfrac{i}{2}f',$

c) $u_{\bar z} = (\operatorname{Re} f)_{\bar z} = \dfrac{1}{2}\overline{f'}, \qquad v_{\bar z} = (\operatorname{Im} f)_z = \dfrac{i}{2}\bar{f}'.$

We saw above that the real differentiability of $f(z)$ at z_0 means that the mapping defined by $w = f(z)$ can be approximated by a real-linear mapping

$$w = w_0 + \alpha(z - z_0) + \beta(\bar z - \bar z_0)$$

(in the sense of §5.2, (11)); so now in exactly the same way complex differentiability means that an approximation (in the sense of (4)) is possible by means of a complex-linear mapping

$$w = w_0 + \alpha(z - z_0)$$

(see III 2, §2).

5.4. If we have defined an arbitrary, *complex-valued linear differential form* (cf. III 4, §10) $\omega = adx + bdy$ in a domain G, where the functions $a(x, y), b(x, y)$ are defined and continuous in G, then this differential form can always be rewritten as follows

$$\omega = adx + bdy = Adz + Bd\bar z \tag{1}$$
$$\text{with} \quad A = \tfrac{1}{2}(a - ib), \qquad B = \tfrac{1}{2}(a + ib).$$

If $\mathfrak{C}$ is an arbitrary rectifiable curve in G, then by means of such a differential form ω we can construct the *curvilinear integral* $\int_{\mathfrak{C}} \omega$, whose value, for a given curve $\mathfrak{C}$, is independent of the choice of parameter for this curve. If we are dealing with a continuously differentiable curve

$$\mathfrak{C} : z(t) = x(t) + iy(t),$$

i.e., if the derivatives $\dot{x}(t)$, $\dot{y}(t)$ in the usual real sense exist and are continuous, then

$$\int_{\mathfrak{C}} \omega = \int (a\dot{x} + b\dot{y})dt.$$

Such a differential form ω is said to be *exact* if for every $z_0 \in G$ there exist a neighborhood $U_\epsilon(z_0)$ and a function f in $U_\epsilon(z_0)$ such that ω is its real total differential

$$(2) \qquad \omega = df, \qquad a = f_x, \qquad b = f_y, \qquad A = f_z, \qquad B = f_{\bar{z}}.$$

Let us now consider a linear differential form $\omega = adx + bdy$ whose coefficients $a(x, y)$, $b(x, y)$ are real-differentiable in a neighborhood $U_\epsilon(z_0)$; for such a form we have

$$(3) \qquad a_y - b_x = -i(A_{\bar{z}} - B_z).$$

In order that this form ω (with real-differentiable functions a, b) be *exact* in $U_\epsilon(z_0)$ it is *necessary and sufficient that*

$$(4) \qquad a_y - b_x = 0$$

or, by (3):

$$(5) \qquad A_{\bar{z}} = B_z = 0 \qquad (\text{in } U_\epsilon(z_0)).$$

The exactness of a linear differential form is equivalent to the property that the value of the corresponding curvilinear integral is independent "in the small" of the path of integration. For a *real* differential form we have

$$\omega = adx + bdy = Adz + Bd\bar{z}$$

with a, b are real, so that

$$(6) \qquad B = \bar{A},$$

and therefore

$$(7) \qquad \omega = 2 \operatorname{Re} (Adz).$$

If a, b are real-differentiable, then provided that ω is exact:

$$(8) \qquad A_{\bar{z}} = \bar{A}_z = 0$$

or

$$(9) \qquad \operatorname{Im} A_{\bar{z}} = 0.$$

A *real differential form*

$$\omega_2 = c_1 dx + c_2 dy = C dz + \bar{C} d\bar{z} \qquad (C = \tfrac{1}{2}(c_1 - ic_2))$$

is said to be *conjugate to the real differential form*

$$\omega_1 = a_1 dx + a_2 dy = A dz + \bar{A} d\bar{z} \qquad (A = \tfrac{1}{2}(a_1 - ia_2)),$$

if

$$(10) \qquad C = -iA \quad \text{or} \quad c_2 = a_1, \quad c_1 = -a_2,$$

and thus

$$(11) \qquad \begin{gathered} \omega_1 = 2\,\mathrm{Re}\,(A dz), \qquad \omega_2 = 2\,\mathrm{Re}\,(C dz) = 2\,\mathrm{Im}\,(A dz), \\ \omega_1 + i\omega_2 = 2A dz. \end{gathered}$$

If ω_1 and ω_2 are both exact and A is real-differentiable in $U_\epsilon(z_0)$, it follows from (9) that $\mathrm{Im}\, A_{\bar{z}} = 0$ and (see (10))

$$(12) \qquad \mathrm{Im}\, C_{\bar{z}} = -\mathrm{Re}\, A_{\bar{z}} = 0,$$

so that

$$(13) \qquad A_{\bar{z}} = 0.$$

For *a differential* $A dz$ with real-differentiable A to be *exact* in $U_\epsilon(z_0)$ it is *necessary and sufficient* (13) that $A_{\bar{z}} = 0$, *which means that* A *is complex-differentiable in* $U_\epsilon(z_0)$.

6. Holomorphic and Harmonic Functions

6.1. *The Concept of Holomorphic Function*

A function $f(z)$ defined in a $U_\epsilon(z_0)$ is said to be *holomorphic* at z_0 (or *regular* according to Riemann) if $f(z)$ is *complex-differentiable in a* $U_{\epsilon_1}(z_0)$. We say that $f(z)$ is holomorphic in a domain G if $f(z)$ is holomorphic at every point of G, i.e., if it is complex-differentiable at every point of G. *Necessary and sufficient conditions for the holomorphy of* $f(z)$ *at the point* z_0 (or in the domain G) *are* 1. *that* f *is real-differentiable at every point of a* $U_\epsilon(z_0)$ (or of the domain G) *and* 2. *that in* $U_\epsilon(z_0)$ (or in G) *the equation* $f_{\bar{z}} = 0$ *is satisfied, or equivalently that the Cauchy-Riemann differential equations* (see §5.2, D(9)) *are satisfied.* The *concept of a holomorphic function* is *fundamental* for the whole theory of functions of a complex variable. In particular, after Goursat's discovery in 1900 of his important proof of the *Cauchy integral theorem* (III 6, §1.2), *continuity* of the complex derivative is no longer a necessary assumption; it is sufficient to assume merely *complex differentiability in the domain.* Weierstrass *chose an entirely different procedure and defined holomorphy* (or *regularity*) of a function at z_0 by the condition that the function must be representable in a neighborhood

$U_\epsilon(z_0)$ by a convergent power series $\sum_{n=0}^{\infty} a_n(z - z_0)^n$. Holomorphy (or regularity) according to Weierstrass is equivalent to holomorphy according to Riemann, but the equivalence can only be proved by means of a rather long chain of important theorems, constituting much of the basic structure of the theory of functions of a complex variable (see III 6, §1.8).

The assumptions concerning complex differentiability in the domain can be weakened somewhat to provide conditions that are still sufficient for holomorphy (Theorem of Looman-Menchoff, 1933, see Heffter [4]). Other possibilities exist for constructing the theory of functions by means of the curvilinear integral, where we may confine ourselves to paths that are parallel to the coordinate axes (see Heffter [4]). We shall not discuss these possibilities here, since they involve the concept of an integral.

For *holomorphic functions* we have in particular the following rules:

H(1) *Addition, subtraction, multiplication, and division, applied to holomorphic functions, lead again to holomorphic functions, assuming that we have avoided zeros in the denominator.*

H(2) *The constant c and the function z are everywhere holomorphic, and thus all polynomials in z are holomorphic, and likewise all rational functions of z at every point where the denominator does not vanish.*

H(3) *If $f(z)$ is holomorphic at z_0 and $g(w)$ is holomorphic at $w_0 = f(z_0)$, then $h(z) = g(f(z))$ is also holomorphic at z_0.*

H(4) *A function $w = f(z)$ that is holomorphic at z_0 with $f'(z_0) \neq 0$ has an inverse $z = g(w)$ in a neighborhood $U_\epsilon(w_0)$, $w_0 = f(z_0)$, and this inverse function is holomorphic in $U_\epsilon(w_0)$.*

6.2. *Differential and Indefinite Integral of a Holomorphic Function*

A function $f(z)$ that is holomorphic at z_0 possesses (by 5.3), in a suitable neighborhood $U_\epsilon(z_0)$, the complex total differential $df = f'(z)dz$, which we simply call the "differential of the holomorphic function $f(z)$."

Of particular importance is the following theorem:

The complex derivative $f'(z)$ of a holomorphic function is again a holomorphic function, from which follows the well-known theorem that a *holomorphic function is arbitrarily often complex-differentiable.*

Conversely, for a preassigned holomorphic function $g(z)$ in $U_\rho(z_0)$ the equation $df = g(z)dz$, or equivalently $f'(z) = g(z)$, can always be solved for a function $f(z)$ that is holomorphic in $U_\epsilon(z_0)$, and the solution is uniquely determined up to an additive constant. This solution is called the *indefinite (complex) integral of the holomorphic function* $g(z)$, or in symbols: $f(z) = \int g(z)dz$.

If we are dealing with a simply connected domain instead of the elementary neighborhood $U_\epsilon(z_0)$, the same statement remains valid. The desired function $f(z)$ is found in the form of a curvilinear integral $\int_{\mathfrak{C}} g(z)dz$, taken over a curve

$\mathfrak{C}$ from z_0 to z in G, e.g., over a polygonal line $\sum \sigma_\nu$; as an example of such a curvilinear integral over a segment σ:

$$z = z_1 + t(z_2 - z_1), \qquad 0 \leqq t \leqq 1,$$

we may take, since

$$\frac{dz}{dt} = z_2 - z_1:$$

$$\int_\sigma g(z)dz = \int_0^1 g(z_1 + t(z_2 - z_1)) \frac{dz}{dt} dt = (z_2 - z_1) \int_0^1 g(z_1 + t(z_2 - z_1))dt.$$

Since the integrand is exact, we may choose an arbitrary path in the interior of the simply connected domain G (from z_0 to z), e.g., a polygonal line with segments parallel to the axes.

6.3. *Geometric Properties of the Mapping Defined by Holomorphic Functions*

Every complex-valued function $f(z)$ defined in a $U_\epsilon(z_0) = U$ determines a mapping $w = f(z)$ of U onto an image set in the w-plane, namely the set of values of f in U.

If $f(z)$ is *holomorphic* in U, the mapping has the following properties:

1. It is *continuous*.
2. It is *domain-continuous*.
3. If $f'(z_0) \neq 0$, the mapping is *locally univalent* (*schlicht*), i.e., there exists a $U_{\epsilon_1}(z_0) = U_1$ that is mapped one-to-one onto a domain U_2 containing w_0, in which the inverse mapping $z = g(w)$ is uniquely defined and is holomorphic (see §6.1, (4)).
4. Let $f'(z_0) = A \neq 0$ and $A = re(\varphi)$, and also let $\mathfrak{M}_0 = \underset{z_0}{\mathfrak{M}\{\mathfrak{C}\}}$ be the set of all continuously differentiable curves $z(t)$ through $z_0 = z(t_0)$ for which z_0 is an ordinary point (i.e., $\dot{z}(t_0) \neq 0$); then it follows that the image curve $w = f(z(t)) = W(t)$ of an arbitrary curve $\mathfrak{C} \in \mathfrak{M}_0$ is a curve $\mathfrak{C}^* = \underset{w_0}{\mathfrak{M}\{\mathfrak{C}\}} = \mathfrak{M}_0^*$. The ratio of the elements of arc of the curve and its image at z_0 and w_0,

$$\left(\frac{|dW|}{|dz|}\right)_{t=t_0} = \left|\frac{\dot{W}}{\dot{z}}\right|_{t=t_0} = |f'(z_0)| = r,$$

does not depend on the direction at the point z_0 but has the same value for all directions through z_0. The infinitesimal distortion at z_0 is the same in all directions; thus the mapping is said to be locally *segment-preserving*, although perhaps *scale-preserving* would be a better term.

Moreover:

$$\left(\frac{dW}{dt}\right)_{t=t_0} = re(\varphi)\left(\frac{dz}{dt}\right)_{t=t_0}$$

or

$$\arg \dot{W}_0 = \arg \dot{z}_0 + \varphi(\text{mod } 2\pi),$$

i.e., the tangent (at w_0) to the image curve $\mathfrak{C}^*$ is turned through the angle φ (in the positive sense) in comparison with the tangent (at z_0) to the original curve $\mathfrak{C}$.

The functional determinant of the mapping at z_0 is $|f'(z_0)|^2 = r^2 > 0$, so that the orientation is preserved. Moreover, if we are given two curves $\mathfrak{C}_1, \mathfrak{C}_2 \in \mathfrak{M}_0$: $z_j(t)$ with $z_j(t_0) = z_0$, whose tangents at z_0 form the angle α with each other, it follows from the two equations

$$w_j(t) = f(z_j(t)), \qquad \dot{w}_{j0} = A\dot{z}_{j0} \qquad \text{with} \qquad A = f'(z_0)$$

(the index 0 denotes $t = t_0$) that

$$\frac{\dot{w}_{20}}{\dot{w}_{10}} = \frac{\dot{z}_{20}}{\dot{z}_{10}}, \qquad \arg\frac{\dot{w}_{20}}{\dot{w}_{10}} = \arg\frac{\dot{z}_{20}}{\dot{z}_{10}} = \alpha,$$

which means that the image curves $\mathfrak{C}_1^*, \mathfrak{C}_2^*$ (which belong to $\underset{w_0}{\mathfrak{M}}\{\mathfrak{C}\}$) form the same angle α at w_0. Such a mapping is said to be *angle-preserving* or *conformal* (at z_0).

5. If $f'(z_0) = 0$ and if z_0 is a zero of $f'(z_0)$ of multiplicity $(n - 1)$, i.e., if $f(z) - f(z_0) = (z - z_0)^n g(z)$ with $g(z_0) \neq 0$ and $g(z)$ holomorphic at z_0 so that $f'(z) = (z - z_0)^{n-1}h(z)$ with $h(z_0) \neq 0$ and $h(z)$ holomorphic at z_0, then it is still true, of course, that $\left|\frac{dw}{dz}\right| = 0$ at z_0 and we can prove the following generalization of preservation of angles: if two curves in $\underset{z_0}{\mathfrak{M}}\{\mathfrak{C}\}$ form an angle α at z_0, their image curves form the angle $n\alpha$ at w_0. To a single passage around z_0 in an arbitrarily small neighborhood $U_\epsilon(z_0)$ there corresponds in the image plane an n-fold passage around w_0 (*n-sheeted branch point* w_0 for the *image surface*). A point sufficiently close to w_0 is the image of exactly n distinct points $z_1, \ldots, z_n \in U_\epsilon(z_0)$. Thus the inverse mapping is n-valued at w_0 and z is a holomorphic mapping of $\sqrt[n]{w - w_0}$.

6. The *surface-element* $dx \wedge dy = \frac{i}{2}(dz \wedge d\bar{z})$ undergoes the change

$$du \wedge dv = \frac{i}{2}(dw \wedge d\bar{w}) = w_0'\bar{w}_0'\,\frac{i}{2}(dz \wedge d\bar{z}) = |f'(z_0)|^2\,\frac{i}{2}(dz \wedge d\bar{z}).$$

7. Conversely, if for a complex-valued function $w = f(z)$ that is real-differentiable in $U_\epsilon(z_0)$ we require preservation of angles at z_0, it follows that at z_0 we must have

$$\frac{dw_2}{dw_1} = \frac{A\,dz_2 + B\,d\bar{z}_2}{A\,dz_1 + B\,d\bar{z}_1}, \qquad \frac{\dot{w}_2}{\dot{w}_1} = \frac{A\dot{z}_2 + B\dot{\bar{z}}_2}{A\dot{z}_1 + B\dot{\bar{z}}_1},$$

(for curves $\mathfrak{C}_j \in \underset{z_0}{\mathfrak{M}}\{\mathfrak{C}\}$). In particular, if two such curves $\mathfrak{C}_1, \mathfrak{C}_2$ are orthogonal to each other at z_0, then

$$\frac{\dot{z}_2}{\dot{z}_1} = i, \qquad \frac{\dot{w}_2}{\dot{w}_1} = \pm i,$$

according to whether the mapping is sense-preserving or not (where we have made the permissible normalization $|\dot{z}_2/\dot{z}_1| = 1$ for the parameters of the curves at z_0). Then it follows from

$$\frac{Ai\dot{z}_1 - Bi\dot{\bar{z}}_1}{A\dot{z}_1 + B\dot{\bar{z}}_1} = \pm i,$$

that

$$A\dot{z}_1 - B\dot{\bar{z}}_1 = \pm A\dot{z}_1 \pm B\dot{\bar{z}}_1 \qquad \text{or} \qquad A\dot{z}_1(1 \mp 1) = B\dot{\bar{z}}_1(1 \pm 1)$$

with $\dot{z}_1 \neq 0$, i.e., either $B = f_{\bar{z}}^0 = 0$ in the case of a *sense-preserving conformal mapping*, or $A = f_z^0 = 0$ in the case of a *sense-reversing conformal mapping*. Thus *a sense-preserving conformal real-differentiable mapping of a $U_\epsilon(z_0)$ is always represented by a holomorphic function with nonvanishing (complex) derivative at z_0*, and a sense-reversing conformal mapping of $U_\epsilon(z_0)$ arises from such a mapping by interchange of z with $\bar{z}$.

6.4. *Relationship to the Theory of Harmonic Functions*

If $f(z) = u + iv$ is *holomorphic* at z_0 ($u = \operatorname{Re} f$, $v = \operatorname{Im} f$), then u and v are arbitrarily often continuously differentiable with respect to x and y, and in particular it follows from $f_{\bar{z}} = 0$, or equivalently from the system of *Cauchy-Riemann differential equations* $u_x = v_y$, $u_y = -v_x$, that:

$$u_{xx} + u_{yy} = 0 \qquad \text{and} \qquad v_{xx} + v_{yy} = 0;$$

i.e., $u = \operatorname{Re} f$ *and* $v = \operatorname{Im} f$ *satisfy the Laplace differential equation* $\Delta u = 0$ and thus are "*harmonic*" functions.

Conversely, if u is a solution of the *Laplace differential equation* in $U_\epsilon(z_0)$, we consider the above *Cauchy-Riemann system* as a system of equations for the unknown function v; since the integrability condition $\Delta u = 0$ of this system is satisfied, there exists a solution v in $U_\epsilon(z_0)$, uniquely determined up to a real additive constant. For example, it can be set up as a curvilinear integral, independent of the path in $U_\epsilon(z_0)$:

$$v(x, y) = \int_{(x_0,y_0)}^{(x,y)} (v_x dx + v_y dy) = \int_{(x_0,y_0)}^{(x,y)} (-u_y dx + u_x dy) = 2 \operatorname{Im} \int_{z_0}^{z} u_z dz,$$

since the condition for exactness is identical with the *Laplace equation*. Thus there exists in $U_\epsilon(z_0)$ a holomorphic function $f(z)$, uniquely determined up to an additive purely imaginary constant, with

$$\operatorname{Re} f(z) = u.$$

The function $v = \operatorname{Im} f(z)$ is called an *harmonic conjugate to u*. Since $-if = v - iu$, we see that $-u$ is then conjugate to v. For the *absolute value* $B = |f|$ of a holomorphic function f ($f \neq 0$) we have:

$$h = \log B = \tfrac{1}{2} \log f\bar{f},$$

so that

$$h_z = \frac{1}{2}\frac{f'\bar{f}}{f\bar{f}} = \frac{1}{2}\frac{f'}{f}$$

and

$$h_{z\bar{z}} = \frac{1}{4}\Delta h = \frac{1}{2}\left(\frac{f'}{f}\right)_{\bar{z}} = 0.$$

Consequently $h = \log B$ is a harmonic function, and the function B itself satisfies the differential equation

$$BB_{z\bar{z}} - B_z B_{\bar{z}} = 0.$$

6.5. *Harmonic Polynomials in Two Real Variables*

For an arbitrary *polynomial solution* $P(x, y)$ of $\Delta P = 0$ we have $P_{z\bar{z}} = 0$, and in view of the formal significance of the operators $\frac{\partial}{\partial z}, \frac{\partial}{\partial \bar{z}}$, we obtain

$$P = A(z) + B(\bar{z}) + C;$$

here C is a complex constant and A, B are polynomials in the one variable z or $\bar{z}$ respectively, with $A(0) = 0$ and $B(0) = 0$.

For a *real* polynomial P we must have: $B(\bar{z}) = \overline{A(z)}$, so that

$$P = \operatorname{Re}(C + 2A(z)) = \operatorname{Re} \sum_{\nu=0}^{p} a_\nu z^\nu,$$

i.e., *every real harmonic polynomial can be built up as a sum of the real parts of* $a_\nu z^\nu$, $\nu = 0, 1, \ldots$.

Bibliography

[1] BEHNKE, H. and SOMMER, F.: Theorie der analytischen Funktionen einer komplexen Veränderlichen. Springer-Verlag, Berlin-Göttingen-Heidelberg (1955).

[2] CARATHÉODORY, C.: Funktionentheorie I = Lehrbücher und Monographien aus dem Gebiete der Exakten Wisssenschaften, Mathematische Reihe, Vol. 8. Birkhäuser, Basel/Stuttgart (1960).

[3] GAUSS, C. F.: Die 4 Beweise der Zerlegung ganzer algebraischer Funktionen . . . (1799–1849). Ed. by E. NETTO = Ostwalds Klassiker der Exakten Wissenschaften, No. 14.

[4] HEFFTER, L.: Begründung der Funktionentheorie auf alten und neuen Wegen. Springer-Verlag, Berlin-Göttingen-Heildelberg (1955).

[5] KNESER, H.: Lehrbuch der Funktionentheorie. Verlag Vandenhoeck & Ruprecht, Göttingen (1958).

[6] KNOPP, K.: Funktionentheorie I = Sammlung Göschen, Vol. 668 (1957).

[7] Knopp, K.: Elemente der Funktionentheorie = Sammlung Göschen. Vol. 1109 (1959).

[8] v. Mangoldt, H. and Knopp, K.: Einführung in die höhere Mathematik. Hirzel-Verlag, Stuttgart, Vol. I (1958), Vol. II (1958).

[9] Ostrowski, A.: Vorlesungen über Differential- und Integralrechnung II = Lehrbücher und Monographien aus dem Gebiete der Exakten Wissenschaften, Mathematische Reihe, Vol. V. Birkhäuser, Basel (1951).

[10] Peschl, E.: Die Rolle der komplexen Zahlen in der Mathematik und die Bedeutung der komplexen Analysis = Veröffentlichungen der Arbeitsgemeinschaft für Forschung des Landes Nordrhein-Westfalen, No. 59 (1958).

Bibliography added in Translation

Hamilton, H. J.: A primer of complex variables, with an introduction to advanced techniques. Wadsworth Publ. Co., Belmont, Calif. 1966.

Jaglom, I. M.: Complex numbers in geometry. (Translated by E. J. Primrose.) Academic Press, New York 1968.

Knopp, K.: Elements of the theory of functions. (Translation of [7] above, by F. Bagemihl.) Dover Publ. Inc., New York 1953.

Schwerdtfeger, H.: Geometry of complex numbers. Mathematical Expositions, No. 13. University of Toronto Press, Toronto 1962.

CHAPTER 6

Functions of a Complex Variable

Introduction

At the head of the historical development of the concept of a function stand the *elementary functions*, a class of functions which is closed under the four elementary operations and also, if taken in a wide enough sense, under differentiation and integration. But even then, this class was soon seen to be too narrow. Mathematicians were forced to deal with functions arising from the elementary functions by *limit processes*. At first they believed that such functions would behave just as "reasonably," from an analytic point of view, as the elementary ones, but they soon realized that the limit processes, e.g., formation of a Fourier series, can give rise to extremely "arbitrary" functions, which may very well fail to have many of the properties, such as differentiability or even continuity, that were formerly regarded as self-evident.

Yet these limit functions can hardly be denied citizenship in the domain of analysis, and it therefore became necessary, on the one hand, to define the extent of the new, bewilderingly extensive class of functions and, on the other, to give an abstract definition of the properties, no longer recognizable a priori, that characterize a function. This situation led to the *Dirichlet concept of a function* and to the setting up of a hierarchy in the all-inclusive class of functions thus defined, a hierarchy established by determining certain subclasses of functions, and the relationships among them, that are defined by special properties.

A comparison of this situation with the earlier one gives rise to the question whether we cannot find a subclass of functions general enough, on the one hand, to ensure that "reasonable" limit processes will not lead outside of it, and special enough, on the other, that its functions will have all those properties which in the earlier days accounted for the privileged position of the elementary functions. As desirable properties we will require that such functions can be represented by convenient analytic expressions and can be differentiated arbitrarily often.

Thus there come to mind at once the *real-analytic functions*, namely functions that can be represented by *power series*. However, it is then necessary to introduce a rather artificial notion of convergence if the limit function of a sequence of real-analytic functions is again to be real-analytic, which is not the case with the usual concept of uniform convergence. In fact, the *Weierstrass approximation theorem* states that every continuous function can even be regarded as the uniform limit of polynomials.

In addition to their unattractive behavior with respect to uniform convergence, the real-analytic functions are hard to characterize in terms of any of their properties. As a subclass of the infinitely often differentiable functions, they consist precisely of those functions that can be represented by their Taylor series; but the decision as to whether a given function can be so represented depends on the behavior of the complicated remainder term in the Taylor formula.

In light of these facts the *passage to complex numbers* is seen not merely as a desirable extension but rather as the natural way of considering the whole problem since, by the simple requirement of *differentiability*, it allows us to characterize exactly those functions that can be represented by power series, and at the same time the usual concept of uniform convergence, in domains of the complex plane rather than in intervals of the real line, is seen to be altogether suitable, since now the limit functions of analytic functions are also analytic.

These basic results serve as the foundation for the theory of such functions, a theory which because of its importance has often been called simply "function theory" and has grown into an imposing edifice of broad and harmonious proportions.

1. Holomorphic Functions in the Complex Plane

By definition a *function* f is a mapping

$$\mathfrak{M} \to \mathfrak{N}$$

of a set of "arguments" $\mathfrak{M}$ into a set of "values" $\mathfrak{N}$. If the set of values has an algebraic structure, the operations defined in it can be transferred to the functions. If the set of arguments and the set of values are also subsets of a field, we can calculate the difference-quotients of a function and thus define differentiability, provided that a limit concept has been defined in the field.

1.1. In the next few sections we consider functions whose arguments and values lie in the field **C** of complex numbers; thus it is meaningful to say of such a function that it is *differentiable* at a point (cf. III 5, §5.3).

Definition. *A function f is said to be holomorphic in a domain*[1] $\mathfrak{G}$ *if it is defined and differentiable at every point* $z \in \mathfrak{G}$; *in particular, it is holomorphic at the point* z_0 *if it is holomorphic in a neighborhood of* z_0.

The actual existence of holomorphic functions is shown by the example of the powers z^n and, more generally, by the *power series with nonvanishing radius of convergence.*

For just as in the real field we can show that every power series

$$f(z) = \sum_{n=0}^{\infty} a_n z^n \tag{1}$$

determines a number R with $0 \leqq R \leqq \infty$ such that the series converges for all $|z| < R$ and diverges for all $|z| > R$, for which reason R is called the *radius of convergence;* the number R can be calculated from the coefficients of (1) by the *Cauchy formula*

$$\frac{1}{R} = \overline{\lim}\, \sqrt[n]{|a_n|}. \tag{2}$$

In the interior of the circle of convergence[2] the convergence of (1) is (absolute and) uniform. If (1) is differentiated (for the time being in a purely formal way, i.e., termwise) we obtain

$$\sum_{n=0}^{\infty} (n + 1)a_{n+1}z^n,$$

which is again a power series whose radius of convergence coincides by (2) with the radius of convergence of (1). Since the differentiated series is also uniformly convergent in the interior of the circle of convergence of (1), this new series can be proved, in exactly the same way as in the real case, to be the derivative of the original series with the result that every power series with $R > 0$ represents a holomorphic function in its circle of convergence. Obviously, we have also shown that all *the derivatives of a power series exist and are holomorphic.* Moreover, it is a trivial but important remark that a power series (1) represents a constant function if and only if all the a_n with $n > 0$ vanish.

Since there exist power series with $r = \infty$, e.g.,

$$\sum_{n=0}^{\infty} \frac{z^n}{n!},$$

[1] A domain is an open connected set of points.

[2] A property is said to hold *in the interior of a domain* if it holds for every compact part of the domain; in the plane a compact set of points is bounded and closed, and conversely.

it follows that for every domain $\mathfrak{G}$ there exist nonconstant functions that are holomorphic in $\mathfrak{G}$.

In particular, we have also shown that for every *real-analytic function*

$$f(x) = \sum_{n=0}^{\infty} a_n x^n$$

represented by its Taylor series in the interval $|x| < R$ has a *holomorphic continuation*

$$f(z) = \sum_{n=0}^{\infty} a_n z^n$$

in the circle $|z| < R$.

1.2. Since it is now clear that the concept of holomorphy leads to interesting functions, it is natural to look for properties of these functions that will give us extensive information about their behavior.

To begin with, it follows as in the real case that *the functions holomorphic in a domain form a subring*[3] *of the ring of functions continuous in the given domain* and that the quotient of holomorphic functions is defined and holomorphic apart from zeros of the denominator.

But now our investigation takes a turn far different from what would be expected from the study of real functions. This surprising turn is due chiefly to two facts: first, the classification of differentiable real functions into functions twice differentiable, three times differentiable, and so forth now becomes meaningless since, as we shall see, every once (complex-) differentiable function necessarily possesses all higher derivatives; and secondly, we cannot immediately undertake the proof of the above statement (namely, that the derivative of a holomorphic function is again holomorphic) but are necessarily forced, as will become clear below, into a peculiar detour, which will prove to be a gateway to the immense riches of the theory of functions. We begin with the following theorem.

Cauchy integral theorem. *Let the function f be holomorphic in the domain G. Then*

$$\int_{\mathfrak{C}} f(z)dz = 0 \tag{3}$$

for every system $\mathfrak{C}$ *of paths*[4] *that is bounding*[5] *in* $\mathfrak{G}$.

As for the curvilinear integral in (3), let us only remark that it can be defined, in analogy with the real case, as the limit of Riemann sums $\sum_{\nu=1}^{n} f(z_\nu)(z_\nu - z_{\nu-1})$ with $z_\nu \in \mathfrak{C}$ (and $z_n = z_0$, if as in the present case $\mathfrak{C}$

[3] The sum, difference and product of holomorphic functions are holomorphic.

[4] A path is a rectifiable curve.

[5] I.e., that consists of finitely many closed oriented Jordan curves forming the oriented boundary of a subdomain of $\mathfrak{G}$.

is closed) where $z_0, z_1, \ldots, z_n$ lie in the positive order on $\mathfrak{C}$. Since a function is necessarily integrable on $\mathfrak{C}$ if it is continuous there, and since the holomorphic function f is continuous in $\mathfrak{G}$, and therefore on $\mathfrak{C}$, the existence of the integral (3) is guaranteed.

For the proof of the theorem, which we will now give in outline, we first replace the paths $\mathfrak{C}$ by polygons $\mathfrak{P}$ formed from chords in such a way that $\int_{\mathfrak{C}}$ and $\int_{\mathfrak{P}}$ differ arbitrarily little from each other, as is possible in view of the continuity of the integrand. Then by means of diagonals the subdomain bounded by these oriented polygons can be divided into triangles, whereupon we see that the integral over the polygons is equal to the sum of the integrals over the sides of all these triangles, since the new sides are traversed exactly twice, once in each direction, so that the corresponding integrals cancel each other. Thus it remains to prove the theorem only for the case that $\mathfrak{C}$ is the perimeter $\dot{\Delta}$ of a triangle Δ. By joining the midpoints of the sides we divide Δ into four congruent triangles, each with perimeter $\frac{1}{2}l$, where l is the perimeter of Δ, i.e., the length of $\dot{\Delta}$. For the same reason as before, the integral over $\dot{\Delta}$ is equal to the sum of the integrals over the four subtriangles. Let us denote by Δ_1 a subtriangle whose integral has the greatest absolute value among the four subintegrals; then we have

$$J \leqq 4J_1,$$

where $J = \left|\int_{\dot{\Delta}} f dz\right|$ and $J_1 = \left|\int_{\dot{\Delta}_1} f dz\right|$. If we now subdivide Δ_1 in the same way, and so forth, we obtain a sequence of nested triangles,

$$\Delta_1, \Delta_2, \ldots, \Delta_n, \ldots,$$

whose common limit point we shall denote by z_0. The triangle Δ_n has the perimeter $l/2^n$, and for

$$J_n = \left|\int_{\dot{\Delta}_n} f dz\right|$$

we have

$$J \leqq 4^n J_n.$$

Since f is differentiable at z_0, it follows that for arbitrarily preassigned $\epsilon > 0$ there exists a $\delta > 0$ such that for

$$\eta(z) := \frac{f(z) - f(z_0)}{z - z_0} - f'(z_0)^{6}$$

we have

$$|\eta(z)| < \epsilon \qquad \text{in the circle} \qquad K_\delta := \{z; |z - z_0| < \delta\}.$$

[6] The symbol $:=$ means that the left-hand side is defined by the right-hand side.

Now let us choose n_0 such that $\Delta_{n_0} \subset K_\delta$, and for $n \geqq n_0$ let us construct the expression

$$J_n = \left|\int_{\dot{\Delta}_n} f(z)dz\right|$$
$$\leqq \left|f(z_0)\int_{\dot{\Delta}_n} dz\right| + \left|f'(z_0)\int_{\dot{\Delta}_n} (z - z_0)dz\right| + \left|\int_{\dot{\Delta}_n} (z - z_0)\eta(z)dz\right|.$$

The first two (elementary) integrals are seen by direct calculation to be equal to zero; for the third integral we have the estimate

$$\left|\int_{\dot{\Delta}_n} (z - z_0)\eta(z)dz\right| \leqq \operatorname*{Max}_{z \in \dot{\Delta}_n} \{|z - z_0|\ |\eta(z)|\}\cdot\text{length of } \dot{\Delta}_n \leqq \epsilon\left(\frac{l}{2^n}\right)^2,$$

since $|z - z_0|$ is not greater than the perimeter of the triangle. Consequently for every $n \geqq n_0$

$$J \leqq 4^n J_n \leqq \epsilon l^2.$$

since this estimate of J holds for every $\epsilon > 0$, it follows that $J = 0$, so that the proof of the theorem is complete.

1.3. The hypothesis that $\mathfrak{C}$ is bounding in $\mathfrak{G}$ is essential for the intermediate inference that the Δ_n reduce to an interior point of $\mathfrak{G}$, and in fact, without this hypothesis the theorem is false, as is shown by the following important example. The function $f(z) = 1/z$ is holomorphic everywhere except at $z_0 = 0$; let $\mathfrak{C}$ be a simple closed path[7] enclosing this point; if $\rho > 0$ is so small that the circle $K_\rho = \{z; |z| = \rho\}$ is also enclosed by $\mathfrak{C}$, the above theorem is applicable to the system of curves $\mathfrak{C} - K_\rho$, and we obtain

$$\int_{\mathfrak{C}} \frac{dz}{z} = \int_{K_\rho} \frac{dz}{z}.$$

The integral on the right can be calculated directly, since on K_ρ we have $dz/z = id\varphi$ with $z = \rho(\cos\varphi + i\sin\varphi)$, and thus

$$\int_{\mathfrak{C}} \frac{dz}{z} = \int_{K_\rho} \frac{dz}{z} = i\int_0^{2\pi} d\varphi = 2\pi i.$$

1.4. However, for functions f, holomorphic in the neighborhood of a point z_0 except possibly at that point itself, there is a special case in which the Cauchy integral theorem still holds even for curves enclosing z_0: namely, the case in which $|f(z)| \leqq \mu$ in the neighborhood. For let $\mathfrak{C}$ again denote the given curve and let K_ρ be a circle about z_0 contained in $\mathfrak{C}$; then again

$$\int_{\mathfrak{C}} f dz = \int_{K_\rho} f dz.$$

[7] A curve is said to be simple if it has no double points.

But, since f is bounded,

$$\left|\int_{K_\rho} f dz\right| \leqq \mu 2\pi\rho,$$

and since this inequality holds for arbitrary $\rho > 0$, we have the desired assertion

$$\int_{\mathfrak{C}} f dz = \int_{K_\rho} f dz = 0.$$

The above result can now be applied as follows. Let f be holomorphic in $\mathfrak{G}$, let z_0 be a point in $\mathfrak{G}$ and let $\mathfrak{C}$ be a simple closed path bounding a neighborhood $\mathfrak{U}$ of z_0 in $\mathfrak{G}$; the difference quotient $\dfrac{f(z) - f(z_0)}{z - z_0}$, regarded as a function of z, is holomorphic and bounded in $\mathfrak{U} - \{z_0\}$, since it can be continuously extended to z_0 in view of the differentiability of f. Thus

$$\int_{\mathfrak{C}} \frac{f(z) - f(z_0)}{z - z_0}\, dz = 0$$

or

$$\int_{\mathfrak{C}} \frac{f(z)}{z - z_0}\, dz = f(z_0) \int_{\mathfrak{C}} \frac{dz}{z - z_0} = 2\pi i f(z_0),$$

and the following fundamental result has been proved.

Let $\mathfrak{C}$ be a simple closed path, defined and bounding in a domain $\mathfrak{G}$. Every function f that is holomorphic in $\mathfrak{G}$ can be represented in the interior $I(\mathfrak{C})$ of $\mathfrak{C}$ by the Cauchy integral formula

(4) $$f(z) = \frac{1}{2\pi i} \int_{\mathfrak{C}} \frac{f(t)}{t - z}\, dt; \qquad z \in I(\mathfrak{G}).$$

1.5. Now let f again be holomorphic and bounded in the neighborhood of a point z_0, except possibly at z_0 itself, and let $\mathfrak{C}$ be a simple closed path around z_0; by the above remark, the formula (4) holds for all $z \in I(\mathfrak{C})$, $z \neq z_0$. But the right-hand side of (4) is also defined for $z = z_0$ and is holomorphic there, since it can obviously be differentiated under the integral sign. Thus we have the following lemma.

Lemma of Riemann. *Let f be holomorphic and bounded in the neighborhood of z_0, except possibly at z_0 itself; then f can be holomorphically extended to z_0.*

1.6. The fact that the right-hand side of (4) can be differentiated under the integral sign leads not only to the existence of f', which was merely assumed before, but also provides the *integral representation*

(4′) $$f'(z) = \frac{1}{2\pi i} \int_{\mathfrak{C}} \frac{f(t)}{(t - z)^2}\, dt.$$

Since it is clear that this expression can again be differentiated (under the integral sign), it follows that f' is holomorphic at the same points as f. Thus we have proved one of the two fundamental results stated above, namely:

Every function f that can be differentiated once in a domain has derivatives of every order in that domain; the successive derivatives are given by

$$f^{(n)}(z) = \frac{n!}{2\pi i}\int_{\mathfrak{C}} \frac{f(t)}{(t - z)^{n+1}}\,dt, \qquad z \in I(\mathfrak{C}). \tag{4''}$$

1.7. As an example of an application of the Cauchy integral representation we now prove the following theorem.

Theorem of Liouville. *Let f be holomorphic and bounded in the entire plane; then f is constant.*

Proof. Let $|f| \leqq M$ for all z and let K_R be a circle of radius R about z_0. Then it follows from (4′) that

$$|f'(z_0)| = \left|\frac{1}{2\pi i}\int_{K_R} \frac{f(t)}{(t - z_0)^2}\,dt\right| \leqq \frac{M}{2\pi}\frac{2\pi R}{R^2} = \frac{M}{R};$$

since this inequality holds for every R, we have $f'(z_0) = 0$, and since z_0 was arbitrary, it follows that $f' = 0$, so that

$$f(z) = f(0) + \int_0^z f'(t)dt = f(0), \qquad \text{q.e.d.}$$

1.8. The most important application of this integral representation of the derivatives is the proof that *the power series not only provide examples of holomorphic functions, but actually represent all possible holomorphic functions.*

For let f be holomorphic in $\mathfrak{G}$, $z_0 \in \mathfrak{G}$, and let the closed circular disk $|z - z_0| \leqq R$ lie entirely in $\mathfrak{G}$; then for any z (considered for the moment as fixed) such that $|z - z_0| < R$, and for arbitrary t, we see that

$$\left|\frac{z - z_0}{t - z_0}\right| < 1$$

on the circle $|t - z_0| = R$, so that the geometric series

$$\frac{1}{t - z} = \frac{1}{t - z_0}\,\frac{1}{1 - \dfrac{z - z_0}{t - z_0}} = \sum_{n=0}^{\infty} \frac{(z - z_0)^n}{(t - z_0)^{n+1}}$$

is uniformly convergent in t. But then, since $f(t)$ is bounded on $|t - z_0| = R$,

$$\frac{f(t)}{t - z} = \sum_{n=0}^{\infty} \frac{f(t)}{(t - z_0)^{n+1}}(z - z_0)^n$$

is uniformly convergent in t and can therefore be integrated termwise.

Thus from (4), with $\mathfrak{C} = \{t; |t - z_0| = R\}$,

$$f(z) = \sum_{n=0}^{\infty} a_n(z - z_0)^n \qquad \text{for} \qquad |z - z_0| < R,$$

where

$$a_n = \frac{1}{2\pi i}\int_{|t-z_0|=R} \frac{f(t)dt}{(t - z_0)^{n+1}},$$

for which by (4″) we can also write

$$a_n = \frac{1}{n!} f^{(n)}(z_0).$$

Thus we have the following result:

Every function f that is holomorphic in $\mathfrak{G}$ can be represented by a power series about any point $z_0 \in \mathfrak{G}$, and this series converges in the largest open circular disk centered on z_0 that is contained in $\mathfrak{G}$.

It follows, in particular, that the radius of convergence of a power series is uniquely determined: i.e., in any *concentric circle properly including the circle of convergence, there cannot exist a function which is holomorphic in the larger circle and coincides in the circle of convergence with the function defined by the power series.* This fact provides an explanation of a phenomenon that remained incomprehensible in real analysis; e.g., the representation

$$\frac{1}{1 + x^2} = \sum_{n=0}^{\infty} (-1)^n x^{2n}$$

no longer holds for $|x| \geqq 1$, although the function $\dfrac{1}{1 + x^2}$ does not show any kind of irregular behavior at the two real points $x = \pm 1$. But the complex continuation $\dfrac{1}{1 + z^2}$ is unbounded in the neighborhood of $\pm i$, and therefore cannot be extended holomorphically. On the circle of convergence of the series there lie nonreal points at which the function is no longer holomorphic.

1.9. *Cauchy Integrals and Power Series* are the two most important instruments for the representation and investigation of holomorphic functions. The Cauchy integrals determine the functions by their values on a closed path, and the power series determine them by the values of their derivatives at one point. Thus a holomorphic function is already completely determined by relatively "few" of its values. We shall state this result in a precise form just below and shall sharpen it to the so-called identity theorem (§1.10), which will lead us later to the concept of an *analytic function.* Let us now make use of these two instruments to draw further important conclusions.

The Cauchy integral formula immediately provides the following result, which was seen in the introduction to be very desirable.

Let the sequence of holomorphic functions f_n, defined in the domain G, be uniformly convergent in the interior[8] *of G; then the limit function is again holomorphic in G and the sequences of derivatives of arbitrary order converge uniformly in the interior of G to the derivatives of the limit function.*

1.10. On the other hand, the power series expansion for holomorphic functions at once gives us the important result that *holomorphic functions assume their values only to integral orders;* by this we mean the following.

For every holomorphic point z_0 of a function f which is not constant[9] at z_0 there exists a well-defined natural number k, the order of multiplicity of f at z_0, such that in a neighborhood of z_0

$$f(z) = f(z_0) + (z - z_0)^k g(z), \qquad g(z_0) \neq 0,$$

where the function g is holomorphic at z_0; *the points with $k > 1$ are isolated,* as is easily shown.

Since $g(z)$ is continuous and $g(z_0) \neq 0$, there exists an entire neighborhood of z_0 in which $g(z) \neq 0$; thus if z_0 is a limit point of points where f takes on the same value, then $f(z) = f(z_0)$ in an entire neighborhood of z_0; in particular, if z_0 is a zero of f, this result states that the nonisolated zeros of a holomorphic function form an open set. On the other hand, in view of the continuity of holomorphic functions in the domain of definition $\mathfrak{G}$, this set is closed. But since $\mathfrak{G}$ is connected, every set that is both open and closed in $\mathfrak{G}$ is either empty or identical with $\mathfrak{G}$. Thus we have proved the following result.

The zeros of a (not identically vanishing) holomorphic function are isolated.

Another statement of the same result is the identity theorem announced above:

Identity theorem. *Two functions are identical with each other if they are holomorphic in $\mathfrak{G}$ and coincide on a set of points with a limit point in $\mathfrak{G}$.*

In particular, the *holomorphic continuation of a real-analytic function,* already seen to be possible in a suitably assigned region, *is possible in only one way.*

[8] Compare §1.1, footnote 2.

[9] A function is said to be constant at a point z_0 if it is constant in a neighborhood of z_0.

An algebraic consequence of this result sharpens the trivial statement that holomorphic functions form a ring:

The functions holomorphic in $\mathfrak{G}$ form an integral ring.[10]

1.11. The argument that led us from the Cauchy integral theorem to the Cauchy integral formula can be generalized as follows (see Figure 1).

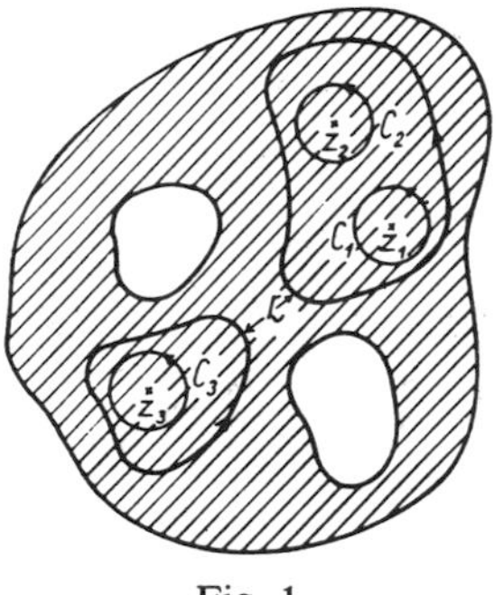

Fig. 1

Let the domain $\mathfrak{G}^*$ be formed from the domain $\mathfrak{G}$ by the exclusion of a finite set of points $\{z_1, \ldots, z_k\}$, and let $\mathfrak{C}$ be a system of paths that is bounding in $\mathfrak{G}$ and contains in its interior all the points of the excluded set. If around each point z_j we now choose a positive circuit C_j containing no excluded point other than z_j, then for every function f holomorphic in $\mathfrak{G}^*$, it follows from the Cauchy integral theorem that

$$\frac{1}{2\pi i}\int_{\mathfrak{C}} f(z)dz = \sum_{j=1}^{k} \frac{1}{2\pi i}\int_{C_j} f(z)dz$$

and, on the other hand, that every integral on the right-hand side is independent of the choice of the circuit C_j, provided it satisfies the above topological assumption. Since such an integral depends only on the behavior of the function in the neighborhood of the point z_j we can therefore define the expression

$$\operatorname*{Res}_{z_j} f := \frac{1}{2\pi i}\int_{C_j} f(z)dz$$

as the *residue of f at the point z_j*. This formula leads to the following generalization of the Cauchy integral theorem.

Residue theorem. *Let f be holomorphic in $\mathfrak{G}$ except at a finite number of points; let $\mathfrak{C}$ be a system of paths that is bounding in $\mathfrak{G}$ and contains none*

[10] A ring is said to be an integral ring, or to be without divisors of zero (cf. IB5, §1.9, footnote 1) if from $ab = 0$, $a \neq 0$ it follows that $b = 0$. In every ring of functions the zero element is the function ν: $\nu(z) = 0$ for all z. Thus we must distinguish between the vanishing of a function f, in which case $f = \nu$, and the vanishing of f at a point z_0, in which case only the value of the function $f(z_0) = 0$.

of these exceptional points. Then

$$\frac{1}{2\pi i}\int_{\mathfrak{C}} f(z)dz = \sum_{j=1}^{k} \operatorname{Res}_{z_j} f,$$

where $z_1, \ldots, z_k$ *are the exceptional points enclosed by* $\mathfrak{C}$.

1.12. We now apply this theorem in the following way: let f be holomorphic and nonconstant in $\mathfrak{G}$, let $\mathfrak{C}$ be a system of paths that is bounding in $\mathfrak{G}$ and let a be a number that is not a value of f on $\mathfrak{C}$; *then*

$$(5) \qquad \frac{1}{2\pi i}\int_{\mathfrak{C}} \frac{f'(z)}{f(z) - a}\,dz$$

is equal to the number of a-points of f enclosed by $\mathfrak{C}$, *each counted in its multiplicity.*

Proof. The integrand is holomorphic inside $\mathfrak{C}$ and on $\mathfrak{C}$ except at a-points of f, i.e., except for solutions of $f(z) = a$. But these points are isolated in $\mathfrak{G}$, so that only finitely many of them are enclosed by $\mathfrak{C}$; thus the above integral is equal to the sum of the residues of the integrand at these points. Let z_1 be such a point. In a neighborhood of z_1

$$f(z) - a = (z - z_1)^{k_1} g_1(z),$$
$$f'(z) = k_1(z - z_1)^{k_1 - 1} g_1(z) + (z - z_1)^{k_1} g_1'(z),$$

where g_1 is holomorphic at z_1 and $g_1(z_1) \neq 0$. Thus in a deleted[11] neighborhood of z_1

$$\frac{f'(z)}{f(z) - a} = \frac{k_1}{z - z_1} + \frac{g_1'(z)}{g_1(z)};$$

but since $g_1(z_1) \neq 0$, the second summand can be holomorphically extended to z_1, so that by §1.3

$$\operatorname{Res}_{z_1} \frac{f'}{f - a} = \frac{1}{2\pi i}\int_{C_1} \frac{k_1 dz}{z - z_1} = k_1, \qquad \text{q.e.d.}$$

1.13. If we now let a vary, we see that (5) represents a continuous (even holomorphic) function of a, as long as we restrict ourselves to values of a that are not assumed by f on $\mathfrak{C}$; but by the last theorem the values of this function are integers, so that the function is a constant and we have proved the following result.

Let f be holomorphic in $\mathfrak{G}$ *and let* $\mathfrak{C}$ *be a system of curves that is bounding in* $\mathfrak{G}$*; then for every value that is not assumed by f on* $\mathfrak{C}$ *there exists in the plane of values of f a neighborhood such that all the values in this neighborhood are assumed equally often by f in the interior of* $\mathfrak{C}$.

In particular, every value assumed by a function that is holomorphic in $\mathfrak{G}$ is an inner point of the set of images onto which $\mathfrak{G}$ is mapped by f. This result can be expressed in the following form.

[11] A neighborhood of z_1 from which z_1 has been removed.

Theorem on the preservation of domains. *The mapping determined by a nonconstant holomorphic function is open, i.e., the images of open sets are open; then it follows by continuity that the images of a domain are connected and thus are again domains.*

The importance of this theorem is shown by one of its immediate consequences; if $|f|$ were to assume its maximum in an interior point z_0 of $\mathfrak{G}$, then $f(z_0)$ would obviously be a boundary point of the set of images of $\mathfrak{G}$ and would be contained in it, which is impossible by the present theorem; thus we have proved the following principle.

Maximum principle. *The absolute value of a nonconstant function that is holomorphic in $\mathfrak{G}$ does not assume a maximum in $\mathfrak{G}$ and assumes a minimum only at the zeros of the function.*

Since the image $f(\mathfrak{G})$ formed from $\mathfrak{G}$ by a function f holomorphic in the domain $\mathfrak{G}$ is again a domain, it is natural to consider an arbitrary function g holomorphic in $f(\mathfrak{G})$ and by its means to define the *composite mapping* $g(f(z))$ defined on $\mathfrak{G}$; we see at once that *such a composition of holomorphic functions is itself holomorphic.* Moreover:

If f assumes every value at most once in $\mathfrak{G}$, then the inverse mapping $f(\mathfrak{G}) \to \mathfrak{G}$ exists and is holomorphic.

2. Meromorphic Functions in the Complex Plane

2.1. Let $\mathfrak{G}$ be a domain in the Gauss plane. The functions holomorphic in $\mathfrak{G}$ form an integral ring. Thus it is natural to make use of the purely algebraic fact that an integral ring always has a quotient field (cf. IB5, §1.12). Then from the elements of this quotient field, which are up to now not functions but purely algebraic objects, we may construct functions by the natural definition

$$\frac{f}{g}(z) := \frac{f(z)}{g(z)},$$

but for the time being we must delete from $\mathfrak{G}$ the zeros of the denominator. Since these zeros are isolated in $\mathfrak{G}$ the remaining set of points is again a domain and the function $\frac{f}{g}$ thus defined is holomorphic in this domain. But we can now restore those zeros of g at which f has a zero of at least the same order as g; for in the neighborhood of such a point $\frac{f}{g}$ remains bounded and can therefore, by the Riemann lemma, be holomorphically extended to this point. Let us denote by $\mathfrak{G}^*$ the domain thus obtained; it is formed from $\mathfrak{G}$ by the deletion of the points at which g vanishes to a higher order than f; at these exceptional points it is clear that $\frac{f}{g}$ cannot be holomorphically defined, since near them it is unbounded. For the reciprocal function g/f, on the other hand, these points are points of holomorphy, and are in fact zeros.

2.2. Thus the domain $\mathfrak{G}^*$ is mapped by the function $w = q(z) := \dfrac{f(z)}{g(z)}$ onto a domain $q(\mathfrak{G}^*)$ of the w-plane. Now if $h(w)$ is holomorphic in $q(\mathfrak{G}^*)$, the composite function $h(q(z))$ is holomorphic in $\mathfrak{G}^*$. But it is quite possible that this function can be holomorphically extended to certain other points that do not belong to $\mathfrak{G}^*$; for example,[12] this is the case for $t = \dfrac{1}{w}$, since here the composite function is simply the reciprocal $\dfrac{g}{f}$, which was seen above to be holomorphic at all the exceptional points of $\dfrac{f}{g}$. This example is typical: for if we know that the composite function $h(q(z))$ formed from $h(w)$ with $w = q(z)$ can be holomorphically extended to an exceptional point of q, then by interpolation of the mapping $t = \dfrac{1}{w}$ we see that

$$k(t) := h\left(\frac{1}{t}\right) = h(w)$$

is holomorphic at $t = 0$, and conversely. Thus *$h(q(z))$ can be holomorphically extended at an exceptional point of q if and only if this extension is possible for $h\left(\frac{1}{t}\right)$ at $t = 0$.*

Now it is desirable to regard the holomorphic extendability of $h(q(z))$ to an exceptional point z_0 as a property of the function $h(w)$. The only difficulty is that the point z_0 does not correspond under the mapping $w = q(z)$ to any point in the w-plane. We can remove this difficulty by adjoining to the w-plane an *ideal point* ∞, which we then assign to z_0 as the image $q(z_0) := \infty$. As a result of this *extension of the set of values* every element of the quotient field of the functions holomorphic in $\mathfrak{G}$ is now defined as a function in the whole of $\mathfrak{G}$, and on the other hand, the same *extension of the set of arguments* naturally leads to the following definition.

Definition. *$h(w)$ is holomorphic at $w = \infty$ if $h\left(\frac{1}{t}\right)$ is holomorphic at $t = 0$.*

Then we may say:

$h(q(z))$ can be holomorphically extended to a point z_0 with $q(z_0) = \infty$ if and only if $h(w)$ is holomorphic at $w = \infty$.

[12] By making $\mathfrak{G}$ smaller we can, if we wish, insure that $q(\mathfrak{G}^*)$ does not contain the point $w = 0$.

2.3. Our definition $q(z_0) = \infty$ made use of the fact that q and $\frac{1}{q}$ are holomorphic in a deleted neighborhood of z_0 and that $\frac{1}{q}(z_0) = 0$. This situation represents a generalization of holomorphy, which it is customary to call *meromorphy* and to define as follows.

Definition. *A function f is said to be meromorphic at the point z_0 if it is either holomorphic at that point or is the reciprocal of a function that is holomorphic there; the function is said to be meromorphic in the domain $\mathfrak{G}$ if it is meromorphic at every point $z \in \mathfrak{G}$. If f is meromorphic at z_0 but not holomorphic, or in other words if $f(z_0) = \infty$, $\frac{1}{f}(z_0) = 0$, then f is said to have a pole of order k at z_0, where k is the order of the zero of $\frac{1}{f}$ at z_0.*

From this definition it is at once clear that *the functions meromorphic in a region $\mathfrak{G}$ form a field*, which contains as a subfield the quotient field of the integral ring of functions holomorphic in $\mathfrak{G}$; in the case considered up to now, namely that $\mathfrak{G}$ is a region of the plane of complex numbers, the two fields of functions coincide.[13]

2.4.[14] Since meromorphy represents a proper generalization of holomorphy it is not surprising that many of the theorems about holomorphic functions are no longer true for meromorphic functions; for example, we already know (§1.3) that $\operatorname{Res}_0 \frac{1}{z} = 1$, so that the Cauchy integral theorem no longer holds for the meromorphic function $\frac{1}{z}$ but must now be stated in the more general form (cf. §1.11 above) that $\frac{1}{2\pi i}\int_C f\,dz$, taken over a simple closed curve C avoiding the poles of f, is equal to the sum of the residues of f at the poles enclosed by C.

Since the poles of a meromorphic function are zeros of its holomorphic denominator, they are isolated, and in the neighborhood of a pole the absolute value of the function increases uniformly beyond all bounds. It follows that a pole cannot be a limit point of zeros of the function. From this remark we see at once that the *identity theorem* (§1.10), stated above for holomorphic functions, *remains true without change for meromorphic*

[13] This result is easily proved but is by no means trivial; for it is to be noted that by definition a meromorphic function is required to be only *locally* a quotient of holomorphic functions, whereas the above statement asserts precisely that this is *globally* the case. Moreover, as we shall soon see, the theorem is false if $\mathfrak{G}$ is, for example, the closed plane (cf. §3.2).

[14] For simplicity we consider in §2.4 only functions defined in domains in the plane of complex numbers; in other words, we omit the point ∞ from the set of arguments.

functions. Also, the fact that holomorphic functions can be represented by power series (§1.8) can be extended to meromorphic functions in the following way.

Let f have a pole of order k at z_0, so that $\frac{1}{f}$ has a zero there of order k. Thus

$$\frac{1}{f}(z) = (z - z_0)^k g(z) \qquad \text{with} \qquad g(z_0) \neq 0.$$

Consequently, if g is holomorphic at z_0, so is $\frac{1}{g}$, and $\frac{1}{g}$ can be represented as

$$\frac{1}{g}(z) = \sum_{\nu=0}^{\infty} a_\nu (z - z_0)^\nu, \qquad a_0 \neq 0,$$

from which, in view of

$$f(z) = \frac{1}{(z - z_0)^k} \frac{1}{g}(z),$$

it follows, if we rename the coefficients $b_\nu := a_{\nu+k}$, that

$$f(z) = \frac{b_{-k}}{(z - z_0)^k} + \cdots + \frac{b_{-1}}{z - z_0} + \sum_{\nu=0}^{\infty} b_\nu (z - z_0)^\nu, \tag{6}$$

and this expansion, called a *Laurent series*, holds uniformly in the interior of the largest deleted circle about z_0 in which f is holomorphic. The power series at the right is called the *holomorphic part of f*, and the other terms are the *principal part of f*.

From (6) we can draw a conclusion of great importance for calculation of the residue of f at z_0. The residue is defined by an integral which, if (6) is the integrand, can be integrated termwise; but, except for the term $\frac{b_{-1}}{z - z_0}$, every summand $b_\nu(z - z_0)$ is the continuous derivative of a function $\left(\text{namely } \frac{b_\nu}{\nu + 1}(z - z_0)^{\nu+1}\right)$, so that the corresponding integral vanishes over closed paths. Thus there remains

$$\operatorname*{Res}_{z_0} f = \frac{1}{2\pi i} \int_C \frac{b_{-1} dz}{z - z_0} = b_{-1};$$

so that *the residue of f at the pole z_0 is equal to the coefficient of* $(z - z_0)^{-1}$ *in the Laurent expansion of f about z_0*.

3. The Theory of Functions on the Closed Plane

In the preceding section we have described the consequences of extending the *set of values* by adjoining the point ∞. But in that section we intentionally avoided more than a mere mention of the corresponding extension of the *set of arguments.* The reason is that the principal method, namely integration, by which we have obtained our results up to now, is not immediately available in the extended plane; for example, we have not yet determined what is meant by a path through the point ∞, to say nothing of an integral along such a path; moreover, the Cauchy integral theorem, in the form in which we have expressed it up to now, is false, as is shown by the example $\int_{|z|=1} \frac{dz}{z}$. The unit circle, taken here for the path of integration, represents a circuit around the point ∞ in which the function $\frac{1}{z}$ is holomorphic, and yet the integral does not vanish.

Thus we must first of all make a somewhat closer study of the set of points obtained as a result of this extension of the plane of complex numbers; we shall then find that in the definition of integration we shall necessarily be led to the concept of a *differential,* by means of which we can reconstruct our theory for the extended set of arguments. At the same time we shall find that the new concepts and new results will form a bridge to the ideas and problems of the theory of functions in its widest sense.

3.1. In motivating the extension of the complex z-plane[15] by adjunction of the point ∞ we were led to define the behavior of a function f at this point in terms of the behavior of $f\left(\frac{1}{t}\right)$ at the point $t = 0$. Thus it is natural to consider the mapping $z = \frac{1}{t}$ as a correspondence not only between the points $t = 0 \leftrightarrow z = \infty$, but also between the neighborhoods of $t = 0$ and the corresponding sets of points on the z-plane, in order to define the latter as *neighborhoods of the point* ∞. Then the extended plane becomes a *topological space* **P**.

This new space can be visualized, for example, as the surface of the Riemann "number sphere" in which the plane is "embedded" by stereographic projection. But in any attempt to "visualize" **P**, it must be remembered that **P** is not a planar set of points and cannot be so represented. Nevertheless, it is possible to consider **P** as made up of two planar domains, and it is to this conception of **P**, as being more easily generalized than the Riemann sphere and therefore more fruitful, that we now give preference.

The representation of **P** as the union of two planar domains is obtained by taking a sufficiently large neighborhood of each of the points 0 and ∞; for example, we may take the interior U_0 of the circle $|z| = 2$ and the exterior

[15] Here we write z instead of w, as above.

U_∞ of the circle $|z| = \frac{1}{2}$ including the point ∞; then, although the neighborhood U_∞ is not a planar domain, it is mapped topologically by $t = \frac{1}{z}$ onto such a domain; and *in any case* **P** *is covered by two open sets* U_0, U_∞ *for each of which there exists a topological mapping* ($z = z$ for U_0, and $t = \frac{1}{z}$ for U_∞) *onto a planar domain.* These statements are summed up by saying that P is a *two-dimensional manifold.* The images in the above-mentioned mappings are called *local parameters*; thus in our case z is a local parameter in U_0, and $t = \frac{1}{z}$ is a local parameter in U_∞. In the modern, extremely graphic terminology, a system of covering open sets together with the corresponding local parameters is called an *atlas* for the manifold, and this atlas is said to consist of *charts*: a chart is simply one of the open sets of the given covering system together with the corresponding local parameter; thus *our atlas for* **P** *consists of two charts.*

The reason why we must pay attention to the local parameters, although they are unnecessary for a description of the point set **P** itself, is that they alone enable us to answer the important question whether a given function is holomorphic (or meromorphic) at a given point $z \in \mathbf{P}$. By means of the local parameters every function defined on **P** is transferred to planar domains, with the result that we can at once say whether a function is holomorphic in a given local parameter. Thus we can assemble our former definitions in the following simple way: *the function f is said to be holomorphic (meromorphic) at the point* $z \in \mathbf{P}$ *if f is holomorphic (meromorphic) in the local parameters.*

But this definition could still give rise to contradictions: for if z lies in two charts, and thus in U_0 and U_∞, i.e., if $\frac{1}{2} < |z| < 2$, it is conceivable that f might be holomorphic in one parameter but not in the other. In order to be certain that this case does not arise, or in other words to ensure that holomorphic dependence on one parameter implies holomorphic dependence on the other, we must verify that these parameters themselves depend holomorphically on each other. In the present situation (cf. Figure 2) this is obviously the case.

For if u_0, u_∞ are the parametric mappings of our two charts, then for a point z in the intersection $U_0 \cap U_\infty$ the corresponding values z and t of the parameters satisfy the relation:[16]

$$t = u_\infty(z) = u_\infty(u_0^{-1}(z)) = : \gamma(z) = \frac{1}{z},$$

[16] Here the entities in question are functions, so that u^{-1} denotes the *inverse function* (if it exists) of u; to make a distinction, we have denoted the *reciprocal function* of f by $\frac{1}{f}$.

which is obviously holomorphic; but then every function that is holomorphic (meromorphic) in z for the first chart is also holomorphic (meromorphic) for the second; for if $f(u_0^{-1}(z))$ is holomorphic in z, then $f(u_\infty^{-1}(t)) = f(u_0^{-1}(\gamma^{-1}(t))$ is also holomorphic (see the end of §1).

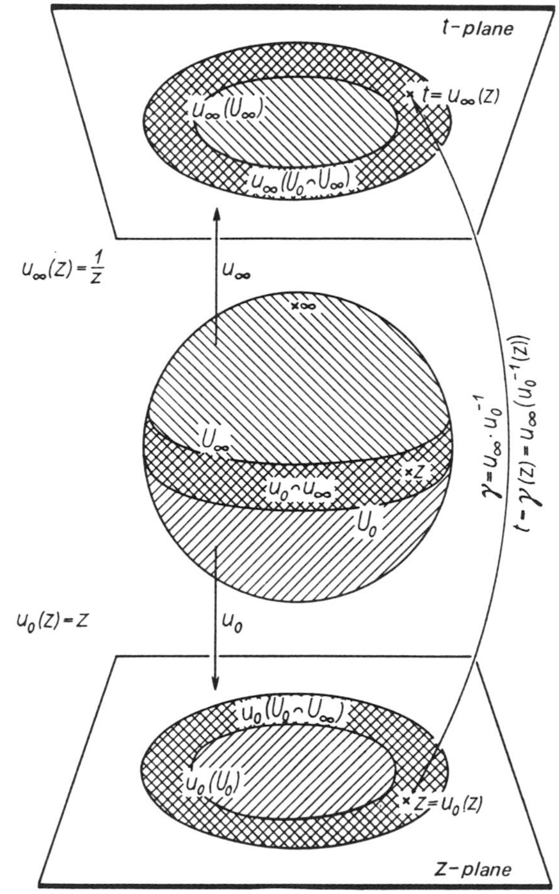

Fig. 2

This property, namely that the local parameters depend holomorphically on each other, is expressed by saying that the charts are *holomorphically related* to each other and that their totality, namely the atlas, defines a *complex structure* on the set of points **P**.

3.2. In addition to these statements, which we have emphasized here because they are suitable for later generalization, we must now add the following special statement: **P** is *compact*,[17] for which reason we shall refer to **P** as the *closed plane*. This property leads to profound differences between the theory of functions on **C** ond on **P**.

For example, the theorem of Liouville (§1.7) leads to a far-reaching result:

Every function that is holomorphic on the closed plane **P** *is constant.*

It is now possible to give a complete characterization of the functions that are meromorphic on **P**. Since the poles of such a function f are isolated, and since in a compact set of points every infinite set of points has a limit point, the function f has only finitely many poles, say $z_1, \ldots, z_n$. If $z_i = \infty$, then in view of the fact that $t = \frac{1}{z}$ the function f has as its principal part (cf. §2.4, (6)) the polynomial:

$$c_1 \frac{1}{t} + \cdots + c_k \frac{1}{t^k} = c_1 z + \cdots + c_k z^k.$$

On the other hand, if z_i is finite, the principal part has the form

$$\frac{c_{i1}}{z - z_i} + \cdots + \frac{c_{ik_i}}{(z - z_i)^{k_i}},$$

which is holomorphic everywhere except at z_i (including $z = \infty$, as is shown by the Riemann lemma (§1.5)). Consequently, f differs from the sum of all its principal parts by a function which is holomorphic everywhere in **P**, i.e., by a constant c_0. Our result then reads:

Every function that is meromorphic on the closed plane **P** *admits a decomposition into partial fractions*

$$f(z) = c_0 + c_1 z + \cdots + c_k z^k + \sum_{i=1}^{n} \sum_{l=1}^{k_i} \frac{c_{il}}{(z - z_i)^l},$$

and is thus rational (*and conversely*).

Thus the field of functions meromorphic on **P** is considerably more inclusive than the quotient field of the holomorphic functions, which contains only constants (cf. §2.3, footnote 1).

It will be noted that we have not yet discussed any extension to the closed plane **P** of the integral theorems by which we were led to the method of power series and Laurent serics; nevertheless, it was permissible for us to use these series in the above results, since in each case we applied them only "locally," i.e., in one chart.

[17] Every covering with open sets contains finitely many open sets that already constitute a covering. That this property holds for **P** can be seen as follows: an arbitrary covering of **P** contains a neighborhood V_∞ of the point ∞; the set $P - V_\infty$ is then a bounded closed set of the complex plane, so that it has the desired property by the theorem of Heine-Borel.

3.3. Yet the extension of these integral theorems to the closed plane is interesting in itself.

A path C in $\mathbf{P}$ is defined as a curve such that its part in U_0 and its part in U_∞ are both rectifiable. Now we can see at once that it makes no sense to integrate a function along a curve C which does not lie entirely in one chart; for such an integral would not be defined until we had determined which of the local parameters in the intersection $U_0 \cap U_\infty$, in this case z and $\frac{1}{z}$, is to be used as the variable of integration. But to make such a choice would introduce a very unsatisfactory element of arbitrariness. Consequently, we must integrate not functions but "magnitudes" which, under a change of the variable of integration, are transformed in such a way that the change caused by the transformation of variables is exactly compensated by the change in the magnitude. Such magnitudes, called *differentials*,[18] are provided, for example, by *the local derivatives of holomorphic functions, i.e., by their derivatives with respect to the local parameters;* for let f be holomorphic in $U_0 \cap U_\infty$; then by definition (§3.1) the derivative $\frac{df}{dz}$ exists (where z is the local parameter in U_0), and the derivative $\frac{df}{dt}$ (where $t = z^{-1}$ is the local parameter in U_∞) also exists; and obviously, if C is a path in $U_0 \cap U_\infty$ with the initial point z_0 and the end point z_1, the integral

$$\int_C df := \int_C \frac{df}{dz}\,dz = \int_C \frac{df}{dt}\,dt = f(z_1) - f(z_0)$$

actually has a value independent of the choice of parameter.

In general, a differential dw is understood to be a collection of functions such that to each choice of parameter there corresponds exactly one of these functions and the functions are related to one another in the same way as the local derivatives of holomorphic functions; i.e., if t and τ are parameters[19] *holomorphically related to each other, and if W_t, W_τ are the corresponding functions of the differential dw, then*

$$W_\tau = W_t \frac{dt}{d\tau}. \tag{7}$$

In the theory of functions we are naturally interested in *holomorphic or meromorphic differentials,* namely differentials whose corresponding

[18] Here we are dealing only with line-integrals. In view of the fact that surface-integrals also have a meaning and define a different kind of differential, it would be more precise to distinguish between linear and quadratic differentials.

[19] For example, $\tau = z$, $t = \frac{1}{z}$.

functions are holomorphic or meromorphic;[20] then it is meaningful to speak of *a zero or a pole of given order for a differential*; on the other hand, it is meaningless to say that the differential takes the value 1 at a certain point, since the factor in the transformation (7) will in general cause this value to be changed under a change of parameter.

At a point $z \neq \infty$ a differential dw is obviously holomorphic or meromorphic together with the corresponding function w_z. On the other hand, the behavior of dw at the point ∞ is determined by its dependence on $t = \frac{1}{z}$. But if we also describe dw at ∞ by setting $dw = w_z dz$, or in other words if we describe dw in terms of the variable z (which is *not* a local parameter at ∞), then in view of the fact that

$$\frac{dz}{dt} = -\frac{1}{t^2},$$

we obtain the relation

$$dw = w_z dz = -w_z \frac{1}{t^2} dt = w_t dt, \qquad \text{so that} \qquad w_t = -\frac{1}{t^2} w_z.$$

Thus we see, in particular, that the differential dz is holomorphic everywhere on **P** except for its pole of order two at ∞; furthermore, a differential dw is holomorphic at ∞ if and only if w_z has a zero there of at least second order.

It also follows that, *except for zero, there does not exist any differential dw which is holomorphic in the whole of* **P** *or is meromorphic with at most one simple pole;* for if such a pole were situated say at infinity, then w_z would be a function everywhere holomorphic on **P** and vanishing, except at ∞, to the first order at least, so that w_z would be identically zero. The differential $\frac{dz}{z}$ is an example of the fact that not every differential is the local derivative of a function.

3.4. Our purpose in defining differentials was to be able to define curvilinear integrals invariantly, i.e., independently of the choice of parameter: thus if we let C be a path that lies entirely in the domain of holomorphy of the differential dw, the integral

$$\int_C dw := \int_C w_t dt = \int_C w_t \frac{dt}{dz} dz = \int w_z dz$$

has in fact the desired invariant property.

[20] It is sufficient to require this property for *one* corresponding function, since the rule for transformation then guarantees the same property for the other functions of the differential.

The Cauchy integral theorem (§1.2) can now again be stated, this time for differentials:

In the domain $\mathfrak{G}$ *let the differential dw be holomorphic; if* $\mathfrak{C}$ *is a system of paths that is bounding in* $\mathfrak{G}$, *then*

$$\int_{\mathfrak{C}} dw = 0.$$

The proof depends on the fact that the subdomain of $\mathfrak{G}$ that is bounded by $\mathfrak{C}$ can be divided by transverse cuts into segments such that each segment lies either entirely in the chart U_0 or entirely in the chart U_∞: if the boundary of such a segment is taken as the path of integration, the theorem is a simple application of its former statement (§1.2); and by addition of these individual integrals, since the integration over the transverse cuts is carried out twice in opposite directions, we obtain the desired theorem.

Thus we can now define the residue of a differential[21] in analogy with our previous (§1.11) definition of a residue:

If z_0 *is an isolated boundary point of a domain* $\mathfrak{G}$ *in which the differential dw is holomorphic, and if C is an arbitrary path in* $\mathfrak{G}$ *around this point, the expression*

$$\operatorname*{Res}_{z_0} dw := \frac{1}{2\pi i}\int_C dw$$

is called the residue of dw at the point z_0.

In the same way as before (§2.4) we see that the residue of dw at the point z_0 is given by the coefficient of $\dfrac{1}{t - t_0}$ in the Laurent series of w_t, where $t = z$ or $t = \dfrac{1}{z}$ according to which local parameter we choose, or are forced, to use.

The residue theorem (§1.11) now reads (see Figure 1, §1.11):

Let dw be holomorphic in $\mathfrak{G}$ *except at finitely many points* $z_1, \ldots, z_k$, *and let* $\mathfrak{C}$ *be a system of paths that is bounding in* $\mathfrak{G}$ *and avoids the points* z_j; *then*

$$\frac{1}{2\pi i}\int_{\mathfrak{C}} dw = \sum_{j=1}^{k} \operatorname*{Res}_{z_j} dw.$$

[21] The concept of a residue is defined for a differential; for a function it is, in general, meaningless; only in the case of the finite complex plane were we able to speak of the residue of a function, since that plane could be described by the one natural chart $(\mathbf{C}, z)$ and the distinction between functions and differentials was therefore unnecessary.

Particularly important, therefore, among the differentials meromorphic in $\mathfrak{G}$ are the ones whose residues all vanish $\left(\text{example: } dz = -\frac{1}{t^2}\,dt \text{ at } \infty\right)$; they are called *differentials of the second kind.* If $\mathfrak{G}$ is simply connected, i.e., if every closed path C in $\mathfrak{G}$ is bounding, then for such differentials we have $\int_C dw = 0$. Thus the integral $\int_{z_0}^{z} dw$ has a value that is independent of the choice of path of integration with initial point z_0 and end point z and therefore this integral is a function of the upper limit z:

$$f(z) = \int_{z_0}^{z} dw.$$

It follows that $dw = df$:

In simply connected domains of **P** *the differentials of second kind are identical with the differentials of meromorphic functions.*

If we apply the residue theorem to a differential dw that is meromorphic in **P** and choose as the path of integration a circuit C around a holomorphic point z_0, so that the poles all lie in the other part of **P** that is bounded by C, then as the result of one passage around z_0 we obtain on the one hand $\frac{1}{2\pi i}\int_C dw = 0$, and on the other the sum of all the residues:

The sum of the residues of a differential meromorphic in **P** *is equal to zero.*

If the function $f \neq$ const is meromorphic in **P**, and therefore rational, the differential $\frac{df}{f - a}$ is meromorphic. In analogy with §1.12 we can then prove that the sum of the residues of this differential is equal to the difference between the number of a-points of f and the number of poles of f; since this sum vanishes, we have the result (for which it would also be easy to give a direct proof):

A function that is meromorphic in **P** *assumes every value equally often in* **P**.

If the differential dw is meromorphic in **P** the quotient $\frac{dw}{dz}$ is a meromorphic function whose zeros and poles coincide with those of dw, except at ∞, where the order of the zero of $\frac{dw}{dz}$ is 2 greater than that of dw; thus it follows that *for every differential meromorphic in* **P** *the total order* (cf. §3.3, p. 235) *of the poles is* 2 *greater than that of the zeros.*[22]

Let us now make a remark about the validity of the *Cauchy integral formula* (4) in a domain that includes the point ∞.

Let C be a simple closed path, let $A(C)$ be the part of **P**, bounded by C, that includes the point ∞ and let f be a function holomorphic on $A(C)$

[22] Example: dz has no zeros but has a double pole at ∞.

and C; we assume that C is positively oriented with respect to ∞ and form the integral

$$\frac{1}{2\pi i}\int_C \frac{f(y)}{y - z}\,dy, \qquad z \in A(C);$$

the poles of the integrand are situated at $y = z$ and $y = \infty$; the residue at the first of these poles is $f(z)$ and at the second $-f(\infty)$; consequently, the *Cauchy integral representation is valid in the neighborhood of* ∞ *only for such functions as are holomorphic at* ∞ *and vanish there.*

4. Riemann Surfaces

In the preceding section we were led in a natural way to extend both the argument-domain and the value-domain of holomorphic functions; we made this extension in order that the quotient field of the functions holomorphic in a given domain might be brought within the scope of the theory of functions. Consequently, in that section we had a general reason for extending the set of complex numbers; but now we shall find that *each of the individual functions gives us a particular reason to extend the domain of values and of arguments in a manner that is specifically determined by the function itself.* This undertaking is desirable for two reasons; first, on these extended, more general sets of points, which because of their common properties can all be subsumed under the one concept of *Riemann surface*, we can again define holomorphy in a natural way and thus confer the rights of citizenship in the theory of functions upon the "many-valued functions" that are of such great importance in practice; in fact, the original motive of Riemann in introducing the surfaces named for him was to resolve the paradox inherent in the very concept of a many-valued function;[23] and secondly we shall find that an extension of this sort opens up to us the most "reasonable" method of dealing with the theory of functions as a whole.

4.1. We begin with a nonconstant function $w = f(z)$, meromorphic in a domain $\mathfrak{G} \subset \mathbf{P}_z$.[24] Since the mapping determined by f

$$f : \mathfrak{G} \to f(\mathfrak{G})$$

from $\mathfrak{G}$ onto the set of values $f(\mathfrak{G})$ has very desirable properties (conformality, preservation of domains), we must look with regret on the fact that in general it cannot be inverted. If we wish to remove this disadvantage, it is clear that we must count each value $w \in f(\mathfrak{G})$ as often as it is assumed by f in $\mathfrak{G}$. In other words, the individual points of the set $\mathfrak{G}$ must be

[23] By definition, a function assigns to each argument *only one* value.

[24] We now denote the closed z-plane by $\mathbf{P}_z$.

distributed over the domain $f(\mathfrak{G})$ in such a way that each point $z \in \mathfrak{G}$ lies exactly "over" the point $f(z) \in f(\mathfrak{G})$; to this "set of points," which may be regarded as being the set of pairs $(z, f(z))$ and is therefore in one-to-one correspondence with the points of $\mathfrak{G}$, we now transfer the topological properties of $\mathfrak{G}$. But the set of points in question is also in correspondence with $f(\mathfrak{G})$, and this latter correspondence, the so-called *projection mapping*, exhibits precisely the most interesting properties of the original mapping f. It is this set of points, taken together with the projection mapping, that is called the *Riemann surface* $\mathfrak{R}$ *belonging to* f *over the plane of values* $\mathbf{P}_w$.

The terminology here is the result of a visualization connecting $\mathfrak{R}$ with $f(\mathfrak{G})$ which we obtain from the fact, unused up to now, that f is meromorphic.

For let $z_0 \in \mathfrak{G}$ be chosen arbitrarily and let $w_0 = f(z_0)$ be a value, of order k, of the function f at the point z_0; then in a neighborhood of z_0 we have[25]

$$f(z) = w_0 + (z - z_0)^k g(z),$$

where g is holomorphic in the given neighborhood and $g(z_0) \neq 0$; thus there exists a neighborhood of z_0 in which, for a certain function h holomorphic in that neighborhood,

$$g(z) = (h(z))^k, \qquad h(z_0) \neq 0.$$

Consequently the function

$$t(z) = (z - z_0)h(z) \tag{8}$$

is holomorphic and *invertible* (one-to-one) at z_0, and we have

$$w = f(z) = w_0 + (t(z))^k.$$

Since the points z_0 with $k > 1$ are isolated in $\mathfrak{G}$, the normal case is $k = 1$. For this case the situation just described means that the point w_0 on $f(\mathfrak{G})$ onto which the "ordinary" point z_0 on $\mathfrak{R}$ is projected has a neighborhood that is in one-to-one correspondence, under the projection mapping, with a neighborhood of z_0. This fact is concisely described by saying: *at an ordinary point the Riemann surface lies univalently* (*schlicht*) *over* $\mathbf{P}_w$. In the case $k > 1$, on the other hand, there is a neighborhood of w_0 over each of whose points $\neq w_0$ there lie on $\mathfrak{R}$ exactly k distinct points in a neighborhood of z_0, whereas over w_0 itself there lies the single point z_0; such a point is called a *branch point of order* $k - 1$, since at that point the surface $\mathfrak{R}$ is k-sheeted over $\mathbf{P}_w$. But now our powers of visualization encounter a difficulty: over a branch point we cannot visualize the surface $\mathfrak{R}$ without self-intersections. This difficulty arises from the fact that we visualize the relation between $\mathfrak{R}$ and $f(\mathfrak{G})$ in a three-dimensional space, but this space

[25] For simplicity we assume that z_0 and w_0 are both finite.

has nothing to do with the formal description of the relation between $\mathfrak{R}$ and $f(\mathfrak{G})$. The self-intersections of $\mathfrak{R}$ are the price we pay in order to satisfy an inordinate craving for visualization (cf. the example $w = z^2$ in Figure 3).

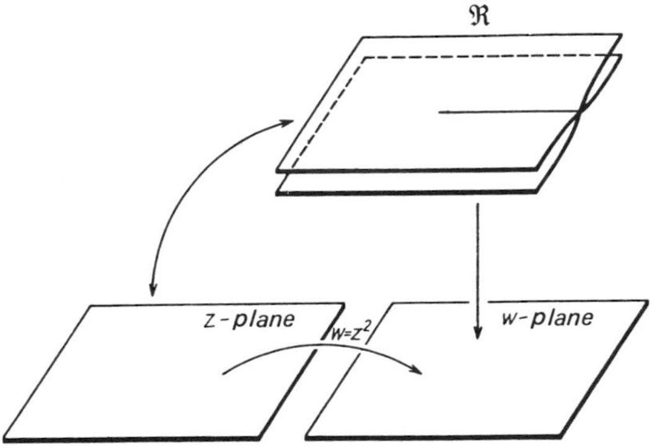

Fig. 3

But we can now investigate the theory of functions on $\mathfrak{R}$; for $\mathfrak{R}$ is topologically related to $\mathfrak{G}$ and we will therefore regard *a function as being holomorphic on $\mathfrak{R}$ if it is holomorphic in the old sense as a function of $z \in \mathfrak{G}$.* In particular, the projection-mapping of $\mathfrak{R}$ onto $f(\mathfrak{G})$ will then be meromorphic: for when referred back to $\mathfrak{G}$ it simply becomes the (meromorphic) mapping $w = f(z)$ by means of which $\mathfrak{R}$ was defined; it is equally obvious that the (invertible) mapping $\check{f}$ is meromorphic, where $\check{f}$ is defined as the function which to each point of $\mathfrak{R}$ assigns the point z from which it originated. *In general, the theory of functions on $\mathfrak{R}$ is completely identical with the theory of functions on $\mathfrak{G}$.*

This discussion, rather trivial up to now, gains in interest if we note that *there is no need to characterize meromorphy on $\mathfrak{R}$ by returning to the z-plane $\mathbf{P}_z$, since this meromorphy can also be characterized in terms of the points on the underlying w-plane $\mathbf{P}_w$*: for in (8) we have found, for each point z_0, a special function $t(z)$ which is holomorphic and invertible in a neighborhood of z_0; thus a function on $\mathfrak{R}$ is holomorphic at the point corresponding to z_0 if and only if it is holomorphic as a function of $t = t(z)$, where for this t the above argument shows that $t = \sqrt[k]{w - w_0}$; thus in the case $k = 1$, the function is holomorphic if and only if it is holomorphic as a function of w at w_0, and on the other hand for $k > 1$ it is in general not a function of w (or rather it is what is paradoxically called a "many-valued" function) but is a function of $\sqrt[k]{w - w_0}$, and *holomorphy for it is defined with respect to this parameter*. In this formulation the original function f is nowhere explicitly mentioned, so that we can investigate the theory of functions on $\mathfrak{R}$ even if we know nothing, or wish for some reason to act as though we knew nothing, about the prehistory of $\mathfrak{R}$ as determined by f; for it may occasionally happen that $\mathfrak{R}$ has been constructed in some other way, so that we can no longer trace its prehistory in the manner just described. The following result is of special interest from this point of view.

Functions that are meromorphic in the region of the w-plane lying below $\mathfrak{R}$, and thus in particular the projection-mapping w itself, are also meromorphic on $\mathfrak{R}$, but in general there exist more meromorphic functions on $\mathfrak{R}$ than on the region lying below $\mathfrak{R}$ on the w-plane: for example, among such functions is the inverse function $\check{f}$ of f as constructed above.[26]

In general, several $z \in \mathfrak{G}$ are assigned by $\check{f}$ to each $w \in f(\mathfrak{G})$; but if w is not a branch point and z is a corresponding value, then f sets up a one-to-one correspondence between certain neighborhoods of z and w: if a pair of corresponding values z, w is preassigned (and w is not a branch point), a particular *function element*, representing z as a meromorphic function of w, is thereby singled out from $\check{f}$. But although the totality of these function elements cannot, in general, be assembled to form a (one-valued) function of w, the function $\check{f}$, defined and meromorphic on the Riemann surface $\mathfrak{R}$, is the "essential feature of the situation," for which each of the function elements provides only a partial view.

4.2. Of basic importance, however, is the inverse situation already mentioned, namely, that the connection (provided up to now by the mapping f) between the Riemann surface and a plane domain $\mathfrak{G}$ is unknown to us; i.e., we know neither the function f nor the corresponding Riemann surface $\mathfrak{R}$ over the w-plane; in this situation we do not know the totality of all the function elements (each of which provides a partial description of f) but *merely a single function element*, i.e., a function that is meromorphic in a plane domain. Then we have the problem of *meromorphic continuation* of this function, i.e., the search for the largest possible set of points that includes the given domain and is the domain of definition for a meromorphic function φ coinciding with the given function in the given subdomain. In view of the remarks in the preceding section we will not expect, in general, that the desired set of points will exist in the original plane, since the function may become "many-valued."[27]

[26] The terminology here is not quite correct; $\check{f}$ does not invert f, but rather the mapping $z \to (z, f(z))$ of $\mathbf{P}_z$ onto $\mathfrak{R}$; thus we write $\check{f}$ instead of f^{-1}.

[27] Of course, it is also true that only in special cases will the given function be the inverse function $\check{f}$ of a function f defined in a plane domain.

On the contrary, we again obtain a Riemann surface, this time over the argument plane, and a meromorphic function on it, namely *the "analytic function" defined by the given function element and the corresponding Riemann surface*. We use the definite article here because it is a fact that the analytic function and the Riemann surface are uniquely determined by the preassigned element, independently of the manner in which they are constructed.

In practice, the meromorphic continuation of the given function element is undertaken by searching for a domain $\mathfrak{G}^*$, overlapping the given domain $\mathfrak{G}$ in which the function φ is already known,[28] such that in $\mathfrak{G}^*$ there is a meromorphic function φ_1^* coinciding with φ in *one* component of the intersection $\mathfrak{G} \cap \mathfrak{G}^*$. Here we must make two remarks: first we see from the identity theorem that for $\mathfrak{G}^*$ there can exist at most one such φ_1^*; and secondly we cannot, in general, sharpen the above requirement by insisting that φ_1^* shall coincide with φ in the whole intersection $\mathfrak{G} \cap \mathfrak{G}^*$, as is shown by the example $\varphi = \sqrt{z}$ in Figure 4; It is precisely because of

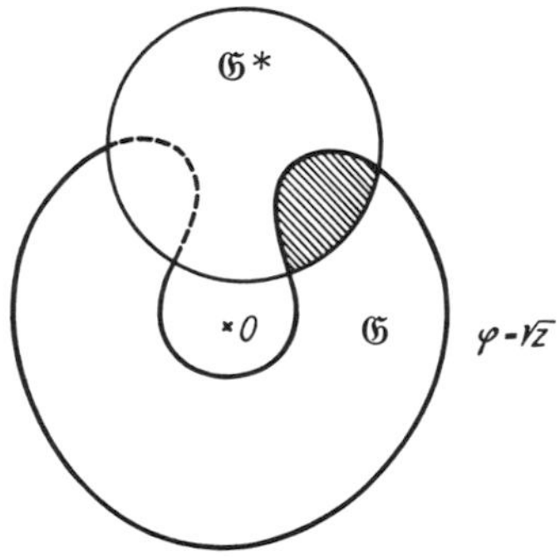

Fig. 4

this possibility that *multivalent* (non-*schlicht*) *domains*, i.e., Riemann surfaces, must be chosen as the domains of definition in the theory of functions. But if two continuations obtained in this way are found to agree in a domain, then naturally we do not count such a domain more than once. For example, if the continuation leads back to the original function after a k-fold circuit of a point z_0, we obtain a closed chain of domains. By means of the subsidiary variable t, obtained from the relation[29] $z - z_0 = t^k$, this chain is mapped holomorphically and one-to-one into a univalent domain of the t-plane containing the origin, and the function φ is transferred to that domain as a meromorphic function of t; furthermore, if φ is

[28] The z-plane here takes the place of the w-plane in the previous section; and the extended function is again denoted by φ.

[29] Again we assume that z_0 is finite.

meromorphic in every sheet of a Riemann surface over a whole neighborhood of z_0, apart from z_0 itself, and if the function φ or its reciprocal is bounded there, then in the t-plane this function can be extended meromorphically (§1.5) to $t = 0$; in this case we adjoin to the Riemann surface a point situated "over z_0" and assign as a neighborhood to this point the chain of domains as extended by it; finally we define the value of the function at this point by the value of the extension of φ; then φ as a function of the parameter t is meromorphic at $t = 0$, and consequently this point is a branch point of order $(k - 1)$ for the function φ or for its Riemann surface over z_0.

In order to proceed with the theory of functions on such a Riemann surface we must again determine when we are to consider a complex-valued function defined on $\mathfrak{R}$ as being holomorphic at a given point of $\mathfrak{R}$. Precisely this question was answered in a natural way in §4.1 for certain special Riemann surfaces; but we saw at that time that the definition given there could be so formulated as to refer only to the projection-mapping of $\mathfrak{R}$, and since this mapping is obviously available in the present case, we can take over the former definition: a function on $\mathfrak{R}$ is holomorphic at a branch point[30] of order $k - 1$ lying above the point z_0 if the given function can be written as a function of $t = \sqrt[k]{z - z_0}$ holomorphic at $t = 0$.

As a direct consequence we see that *the following two meromorphic functions exist on* $\mathfrak{R}$: *the projection-mapping z and the defining analytic function* φ.

4.3. Now it is clear that we can proceed further with the argument of §4.1:

Let $\mathfrak{R}_z$ be a Riemann surface over the z-plane $\mathbf{P}_z$ and let f be a meromorphic function on $\mathfrak{R}_z$. Then over the value-plane $\mathbf{P}_w$ we can construct a Riemann surface $\mathfrak{R}_w$ such that f appears there as an invertible mapping of $\mathfrak{R}_z$ onto $\mathfrak{R}_w$, and on $\mathfrak{R}_w$ a function is obviously holomorphic at a point if and only if when transferred to $\mathfrak{R}_z$ it is holomorphic at the corresponding point; in particular, the projection-mapping w and the inverse $\check{f}$ of f are therefore meromorphic functions on $\mathfrak{R}_w$.

But if we take into account that the given function f represents the transfer of the projection-mapping w from $\mathfrak{R}_w$ to $\mathfrak{R}_z$, and that the projection-mapping z plays the same role for the function $\check{f}$, and that therefore a projection-mapping and its defining function are interchanged if the two Riemann surfaces (on each of which the theory of functions is the same as on the other) are interchanged, we realize the complete irrelevance in our present theory of functions of the fact that our Riemann surfaces are covering surfaces over the underlying planes; in other words: *behind*

[30] It is convenient to regard an ordinary point as a branch point of order zero.

the concept of a Riemann surface as we have presented it up to now there lies a more general definition, which no longer depends on coverings.

4.4. We now recall the description given in §3.1 for the closed plane **P**. We had covered **P** with two domains, which were then superimposed on complex planes, namely the *parameter planes*, by means of *invertible* mappings; and at that time we saw that the one-to-one relation between parts of the parameter planes, arising from the fact that the domains on **P** necessarily overlap, is holomorphic; but this statement is merely another expression of the fact that the holomorphy of functions on **P** can be described as holomorphy in terms of a local parameter, since the holomorphy is independent of the choice of parameter.

In §4.1 and §4.2 we saw that the holomorphy of functions at a point of a Riemann surface can also be described in terms of a suitable parameter t; if the point in question is an ordinary point, this parameter merely determines the projection-mapping of a neighborhood of the point into the complex plane over which the Riemann surface lies; but when we were dealing with a branch point, we made use of a different parameter, which mapped a neighborhood of the given point in a *one-to-one* way onto a neighborhood of its image point, while the projection-mapping depended holomorphically on this parameter (cf. the example in Figure 3).

Thus in order to be able to define holomorphy in a meaningful way it is sufficient to know that every point of a Riemann surface belongs to an open set for which there is given a topological mapping, called a *local uniformizing parameter*, into the complex plane such that for every inter-

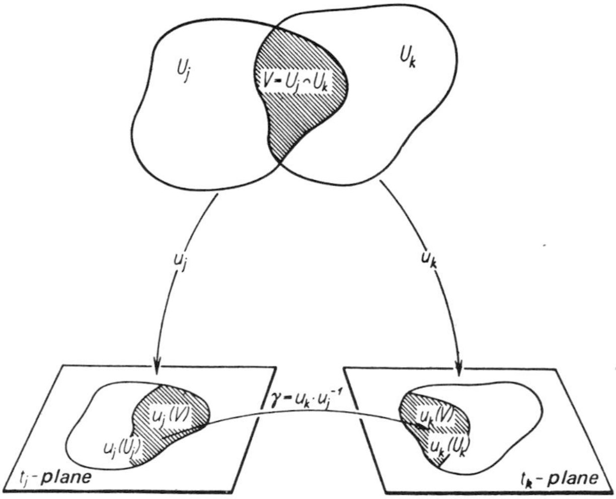

Fig. 5. u_j, u_k are topological; $j = u_k u_j^{-1}$ is holomorphic.

section of these open sets (which are mapped into different planes) the corresponding parts of the image domains correspond holomorphically to each other (Figure 5). Again we simply say: the surface is covered by neighborhoods from an *atlas of holomorphically related charts.*

We can also express this property by saying that *the surface carries a complex-analytic structure*; this is the precise condition under which a theory of functions is possible. Surfaces that satisfy this condition are called *abstract Riemann surfaces,* which therefore include all those surfaces,[31] and only those surfaces, on which a theory of functions is meaningful.

Our earlier Riemann surfaces, which we now call *concrete* in distinction to the abstract surfaces, are abstract Riemann surfaces with the added property that two meromorphic functions f and g exist on them: namely, the projection-mapping[32] and the function which defines the Riemann surface. The existence of these two functions makes the definition of a concrete Riemann surface somewhat narrower than that of an abstract Riemann surface, since for the latter there is no a priori necessity for the existence of any functions, apart from the constants, that are meromorphic on the whole surface.

But now that we have discovered the structure of the sets of points on which it is possible for meromorphic functions to exist at all, it is our task to develop their "theory of functions" far enough to provide precise statements of existence sufficient for the following fundamental conclusion: every abstract Riemann surface can in fact be regarded as a concrete Riemann surface; in other words, the definition of the concrete Riemann surface includes properties that are superfluous in the sense that they can be proved from other properties, namely those that define an abstract Riemann surface.

The general function-theoretic resources for this purpose will be obtained below in analogy with the developments in §3.

Among these is included, to begin with, the concept of a *holomorphic or meromorphic differential,* which will be defined exactly as in §3.3; here again the most important examples of differentials will be the derivatives of meromorphic functions with respect to the local uniformizing parameters. For holomorphic differentials the statement and proof of the Cauchy integral theorem are the same as in §3.4; the proof depends on the fact

[31] Instead of "*surface*" it is also customary to speak of a "*two-dimensional manifold,*" i.e., a *connected Hausdorff-space* which is locally Euclidean and two-dimensional; in other words, every point has a neighborhood homeomorphic to an ordinary circular disk.

[32] The branch points are a property of the projection-mapping: their dependence on the local uniformizing parameter is of higher than the first power. On the other hand, it is clearly meaningless to speak of branch points on an abstract Riemann surface.

that the Riemann surface is *oriented* by the orientation of the parameter planes, so that in a positive passage around the segments of a decomposition of the domain in question the parts of the boundary that are common to two subdomains are traversed in both directions. The concept of the *residue* of a meromorphic differential at a pole and the resulting residue theorem follow here in the same way as in §3.4.

5. Functions on a Riemann Surface

The definitions of the preceding section determine the behavior of a holomorphic function *everywhere in the small* on a Riemann surface $\mathfrak{R}$. Thus it is natural to ask: which global properties necessarily follow from the local properties? If we consider this question as being fully answered only when we have a clear view of all meromorphic functions on $\mathfrak{R}$, we are still a long way, in the present state of mathematics, from being able to answer it for every Riemann surface. Only for two special classes of Riemann surfaces, *compact* and *simply connected*, do we have far-reaching results. For example, for the closed z-plane $\mathbf{P}$, which belongs to both these classes, we were able to characterize the meromorphic functions as being precisely those functions that are rational in z; this result was based in an essential way on the fact that a function everywhere holomorphic on $\mathbf{P}$ is necessarily constant.

5.1. The argument for the latter result was based in turn on the compactness of $\mathbf{P}$ and can be carried out equally well for *every compact Riemann surface* $\mathfrak{R}$. For if the function f is holomorphic on $\mathfrak{R}$, then $|f|$ is real and continuous on $\mathfrak{R}$ and therefore, in view of the compactness of $\mathfrak{R}$, assumes its maximum at some point x_0; thus the local uniformizing parameter at x_0 transplants the function f as a holomorphic function to a planar domain in which the maximum of its absolute value is assumed at the *interior* point corresponding to x_0; consequently, by the maximum principle (§1.13) this absolute value is constant, and then the same statement can be made for f in a neighborhood of x_0 and therefore on the whole of $\mathfrak{R}$ (since the statement and proof for the identity theorem (§1.10) remain unchanged for Riemann surfaces). *Thus on a compact Riemann surface every holomorphic function is constant, and therefore the meromorphic functions are completely determined, up to an additive constant, by their (finitely many) poles.*

5.2. We now wish to derive some general results for *meromorphic functions on a compact Riemann surface*, for which purpose we assume that there exists a nonconstant meromorphic $z = z(x)$ on $\mathfrak{R}$. Then $\dfrac{dz}{z - a}$ is a meromorphic differential, so that the sum of all its residues must be equal to zero, as can be proved on compact Riemann surfaces in exactly the same way as for the closed plane **P** (cf. §3.4); for our special differential this sum of residues is again equal to the difference between the number of poles and the number of a-points of z; thus we again have the following theorem.

Every nonconstant function that is meromorphic on a compact Riemann surface $\mathfrak{R}$ *assumes every value equally often.*[33] The points at which a meromorphic function z assumes its values to higher than the first order are the zeros of the differential dz and thus are isolated. In the present case this means: *to every function z that is meromorphic on a compact Riemann surface* $\mathfrak{R}$ *there corresponds a natural number n_z such that every value of z, apart from finitely many exceptional values, is assumed, and in fact to the first order, at precisely n_z points on* $\mathfrak{R}$. In other words: *the function z maps* $\mathfrak{R}$ *like an n_z-sheeted concrete Riemann surface* $\mathfrak{R}_z$ *over the closed z-plane*; the n_z sheets are everywhere univalent (*schlicht*) over the z-plane except for the finitely many branch points of $\mathfrak{R}_z$.

Now let w be a second meromorphic function on $\mathfrak{R}$. By means of z the function w is transferred to $\mathfrak{R}_z$. For every value a of z which is not a branch point there thus correspond exactly n_z values $w_1, \ldots, w_{n_z}$ over a. If these values are all finite, they satisfy the equation

$$(w - w_1)(w - w_2)\cdots(w - w_{n_z}) = w^{n_z} + \sum_{i=1}^{n_z} (-1)^i \sigma_i(w_1, \ldots, w_{n_z}) w^{n_z - i} = 0,$$

where the elementary symmetric functions[34] σ_i are well-defined functions of a. Since it is seen at once that the holomorphy of w is also transferred, the σ_i are holomorphic functions of a, which can be meromorphically extended (as is shown by means of the Riemann lemma (§1.5)), to the finitely many exceptional points; thus (§3.2) the σ_i are rational functions of a. Consequently there exists a polynomial $P(w, z)$ that vanishes everywhere on $\mathfrak{R}$ except perhaps at finitely many points; but since this polynomial is meromorphic on $\mathfrak{R}$, the identity theorem (§2.4) implies that it must vanish everywhere on $\mathfrak{R}$. If the polynomial is reducible, our result holds for an irreducible factor (cf. IB5, §§4.1 and 2.3). Thus we have proved the following statement.

If z is a nonconstant function meromorphic on the compact Riemann surface $\mathfrak{R}$, *then every function w meromorphic on* $\mathfrak{R}$ *is connected with z by an irreducible algebraic relation $P(w, z) = 0$; here P is a polynomial in z and w whose degree in w is at most n_z.*

[33] This result again implies that the constants are the only functions holomorphic on the whole of $\mathfrak{R}$.

[34] Cf. IB4, §2.4.

Thus in the field $K_{\mathfrak{R}}$ of functions meromorphic on $\mathfrak{R}$ there exists an element v of maximal degree, which is necessarily a primitive element[35] of $K_{\mathfrak{R}}$ over the field $\mathbf{C}(z)$ of functions rational in z. For if there existed a $w \in K_{\mathfrak{R}}$ not rational in z and v, then the field $\mathbf{C}(z)(v, w)$ would be of higher degree (cf. IB7, §2) than $\mathbf{C}(z)(v)$ and a primitive element in it would be of higher degree than v. Thus we have proved the following result.

The field $K_{\mathfrak{R}}$ of functions meromorphic on the compact Riemann surface $\mathfrak{R}$ is an algebraic function field (unless it consists merely of the field of constants $\mathbf{C}$). *In other words, there exist in $K_{\mathfrak{R}}$ two functions z and v with the following properties:*

1. $K_{\mathfrak{R}} = \mathbf{C}(z, v)$, *so that every $w \in K_{\mathfrak{R}}$ is a rational function in z and v;*
2. *The functions z and v are connected by an irreducible algebraic equation.*

Conversely, it is easy to show that *every irreducible equation $P(v, z) = 0$* in the sense of §4.2 *defines an analytic function whose corresponding Riemann surface is compact.*[36]

The compact Riemann surfaces are identical with the Riemann surfaces of algebraic functions.

For the proof of this theorem we still lack the following two details: *first, the existence of a nonconstant meromorphic function z on $\mathfrak{R}$; and secondly, the proof that $K_{\mathfrak{R}}$ is exactly of degree n_z over* $\mathbf{C}(z)$; for if this degree were smaller than n_z, then the Riemann surface $\mathfrak{R}_z^*$ corresponding to a generator $v \in K_{\mathfrak{R}}$ would have fewer sheets than $\mathfrak{R}_z$, so that the latter would not be the Riemann surface of an algebraic function; but in the present case $\mathfrak{R}_z$ is identical with $\mathfrak{R}_z^*$.

Both these assertions are direct results of the following theorem.

Every Riemann surface is "meromorphically separable;" i.e., for every two points x_1, x_2 there exists a meromorphic function v_{12} with $v_{12}(x_1) = 1$, $v_{12}(x_2) = 0$: this function separates the points x_1, x_2.

Since such a function is not constant, the first of the two assertions follows at once. But the second also follows: for let a be a nonexceptional value of a nonconstant function z and let $x_1, \ldots, x_{n_z}$ be the (consequently distinct) points on $\mathfrak{R}$ at which z takes the value a, and let us then form the functions

$$v_{ik}, \qquad 1 \leqq i, \quad k \leqq n_z, \quad i \neq k$$

with the property that

$$v_{ik}(x_i) = 1, \qquad v_{ik}(x_k) = 0$$

and from them the functions

$$v_i := \prod_{\substack{k=1 \\ k \neq i}}^{n_z} v_{ik};$$

[35] Cf. IB7, §6, theorem 3.

[36] See, e.g., Behnke-Sommer [1*], p. 418.

for these latter functions we have

$$v_i(x_k) = \delta_{ik},$$

so that the function

$$v := \sum_{j=1}^{n_z} j v_j$$

has the values

$$v(x_j) = j, \qquad 1 \leqq j \leqq n_z;$$

thus for $z = a$ the algebraic equation connecting z and v has at least n_z distinct solutions in v, so that its degree in v cannot be smaller than n_z.

The foregoing theorem on the existence of separating functions can easily be reduced to a theorem in which only one arbitrary point occurs:

There exists a meromorphic function φ with a single arbitrarily prescribed pole x_1 with suitable principal part.

For then this point x_1 will be separated from a distinct point x_2 by the meromorphic function

$$v_{12}(x) := 1 - \frac{1}{1 - \varphi(x_2) + \varphi(x)},$$

since $v_{12}(x_1) = 1$, $v_{12}(x_2) = 0$.

The functions occurring in this theorem are called *elementary functions*; on $\mathfrak{R}$ they obviously play the role of polynomials on the closed plane, since the latter are characterized by the property of having a single pole at ∞. The importance of this analogy is due to the fact that the ordinary interpretation of a polynomial as a linear combination of powers cannot be extended to the case of Riemann surfaces; for on them there does not exist, in general, any distinguished function which, like z on $\mathbf{P}$, describes the points of $\mathfrak{R}$.

We shall not further pursue the question of proving the existence, asserted above, of elementary functions; let us merely say that the usual "potential-theoretic" method[37] consists of embedding the functions that are meromorphic in $\mathfrak{R}$ in classes of more general entities, in which we can step by step create the desired properties.

Here we have merely attempted to prove that the theory of functions on compact Riemann surfaces is identical with the theory of algebraic functions, which is an extremely well-developed discipline.[38]

5.3. Of quite a different kind are the results concerning *simply connected Riemann surfaces*:

There exist exactly three distinct types of simply connected Riemann surfaces: the unit circle, the plane of complex numbers and the closed plane.

[37] See, e.g., A. Pfluger [3*], §§30 and 38; a method depending purely on function-theoretical arguments is given in Behnke-Sommer [1*], p. 523 ff.

[38] Behnke-Sommer [1*], chapter VI, §§3–5.

The precise meaning of this statement is as follows:

On every simply connected Riemann surface $\mathfrak{R}$ there exists a meromorphic function which maps $\mathfrak{R}$ in a one-to-one way onto one of the above-mentioned "normal domains;" as a result, all the values of the function are assumed exactly once.

In other words: on a simply connected *Riemann surface* $\mathfrak{R}$ there exists a *global uniformizing parameter*, with which the set of local uniformizing parameters may be replaced.

This famous theorem, known as the "principle of uniformization," is proved in somewhat the following way;[39] we first show, by a process of exhaustion[40] of $\mathfrak{R}$, that there exists a univalent (*schlicht*) meromorphic function, i.e., a function which assumes every value at most once, and then we make use of the Riemann mapping theorem,[41] according to which every simply connected domain in the plane, and thus the image domain of $\mathfrak{R}$, can be mapped meromorphically and one-to-one onto one of our three normal domains.

The uniformization theorem states, for example, that every compact, simply connected Riemann surface is equivalent to the closed plane. For surfaces of higher connectivity there is no general principle of equivalence.[42]

The question whether a noncompact, simply connected Riemann surface is of hyperbolic or parabolic type, i.e., is equivalent to the circle or to the plane,[43] forms the so-called *type problem.*

The theory of functions on an arbitrary Riemann surface is still in a state of rapid development. The methods by which its problems are attacked are in every case natural generalizations of the methods effective in the two cases dealt with above. At any rate, to give one important example, it has been proved that *on arbitrary, noncompact Riemann surfaces* (which can be exhausted by means of compactifiable surfaces) *there exist nonconstant holomorphic* (and thus not merely meromorphic as in the compact case) *functions* guaranteeing that the *surface can be regarded as concrete surface 2* [44] in the sense of §4.2.

[39] The clearest and most elegant proof is the one given by B. L. van der Waerden [5].

[40] That such an exhaustion is always possible is a nontrivial result due to T. Radó [4].

[41] See, e.g., Behnke-Sommer [1*], p. 336 ff.

[42] This problem is called the *modulus problem*, since for the simplest case of the elliptic configuration, i.e., the *Riemann surface* for

$$w^2 = a_0 z^4 + a_1 z^3 + a_2 z^2 + a_3 z + a_4,$$

it is solved by the quotient of two characteristic magnitudes called "moduli."

[43] In any event the two cases are mutually exclusive; for a holomorphic mapping of the plane onto a circle would be bounded at ∞ and could therefore be extended holomorphically in such a way as to be holomorphic on the closed plane, and thus constant and not invertible.

[44] Behnke-Stein [2]; cf. also A. Pfluger [3*], §40.

We have now erected the scaffolding for a *general theory of functions* and have tried to show how it naturally depends on the concept of holomorphy. We should now proceed to build up the *theory of functions proper* with its numerous individual results, all of them not only elegant from the mathematical point of view but also of great utility in practical life. In the foregoing text we have not discussed such special questions, in order not to obscure the abstract development of the argument; and lack of space now prevents us from adding to this general theory even a slight indication of the richness of the special theory.

Bibliography

[1] BEHNKE-SOMMER: Theorie der analytischen Funktionen einer komplexen Veränderlichen. Springer, Berlin-Göttingen-Heidelberg 1955. (Textbook.)

[2] BEHNKE-STEIN: Entwicklungen analytischer Funktionen auf Riemannschen Flächen. Math. Ann. **120** (1948).

[3] PFLUGER, A.: Theorie der Riemannschen Flächen. Springer, Berlin-Göttingen-Heidelberg 1957. (Textbook.)

[4] RADÓ, T.: Über den Begriff der Riemannschen Fläche. Acta Szeged **2** (1925).

[5] VAN DER WAERDEN, B. L.: Topologie und Uniformisierung der Riemannschen Flächen. Sitzungsber. sächs. Akad. Wiss. **93** (1941).

Bibliography added in Translation

AHLFORS, L. V.: Complex analysis. An introduction to the theory of analytic functions of one complex variable. McGraw Hill, New York 1953.

AHLFORS, L. V. and SARIO, L.: Riemann surfaces, Princeton Mathematical Series, No. 26, Princeton Univ. Press., Princeton, N.J. 1960.

BOAS, R. P.: Entire Functions. Academic Press, New York 1954.

HILLE, E.: Analytic Function Theory. 2 vols. Ginn and Co., Boston 1959–1962.

KAPLAN, W.: Introduction to analytic functions. Addison-Wesley Publ. Co., Reading, Mass. 1966.

NEVANLINNA, Rolf; PAATERO, V.: Einführung in die Funktionentheorie. Birkhauser Verlag, Basel–Stuttgart 1965.

SPRINGER, G.: Introduction to Riemann surfaces. Addison-Wesley Publ. Co., Reading, Mass. 1957.

CHAPTER 7

Points at Infinity

Introduction

Among the very few questions with which the general public is likely to approach a mathematician, the intersection of parallel lines at infinity seems to be regarded as considerably less important than such questions as the quadrature of the circle, the *Fermat theorem*, or the trisection of an angle. A person with a mathematical education is rather inclined not to take this particular question very seriously. If the questioner insistently requires an explanation of the intersection of parallel lines at infinity, he is likely to say something about convention or fiction. But such an answer is just as wrong as it would be, for example, for the question "Is $\sqrt{5}$ really a number?" All mathematical objects are abstract entities. To be sure, some of them stand in idealized analogy with material objects. (For example, the point, line and plane of finite Euclidean geometry.) But as a mathematical object a point at infinity exists in exactly the same sense as other mathematical objects, e.g., the finite points. This "existence" of the points at infinity will be discussed below in §2. It is a familiar fact that they first turned up in elementary geometry, where they should be introduced as soon as the intersection of straight lines is discussed. Moreover, it can hardly be said that the concept of a point at infinity is very difficult. Thus there have always been teachers, particularly those who are enthusiastic about projective geometry, who have insisted that the course of instruction in secondary schools should include a detailed treatment of points at infinity, since they believe that the abstract concepts so characteristic of mathematics can be understood better and more quickly by means of points at infinity than in any other way.

From the scientific point of view it must be admitted that in Euclidean geometry, in real analysis and even in the theory of functions of a complex variable it would be quite possible to avoid the introduction of points at infinity. But as we shall see below, they enable us to state many theorems in a much simpler form. But the question as to what sort of points at infinity we should introduce, whether the single point in the *Gauss plane* of the theory of functions, or the whole line of points in *Euclidean geometry*, must be decided by mathematical arguments. Under certain quite natural conditions the usual choice of points at infinity is the only possible one. For example, no one would ever study the theory of functions of a complex variable on the projective plane, or real projective geometry on the plane extended by a single point at infinity.

In higher dimensions, however, the decision is by no means unique. For the space C^n of n complex variables there exist infinitely many possibilities, all of them equally justifiable from the point of view of the theory of functions. The passage from one compactification of C^n to another has given rise to the concept of a modification, which plays an important role in the modern theory of functions of several complex variables. The concepts "*compactification*" and "*modification*" will be defined below.

I. The Usefulness of Points at Infinity

A) *Let us first consider the Euclidean plane.* We must above all remember that any question concerning properties of lines is meaningful only as part of a given mathematical system. The lines in elementary geometry, i.e., the straight lines described by the *Euclid-Hilbert system of axioms*, have properties different from those of the lines in a non-Euclidean geometry. This fact is particularly important for the theorems about intersections. The lines through a point P that is not on a given line g behave differently in elliptic, Euclidean, and hyperbolic geometry. In the first case all these lines have a point of intersection with g in the finite plane (consider the great circles on a sphere). In the second case there is exactly one exceptional line, and in the third case infinitely many. Thus we must first state our system of axioms. Only then can we discuss the intersection of parallel lines (i.e., lines not intersecting in the finite plane) at a point at infinity.

If we are simply asked about the intersection of parallel straight lines, it is natural to think of Euclidean geometry and its axioms, by which, in accordance with the general practice of the last sixty years, we mean the *Hilbert axioms*. But in these *Hilbert axioms* the concept of an infinitely distant element, and in particular of an infinitely distant point, does not occur. Also, if we look at the theorems that follow immediately from these axioms (theorems that form the basic introduction to mathematical rigor in the schools) here too we will search in vain for the concept of an infinitely distant element. In fact, the older textbooks, as long as they were not dealing with projective geometry, often avoided these infinitely distant elements altogether, although some mention of the intersection of parallel lines can be found as far back as Kepler.[1] But then, towards the end of the 19th century, infinitely distant elements (points, lines, planes were systematically introduced[2] as one of the fruits of the new insight into projective geometry. However, as long we do not wish to proceed from Euclidean geometry to projective, it is not absolutely necessary to introduce points at infinity. Although it is only for convenience that we extend the plane by adjoining points at infinity, still it is well known that we thereby obtain a remarkable simplification in elementary geometry. This simplification already applies to the fundamental statement: two lines either intersect in a point or else are parallel. For after introduction of the line at infinity the corresponding statement reads: two distinct lines intersect in exactly one point. Then there is the dual statement: through two given points there passes exactly one line. It is true that the wording of this statement is exactly the same as in the axioms of Euclidean geometry. But now the statement is more inclusive, since one of the points, or even both of them, may lie at infinity. In view of the fact that an infinitely distant point can be represented by a direction, the modified statement also contains the following information: through a given point there passes exactly one straight line in a given direction. And for three-dimensional Euclidean geometry the simplification is even more remarkable. Consider, for example, the following statement. Three given planes either have a line in common, or they have a single point in common, or they intersect pairwise in three parallel lines, or two of the planes are parallel and cut the third plane in two parallel lines, or all three planes are parallel. As soon as we have closed the space by adjoining the infinitely distant plane, this complicated statement becomes simply: three planes have either a line or a point in common.

These examples show the convenience of points at infinity in the fundamental statements of *Euclidean geometry*. But if we proceed further into the subject, so that the basic theorems must be employed many times, the simplification becomes even more noticeable. Consider, for example, the theorem of Pappus and Pascal (Figure 1). The hypothesis of this well-known theorem runs as follows: "On each of two lines g and g' intersecting at a point P there are chosen three points A_1, A_3, A_5 and A_2, A_4, A_6, all distinct from P. The point S_1 is the intersection of the line through A_1, A_2 with the line through A_4, A_5; correspondingly, S_2 is the intersection of the line through A_2, A_3 with the line through A_5, A_6 and S_3 is the intersection of the line through A_3, A_4 with the line through A_6, A_1." Then if we are restricted to finite points on the plane, the conclusion of the

[1] See Johannes Tropfke, Geschichte der Elementar-Mathematik, Band 4, 2. Auflage (1923), pp. 53–54.

[2] See, e.g., Theodor Reye, Die Geometrie der Lage, 2. Auflage (1877), pp. 15 ff.

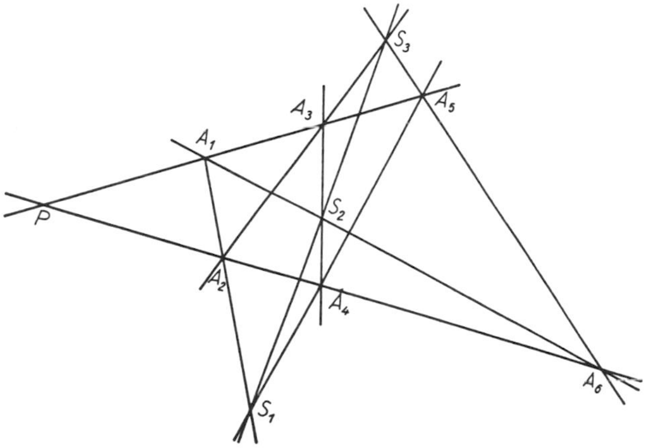

Fig. 1. Theorem of Pappus and Pascal

theorem reads: "The points S_1, S_2, S_3 all lie on one line, or the three pairs of lines are all parallel (so that there are no points of intersection S_1, S_2, S_3), or the lines in one of the pairs are parallel and the line s_1 joining the points of intersection of the other two pairs is parallel to the lines in the first pair." But if the improper elements of the *Euclidean plane* have already been introduced, the theorem of Pascal and Pappus can be stated very simply "S_1, S_2, S_3 are collinear." Moreover, the statement not only becomes much more concise; it includes several additional cases, which would otherwise require separate treatment, namely: 1. the lines g and g' are parallel, so that P is at infinity; 2. either A_1 or A_2, or both, are at infinity.

The reason for these advantages is quite obvious. The position of two parallel lines is no longer distinguished from any other position for two lines. The simplification arises from the fact that now there are no exceptions to the rule: "Two lines in a plane intersect in a point."

Thus the introduction into Euclidean geometry of infinitely distant points or, more precisely, of the points at infinity in the projective sense, is unquestionably useful, and after a little practice we are able, not only to argue logically points are about these points at infinity, but to visualize them as well. However, the student may well become confused when he subsequently learns that in the theory of functions only one infinitely distant point is allotted to the complex plane.

Of course, the *Gauss complex plane* is no different, before the adjunction of points at infinity, from the Euclidean plane with rectangular coordinates. The combination of the two coordinates x and y to form a single complex coordinate z naturally makes no difference in this respect. But after closure the two planes have quite different properties, as will be explained in detail in the following sections.

B) Here we must discuss the reasons for introducing *points at infinity in analysis*. In the present section we shall deal with *real analysis*, about which there is much less to be said than about complex analysis, which will be discussed in the next section.

First of all, it is important to understand the following fact: *we do not introduce any number* ∞, since ordinarily an element is not called a number unless it belongs to a field that includes the rational numbers. But since subtraction $\infty - \infty$ cannot be uniquely defined (as a consequence of the equation $\infty + a = \infty$ for all $a \neq \infty$), we cannot speak here of a number ∞. Admittedly, however, the points $+\infty$ and $-\infty$ do occur in analysis, as the "end points" of the half-lines $a < x$ and $x < a$.

Thus we interpret:

1. $\lim_{n\to\infty} a_n = +\infty$: for every number N, no matter how large, there exists a n_0, such that $a_n > N$ for $n > n_0$.

2. $\lim_{x\to\infty} f(x) = A$: for every $\epsilon > 0$ there exists an x_0 such that

$$|f(x) - A| < \epsilon \qquad \text{for} \qquad x > x_0.$$

3. $\lim_{x\to x_0} f(x) = +\infty$: for every M, however large, there exists a $\delta > 0$ such that $f(x) > M$ for $|x - x_0| < \delta$.

4. $-\infty < f(x) < +\infty$ for $x \in J$: to every $x \in J$ the function f assigns a real number.

5. $-\infty < f(x) \leqq +\infty$ for $x \in J$ and $f(x)$ continuous in J: to every $x \in J$ the function f assigns a real number, or else $x \in J$ is an exceptional point for which 3 is valid.

The convenience of the symbol ∞ in such formulas is immediately obvious, and since every student of the infinitesimal calculus can easily rewrite these statements in such a way as to avoid the symbol ∞, there is no need for us to discuss them further.

But we must note that a sequence can now be convergent, namely to $+\infty$ (or $-\infty$), even though it was called divergent before (e.g., the sequence of natural numbers). Thus it is only in the more advanced parts of the infinitesimal calculus that systematic use is made of the above formulas. But then it is to be noted that in the Weierstrass-Bolzano theorem the condition of boundedness can be omitted. The theorem then runs: "Every infinite sequence of real numbers has at least one limit point." (We say $+\infty$ is a limit point of the sequence a_n if for every M, however great, there exist infinitely many a_n for which $M < a_n$.) This elimination of an exception for unbounded sequences is naturally of great value, since the Weierstrass-Bolzano is the most important theorem in real analysis.

If in a (Hausdorff) space R the Heine-Borel theorem is valid, i.e., if in *every covering $\{W_\iota, \iota \in J\}$ of R, where J is a (finite or infinite) index set and the sets W_ι are open, there exist finitely many $W_{\iota_1}, \ldots, W_{\iota_\zeta}$ already covering R, then R is said to be compact* (formerly also called bicompact). As will be shown in §4, *every infinite set of points in a compact space has at least one limit point.* Thus the Weierstrass-Bolzano theorem is valid in such a space without exception. If a given space R is extended in such a way as to become compact, we speak of a compactification of R (often called a closure in the older literature). This compactification must satisfy the usual axioms for such concepts, as discussed below in §4.

The introduction of the two points at infinity on the real line is such a compactification, and the same remark holds for the points at infinity in the projective plane (projective space).

C) *We now come to complex analysis.* So far as numbers are concerned, the same remark can be made as in real analysis: there cannot exist any complex number ∞. But this fact only makes it of greater importance that we compactify the space in such a way that any infinite sequence of points has at least one limit point. Here we adjoin, not a line as in the projective geometry of the plane, but a *point* ∞. For this procedure there are many reasons. The reason usually given in the textbooks is that only then can the transformation $w = \frac{1}{z}$, and consequently all transformations

$$(1) \qquad w = \frac{az + b}{cz + d} \quad \text{with} \quad ad - bc \neq 0,$$

map the extended plane one-to-one onto itself without exception. This explanation is completely satisfactory only if we confine our attention to the linear transformations (1), which admittedly are of outstanding importance but are certainly not the only transformations. Nevertheless, in the theory of functions of a complex variable there exists only *one* closure. Now this theory is in fact the theory of holomorphic functions and their meromorphic extensions (see III 6). Therefore it deals only with domains such that for every point a criterion of holomorphy can be set up; in other words, the domain must be given a complex structure. In the case of one variable the domains for which this can be done are precisely the *Riemann surfaces.* Thus we must demand that the plane be extended by points at infinity in such a way that the extended plane is a *Riemann surface.* But this requirement already puts very strict limitations on the possibility of introducing points at infinity, in view of the following theorem (Radó-Behnke-Stein-Cartan):[3] *If $\mathfrak{R}$ is a Riemann surface and f is*

[3] Formulated here only for $n = 1$. But it holds in general for functions of n complex variables. See Erhard Heinz: Ein elementarer Beweis des Satzes von Radó-Behnke, Stein-Cartan. Math. Ann. 131 (1956), pp. 258–259. For $n = 1$ see C. Carathéodory, Conformal representation, Cambridge Tracts 1952.

a function which is continuous on $\mathfrak{R}$ *and is holomorphic at all points* $z \in \mathfrak{R}$ *for which* $f(z) \neq 0$, *then f is holomorphic everywhere in* $\mathfrak{R}$. Let us now consider the z-plane compactified to the *Riemann surface* $\mathfrak{R}$ and remove the points $|z| \leqq 1$. The remaining $\mathfrak{R}'$ is again a *Riemann surface*. On it we consider the function $w = \frac{1}{z}$, to which at the points at infinity we assign the value zero. Then $f(z)$ is certainly continuous at all points of $\mathfrak{R}'$. It follows from the above theorem that $f(z)$ is holomorphic at all points of $\mathfrak{R}'$. All the points at infinity on $\mathfrak{R}'$ (and thus also on $\mathfrak{R}$) are zeros of $f(z)$. Since $f(z)$ does not vanish identically, its zeros are isolated. In a sufficiently small neighborhood of a point P at infinity there thus exist, apart from P itself, only finite points. Such a neighborhood $U(P)$ is mapped by $w = \frac{1}{z}$ onto a neighborhood of the origin. Thus there exists an $\epsilon > 0$ such that all complex numbers $|w| < \epsilon$ are assumed in $U(P)$. But since $w = \frac{1}{z}$ takes on each value only once in the finite part of the plane, there cannot exist more than one point at infinity. On the other hand, this compactification by one point is in fact possible (see §3).

Now since $f(z) = \frac{1}{z}$ maps the compactified plane one-to-one and holomorphically (i.e., *biholomorphically*) onto itself, it follows that the points $|z| > N$ for sufficiently large N are neighborhoods of the point ∞. (Their images are in fact the neighborhoods $|w| < \frac{1}{N}$ of the image $w = 0$.)

This point ∞, together with a neighborhood of it, is mapped biholomorphically by $w = \frac{1}{z}$ onto a neighborhood of $w = 0$. Thus $g(z)$ is holomorphic at $z = \infty$ if $g\left(\frac{1}{w}\right) = h(w)$ is holomorphic at $w = 0$ (in accordance with the definition of a *Riemann surface*).

If $g(z)$ is holomorphic at $z = \infty$ (or meromorphic), then $g(\infty)$ is defined. "Aha! So infinity is after all a number," the Incorrigibles will say. But, although we cannot prevent anyone from saying that what turns up as the argument of a function (i.e., of a mapping) is a number, "infinity" does not thereby acquire "numerical properties" (the points on a Riemann surface are not numbers either), so that we ourselves will adhere to the above definition of a number as an element of a field containing the rationals, so that ∞ is a point in the plane but is not a number. Then if $w = f(z)$ has a pole at z_0, its image in the w-plane is the point $w_0 = \infty$, since $w' = \frac{1}{w} = \frac{1}{f(z)} = f^*(z)$ has the image point $w_0' = 0$ for z_0. (But if $f(z)$ is essentially singular at z_0, then z_0 has no image point in the closed w-plane, since $f^*(z)$ is also essentially singular at z_0.)

We now have the generalized chain rule: if $w = f(z)$ is meromorphic at z_0 and $w^* = g(w)$ is meromorphic at $w_0 = f(z_0)$, then $g(f(z)) = h(z)$ is also meromorphic at z_0 (see III 6, §2).

2. Existence and Properties of the Projective Plane

We now carry out the procedure of extending the *Euclidean plane* E_0 to the projective plane E_p; we do this first in a purely geometric way, and thus in particular without the use of coordinates, and then analytically, i.e., by means of homogeneous coordinates.

I. Let E_0 be a plane for which the *Hilbert system of axioms* is assumed. Consequently, by points and lines we shall always mean points and lines of E_0.

The property of two straight lines g_1 and g_2 in E_0 of being parallel to each other, in symbols $g_1 \| g_2$, is reflexive, symmetric and transitive. Thus we can form classes of parallel lines (cf. the classes of numbers congruent mod m in the theory of numbers, or the classes of equivalent pairs of integers in the construction of rational numbers. For the concept of a class see also IA, §8.5 and IB6, §4.1.) It is clear that there exist infinitely many classes of parallel lines and that every line belongs to exactly one class.

Each of the classes thus introduced is now called a *point at infinity*, so that there exist infinitely many points at infinity. The totality of all the points at infinity is again called the "*line at infinity.*" Further, an expression like "*g passes through the infinitely distant point K*" means that g lies in the class K. Then every line g passes through an infinitely distant point and, since g belongs to only one class, through exactly one infinitely distant point. Further, by a "point" we mean either a point in the plane in the former sense or else an infinitely distant point (and similarly for the lines), so that the *incidence theorem* follows at once: "*Two distinct lines g_1 and g_2 have exactly one point in common*". The set of points on a line "closed" in this way, together with the above relations among them, is now called the *projective plane E_p*.

II. By a number triple $p = (x_1, x_2, x_3)$ we mean three real numbers, not all of which vanish. Two number triples $\tau = (x_1, x_2, x_3)$ and $\tau' = (x_1', x_2', x_3')$ are said to be equivalent, in symbols $\tau \sim \tau$ if and only if there exists a $\lambda \neq 0$ such that $x_i' = \lambda x_i$ $(i = 1, 2, 3)$. Then for arbitrary number triples τ_1, τ_2, τ_3:

1. $\tau_1 \sim \tau_1$;
2. from $\tau_2 \sim \tau_1$ follows $\tau_1 \sim \tau_2$;
3. from $\tau_3 \sim \tau_1$, $\tau_1 \sim \tau_2$ follows $\tau_3 \sim \tau_2$.

The sets of equivalent number triples can thus be combined into classes K (every number triple belongs to exactly one class K), and the classes containing a number triple of the form $(a, b, 0)$ are said to be *singular*. In the same way the homogeneous linear equations $a_1x_1 + a_2x_2 + a_3x_3 = 0$ that can be transformed into one another by multiplication with a constant $\lambda \neq 0$ are combined into classes G, and the class of equations $a_3x_3 = 0$ is said to be singular. The class K is said to lie on the class G if for a number triple in K and an equation in G, and consequently for all number triples in K and all equations in G, the given equation is satisfied by the given number triple.

The totality of the nonsingular classes K and G is monomorphic to the totality of the points P and the lines g of the given Euclidean plane E_0 (see IA, §4.6).

In other words: every statement about points and lines in a Euclidean plane E_0 holds equally well for the (nonsingular) classes K and G, and conversely.

This monomorphy is almost self-evident if we make use of the fundamental theorem that in the plane E_0 it is possible to introduce rectangular coordinates x, y over the real field. If P has the coordinates x, y, then the corresponding class K is the class of all triples $(\lambda x, \lambda y, \lambda)$ with arbitrary $\lambda \neq 0$, and conversely the class containing the number triple (x_1, x_2, x_3) corresponds to the point P with coordinates $x = x_1/x_3$, $y = x_2/x_3$.

But we must note that it is not the entire set of classes K and G that is monomorphic to the points P and the lines g, but only *a subset* of it, namely the nonsingular classes. *Thus the singular classes K are called points at infinity*, and their totality constitutes the line at infinity; from now on we speak, not of the classes K and G, but of the corresponding P and g. The P, g and the infinitely distant elements, together with the relations subsisting among them, form a projective plane E_p.

These two ways of introducing the points at infinity have been described in some detail in order to emphasize their analogy with the usual methods of extension in arithmetic and algebra. We see that the points at infinity have a no less valid claim to existence than say the algebraic numbers. It would not occur to anyone to refuse to deal with algebraic numbers on the ground that they "do not really exist." The points at infinity are on exactly the same footing as the finite points.

Some of the properties of the projective plane have already been described in §1. Here let us merely make some additional remarks relevant to our present purposes.

The totality of the transformations with $|a_{ik}| \neq 0$

$$T: \begin{cases} x' = \dfrac{a_{11}x + a_{12}y + a_{13}}{a_{31}x + a_{32}y + a_{33}} \\[2ex] y' = \dfrac{a_{21}x + a_{22}y + a_{23}}{a_{31}x + a_{32}y + a_{33}} \end{cases}$$

or in homogeneous coordinates:

$$x_1' = a_{11}x_1 + a_{12}x_2 + a_{13}x_3,$$
$$x_2' = a_{21}x_1 + a_{22}x_2 + a_{23}x_3,$$
$$x_3' = a_{31}x_1 + a_{32}x_2 + a_{33}x_3$$

sets up a one-to-one correspondence between 1. the points and 2. the lines of E_p. Before the introduction of the points at infinity, this one-to-one correspondence obviously did not hold for transformations T with non-constant denominator. Thus the group of one-to-one linear mappings has been significantly extended by the introduction of the points at infinity.

Another property displayed by the plane E_p after the adjunction of points at infinity is the one which has given a name (compactification) to our process of extension. *The plane E_p is compact.* We shall speak more in detail on this subject in §4. Here let us merely say that in E_p (as in every compactified plane) the Weierstrass-Bolzano theorem is valid without the condition of boundedness; in other words, every infinite sequence of points has at least one limit point. Of course, before stating this theorem we must make some suitable agreement as to when a point at infinity is to be called a limit point for a sequence a_n; such a definition can easily be given if for a point at infinity we make use of the representation of its singular class by real number triples $(x_1, x_2, 0)$. Then a set M in E_p is said to be *closed* if it contains all its limit points, and *open* if the complement of M in E_p is closed.

A set of points G is called a *domain* if it is open and connected. There exist infinite domains, namely those that contain points at infinity. The finite points $(x, y) \neq (0, 0)$, taken together with the points at infinity $(x, y, 0)$, form an infinite domain G if in both cases the condition $\left|\frac{y}{x}\right| < c$, with $c > 0$, is satisfied; in particular, this domain contains the point at infinity on the x axis $(x, 0, 0)$.

The domain G just defined is divided into four subdomains by the line at infinity and the x axis (see Figure 2).

Here the line at infinity separates Part I from Part IV, and Part II from Part III, but we can pass from Part I to Part IV without crossing

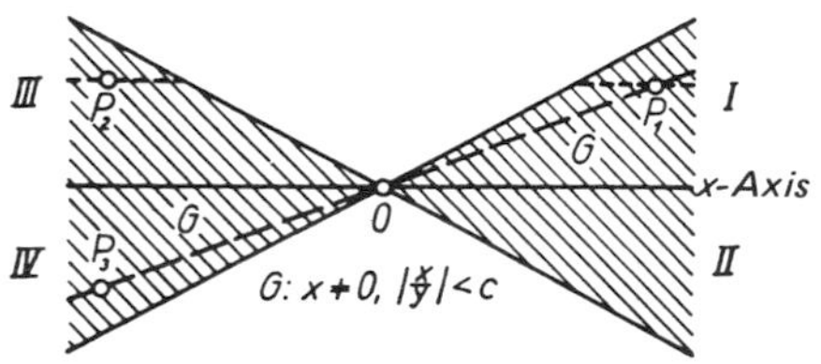

Fig. 2. Projective closure

the x-axis, say by way of the point at infinity on the line $\frac{y}{x} = \frac{c}{2}$. On the other hand, Parts I and II are separated from each other by the x-axis. Parallels to the x-axis running through I and III in G join points of I and III, but in the meantime they cross the x-axis (namely at the point at infinity).

The plane E_p has very surprising properties with respect to direction and orientation, as we can well recognize by just dividing G into the two subdomains G_1, consisting of II and III, and G_2, consisting of I and IV. If we traverse the x-axis in a positive direction beginning with the origin, at first G_1 is on the "left," but after passage through the point at infinity, the same domain is on the "right." We say E_p is a one-sided, i.e., nonorientable, surface. (The most familiar example of a nonorientable surface is the *Möbius strip* in Figure 3.) Now the points at infinity also have (open)

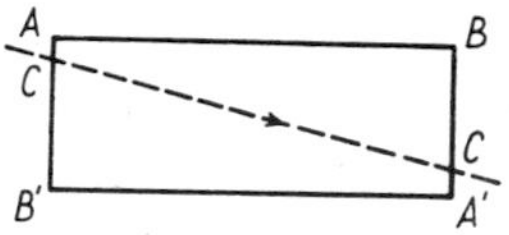

Fig. 3. Möbius strip

neighborhoods which are the projective (and therefore certainly topological) images of neighborhoods of the origin. To such neighborhoods (corresponding say to circles about (x_0, y_0)) belong the sets of points: $\left|\frac{y}{x} - \alpha\right| < \epsilon$, $x^2 + y^2 > \frac{1}{\epsilon^2}$, together with the points at infinity $(x, y, 0)$ with $\left|\frac{x}{y} - \alpha\right| < \epsilon$. This special property is of fundamental importance in topology and will be discussed further in §4. We describe it here by saying that the projective plane is *closed.* As fundamental properties obtained from the projective closure of the plane we have already recognized:

1. The general validity of the incidence theorem.
2. The one-to-one mappings of the group of projective transformations.
3. The compactification of the plane to a *locally Euclidean Hausdorff space* (i.e., to a manifold).

3. The Function-Theoretic Closure of the Euclidean Plane

Instead of the compact z-plane E_K we often speak of the number sphere, for the reason that these two manifolds can be mapped onto each other one-to-one and conformally by means of the stereographic projection. The essential feature of this projection, discussed in every textbook on the theory of functions, is that the points of the finite plane are assigned one-to-one to the points of the sphere with the exception of a single point N, called the north pole.

The mapping is topological, and even conformal. In particular, a sequence of points z_n in the plane E_K converges to ∞ if and only if its images P_n converge to N. Thus the mapping remains topological if the plane is extended by $z = \infty$, the image of this point being the point N on the sphere. Instead of the compactified plane it may often seem preferable to speak of the Riemann sphere, since then the existence of the point at infinity is no longer contested (it is just another name for the north pole of the sphere). The advantage of the sphere is that it provides a metric which, unlike the Euclidean metric, does not fail for the point ∞. By the *chordal distance* χ of two points P_1, P_2 on the sphere (and thus between the corresponding complex numbers z_1, z_2) we mean the distance between P_1 and P_2 in the three-dimensional Euclidean space. For two finite points z_1, z_2 this distance is

$$\chi(z_1, z_2) = \frac{|z_1 - z_2|}{\sqrt{(1 + |z_1|^2)(1 + |z_2|^2)}},$$

and for z_1 and ∞ it is

$$\chi(z_1, \infty) = \frac{1}{\sqrt{1 + |z_1|^2}}.$$

Since the sphere is oriented (right and left are never interchanged by a circuit around a closed curve, and this property is transferred by the stereographic projection onto the compactified plane E_K) the plane E_K is also oriented. In Figure 2 the orientability is made clear by the fact that I, II, III, IV are four separated domains with only the boundary point ∞ in common. During a passage along the x-axis it can no longer happen that a domain which was at first on the left will later be on the right. Also, the incidence theorem for lines is also no longer valid. For now two lines intersect in two points (S and ∞) if they are not parallel or in one point (namely ∞) if they are parallel. On E_K the projective transformations are in general not one-to-one (a fractional transformation transforms the point at infinity into a line). Conversely, the transformations $w = \dfrac{az + b}{cz + d}$ with $ad - bc \neq 0$ are not, in general, one-to-one on E_p.

The surface E_K is a *Riemann surface*. As a system of neighborhoods about a finite point z_0 (basis of neighborhoods in the sense of the *Reimann surface*) we may take all the circles around that point; as a system of neighborhoods for infinity we may take the exteriors of all circles about the origin (namely the points z for which $\chi(z, \infty) < \epsilon$). Two charts (cf. III 6, §3.1) are sufficient to cover E_K: for example, the unit circle, and the exterior of the circle $|z| = \frac{1}{2}$.

4. Neighborhoods and Compactification

In order that the Weierstrass-Bolzano theorem may be valid without exception in the extended plane, it is not sufficient merely to introduce points at infinity in one way or another. It is also necessary to define what is meant by saying that a point at infinity is a limit point of a set or of a sequence. This is exactly what we have just done for E_p (§2) and E_K (§3), thereby obtaining the concepts of closed and open sets (and consequently also the concept of neighborhood of a point at infinity). Only then can we raise the question of the compactness of E_p and E_K and ask whether the plane E_0 extended in this way has been compactified by E_p and E_K. However, in order to define neighborhoods in a given space, it is certainly not necessary to begin by introducing limit points. Nowadays it is much more usual to proceed in the opposite direction. We first introduce neighborhoods, which must satisfy the *Hausdorff axioms*. In this way we obtain a *Hausdorff space*. For such a space we have the usual definitions of open sets and limit points, and then we can again ask about the compactness of the space (see §1). In the case of E_p and E_K this second procedure is just as natural as the first, since we have already stated what we mean by neighborhoods of a point at infinity. In E_K they were the sets of points $|z| > N$ including $z = \infty$ and in E_p, for the point at infinity $(x_1, y_1, 0)$, they were the sets of points:

$$x^2 + y^2 > \frac{1}{\epsilon^2} \qquad \text{with} \qquad \left|\frac{y}{x} - \frac{y_1}{x_1}\right| < \epsilon$$

and

$$x^2 + y^2 > \frac{1}{\epsilon^2} \qquad \text{with} \qquad \left|\frac{x}{y} - \frac{x_1}{y_1}\right| < \epsilon.$$

We have already asserted that E_p and E_K are compact. Let us now prove this assertion. The proof will be in two parts, I and II.

I. The spaces E_p and E_K have *countable topologies*, i.e., in both planes every open set can be represented as the union of elements of a fixed countable system of special open sets. Such a system is called a *base for the topology* of E_p and E_K respectively. A *countable base for the open plane* E_0 is obtained by taking all the circles with radii $\frac{1}{n}$ about the points $r_1 + ir_2$; r_1, r_2 rational. For E_K we must take, as further base elements, the exterior of the circles $x^2 + y^2 = N^2$, where N runs through all natural numbers. For E_p we must add the following sets.

1. The sets of points, $\left|\frac{y}{x} - r_3\right| < \frac{1}{N}$, $x^2 + y^2 > N^2$, together with the points at infinity $(1, r, 0)$, with $|r - r_3| < \frac{1}{N}$. Here again r_3 runs through all rational numbers and N through all natural numbers.

2. Neighborhoods of the point at infinity on the y-axis, say $\left|\frac{x}{y}\right| < \frac{1}{N}$, $x^2 + y^2 > N^2$, together with the points at infinity $(\epsilon, 1, 0)$ with $|\epsilon| < \frac{1}{N}$. Obviously the totality of all these neighborhoods is countable and forms a base. In §§2 and 3 we saw that in E_p and E_K every infinite sequence of points has at least one limit point. A space (here E_p and E_K) in which this property holds for every infinite sequence is said to be *sequentially compact* or semicompact (in the older literature, compact).

II. We now have the following fundamental theorem.

Theorem A. *If the sequentially compact Hausdorff space T has a countable base, then T is compact.*

Proof. Let $V = \{V_k, k = 1, 2, 3, \ldots\}$ be a countable base for the open sets of T. Let U be any open covering of T. Since V is countable and every element of U is the union of elements of V, it follows that U contains a countable subcovering $U' = \{U_1', U_2', \ldots\}$ of the whole of T. If finitely many elements of U' already cover the space T, nothing remains to be proved. But if not, let us choose an infinite sequence P_ν from T for which

$$P_\nu \in T - U_1' \cup U_2' \cup \cdots \cup U_\nu'.$$

By hypothesis, the P_ν have a limit point $P \in T$. But P does not lie in any of the open sets $U_1' \cup \cdots \cup U_\nu'$ and therefore not in T, which is a contradiction. Consequently T is already covered by finitely many of the U_i'. Since we have already seen in I that E_p and E_K have countable topologies, the proof of the assertion that *E_p and E_K are compact* is now complete.

Not every sequentially compact space is compact, but the converse is true.

Theorem B. *Every compact space is sequentially compact.*

On account of the importance of this theorem, let us give an outline of the proof. If the statement were not true, there would exist in T an infinite sequence P_ν without limit points. Thus for every $P \in T$ there would exist a neighborhood $U(P)$ containing at most one point of the sequence P_ν. If T is compact, a finite subcovering can be selected from the covering $\{U(P), P \in T\}$. This subcovering can contain only finitely many of the P_ν, but it covers the whole of T, which is the desired contradiction.

It is now natural to ask whether a compact space can be compactified a second time. The answer is in the negative, as we shall prove by first laying down a *system of axioms for the concept of compactification* (thereby making the above definition precise).

Let $\mathfrak{R}$ and $\mathfrak{R}'$ be Hausdorff spaces. Then $\mathfrak{R}'$ is said to be a compactification of $\mathfrak{R}$ if: 1. *$\mathfrak{R}$ is an open subset of $\mathfrak{R}'$;* 2. *every open neighborhood of $\mathfrak{R}' - \mathfrak{R}$ has a nonempty intersection with $\mathfrak{R}$;* 3. *$\mathfrak{R}'$ is compact.*

The requirements 1 and 3 are obviously necessary, and 2 means that the newly adjoined ("infinitely distant") points are to be limit points of $\mathfrak{R}$.

The compactification defined in this way has the following characteristic properties, which justify the name:

a) $\mathfrak{R}$ *is a subspace of* $\mathfrak{R}'$.
b) *The topology of* $\mathfrak{R}$ *is not changed by the compactification* (as follows from the axioms for a Hausdorff space).
c) *If* $\mathfrak{R}''$ *is a compactification of* $\mathfrak{R}'$, *then* $\mathfrak{R}'' = \mathfrak{R}'$.

It remains to prove c). We assume that $N = \mathfrak{R}'' - \mathfrak{R}'$ is not empty and let $\mathfrak{U} = \{U\}$ be the system of open neighborhoods of $N = \mathfrak{R}'' - \mathfrak{R}'$. Then $\mathfrak{U} \cap \mathfrak{R}' = \{U \cap \mathfrak{R}'\}$ has the following properties. 1. The empty set does not belong to $\mathfrak{U} \cap \mathfrak{R}'$. 2. If $U_1 \cap \mathfrak{R}'$ and $U_2 \cap \mathfrak{R}'$ are contained in $U \cap \mathfrak{R}'$, then $U_1 \cap U_2 \cap \mathfrak{R}'$ is also contained in $\mathfrak{U} \cap \mathfrak{R}'$.

A system $\mathfrak{U}^*$ of sets satisfying the axioms 1 and 2 is called a *filter base* (see III 1, §5.2). Filter bases are often used in present day topology instead of sequences of points. In the same way as for sequences, we can define limit points of filter bases. A point P_0 is called a *limit point of a filter base* $\mathfrak{W}$ if every neighborhood of P_0 has a non-empty intersection with every element $W \in \mathfrak{W}$. As in theorem B we can then prove the following theorem.

Theorem C. *In a compact space every filter base has at least one limit point.*

Now let the point Q_0 of the space $\mathfrak{R}'$ (which is a compactification of $\mathfrak{R}$) be a limit point of $\mathfrak{U} \cap \mathfrak{R}'$. Then every neighborhood $V(Q_0)$ has a non-empty intersection with every neighborhood of N. On the other hand, $\mathfrak{R}''$ is a Hausdorff space. Consequently, for every point $P \in N$ we can find neighborhoods $U(P)$ and $U(Q_0)$ such that $U(P) \cap U(Q_0) = 0$. Since N, as the complement of an open set, is closed and is therefore compact (as a closed set in a compact space), it is covered by finitely many of the neighborhoods $U(P)$. Let their union be denoted by W. Then there exists a neighborhood $V(Q_0)$ that has a non-empty intersection with W or, in symbols, $V(Q_0) \cap W = 0$, which is a contradiction. Thus N must be empty and consequently $\mathfrak{R}'' - \mathfrak{R}'$. A space that is the result of a compactification cannot be further extended by means of a new compactification.

5. Other Methods of Closing the Plane

In order to make a survey of closures of the plane other than E_p and E_K, we map the open plane E one-to-one and continuously onto the open unit disk S (we shrink the plane onto the disk), say by the transformations:

$$(1) \qquad x' = \frac{1}{1 + R}x, \qquad y' = \frac{1}{1 + R}y,$$

where x, y are coordinates in E and x', y' are coordinates in S. Also, we set $R = \sqrt{x^2 + y^2}$, $R' = \sqrt{x'^2 + y'^2}$, so that $R' = \frac{R}{1 + R}$. (By $\bar{S}$ we mean as usual the disk S together with its boundary points.) If this shrinkage is now extended in any natural way to the closed planes E_p and E_K, the points at infinity will be mapped onto the boundary of S, although no longer one-to-one. Thus in the projective closure, one and the same point at infinity is mapped onto two opposite points of a diameter of $\bar{S}$, and only onto two such points. In the function-theoretic closure, on the other hand, all the points of the boundary of S are images of one and the same point at infinity. Thus the problem of compactifying E can be regarded in particular as the problem of identifying the boundary points of S among themselves. These various compactifications of E differ from one another only in the various identifications of the points on the boundary of S. For example, we obtain the torus-like closure if we divide the bounding circumference into four parts (Figure 4):

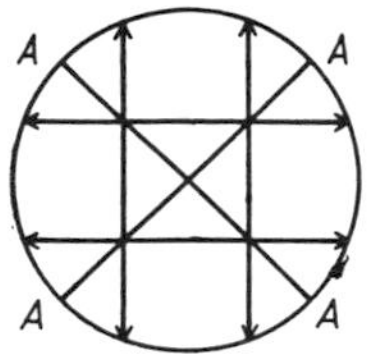

I. $-\frac{\pi}{4} < \varphi < +\frac{\pi}{4}$, III. $\frac{3}{4}\pi < \varphi < \frac{5}{4}\pi$,

II. $+\frac{\pi}{4} < \varphi < \frac{3}{4}\pi$, IV. $\frac{5}{4}\pi < \varphi < \frac{7}{4}\pi \equiv -\frac{\pi}{4}$.

Fig. 4

We now identify the four end points of these circular arcs as "the one boundary point A." Then we identify pairwise the points of Part I with those of Part III, and the points of Part II with those of Part IV, in such a way that the boundary points with amplitudes α and $\pi - \alpha$, and those with amplitudes $\alpha + \pi/2$ and $(3/2)\pi - \alpha$, with α running from $-\pi/4$ to $+\pi/4$, are identified. The plane compactified in this way will be called E_T.

Thus the boundary of the circle is divided into two "closed" curves, each of which runs from A to A and has only the point A in common with the other curve. Each of the pairs of boundary points, or the one quadruple A of boundary points, is then taken to represent *one* point at infinity. Consequently, under this compactification the points at infinity make up two curves K_1 and K_2 with a common point of intersection. (The fact that we thus obtain a torus can be visualized if, before the identification, we map the open disk topologically onto an open square and then undertake the identification of the boundary points in such a way that the four

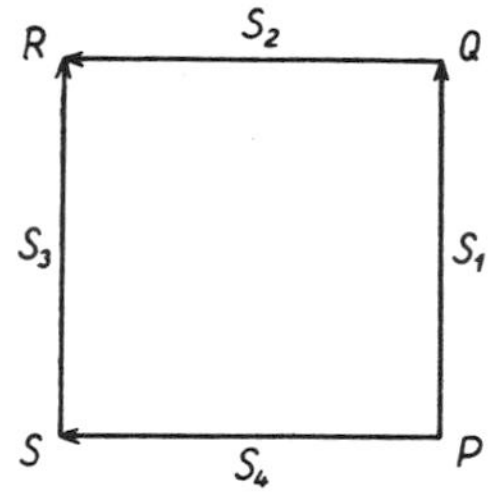

Fig. 5

vertices form the above quadruple. If we visualize the square as being made of paper (Figure 5), we must bring S_1 and S_3 into coincidence in such a way that R falls on Q and S on P. But then we need only cover S_2 with S_4 to obtain the torus.) Let us again look at the set of points G: $x \neq 0$, $|y/x| < c < 1$ (Figure 2). Under the mapping onto S the set of points goes into $x' \neq 0$, $|y'/x'| < c$, $\sqrt{x'^2 + y'^2} < 1$. The points of the circular arcs C_1 and C_2 (see Figure 6) are then assigned pairwise to each other in such a way that if, beginning at B and B', we traverse C_1 and C_2 "with equal velocity," the two points reached at any given time "represent" the same point at infinity, i.e., the points inside G' can be connected with one another by a curve lying inside G'. Thus the same remark holds for the original G of G' (see Figure 6). But now, on account of the correspondence

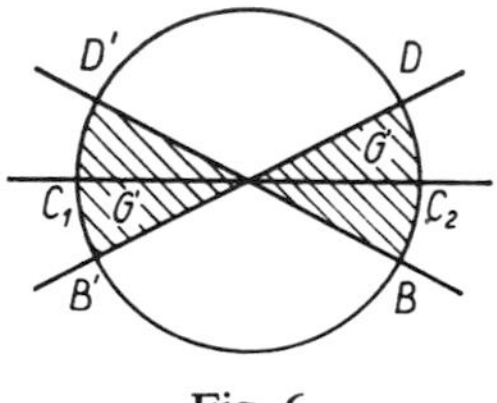

Fig. 6

between C_1 and C_2, it follows that the subdomains I and III (and also II and IV) of G are connected with each other, whereas I and IV, and II and III, are separated from each other by the x-axis. Thus as we traverse the x-axis, the right and left sides will not be interchanged at infinity. In E_T there is no closed curve L that divides E_T into two domains G_1 and G_2 such that in a circuit around L the domain G_1 lies at one time on the right and at another time on the left. We express this fact by saying that E_T is orientable (like E_K). It follows that we cannot map *the closed projective plane E_p one-to-one and continuously (topologically) onto the torus-like closed plane E_T*; for if this were possible, there would exist a domain on E_T, namely the image of G on E_p, that would be divided by a nonself-intersecting curve L into two subdomains, namely the images of G_1 and G_2, in such a way that during a circuit of L each of the two subdomains is at one time on the left and at another time on the right of L. But it is also obvious that E_T *cannot be mapped topologically onto* E_K. For on E_T there exist closed curves C that do not partition E_T. If we take the x-axis as a curve C, then all the points that are not on this axis can be connected with one another by a curve that does not cross the x-axis. But the space E_K certainly does not have this property, since otherwise it would also hold for the sphere, which is the image of E_K.

Like E_p and E_K, the space E_T is *compact* after the introduction of suitable neighborhoods of the points at infinity, since the circular disk $\bar{S}$ is sequentially compact. After the identification, we introduce neighborhoods in the obvious way for the points in $\bar{S}_T$ defined by the identification. Then $\bar{S}_T$ is compact. We obtain E_T by a one-to-one mapping of $\bar{S}_T$. Then the neighborhoods in E_T are defined as the images in $\bar{S}_T$. In this way E_0 is compactified to E_T.

We have now discovered two further properties of closed planes. The plane E_p admits the projective transformations as one-to-one mappings into itself, and correspondingly the plane E_K admits the fractional linear holomorphic transformations. Neither of these two sets of mappings is defined at the points at infinity of E_T. Nevertheless E_T does admit a group of one-to-one continuous mappings, namely the mappings of the torus into itself. The torus admits a two-parameter rotation group. Naturally, the formulas for the transformations in the corresponding group of E_T can no longer be written linearly in the rectangular coordinates x and y. Nevertheless, like E_p and E_K, the plane E_T exhibits a *transitive group of transformations*. For any two points P_1 and P_2 of E_T one of these transformations takes P_1 into P_2. Yet none of these transformations will take all the points at infinity simultaneously into the finite plane, since the image of two closed curves that intersect in only one point (as is the case with the points at infinity and the image of the point A in S_T) cannot lie entirely in the finite plane E_0.

As a particular property of E_p we mentioned the general validity of the incidence theorem (for lines). This theorem fails to hold for E_T, just as it fails for E_K. For example, the two lines $y = +x$ and $y = -x$ intersect twice, at the origin and at the point at infinity that appears on the boundary of S_T as the (four times represented) boundary point A. On the other hand, the two lines $x = 0$ and $y = 0$ intersect only once. A general statement about replacements for the incidence theorem can be made only after we have determined how the lines join one another at infinity. The arrangement here is less obvious than for E_p and E_K, because the lines, except for $x = 0$, $y = 0$, $y = \pm x$ and their parallels, are no longer closed curves. Of course, they can be made into closed curves by pairwise combinations, in which each line g is associated with its mirror image in the x-axis or the y-axis (depending on the inclination of g) (Figure 7).

But then for two such T-lines the number of points of intersection varies from one to four. If we attempted to state the Pappus-Pascal theorem in terms of this complicated incidence, we would obtain an extremely complex and unintelligible result. Here again it is obvious that statements concerning the intersections of lines are most conveniently expressed in E_p. But now at least it is immediately clear how we can obtain other closures of the plane in addition to E_p, E_S, E_T. We divide the boundary of the

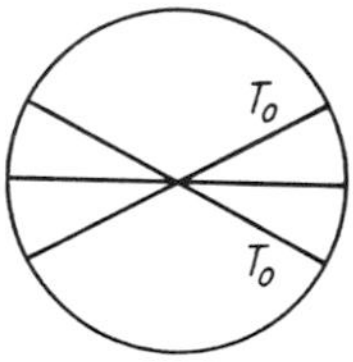

Fig. 7. The torus-line T

image circle S into two subintervals, identify the end points of these subintervals, and then set up a correspondence between the intervals by identifying each point of the interval a with a point of the interval a', where a' is traversed in the same sense as a or in the opposite sense. The first case occurs in E_p, and the second in E_T. If the correspondence for all the pairs is in the opposite sense, then during a circuit around a closed curve K through points at infinity the right and left sides of the curve will not be interchanged; more precisely, a second curve, provided only that it remains in a sufficiently close neighborhood of K and does not intersect K, will remain on one side (right or left). But this will not be the case, at least not for all curves, if the correspondence for a pair of intervals a and a' is in the same sense (as is shown already by E_p).

Now it is a fundamental theorem of topology that every closed two-dimensional surface in R^3 can be mapped one-to-one and continuously onto the unit circle, provided we identify the boundary points in the manner described above. Thus *we can always close the plane in such a way that it is the one-to-one continuous image of a preassigned closed two-dimensional surface.* For every natural number p there exists a closed two-dimensional surface containing p nonintersecting closed curves $C_1, \ldots, C_p$ such that the set of all points not on $C_1, \ldots, C_p$ is a domain. The same remark then holds for the closed plane. In this connection it is to be noted that on E_T the closed line $x = 0$ does not partition the plane E_T. If we close the plane by means of an identification of the boundary provided by a one-sided surface (for example, the infinite Möbius strip), we obtain a nonorientable plane; on the other hand, if the identification is provided by a two-sided surface, the resulting closed plane will be an orientable manifold. From the topological point of view these statements are self-evident, since the property of orientability is preserved under topological mappings. Naturally there are other possibilities for extending the plane by the adjunction of points at infinity. But then it is no longer possible to map the plane E_0 one-to-one and continuously onto the circle $\bar{S}$ (with identification of the boundary) and thereby to describe E_0 in terms of $\bar{S}$.

6. Modifications

1. In addition to its topology, a topological space can carry a local structure; for example, local complex coordinate systems (called local uniformizing parameters) are defined on every *Riemann surface* $\mathfrak{R}$. If the neighborhoods $\mathfrak{U}_1$ and $\mathfrak{U}_2$ with coordinates $z^{(1)}$ and $z^{(2)}$ have an intersection ϑ, then by the definition of a *Riemann surface* there exists on ϑ a one-to-one holomorphic transformation of $z^{(1)}$ into $z^{(2)}$. A function f defined on ϑ and holomorphic there as a function of $z^{(1)}$ is also holomorphic as a function of $z^{(2)}$, so that the concept of a holomorphic function on $\mathfrak{R}$ is well-defined (see III 6, §4.1). We say that a complex structure is thereby defined on $\mathfrak{R}$.

In the same way we see that the (real) projective plane has a local structure. For it is possible, for every point $Q \in E_p$, to find a neighborhood $U(Q)$ that can be mapped one-to-one and projective-linearly onto a domain of the *Euclidean plane.* The coordinates thereby introduced in $U(Q)$ are uniquely determined up to a projective-linear transformation.

The concept of a local structure cannot be described more precisely here, since it depends on an extremely abstract line of thought. Detailed information can be found in the works of MacLane and Eilenberg.

If X is a topological space with a local structure, then in compactifying it to a space Ψ we will naturally wish to extend the special local structure of X. In many cases this wish cannot be met. In other cases we obtain exactly *one* compactification satisfying the condition, and then the extension of the local structure of X to Y will be uniquely determined. For example, we have seen (§1,C) that the complex plane can be closed in *only* one way, namely, to the *Riemann sphere,* if we require that for every point at infinity there be given a local complex structure consistent with the local structure at every other point of the compactified plane. On the other hand, as was shown in the present section, there exist infinitely many topological closures of the plane.

2. We now show that *the projective closure of the plane is uniquely determined* if we assume that the (local) linear structure of the plane is extended to the points at infinity. Here we must pay particular attention

to axiom 2 of compactification (see §4, p. 265) to the effect that the points at infinity are nowhere dense in the extended space (for otherwise we would have to consider closure by the three-dimensional sphere, which may be regarded as a two-sheeted unbranched closed covering surface of the projective plane).

Proof. Let X be a connected compact *Hausdorff space* containing E_0, and let $U = \{U_K, K = 0, 1, \ldots, i_0\}$ be a finite open covering of X. In every open set U_K let two-dimensional real coordinates be defined, i.e., let U_K be mapped by a topological mapping ψ_k onto a domain G of the (x_1, x_2)-plane. Also let the coordinate transformations defined in the intersections $U_{K_1} \cap U_{K_2}$ be projectively linear. Finally, let $U_0 = E_0$ be the finite (x_1, x_2)-plane. Then it is clear that we must consider X as an "admissible" closure of the (x_1, x_2)-plane, since the local linear structure of E_0 has certainly been extended to the whole of X. Thus we must show that X is the projective plane. Let $P \in M = X - U_0$ be an arbitrary point "at infinity." Since all the coordinate transformations in X are projective-linear, the mapping ψ_0 can be extended to a projective-linear mapping into a neighborhood $U(P)$. This extension maps $U(P)$ topologically onto a domain $G \subset E_p$.

If we apply this procedure to every point $P \subset M$, we obtain an extension of ψ_0 to a continuous, locally topological mapping $\bar{\psi}: X \to E_p$. This mapping $\bar{\psi}$ is projective-linear with respect to the local coordinates.

It follows at once that $\bar{\psi}$ is a one-to-one mapping of X. In fact, if $P_1 \neq P_2 \in X$ were two points with $\bar{\psi}(P_1) = \bar{\psi}(P_2) = Q$, there would exist disjoint neighborhoods $U(P_1)$, $U(P_2)$, that would be mapped topologically by $\bar{\psi}$ onto a neighborhood $V(Q)$. Then we would have

$$\bar{\psi}(U(P_1) - M) \cap \bar{\psi}(U(P_2) - M) \neq 0,$$

so that there would exist finite points $P_1' \neq P_2' \in U_0$ with $\psi_0(P_1) = \psi_0(P_2)$, in contradiction to the hypothesis that ψ_0 is a topological mapping.

Furthermore: $\bar{\psi}$ maps X onto E_p.

Proof. On the one hand, $\bar{\psi}(X) \subset E_p$ is a compact subset of E_p, since it is the continuous image of a compact space. On the other hand, $\bar{\psi}(X)$ is also an open subset, since $\bar{\psi}$ is a local-topological mapping. Consequently $\bar{\psi}(X) = E_p$, since E_p is connected.

3. The generalization of *Riemann surfaces* to higher dimensions, namely complex manifolds, has been studied in great detail in recent years. Complex manifolds are defined in exactly the same way as Riemann surfaces. They are *Hausdorff spaces* that are covered by local systems of complex coordinates $z_1, \ldots, z_n$. The coordinate transformations $z^* = f_\nu(z_1, \ldots, z_n)$, $\nu = 1, \ldots, n$ defined in the intersection of any two such coordinate systems are holomorphic, i.e., the functions $f_\nu(z_1, \ldots, z_n)$ occur-

ring in them are holomorphic functions of the complex variables $z_1, \ldots, z_n$. The totality of these local complex coordinate systems is also called a complex structure for X (see III 6, §4).

If we wish to close the space C^n of n-tuples of complex numbers to a space X, it is reasonable to require that the natural complex coordinate system for C^n can be extended by further complex coordinate systems in such a way that X becomes a complex manifold. But now, in contradistinction to the cases discussed under 1 and 2, it is possible to construct *infinitely many closures for* C^n, $n > 1$. Thus we can no longer show, as in the case of $n = 1$ for the z-plane, that there exists exactly one reasonable closure. Let us give some examples for $n = 2$.

a) *The Osgood space* $\overline{C_2}$. We introduce the points at infinity (z_1, ∞) with $z_1 \in C^1$, and (∞, z_2) with $z_2 \in C^1$ and also (∞, ∞), and extend the topology and the complex structure of C^2 to the set thus obtained in such a way that the mappings

$$z_1^* = \frac{az_1 + b}{cz_1 + d}, \qquad z_2^* = \frac{a'z_2 + b'}{c'z_2 + d'}, \qquad \begin{vmatrix} a & b \\ c & d \end{vmatrix} \neq 0, \qquad \begin{vmatrix} a' & b' \\ c' & d' \end{vmatrix} \neq 0$$

are one-to-one holomorphic mappings of $\overline{C^2}$ onto itself. The *Osgood space* is isomorphic to the *Cartesian product* of the *Riemann sphere* with itself. It has two planes at infinity

$$E_1 = \{(z_1, \infty) : z_1 \in C^1 \cup \infty\} \approx E_K,$$
$$E_2 = \{(\infty, z_2) : z_2 \in C^1 \cup \infty\} \approx E_K,$$

which intersect at the point (∞, ∞).

For the theory of functions of two complex variables the *Osgood space* has many disadvantages. For example, the homogeneous linear transformations $z_1^* = az_1 + bz_2$, $z_2^* = cz_1 + dz_2$ are not, in general, one-to-one holomorphic mappings of $\overline{C^2}$ onto itself. Consequently, in recent work on the theory of functions a growing preference has been given to the projective closure of C^2.

b) *The complex-projective space* P^2. Here we introduce the points at infinity (z, ∞), $z \in \overline{C^1}$ and extend the topology and the complex structure of C^2 in such a way that the point (z, ∞) is the point at infinity common to the "complex lines" (real two-dimensional analytic planes) $z_2 = z \cdot z_1 + c$, and the complex-projective mappings are one-to-one holomorphic mappings of P^2 onto itself. The space P^2 has only one "infinitely distant plane," consisting of the points which in complex homogeneous coordinates can be described by $(\Lambda z_1, \Lambda z_2, 0)$. Under the complex-projective transformation

$$z_1' = \frac{1}{z_1}, \qquad z_2' = \frac{z_2}{z_1}$$

this plane is the holomorphic image of $z_1 = 0$.

In addition to the above closures, which have been of great use in the theory of functions, other closures of C^2, quite different in character, are "possible" though not customary.

c) For example, let us consider the holomorphic mapping of C^2 into itself first introduced by Bieberbach. The mapping $\varphi: C^2 \to C^2$ constructed by Bieberbach has the following properties:

1. φ is one-to-one.
2. The functional determinant of φ is everywhere equal to 1.
3. $C^2 - \varphi(C^2)$ contains an open non-empty subset.

Since $\varphi(C^2) \approx C^2$, we may consider that C^2 is represented by the subset $\varphi(C^2)$ of P^2 (or also of $\overline{C^2}$) and that the set $P^2 - \varphi(C^2)$ is the set of points at infinity. In this way we obtain a closure of C^2 that contains *hyperinfinitely distant points.* Every hyperinfinitely distant point is (by definition) surrounded solely by nonfinitely distant points. In the theory of functions, however, the *Bieberbach closure* is not often used. The coordinate functions z_1 and z_2 become essentially singular at the boundary of the domain of finite points and therefore they cannot be extended to the infinitely distant points, in the form of meromorphic functions, as in the Osgood and projective spaces.

It is tempting to compactify C^2 with a single point P, as is possible in topology. But in this case the complex structure of C^2 can no longer be extended to P_∞; for if in a neighborhood $U(P_\infty)$ there existed complex coordinates z_1^*, z_2^* which in $U' = C^2 \cap U(P_\infty)$ were connected with z_1, z_2 by a holomorphic transformation of coordinates $z_1 = f(z_1^*, z_2^*)$, $z_2 = g(z_1^*, z_2^*)$, then the function f, for example, would be holomorphic in $U(P_\infty) - P_\infty$ and yet would be unbounded in every neighborhood of P_∞. By a theorem of Riemann on removable singularities, a holomorphic function of two variables cannot have a nonremovable singularity at an isolated point; but then $f \equiv z_1$ can be extended holomorphically to P_∞ and thus is bounded there, which is a contradiction. The closure of the four-dimensional *Euclidean space* by a single point plays an important role, however, in the theory of the right-regular quaternion functions developed by Rudolf Fueter.

4. In view of the foregoing discussion, it is natural to ask how the various closures of C^n are related to one another. Let X', X be two closures of C^n that are admissible from the point of view of the theory of functions, and let N' and N be the corresponding sets of infinitely distant points. Then $X' - N' = X - N = C^n$, and the identical mapping i maps $X' - N'$ one-to-one and holomorphically onto $X - N$. Furthermore, N, N' are closed sets in X and X' (since their complements are certainly open). If U is an open neighborhood of N, then $N' \cup i^{-1}(U - N)$ is an open neighborhood of N'.

These characteristic properties of the passage from one admissible closure of C^n to another have led to the concept of a *modification*.[4] We now give a definition of this concept.

Let X', X be connected complex manifolds, let $N' \subset X'$, $N \subset X$ be closed sets, and let N be properly contained in X. Further, let φ be a one-to-one holomorphic mapping of $X' - N'$ onto $X - N$, so that $X - N$ and $X' - N'$ are holomorphically equivalent (in the above examples they are even identical). Now let the following axiom be satisfied: if U is a neighborhood of N, then $N' \cup \varphi^{-1}(U - N)$ is a neighborhood of N'. Then the passage from X to X' is called a modification of X.

Thus a modification is a process in which a closed subset $N \subset X$ is replaced by a closed subset N' in such a way that the space X' thus obtained is again a complex manifold (for example, the passage from P_2 to $\overline{C^2}$).

As is shown by the example of the *Bieberbach closure* of C^2, the new set N' can contain inner points, even when N is a nowhere dense subset of X. But this phenomenon cannot occur if the following condition is satisfied.

(*) There exists a neighborhood $U(N)$ and a function f that is holomorphic in $U(N)$, does not vanish identically in any open subset of U and is such that $f(x) = 0$ for $x \in N$. If (*) is satisfied, there cannot be any hyperinfinitely distant points. Of particular importance in the theory of functions are those closures of C^n in which the coordinate functions $z_1, \ldots, z_n$, and consequently all polynomials, can be extended in a unique way to meromorphic functions on the closed space. It turns out that in closures X, X' of this kind the identity $i : X' - N' \to X - N$ can be extended to a one-to-one meromorphic mapping of X' onto X. Modifications in which φ has this property are called *meromorphic modifications*. They are analogues of the classical birational transformations in algebraic geometry. Particularly interesting are the *proper continuous modifications*. In this case φ can be extended to a proper continuous mapping $\hat{\varphi}: X' \to X$. By an analogue of the theorem of Radó it follows that *$\hat{\varphi}$ maps X' holomorphically onto X and the inverse $\hat{\varphi}^{-1}$ is a meromorphic mapping*. The best-known example of a continuous proper modification is the *Hopf σ-process*, in which a point $x \in X$ is replaced by an $(n - 1)$-dimensional complex-projective space P^{n-1} (so that $N = \{x\}$, $N' = P^{n-1}$). On the other hand, the modification of $\overline{C_2}$ to P_2 is *not* properly continuous.

[4] As the first publication on this question, see H. Behnke and K. Stein, Modifikationen komplexer Mannigfaltigkeiten und Riemannscher Gebiete, Math. Ann. 125 (1951). For general orientation see the report H. Behnke and H. Grauert, Analysis in non-compact complex spaces, Princeton University Press 1960.

CHAPTER 8

Ordinary Differential Equations

1. Introduction

1.1. By a *solution* of an ordinary differential equation

$$(1) \qquad F(x, y, y', \ldots, y^{(n)}) = 0$$

we mean any function

$$y = \varphi(x)$$

that is differentiable over a domain I_x and is such that the function

$$F(x, \varphi(x), \varphi'(x), \ldots, \varphi^{(n)}(x))$$

exists and vanishes identically over I_x. Here the variable x can be either complex or real, but in the present chapter we shall always assume that x is real. Then I_x denotes the usual open intervals. We shall arrange the discussion under the following four headings:

1. Under what assumption can we expect that solutions (1) will exist at all?
2. By what methods can we obtain solutions?
3. What properties will a function have, if it is a solution of a differential equation?
4. Under what circumstances does a given differential equation have solutions that satisfy certain preassigned conditions?

1.2. Although the last of these questions necessarily refers to special equations, the differential equations in the first three are arbitrary. Nevertheless, for reasons connected not only with the methods available to us but with the inherent nature of the case, we do not begin our discussion with the most general type (1) of differential equations but proceed gradually from simpler types, in which the essential phenomena stand out more clearly. In this way we will arrive at various classifications of differential equations.

A differential equation in the general form (1) is said to be *implicit*, in contrast to the *explicit* differential equation

$$y^{(n)} = f(x, y, \ldots, y^{(n-1)}),$$

solved for the highest derivative of the desired function.

The order of highest derivative occurring in a differential equation is called the *order* of the differential equation.

The simplest type, on the basis of this classification, namely the explicit differential equation of first order

$$y' = f(x, y), \tag{2}$$

will play the principal role, not only because it constitutes the first non-trivial case, but also because it is not too special to suggest analogies for the general case.

1.3. The differential equation

$$y' = f(x), \tag{3}$$

expressing the elementary problem of quadrature, already shows that some assumptions must be made on the function f if the problem is to be solvable. For example, by a theorem of Darboux f must assume every value intermediate between any two of its values, if it is to be the derivative of a function $y = \varphi(x)$. On the other hand, continuity of f is already a sufficient condition for (3) to be solvable. But again, continuity is certainly not necessary, since discontinuous derivatives exist. We shall not raise the difficult question of the exact conditions for solvability but shall be content with the above remark that the continuity of f guarantees the solvability of (3).

In the more general case (2) it will turn out that continuity of f is again sufficient for the existence of solutions. However, since these solutions may have undesirable properties, we shall from time to time impose further restrictions.

2. Methods of Integration

2.1. Let the differential equation

$$y' = f(x, y) \tag{2}$$

be interpreted in the real plane with Cartesian coordinates x, y in the sense that at each point (x_0, y_0) at which f is defined we draw the straight line

$$y = y_0 + f(x_0, y_0)(x - x_0).$$

This line, considered as consisting of the point (x_0, y_0) and the slope $f(x_0, y_0)$, is called a *line-element*; the totality of the line-elements is the *direction field* of the differential equation. This interpretation obviously leads to the following criterion:

A function $y = \varphi(x)$ is a solution of the differential equation (2) if and only if its graph[1] lies in the domain of definition of f and at every point has a tangent coinciding with the line-element determined by the direction field of the differential equation.

Thus we have the following method for *graphical approximation* of the solutions.

On the x-axis let us introduce a discrete subdivision and at each point choose an arbitrary line-element from the direction field of the differential equation; we now follow it to the right and to the left up to the next point that lies over a point in our subdivision of the x-axis, and then proceed further along the line-element at the new point, and so forth.

In this way we obtain a polygonal line which may be expected to provide an approximation to a solution of the differential equation, an approximation which will be more nearly exact for finer subdivisions of the x-axis; and in fact, this polygonal procedure can be employed for graphical, and also numerical, integration of the differential equation (2); *many of the practical methods of solution in actual use are simply refinements of this method.*[2]

The method is justified by the following convergence theorem of Peano:[3] *for every sequence of subdivisions of the x-axis with norm converging to zero the polygonal method, beginning at an arbitrary point $P_0 = (x_0, y_0)$, provides a sequence of polygonal lines through P_0 from which it is possible to select a subsequence converging uniformly, in a certain x-interval about x_0, to a solution of* (2) *through P_0; here it is assumed that $f(x, y)$ is continuous in a domain containing P_0.*

Although this theorem guarantees the existence of solutions $y = \varphi(x)$ for which $y_0 = \varphi(x_0)$ with arbitrarily preassigned x_0, y_0, its constructive value is very much weakened by the necessity of choosing a certain subsequence. This disadvantage is eliminated if in addition to its continuity the function f satisfies a further restrictive condition (stated below), which will continue to play a very important role throughout the chapter. For if a polygonal sequence does not converge as a whole, it must have at least two subsequences that converge to different solutions, so that in this case there must be more than one integral curve through P_0. Consequently, if we know that only one solution passes through P_0, then each of the polygonal sequences constructed above must converge as a whole to this solution.

[1] I.e., the curve by which $y = \varphi(x)$ is represented in an x, y-diagram.

[2] See, e.g. [7], chapter 10.

[3] For the proof see, e.g. [4], §7.

The condition just mentioned, the so-called *Lipschitz condition*, guarantees precisely that *the solutions are unique*[4] (cf. §3), i.e., that there do not exist distinct solutions through the point P_0, two functions being considered distinct if they are not identical over any interval around x_0.

By definition, the *Lipschitz condition* is satisfied if the difference quotient of f with respect to y is uniformly bounded with respect to x; or more precisely, if for every rectangle R (with sides parallel to the axis) in the domain G_f of definition of f there exists a *Lipschitz constant* M_R satisfying the equation

$$|f(x, y_2) - f(x, y_1)| \leqq M_R|y_2 - y_1|$$

for every two points (x, y_1) and (x, y_2) in R.

Somewhat stronger, but more convenient in view of the mean-value theorem of calculus, is the requirement that f have a partial derivative with respect to y everywhere in G_f.

2.2. Another important method of solution is the Picard-Lindelöf *method of iteration.* This method depends on a connection between differential and integral equations; namely, every solution $y = \varphi(x)$ passing through (x_0, y_0) and lying in the domain of continuity of the differential equation (2)

$$y' = f(x, y)$$

is differentiable and therefore certainly continuous, and thus satisfies the integral equation

$$y = y_0 + \int_{x_0}^{x} f(t, y)dt; \tag{4}$$

and conversely, it is clear that every continuous solution of (4) is also a solution of (2) passing through (x_0, y_0).

We now construct a sequence of continuous functions φ_n, all of which pass through (x_0, y_0), by setting:

$$\varphi_0(x) = y_0,$$

$$\varphi_n(x) = y_0 + \int_{x_0}^{x} f(t, \varphi_{n-1}(t))dt, \qquad n > 0.$$

In a certain x-interval about x_0 these functions all lie in the domain of continuity of f. *If f also satisfies the Lipschitz condition, the sequence converges uniformly*[5] *in this interval to any solution of* (2) *through* (x_0, y_0); as a corollary, it follows at once that such a solution is unique.

[4] A more detailed analysis of uniqueness of solutions is to be found in [4], §12, and [2], §1.5.1.

[5] For the proof see, e.g. [4], §6.

2.3. Up to now we have laid only rather weak conditions on the differential equation (2), but let us now make the assumption that f is *analytic;* more precisely, in a neighborhood of the point (x_0, y_0) through which we wish to determine a solution, let $f(x, y)$ be represented by an absolutely convergent power series in $x - x_0$ and $y - y_0$. *Then there exists exactly one solution through* (x_0, y_0), *and this solution is again analytic;*[6] it can be written as a power series in $x - x_0$ with undetermined coefficients, and the differential equation provides a recursion formula from which the coefficients can be uniquely determined.

2.4. An explicit differential equation of the second order

$$y'' = f(x, y, y') \tag{5}$$

can also be interpreted geometrically in the x, y-plane in a natural way. Through every point and corresponding to every slope let us draw a circular arc with radius determined by the condition that at the point x_0 the second derivative of the function representing the circle has the value determined by (5). In terms of this three-parameter family of circles, which is called the *curvature field* of (5), the solutions of (5) can be described as those curves whose circle of curvature at each point is determined by one of the curvature elements at that point.

However, this interpretation is useful only in special cases;[7] for our general discussion it is too difficult to visualize, especially when extended to differential equations of higher order. In order to avoid such a strain on our powers of visualization let us rather *increase the number of dimensions of the space*, which will enable us to make a different interpretation, *allowing us to handle the explicit differential equation of* n-th *order*

$$y^{(n)} = f(x, y, y', \ldots, y^{(n-1)}) \tag{6}$$

in complete analogy with the equation of first order.

Since not only x and y, but also $y', \ldots, y^{(n-1)}$ are independent variables in (6), we interpret all these magnitudes as Cartesian coordinates in an $(n + 1)$-dimensional space $E^{(n+1)}$. Thus if we write y_1 instead of y, write y_2 instead of y' and finally y_n instead of $y^{(n-1)}$, then for a solution $y = \varphi(x)$ of (6) these new magnitudes are functions $y_i = \varphi_i(x)$ $(i = 1, \ldots, n)$ satisfying the relations

$$\varphi_2 = \varphi_n', \ldots, \qquad \varphi_n = \varphi_{n-1}', \qquad \varphi_n'(x) = f(x, \varphi_1(x), \ldots, \varphi_n(x));$$

[6] For the proof see, e.g. [4], §8.
[7] See, e.g. [1], p. 102.

consequently, $\varphi_1, \ldots, \varphi_n$ satisfy the following system of differential equations of first order:

$$(7) \qquad \begin{aligned} y_1' &= y_2 \\ y_2' &= y_3 \\ &\vdots \\ y_{n-1}' &= y_n \\ y_n' &= f(x, y_1, \ldots, y_n). \end{aligned}$$

Conversely, if $y_i = \varphi_i(x)$ $(i = 1, \ldots, n)$ is a system of solutions of the system (7), it is clear that $y = \varphi_1(x)$ is a solution of (6) to which the given system of solutions of (7) corresponds in the above way; thus *the solutions of* (6) *are in one-to-one correspondence with the systems of solutions of* (7).

It is desirable to eliminate the lack of symmetry in the way in which $y_1, \ldots, y_n$ appear in (7); i.e., instead of (7) we consider the more general system

$$(8) \qquad \begin{aligned} y_1' &= f_1(x, y_1, \ldots, y_n) \\ &\cdots\cdots\cdots\cdots \\ y_n' &= f_n(x, y_1, \ldots, y_n). \end{aligned}$$

A system of solutions $y_i = \varphi_i(x)$ of (8) will be represented in $E^{(n+1)}$ by a curve lying over the x-axis and a tangent at an arbitrary point $(x_0, y_{10}, \ldots, y_{n0})$ to the line

$$y_i = y_{i0} + f_i(x_0, y_{10}, \ldots, y_{n0})(x - x_0) \qquad (i = 1, \ldots, n).$$

Thus, if we assume that the functions f_i are defined in a domain of $E^{(n+1)}$ and then draw the above line through each point $(x_0, y_{10}, \ldots, y_{n0})$ of this domain, we again obtain a family of line-elements whose totality we describe as the direction field of the system of differential equations. Then we can characterize the systems of solutions $y_i = \varphi_i(x)$ of (8) by the property that their graphs are tangent at each point to the line-element prescribed by the direction field belonging to (8).

So for systems of first order, and consequently also for explicit differential equations of n-th order, we have reached a situation completely analogous to the above situation for differential equations of first order. The corresponding results are valid, and can be proved in exactly the same way as before. In fact they do not even need to be proved again, since they hold for the present systems if we merely replace the absolute values in the various inequalities by a corresponding concept in the space of the $y_1, \ldots, y_n$. For example, a function $f(x, y_1, \ldots, y_n)$ which is continuous in the domain G_f of $E^{(n+1)}$ is said to satisfy the *Lipschitz condition* if for every rectangular parallelepiped R in G_f with surfaces parallel to the axes there exists a number M_R such that for any two points $(x, y_1^1, \ldots, y_n^1)$ and $(x, y_1^2, \ldots, y_n^2)$ in R we have the inequality

$$|f(x, y_1^2, \ldots, y_n^2) - f(x, y_1^1, \ldots, y_n^1)| \leqq M_R \max_{1 \leqq i \leqq n} |y_i^2 - y_i^1|$$

in particular, this inequality is satisfied if in G_f the function f is continuously differentiable in all the y_i.

2.5. An implicit differential equation of first order

$$F(x, y, y') = 0 \tag{9}$$

will likewise define a direction field in the x, y-plane, but now a point (x, y) will in general be associated with several line-elements, simply because the equation $F(x, y, p) = 0$ will have several solutions in p. The solutions $y = \varphi(x)$ of (9) will again be functions whose graphs will at every point be tangent to a corresponding line-element of the direction field.

In a certain sense a direction field of this sort can be regarded as a superposition of the direction fields of several explicit differential equations. A simpler direction field will then be uniquely determined, in the small, by a line-element (x_0, y_0, p_0) corresponding to (9), provided that the equation $F(x, y, p) = 0$ is satisfied by exactly one continuous function $p = f(x, y)$ in a neighborhood of (x_0, y_0) with $f(x_0, y_0) = p_0$. A line-element with this property, which we shall call a *regular* line-element, will therefore, at least locally, select from the direction field of (9) the direction field corresponding to the equation $y' = f(x, y)$, whose solutions can be obtained by the method discussed above.

The nonregular line-elements are called *singular*, and the corresponding points are singular points; on the other hand, solutions of (9) are called regular or singular according to whether all their line-elements are regular or singular; the totality of singular points is called the *discriminant manifold*; all the singular solutions lie entirely in this manifold.

If, in a domain of the three-dimensional space of its arguments, the function $F(x, y, p)$ in (9) is continuous, and is continuously differentiable with respect to p, then by the theorem on implicit functions (cf. III 2, §5.6) every line-element (x_0, y_0, p_0) for which

$$F_p(x_0, y_0, p_0) \neq 0$$

is regular.[8] It can be shown[9] that any solution of (9) through this line-element, provided it does not encounter the discriminant manifold, is also a solution of the uniquely determined explicit differential equation $y' = f(x, y)$; it is a regular solution of (9). The totality of line-elements satisfying the equations

$$F(x, y, p) = 0, \qquad F_p(x, y, p) = 0$$

[8] In fact, there exists a neighborhood of (x_0, y_0) which, apart from the solution $p = f(x, y)$, does not contain any line-elements whose p-values differ arbitrarily little from p_0.

[9] Cf. [6].

also form a direction field on the set of the corresponding points, which includes the discriminant manifold. Curves $y = \varphi(x)$ which at every point are tangent to one of these line-elements are also solutions of (9). In this way all singular solutions are obtained, but perhaps also some regular solutions. If we assume that $F(x, y, p)$ is also continuously differentiable with respect to x and y, we can deduce a necessary condition for twice differentiable solutions of this kind: they must satisfy

$$F(x, \varphi(x), \varphi'(x)) = 0$$

identically in x, and thus by the chain rule

$$F_x + F_y\varphi' + F_p\varphi'' = 0;$$

since they must also satisfy $F_p = 0$, it follows that they consist solely of points of (x, y) for which the three equations

$$(10) \qquad F = 0, \qquad F_p = 0, \qquad F_x + pF_y = 0$$

are solvable for p.

Let F_p be everywhere continuously differentiable in all its arguments and let F_{pp} vanish nowhere. Then in a neighborhood of every point of a solution we can solve $F_p = 0$ for p, where p is continuously differentiable (cf. III 2, §5.7). Then if $p = \pi(x, y)$ is such a solution, let $y = \varphi(x)$ be a differentiable function for which the equations

$$F(x, \varphi(x), \pi(x, \varphi(x))) = 0,$$
$$F_p(x, \varphi(x), \pi(x, \varphi(x))) = 0,$$
$$F_x(x, \varphi(x), \pi(x, \varphi(x))) + \pi(x, \varphi(x))\cdot F_y(x, \varphi(x), \pi(x, \varphi(x))) = 0$$

are satisfied identically in x. Differentiating the first equation, and noting the second, we obtain

$$F_x(x, \varphi(x), \pi(x, \varphi(x))) + \varphi'(x)\cdot F_y(x, \varphi(x), \pi(x, \varphi(x))) = 0.$$

If we also assume that F_y nowhere vanishes, it follows from this equation and from the third of the above equations that

$$\pi(x, \varphi(x)) = \varphi'(x).$$

Substituting this result in the first equation, we finally obtain

$$F(x, \varphi(x), \varphi'(x)) = 0.$$

Thus $y = \varphi(x)$ is a solution of (9).

2.6. To bring the present section to a close, we give a brief discussion of the most important methods by which certain types of differential equations may be reduced to simpler forms.[10] Among these simpler forms

[10] An extensive survey of these questions is given in [5].

we include in particular the differential equations that can be solved directly by *quadratures*, i.e., whose solutions can be obtained by integration of known functions.

a) In two special cases the order of a differential equation may be reduced by one:

α) $$F(x, y', \ldots, y^{(n)}) = 0;$$

the substitution $y' = z$ gives $F(x, z, z', \ldots, z^{(n-1)}) = 0$.

β) $$F(y, y', \ldots, y^{(n)}) = 0:$$

if we confine ourselves here to strictly monotone arcs of the solution curves, we may write them in the form $x = \psi(y)$. Thus if we rewrite the differential equation with y as the independent variable we obtain from

$$\frac{dy}{dx} = \frac{1}{\frac{dx}{dy}}, \qquad \frac{d^2y}{dx^2} = -\frac{\frac{d^2x}{dy^2}}{\left(\frac{dx}{dy}\right)^3}, \qquad \text{etc.,}$$

a differential equation of the form

$$G(y, x', \ldots, x^{(n)}) = 0 \qquad \text{with} \qquad x^{(k)} = \frac{d^kx}{dy^k},$$

which is of type α.

b) If $\Phi(x, y)$ is continuously differentiable in a given domain, then every differentiable function $y = \varphi(x)$ for which

$$\Phi(x, \varphi(x)) = \text{const}$$

satisfies the differential equation

$$\Phi_x(x, y) + \Phi_y(x, y)y' = 0,$$

and conversely; such a differential equation is said to be *exact* with the *stem function* Φ.

It is natural to ask how we can tell by looking at a differential equation

(10) $$g(x, y) + h(x, y)y' = 0$$

whether it is exact, and if it is exact, how we can find a stem function. To these questions the following theorem gives a very general answer.

Let G be simply connected and let g and h be continuously differentiable in G; then (10) *is exact if and only if the integrability condition*

$$g_y = h_x \text{ in } G$$

is satisfied. A stem function is obtained by

(11) $$\Phi(x, y) = \int_{x_0, y_0}^{x, y} g(\xi, \eta)d\xi + h(\xi, \eta)d\eta,$$

where the curvilinear integral is independent of the path and is therefore meaningful without further information about the path of integration from the fixed point (x_0, y_0) to (x, y) in G.

The assumption that G is simply connected serves only to ensure that the stem function is unique; this condition is not necessary for the solvability of (10). If it is not satisfied, we will in general obtain from (11) a many-valued function Φ; but this function will still be such that the integrals of (10) are characterized by $\Phi = \text{const.}$

If (10) is not exact, it often possible to find an *integrating factor* (*Euler multiplier*) $M(x, y)$ such that the equation

$$Mg + Mhy' = 0$$

(whose solutions apart from the manifold $M = 0$ are identical with those of (10)) is exact. The integrating factors M are the continuously differentiable solutions of the partial differential equation

$$M_x h - M_y g = M(g_y - h_x),$$

which in certain special cases is easier to solve than the ordinary differential equation given originally.

Thus the following five cases are readily solvable.

α) A differential equation with *separated variables*

$$g(x) + h(y)y' = 0, \qquad g \text{ and } h \text{ continuous},$$

is exact; its solutions are obtained from

$$\int_{x_0}^{x} g(\xi)d\xi + \int_{y_0}^{y} h(\eta)d\eta = \text{const.}$$

β) The equation

$$y' = f(ax + by + r), \qquad b \neq 0,$$

is transformed by the substitution $z = ax + by + r$ into

$$z' = a + bf(z);$$

thus α) is applicable in domains in which

$$a + bf(z) \neq 0.$$

γ) In the equation

$$y' = f\left(\frac{y}{x}\right)$$

we set $\frac{y}{x} = z$, obtaining

$$xz' = f(z) - z.$$

Again the variables can be separated and we obtain solutions in those domains in which $x \neq 0$ and $f(z) \neq z$.

δ) As for

$$y' = f\left(\frac{ax + by + r}{cx + dy + s}\right) \qquad \text{with} \qquad ad - bc = 0, \qquad d \neq 0,$$

we see that since

$$\frac{ax + by + r}{cx + dy + s} = \frac{b}{d} + \frac{dr - bs}{d} \frac{1}{cx + dy + s},$$

we actually have the case β).

ϵ) Finally the case

$$y' = f\left(\frac{ax + by + r}{cx + dy + s}\right) \qquad \text{with} \qquad ad - bc \neq 0$$

can be reduced to γ) by determining x_0 and y_0 from

$$ax_0 + by_0 + r = 0, \qquad cx_0 + dy_0 + s = 0$$

and substituting

$$u = x - x_0, \qquad v = y - y_0.$$

For $u \neq 0$ we obtain

$$\frac{dv}{du} = f\left(\frac{a + b\dfrac{v}{u}}{c + d\dfrac{v}{u}}\right).$$

c) One of the most important classes of differential equations consists of the *linear* equations of the form

$$(12) \qquad y^{(n)} + a_{n-1}(x)y^{(n-1)} + \cdots + a_1(x)y' + a_0(x)y = b(x),$$

linear in the unknown function and its derivatives. We assume that the coefficients $a_i(x)$ and $b(x)$ are continuous over an interval J.

For arbitrarily preassigned x_0 and $y_0, y_0', \ldots, y_0^{(n-1)}$, Peano showed that there exists an integral $y = \varphi(x)$ of (12), defined over the whole of J, with $\varphi^{(\nu)}(x_0) = y_0^{(\nu)}$ $(\nu = 0, \ldots, n - 1)$. Now let y_p be such a solution of (12), defined over the whole of J, and let y be an arbitrary solution. Then in view of the linearity of (12) the function $\eta = y - y_p$ satisfies the corresponding *homogeneous* differential equation

$$(13) \qquad \eta^{(n)} + a_{n-1}(x)\eta^{(n-1)} + \cdots + a_1(x)\eta' + a_0(x)\eta = 0;$$

and conversely, if η is a solution of (13), then $y = y_p + \eta$ is a solution of the nonhomogeneous differential equation (12). Thus we obtain all solutions y of (12) if we can find

(A) all solutions η of the homogeneous equation (13) and
(B) a particular solution y_p of (12) defined on the whole of J.

A. Only in a few special cases (see the examples just below) can we directly determine all the solutions of (13):

α) If the differential equation is *of first order*,

$$\eta' + a(x)\eta = 0,$$

the variables can be separated and we obtain all the solutions in the form

$$\eta = \eta_0 e^{-\int_{x_0}^{x} a(t)dt}.$$

β) If all the coefficients a_ν in (13) are *constants*, $\eta = e^{\lambda x}$ is a solution, provided λ is a root of the *characteristic equation*

$$\lambda^n + a_{n-1}\lambda^{n-1} + \cdots + a_1\lambda + a_0 = 0. \tag{14}$$

This statement means, if $\lambda = \sigma + i\tau$ is complex, that $e^{\sigma x}\sin \tau x$ and $e^{\sigma x}\cos \tau x$ are solutions of (13). If λ is a k-fold root of (14), the functions

$$e^{\lambda x}, xe^{\lambda x}, \ldots, x^{k-1}e^{\lambda x}$$

are solutions of (13). It is clear that in this way we obtain exactly n solutions $\eta_1, \ldots, \eta_n$, and the general theory, sketched just below, then shows that every solution of (13) is obtained as a linear combination

$$\eta = c_1\eta_1 + \cdots + c_n\eta_n.$$

γ) The homogeneous *Euler differential equation*

$$y^{(n)} + \frac{a_{n-1}}{x}y^{(n-1)} + \cdots + \frac{a_0}{x^n}y = 0,$$

which we shall consider for $x > 0$, is reduced by the substitution

$$t = \log x, \qquad z(t) = z(\log x) = y(x)$$

to a linear differential equation with constant coefficients.

δ) In general the equation (13) cannot be solved by quadratures. However, we can deduce some of the properties of the solutions, which depend on the algebraic structure of (13).

Let $\eta_1, \ldots, \eta_n$ be any solutions of (13) defined in a neighborhood of x_0. By the *Wronski determinant* of such a system of functions we mean the expression

$$W(\eta) = \begin{vmatrix} \eta_1 & \cdots & \eta_n \\ \eta_1' & \cdots & \eta_n' \\ \vdots & & \vdots \\ \eta_1^{(n-1)} & \cdots & \eta_n^{(n-1)} \end{vmatrix}$$

If for abbreviation we write this expression in the form $[\eta, \eta', \ldots, \eta^{(n-1)}]$, then from (13) we obtain

$$\frac{dW}{dx} = [\eta', \eta', \eta'', \ldots, \eta^{(n-1)}] + [\eta, \eta'', \eta'', \ldots, \eta^{(n-1)}] + \cdots$$
$$+ [\eta, \eta', \ldots, \eta^{(n-2)}, \eta^{(n)}]$$
$$= [\eta, \eta', \ldots, \eta^{(n-2)}, -a_{n-1}\eta^{(n-1)}] = -a_{n-1}W.$$

Thus the *Wronski determinant* satisfies a homogeneous linear differential equation of first order and therefore by α) has the form

$$W(\eta(x)) = W(\eta(x_0))e^{-\int_x^{x_0} a_{n-1}(t)dt}.$$

It follows that $W(\eta(x))$ either vanishes identically or does not vanish at any point.

A system of solutions $\eta_1, \ldots, \eta_n$ with $W(\eta) \neq 0$ is called a *fundamental system.* A system of functions of this sort, which are even defined on the whole of J, is obtained from the Peano theorem if we simply set the values $\eta_k^{(\nu)}(x_0)$ $(k = 1, \ldots, n; \nu = 0, \ldots, n-1)$ equal to 1 or 0, according to whether $k = \nu + 1$ or not.

Now let η be an arbitrary solution of (13). Since the *Wronski determinant*, constructed for the fundamental system at the fixed point x_0, nowhere vanishes, we see that for every point x the system of equations

$$\begin{aligned} \eta(x) &= c_1(x)\eta_1(x) + \cdots + c_n(x)\eta_n(x) \\ \eta'(x) &= c_1(x)\eta_1'(x) + \cdots + c_n(x)\eta_n'(x) \\ &\vdots \\ \eta^{(n-1)}(x) &= c_1(x)\eta_1^{(n-1)}(x) + \cdots + c_n(x)\eta_n^{(n-1)}(x) \end{aligned}$$

has a unique solution in the $c_\nu(x)$ consisting of functions that are rational in the $\eta_k^{(\nu)}$ and are therefore differentiable.

By comparing these equations with the equations obtained from them by differentiation we see that

$$\begin{aligned} c_1'\eta_1 + \cdots + c_n'\eta_n &= 0 \\ &\vdots \\ c_1'\eta_1^{(n-2)} + \cdots + c_n'\eta_n^{(n-2)} &= 0 \end{aligned}$$

and

$$\eta^{(n)} = c_1\eta_1^{(n)} + \cdots + c_n\eta_n^{(n)} + c_1'\eta_1^{(n-1)} + \cdots + c_n'\eta_n^{(n-1)}.$$

From this last relation it follows, since the $\eta, \eta_1, \ldots, \eta_n$ are solutions of (13), that

$$c_1'\eta_1^{(n-1)} + \cdots + c_n'\eta_n^{(n-1)} = 0.$$

These equations extend the $n - 1$ equations for the c_k' to a system whose determinant is again the nowhere vanishing Wronski determinant of $\eta_1, \ldots, \eta_n$; it follows that $c_1' = \cdots = c_n' = 0$, so that the c_k are constant.

Thus every solution η *of* (13) *is a linear combination*

$$\eta(x) = c_1\eta_1(x) + \cdots + c_n\eta_n(x)$$

of the functions of a fundamental system. Since the converse is also true, it follows that the totality of the solutions of (13) is identical with all the linear combinations of functions of a fundamental system; *in particular, every solution is defined on the whole of J.*

B. The solution of the nonhomogeneous equation (12) can always be found if we already have a fundamental system $\eta_1, \ldots, \eta_n$ for the homogeneous equation (13).

The method, which is known as *variation of parameters*, begins by writing down the expression

$$y = u_1(x)\eta_1(x) + \cdots + u_n(x)\eta_n(x). \tag{14}$$

We then impose on the unknown functions $u_k(x)$ the identities

$$\begin{aligned} u_1'\eta_1 \quad &+ \cdots + u_n'\eta_n \quad = 0 \\ &\vdots \\ u_1'\eta_1^{(n-2)} &+ \cdots + u_n'\eta_n^{(n-2)} = 0, \end{aligned}$$

and thus obtain

$$\begin{aligned} y^{(k)} &= u_1\eta_1^{(k)} + \cdots + u_n\eta_n^{(k)} \qquad (k = 0, \ldots, n-1), \\ y^{(n)} &= u_1\eta_1^{(n)} + \cdots + u_n\eta_n^{(n)} + u_1'\eta_1^{(n-1)} + \cdots + u_n'\eta_n^{(n-1)}. \end{aligned}$$

From the equations (12) for y and (13) for the η it follows that

$$u_1'\eta_1^{(n-1)} + \cdots + u_n'\eta_n^{(n-1)} = b.$$

This equation completes the above conditions to a system of equations for the u_i' whose determinant is again the nowhere vanishing Wronski determinant of the fundamental system, so that the u_i' are uniquely determined as continuous functions on J. Thus the unknowns u_i in the above expression are also determined up to additive constants. Consequently (14) *determines all solutions of* (12), *and these solutions are defined on the whole of J.*

The linear differential equation of first order has been integrated above in an elementary way, namely by quadratures. The next most difficult types are obtained if we either retain the linearity and increase the order:

$$y'' + a_1(x)y' + a_0(x)y = 0$$

(linear differential equation of second order), or retain the order but perturb the linearity by the addition of a quadratic term:

$$z' = \alpha(x)z^2 + \beta(x)z + \gamma(x)$$

(Riccati differential equation). These two types are "equally difficult," since the substitution $z = \frac{y'}{y}$ transforms one of them into the other. For example, the following two equations correspond to each other

$$y'' = x^k y \qquad \text{and} \qquad z' + z^2 = x^k.$$

For the latter equation, Liouville has shown that it can be solved by quadratures only for the values $k = -2$ and $k = -2\left(1 + \frac{1}{\nu}\right)$ with odd ν; *thus the equation*

$$y'' = \frac{y}{x}$$

is a simple example of a linear differential equation that cannot be solved by elementary means.

3. Properties of the Solutions of Explicit Differential Equations

3.1. Since the following discussion can again be extended from systems of first order to explicit differential equations of higher order, we confine our attention to

$$y' = f(x, y). \tag{15}$$

Here again we assume that the function f is continuous in a domain G of the x, y-plane. Then it follows that every solution, regarded as a function of x, is not only differentiable but even continuously differentiable; more generally we can show in the same way that *every solution is $(p + 1)$-times continuously differentiable if f has continuous derivatives up to the p-th order, with respect to its two variables.*

The existence of solutions is guaranteed by the *Peano theorem* (cf. §2.1); but to begin with, this guarantee holds only in the small. Thus we ask whether and to what extent it is possible to combine solutions of this sort so as to form curves in the large. The question is answered in the following way.

If an integral curve comes arbitrarily close to an inner point P of G, then either the curve includes P as an inner point or it unites itself differentiably with every integral curve through P. It readily follows that every integral curve that does not come arbitrarily close to the boundary of G is a proper subarc of an integral curve with a larger interval of definition. Finally, this result implies that *every integral curve can be extended,* in general in more than one way, *to an integral curve that traverses G from edge to edge.*[11]

[11] For the proof see, e.g. [4], §9.

3.2. We now assume that the f of (15) also satisfies the Lipschitz condition in G. Then as mentioned above in §2.1, we have the following result.

Uniqueness theorem. *Through every point of G there passes exactly one integral curve of* (15).

It is easy to give a direct proof, clearly showing the effect of the Lipschitz condition. Let (x_0, y_0) be the point of G in question and let φ_1, φ_2 be two integrals of (15) with $\varphi_1(x_0) = \varphi_2(x_0) = y_0$. Then

$$\varphi_i(x) = y_0 + \int_{x_0}^{x} f(t, \varphi_i(t))dt, \qquad i = 1, 2$$

and therefore, by the Lipschitz condition,

$$|\varphi_2(x) - \varphi_1(x)| \leqq L \int_{x_0}^{x} |\varphi_2(t) - \varphi_1(t)|dt, \qquad \text{if} \qquad x \geqq x_0.$$

But then it follows from the lemma just below that $|\varphi_2(x) - \varphi_1(x)| \leqq 0$, so that $\varphi_2(x) = \varphi_1(x)$ must hold for $x \geqq x_0$; and the proof is the same for $x \leqq x_0$.

Lemma. *Let $\psi(x)$ be continuous in $J = [x_0, a)$, and let there be an $L \geqq 0$ such that for all x in J*

$$\psi(x) \leqq L \int_{x_0}^{x} \psi(t)dt.$$

Then ψ is nowhere positive in J.

Proof. If points x existed in J with $\psi(x) > 0$, they would have a greatest lower bound x_1 such that

$$\psi(x) \leqq 0 \qquad \text{for} \qquad x_0 \leqq x \leqq x_1.$$

On the other hand, there would exist arguments x to the right of x_1, and arbitrarily close to it, for which $\psi(x) > 0$. If x_2 is so chosen that in the interval $\left[x_1, x_1 + \frac{1}{2L}\right]$ the function ψ has a maximum, then

$$\psi(x_2) > 0,$$

is in contradiction to the assumed inequality

$$\psi(x_2) \leqq L \int_{x_0}^{x_2} \psi(t)dt \leqq L \int_{x_1}^{x_2} \psi(t)dt \leqq L\psi(x_2)(x_2 - x_1)$$

$$\leqq L\psi(x_2)\frac{1}{2L} = \frac{1}{2}\psi(x_2). \qquad \text{q.e.d.}$$

3.3. By means of the uniqueness theorem we can describe the behavior of the integral curves of (15) in the following way: *the integral curves cover G simply and without gaps and run from edge to edge.* In particular, an integral curve is thus completely determined by one of its points. In other words, we can characterize the solution by assigning the "initial conditions" $\varphi(x_0) = y_0$. The totality of the solutions can then be described by a single function, the "general solution"

$$y = \varphi(x; x_0, y_0),$$

which for fixed x_0, y_0 is a solution of (15) such that

$$y_0 = \varphi(x_0; x_0, y_0)$$

identically in x_0, y_0. If $\varphi(x; x_0, y_0) = \varphi(x; x_1, y_1)$ for one x, the uniqueness theorem shows this equation holds identically in x and is therefore satisfied if and only if

$$\varphi(x_0; x_1, y_1) = y_0.$$

Consequently, we have the following functional equation, identically in all its variables,

$$\varphi\Big(x; x_0, \varphi(x_0; x_1, y_1)\Big) = \varphi(x; x_1, y_1).$$

We now show that *the general solution depends continuously on the coordinates* x_0, y_0 *of the initial point*. A consequence of this fact, of extreme importance in practical applications, is that two integral curves through neighboring initial points remain close to each other throughout their whole course.

Let us set

$$\varphi_0(x) = \varphi(x; x_0, y_0), \qquad \varphi_1(x) = \varphi(x; x_1, y_1),$$

so that

$$\varphi_0(x_0) = y_0, \qquad \varphi_1(x_1) = y_1.$$

If (x_1, y_1) is sufficiently close to (x_0, y_0), then φ_1 is also defined for x_0 and we can write

$$\varphi_1(x) - \varphi_0(x) = \varphi_1(x_0) - \varphi_0(x_0) + \int_{x_0}^{x} [f(t, \varphi_1(t)) - f(t, \varphi_0(t))]dt.$$

If for abbreviation we set $\rho = |\varphi_1 - \varphi_0|$, the *Lipschitz condition* gives us the integral equation

$$\rho(x) \leqq \rho(x_0) + L\int_{x_0}^{x} \rho(t)dt \qquad \text{for} \qquad x \geqq x_0.$$

But the function

$$\sigma(x) = \rho(x_0)e^{L(x-x_0)}$$

satisfies the integral equation

$$\sigma(x) = \rho(x_0) + L\int_{x_0}^{x} \sigma(t)dt.$$

Thus for $\psi = \rho - \sigma$ we have

$$\psi(x) \leqq L \int_{x_0}^{x} \psi(t)dt \qquad \text{for} \qquad x \geqq x_0.$$

Thus the lemma means that

$$\rho \leqq \sigma \qquad \text{for} \qquad x \geqq x_0.$$

Similarly, removing the restriction on x, we obtain

$$|\varphi_1(x) - \varphi_0(x)| \leqq |\varphi_1(x_0) - \varphi_0(x_0)|e^{L|x-x_0|}$$

for $x \gtreqless x_0$. For the first factor we also have

$$\begin{aligned} |\varphi_1(x_0) - \varphi_0(x_0)| &\leqq |\varphi_1(x_1) - \varphi_1(x_0)| + |\varphi_1(x_1) - \varphi_0(x_0)| \\ &\leqq M|x_1 - x_0| + |y_1 - y_0|, \end{aligned}$$

where M is a bound on $|f|$ in a suitably chosen part of G containing the curves φ_0 and φ_1 for the arguments in question. Thus we finally obtain

$$|\varphi(x; x_1, y_1) - \varphi(x; x_0, y_0)| \leqq (M|x_1 - x_0| + |y_1 - y_0|)e^{L|x-x_0|},$$

which implies the desired continuity.

Let us also mention[12] the following, essentially deeper refinement of this theorem:

If the function f is p-times continuously differentiable in both variables, the general solution has continuous derivatives in all variables up to the p-th order.

3.4. Finally we discuss the question, of equal importance for the applications, of the behavior of the solutions of (15) when the differential equation itself is slightly changed.

Let f and f^* be continuous in G and let f satisfy the Lipschitz condition there; let the difference $\delta = f^* - f$ be uniformly bounded in G:

$$|\delta(x, y)| \leqq \Delta \text{ in } G.$$

Let φ and φ^* be solutions of (15) and of $y' = f^*(x, y)$, respectively, with the common initial condition

$$\varphi(x_0) = \varphi^*(x_0) = y_0.$$

For $x \geqq x_0$ these assumptions imply

$$\begin{aligned} |\varphi^*(x) - \varphi(x)| &= \left| \int_{x_0}^{x} [f^*(t, \varphi^*(t)) - f(t, \varphi(t))]dt \right| \\ &\leqq \int_{x_0}^{x} |f(t, \varphi^*(t)) - f(t, \varphi(t))|dt + \int_{x_0}^{x} |\delta(t, \varphi^*(t))|dt \\ &\leqq L \int_{x_0}^{x} |\varphi^*(t) - \varphi(t)|dt + \Delta(x - x_0). \end{aligned}$$

[12] For the proof see, e.g. [4], §18.

We now form the comparison function

$$\sigma(x) = \frac{\Delta}{L}(e^{L|x-x_0|} - 1),$$

which for $x \geqq x_0$ satisfies the integral equation

$$\sigma(x) = L\int_{x_0}^{x} \sigma(t)dt + \Delta(x - x_0).$$

Thus for $\psi = |\varphi^* - \varphi| - \sigma$ we again have

$$\psi(x) \leqq L\int_{x_0}^{x} \psi(t)dt \qquad \text{for} \qquad x \geqq x_0$$

and consequently the lemma shows that

$$|\varphi^* - \varphi| \leqq \sigma,$$

which can also be proved for $x \leqq x_0$. The relation thus obtained

$$|\varphi^*(x) - \varphi(x)| \leqq \frac{\Delta}{L}(e^{L|x-x_0|} - 1)$$

can be interpreted very concisely by saying that *the solutions of* (15) *depend continuously on the differential equation.*

4. Solutions with Special Properties

For many problems the statements of the general theory are not sufficient; we require solutions with special properties and must look for criteria guaranteeing the existence of such solutions. In the present section we define some of the most important problems of this kind.

4.1. *Boundary Value Problems*

A solution y of a differential equation of second order can in general be characterized by assigning the two initial values $y(x_0)$, $y'(x_0)$ (cf. §2.4). But it is often of greatest interest to find solutions satisfying two conditions of another kind; namely, the value of the solution is preassigned at two *distinct* points x_0, x_1. Such a *boundary value problem* is by no means always solvable, as is shown by the example

$$y'' + y = 0, \qquad y(0) = 0, \qquad y(\pi) = 1.$$

The general solution is $a \sin t + b \cos t$, but the first boundary condition demands $b = 0$ and the second, $b = -1$.

Again it is the linear differential equations that admit an extensive theory of boundary value problems. We shall confine our attention to the (certainly typical) case of the differential equation

(16) $$y'' + q(x)y = r(x), \qquad q \text{ and } r \text{ continuous},$$

and the "*first homogeneous boundary value problem*"[13]

$$(17) \qquad y(0) = 0, \qquad y(1) = 0.$$

We speak here of a *homogeneous problem* if $r(x)$ vanishes identically, and otherwise of a *nonhomogeneous problem.* The homogeneous problem always has the *trivial* solution $y = 0$; every other solution will be called *nontrivial.*

Let y_1, y_2 be a fundamental system for the homogeneous equation corresponding to (16)

$$(16_h) \qquad y'' + qy = 0.$$

The homogeneous problem (16_h), (17) has a nontrivial solution if and only if from

$$c_1 y_1(0) + c_2 y_2(0) = 0,$$
$$c_1 y_1(1) + c_2 y_2(1) = 0$$

it follows that $c_1 = c_2 = 0$, or in other words if the determinant

$$\Delta := y_1(0)y_2(1) - y_2(0)y_1(1)$$

is different from zero. In this case the nonhomogeneous problem (16), (17), is uniquely solvable:

If y_p is a particular solution of (16), then c_1, c_2 are uniquely determined by the equations

$$c_1 y_1(0) + c_2 y_2(0) = -y_p(0),$$
$$c_1 y_1(1) + c_2 y_2(1) = -y_p(1).$$

There remains the question whether under the given conditions the homogeneous and the nonhomogeneous problems are both solvable; information on this point is obtained by making a suitable choice of the fundamental system. Thus let $\Delta = 0$ and let y_1 be a nontrivial solution of the homogeneous problem. If y_2 then completes y_1 to a fundamental system, it certainly follows that $y_2(0) \neq 0$ and $y_2(1) \neq 0$, since otherwise the Wronski determinant would vanish.

For a solution y of the nonhomogeneous problem (see §2.6, c), B) we set

$$y = uy_1 + vy_2$$

with

$$u'y_1 + v'y_2 = 0,$$
$$u'y_1' + v'y_2' = r.$$

[13] In the "second" boundary value problem the values of y' are assigned and in the "third," linear combinations of y and y'.

From the boundary condition (17) we then have

$$v(0)y_2(0) = 0, \qquad v(1)y_2(1) = 0,$$

so that

$$v(0) = v(1) = 0.$$

But for our present differential equation the Wronski determinant is constant:

$$W(y_1, y_2) = \begin{vmatrix} y_1(0) & y_2(0) \\ y_1'(0) & y_2'(0) \end{vmatrix} = -y_1'(0)y_2(0).$$

Thus we obtain

$$v'(x) = -\frac{y_1(x)r(x)}{y_1'(0)y_2(0)},$$

so that

$$v(x) = -\frac{1}{y_1'(0)y_2(0)}\int_0^x y_1(t)r(t)dt.$$

Here we already have $v(0) = 0$, while the other condition $v(1) = 0$ implies

$$\int_0^1 y_1(x)r(x)dx = 0,$$

which in the terminology of functional analysis means that the functions y_1 and r are orthogonal to each other. Thus we have proved the following theorem.

Alternative theorem. *The nonhomogeneous problem is uniquely, or non-uniquely, solvable if and only if the homogeneous problem has no nontrivial solution or, respectively, is solvable and the "perturbation function" r is orthogonal to the solution of the homogeneous problem.*

Thus the entire question has been reduced to the homogeneous problem. Since the above criterion for solvability implies the knowledge of a fundamental system, which is generally difficult to obtain, let us continue our discussion in the following way.

Since in general the homogeneous problem has no nontrivial solution, let us formulate the question somewhat differently. Instead of (16_h), we consider the differential equation

$$y'' + (q(x) + \lambda)y = 0 \tag{18}$$

and ask for what values of λ the boundary problem (17) has a nontrivial solution. Such values of λ are called *eigenvalues* of the problem and the nontrivial solutions are the corresponding *eigenfunctions.*

Although it will not be required below, let us prove the following important theorem.

Eigenfunctions corresponding to distinct eigenvalues are orthogonal.

For let y_i be an eigenfunction belonging to the eigenvalue λ_i $(i = 1, 2)$. Then y_2 can be regarded as a solution of the nonhomogeneous boundary value problem for

$$y'' + (q(x) + \lambda_1)y = (\lambda_1 - \lambda_2)y_2(x).$$

Since $\lambda_1 \neq \lambda_2$, the assertion then follows from the alternative theorem.

Now instead of asking directly for the eigenvalues of the boundary problem (17), (18), let us first consider an arbitrary value of λ and determine the solution, unique up to a numerical factor, of (18) satisfying the first boundary condition $y(0) = 0$; and let us then attempt to make statements about the number of zeros of this solution in the interval $(0, 1)$.

For this purpose we rewrite (18) as a system of differential equations of first order

$$(19) \qquad \begin{aligned} y' &= z, \\ z' &= -(q(x) + \lambda)y, \end{aligned}$$

and interpret a solution $y(x)$ of (18) as the curve in the y, z-plane represented by the pair of functions $y(x)$, $z(x)$ with x as a parameter. Since for a nontrivial solution of (18) the functions y and y' cannot vanish simultaneously, the corresponding curve does not pass through the origin of the y, z-plane. If we introduce the polar coordinates

$$y = \rho \sin \vartheta, \qquad z = \rho \cos \vartheta,$$

the system (19) is transformed into

$$(20) \qquad \begin{aligned} \vartheta' &= (q + \lambda) \sin^2 \vartheta + \cos^2 \vartheta, \\ \rho' &= \rho(1 - q - \lambda) \sin \vartheta \cos \vartheta. \end{aligned}$$

This system has the advantage that the functions are "*separated.*" The first equation contains $\vartheta(x)$ alone and by substituting its solution into the second equation we obtain the corresponding $\rho(x)$, uniquely up to a numerical factor. By confining our attention in this way to the first equation (20), we are not only able to solve the differential equation (18) but also, from the corresponding solution $\vartheta(x)$ of (20), we can read off the zeros of a solution $y(x)$ Since $\rho \neq 0$, these zeros correspond precisely to the values $n\pi$ of $\vartheta(x)$. The first boundary condition $y(0) = 0$ is satisfied by setting $\vartheta(0) = 0$.

At the points $\vartheta = n\pi$ we have $\vartheta' = 1$, from which it follows at once that n is never negative for $x \in [0, 1]$ and that for increasing x the values of ϑ run through increasing values of $n\pi$, so that in particular none of these values can be assumed by ϑ more than once Then we can prove the following, quite natural statements.

A. Assume $\lambda_1 < \lambda_2$ and let ϑ_1 and ϑ_2 be the corresponding solutions of (20) with $\vartheta_1(0) = \vartheta_2(0) = 0$. Then

$$\vartheta_1(x) < \vartheta_2(x) \qquad \text{for} \qquad 0 < x \leqq 1.$$

For the corresponding solutions $y_1(x)$, $y_2(x)$ of (19) we obtain the following theorem.

Comparison theorem. *Let y_1 and y_2 be solutions of* (18) *for $\lambda_1 < \lambda_2$ with $y_1(0) = y_2(0) = 0$; then y_2 has at least as many zeros in* $[0, 1]$ *as y_1.*

B. Let us now consider the value of $\vartheta(x)$ for $x = 1$ as a function $\vartheta(1, \lambda)$ of λ. By §3.4 this function is continuous and by A it is monotone increasing. Furthermore, it can be shown that

$$\lim_{\lambda \to -\infty} \vartheta(1, \lambda) = 0, \qquad \lim_{\lambda \to +\infty} \vartheta(1, \lambda) = \infty.$$

Thus for every natural number n there exists exactly one λ_n such that $\vartheta(1, \lambda_n) = n\pi$. Since $\vartheta(x, \lambda_n)$ assumes each of the values $k\pi$, with $0 \leqq k \leqq n$, exactly once in $[0, 1]$, the corresponding function y has exactly $n - 1$ zeros in the open interval $(0, 1)$. These results can be combined in the following theorem.

Oscillation theorem. *The eigenvalues of the (homogeneous) boundary problem*

$$y'' + (q(x) + \lambda)y = 0, \qquad y(0) = y(1) = 0$$

form a sequence $\{\lambda_n\}$ monotonely increasing to infinity, and the eigenfunction belonging to λ_n has exactly $n - 1$ zeros in $(0, 1)$.

4.2. *Stability*

In practical applications the independent variable is frequently the time t, so that the differential equation describes some process taking place in the course of time. If the process begins at $t = 0$, we are interested in its state at every subsequent time $t > 0$; thus, in particular, the differential equation will be defined for all $t \geqq 0$. But now it is of great importance to know whether a process is stable or not, i.e., whether under slightly altered initial conditions the alteration in the process will remain slight for all $t > 0$. The statement, proved in §3.3, that the solution depends continuously on the initial conditions is not sharp enough for our present purposes; it states only that by a sufficiently small change in the initial conditions we can make the difference in the two corresponding solutions less than any preassigned quantity during the whole of a finite interval $0 \leqq t \leqq t_1$. On the other hand our present question calls for the following definition.

Definition.[14] *Let the differential equation*

$$y^{(n)} = f(t, y, y', \ldots, y^{(n-1)})$$

be defined for $t \geqq 0$. A solution φ_0 is said to be stable if for every $\epsilon > 0$

[14] Other concepts of stability are more suitable for many purposes, but we cannot discuss them here; cf. [3].

there exists a $\delta(\epsilon) > 0$ *such that for every solution* φ *and for all* $t \geqq 0$ *we have*

$$|\varphi(t) - \varphi_0(t)| < \epsilon,$$

provided that $|\varphi^{(i)}(0) - \varphi_0^{(i)}(0)| < \delta$ *for* $i = 0, \ldots, n-1$.

Now it is the task of the theory of stability to set up useful sufficient conditions for the stability of solutions. In the case of a linear differential equation

$$y^{(n)} + a_{n-1}y^{(n-1)} + \cdots + a_0 y = b(t) \tag{21}$$

with constant coefficients a_i ($i = 0, \ldots, n-1$) it is easy to give an exact criterion.

Here $\eta = \varphi - \varphi_0$ is a solution of the homogeneous equation

$$\eta^{(n)} + a_{n-1}\eta^{(n-1)} + \cdots + a_0\eta = 0 \tag{22}$$

with the initial values

$$\eta^{(i)}(0) = \varphi^{(i)}(0) - \varphi_0^{(i)}(0).$$

Thus a solution φ_0 of (21) is stable if and only if the trivial solution of (22) is stable; in particular, it follows either that all solutions of (21) are stable or that all solutions are unstable. Now let the functions $\eta_1, \ldots, \eta_n$, with

$$\eta_i = t^{k_i}e^{\lambda_i t},$$

be the fundamental system for (22) constructed in §2.6, c), A, β). Then an arbitrary solution η has the form

$$\eta = c_1\eta_1 + \cdots + c_n\eta_n,$$

so that small values of c_i obviously correspond to small initial values $\eta(0), \eta'(0), \ldots, \eta^{(n-1)}(0)$, and conversely. Thus $\eta_0 \equiv 0$ is a stable solution of (22) if and only if for every $\epsilon > 0$ there exists a $\gamma(\epsilon) > 0$ such that for all $t \geqq 0$ and $i = 1, \ldots, n$

$$|c_i\eta_i(t)| < \epsilon \qquad \text{if} \qquad |c_i| < \gamma.$$

But this will be the case if and only if all the η_i are bounded for $t \geqq 0$. However, for $\lambda = \alpha + i\beta$ we have

$$t^k e^{\lambda t} = t^k e^{\alpha t}(\cos \beta t + i \sin \beta t).$$

Thus the η_i are bounded for $t \geqq 0$ if and only if the real parts of all the λ_i are nonpositive and every λ_i with vanishing real part occurs only as a simple zero of the characteristic equation

$$\lambda^n + a_{n-1}\lambda^{n-1} + \cdots + a_0 = 0 \tag{23}$$

corresponding to (22). However, we shall restrict ourselves to finding a necessary condition for all the λ_i to have *negative* real parts, not because this condition can be formulated very simply, but because a solution $\cos \beta t + i \sin \beta t$ may become an unbounded function under a slight change in the differential equation. Thus the stability of a given solution of (21) is reduced to the question: when do all the roots of the characteristic polynomial (23) have negative real parts?[15] It is possible to answer this function-theoretic question without calculating the roots of (23). Among the various possible methods the most convenient one is due to I. Schur.

Polynomials of first or second degree (with real coefficients) are Hurwitz polynomials if and only if their coefficients are positive; for quadratic polynomials this assertion follows from the representation of the polynomial in the form

$$(\lambda - \lambda_1)(\lambda - \lambda_2), \qquad \text{if } \lambda_1, \lambda_2 \text{ are real,}$$

$$(\lambda - \lambda_1)(\lambda - \bar{\lambda}_1) = \lambda^2 - \lambda(\lambda_1 + \bar{\lambda}_1) + |\lambda_1|^2, \qquad \text{if } \lambda_1 \text{ is complex.}$$

Since polynomials of n-th degree are products of linear and quadratic polynomials, a Hurwitz polynomial of n-th degree always has positive coefficients; for $n > 2$; however, this necessary property is not sufficient as is shown by the example $\lambda^3 + 1$. For $f(\lambda) = \lambda^n + a_{n-1}\lambda^{n-1} + \cdots + a_0$ we form the polynomial $g(\lambda) = \alpha f(\lambda) - \beta f(-\lambda)$ with $|\alpha| > |\beta|$. *Then $f(\lambda)$ is a Hurwitz polynomial if and only if $g(\lambda)$ is a Hurwitz polynomial.*

Proof. It is only necessary to prove the statement in one direction, since

$$g(-\lambda) = -\beta f(\lambda) + \alpha f(-\lambda),$$

so that

$$f(\lambda) = \gamma g(\lambda) - \delta g(-\lambda) \qquad \text{with} \qquad |\gamma| = \left|\frac{\alpha}{\alpha^2 - \beta^2}\right| > |\delta| = \left|\frac{-\beta}{\alpha^2 - \beta^2}\right|.$$

Let us write $\lambda = u + iv$, and $\lambda_j = u_j + iv_j$ for the zeros of $f(\lambda)$. Then

$$f(\lambda) = \prod_1^n (\lambda - \lambda_j) \qquad \text{and} \qquad f(-\lambda) = \prod_1^n (-\lambda - \bar{\lambda}_j).$$

Now let f be a Hurwitz polynomial, so that all $u_j < 0$. Then we have the decisive inequalities

$$0 < |\lambda - \lambda_j| = |\lambda + \bar{\lambda}_j| \qquad \text{for} \qquad u = 0,$$

$$|\lambda - \lambda_j| > |\lambda + \bar{\lambda}_j| \qquad \text{for} \qquad u > 0.$$

Since $|\alpha| > |\beta|$, it follows that $|\alpha f(\lambda)| > |\beta f(-\lambda)|$, i.e., $g(\lambda) \neq 0$ for $u \geqq 0$, as was to be proved.

[15] Since A. Hurwitz was the first to answer this question, such polynomials are called *Hurwitz polynomials.*

Now let us make the special choice

$$\alpha = f(1) \qquad \text{and} \qquad \beta = f(-1).$$

Then $\alpha = 1 + a_{n-1} + \cdots + a_0$, $\beta = \pm 1 \pm a_{n-1} \pm \cdots \pm a_0$, and since f is a Hurwitz polynomial, it follows that all $a_i > 0$, so that in fact $|\alpha| > |\beta|$. But the polynomial

$$g(\lambda) = f(1)f(\lambda) - f(-1)f(-\lambda)$$

constructed in this way has a zero at -1, so that

$$f_1(\lambda) = \frac{f(1)f(\lambda) - f(-1)f(-\lambda)}{\lambda + 1} \tag{24}$$

is *a polynomial of degree* $n - 1$ *and is a Hurwitz polynomial if and only if* g *is a Hurwitz polynomial.*

This result leads to the recursive procedure of I. Schur. *For the polynomial* f *in question we construct the polynomial* f_1 *in* (24), *and then for* f_1 *we construct an* f_2 *and so forth up to the constant* f_n. *Then* f *is a Hurwitz polynomial if and only if all the polynomials* $f_1, f_2, \ldots, f_n$ *have positive coefficients.*

For nonlinear differential equations the question of stability has been extensively investigated.[16] The theory began with Ljapunov, whose chief result we will now present for the simplest case of the differential equation

$$y' = f(t, y). \tag{25}$$

Here we assume that the integral in question is again $y_0 \equiv 0$.

The Ljapunov criterion for stability. *In the decomposition*

$$f(t, y) = -ay + F(t, y)$$

assume that $a > 0$, $F(t, 0) = 0$, *and that there exists a* $\delta > 0$ *such that* $\left|\dfrac{F(t, y)}{y}\right| < \frac{1}{2} a$ *for all* $t \geqq 0$ *and* $|y| \leqq \delta$. *Then* $y_0 \equiv 0$ *is a stable solution of* (25).

Proof. Let $y(t)$ be a solution of (25) with $|y(0)| < \delta$. Then

$$\frac{dy^2}{dt} = 2y(t)(-ay(t) + F(t, y(t))) = -2y^2(t)\left(a - \frac{F(t, y(t))}{y(t)}\right). \tag{*}$$

Now if there exists a $t > 0$ with $|y(t)| \geqq \delta$, let t_0 be the smallest such t. Then

$$0 < y^2(t_0) - y^2(0) = t_0\left[\frac{dy^2}{dt}\right]_{t=t_1} \qquad \text{with} \qquad 0 \leqq t_1 \leqq t_0.$$

[16] See the monograph [3] quoted above.

But $|y(t_1)| \leq \delta$ by the definition of t_0, so that by (*) we have

$$\left[\frac{dy^2}{dt}\right]_{t=t_1} \leqq 0,$$

which is impossible. Thus

$$|y(t)| < \delta \qquad \text{for all} \qquad t \geqq 0,$$

provided this inequality holds for $t = 0$. The proof is therefore complete.

4.3. *Periodic Solutions*

To illustrate the subject of the present section let us first consider a linear differential equation of second order with constant coefficients

$$y'' + ay' + by = g(t).$$

For this equation to have periodic solutions it is necessary either that $g(t)$ is periodic or that $g(t) = 0$ and $a = 0$.

If $g(t)$ is periodic we speak of a *forced oscillation*, and if $g(t) = 0$, of a *free oscillation*; but the latter does not occur if the so-called damping factor a is present. Now it is a fact of extraordinary importance for technical applications that nonlinear differential equations can also have periodic solutions, both forced (in which case we speak of *synchronization*) and free, where the latter, in contrast to the linear case, can also occur in a damped system (so-called *relaxation*).

The discovery of these phenomena for certain nonlinear differential equations was followed by many extremely ingenious investigations[17] of the existence and behavior (in particular with respect to stability) of periodic solutions of other equations.

To conclude the present chapter we give an example which (as is characteristic for this type of question) depends in an essential way on topological arguments. We are dealing here with a special *quasilinear* differential equation (linear in the derivatives of y)

$$y'' + a(y)y' + b(y) = g(t), \tag{26}$$

which in the "unperturbed" case $g = 0$ is stationary (i.e., does not depend on the time); *in the perturbed case g is assumed to be periodic with period T*, and the functions a, b, g are defined and continuously differentiable for all values of the variables. A more convenient interpretation is obtained by replacing (26) with the system

$$\begin{aligned} y' + A(y) &= z \qquad \left(\frac{dA(y)}{dy} = a(y)\right) \\ z' + b(y) &= g(t). \end{aligned} \tag{27}$$

[17] Some quite subtle examples are given in [2], §§3–4.

Then every solution $y(t)$, $z(t)$ is represented by a curve (a so-called *characteristic*) in the z, y-plane (the *phase plane*). If y is periodic, z is also periodic with the same period; thus the characteristic is closed.

Conversely, if a characteristic has a closed arc, so that there exist values t_0 and $T > 0$ with

$$(28) \qquad y(t_0 + T) = y(t_0), \qquad z(t_0 + T) = z(t_0),$$

it follows from (27) that $y'(t_0 + T) = y'(t_0)$ also. Now if g is periodic with the period T, and if y is a solution (26), the function

$$x(t) := y(t + T)$$

is also a solution of (26). Since its values for t_0 are

$$x(t_0) = y(t_0), \qquad x'(t_0) = y'(t_0),$$

it is identical, by the uniqueness theorem, with $y(t)$:

$$y(t + T) = y(t) \qquad \text{for all } t,$$

and, provided g is not constant, we see from (26) that this periodic function is certainly not constant. If for a nonconstant perturbation g, periodic with the period T, we can prove (28), we have thereby proved the existence of a nonconstant periodic solution of (26).

At time t_0 exactly one characteristic passes through each point (y, z) of the phase plane; on it we mark the point (y^T, z^T) corresponding to the time $(t_0 + T)$. Since the integrals depend continuously on the initial conditions, the correspondence

$$(29) \qquad (y, z) \rightarrow (y^T, z^T)$$

is a continuous mapping of the phase plane into itself. Now if we can find a compact, simply connected domain that is also mapped into itself, the Brouwer fixed point theorem states that this domain must contain a point (y_0, z_0) left unchanged by the mapping. Thus (28) holds for the characteristic through this point, which therefore defines a nonconstant periodic solution of (26).

Now let $u(y, z)$ be a continuously differentiable function whose level curves are simple closed curves covering the phase plane in such a way that the compact domains G_c bounded by the level curves $u = c$ form an increasing chain for increasing c:

$$G_{c_2} \supset G_{c_1} \qquad \text{for} \qquad c_2 > c_1.$$

Then along a characteristic $y(t)$, $z(t)$ the function u depends only on t. If there is a value c_0 such that

$$\frac{du(y(t), z(t))}{dt} < 0$$

whenever

$$u(y(t), z(t)) \geqq c_0,$$

then under the transformation (29) the domain G_{c_0} will be mapped into itself for arbitrary t_0. For otherwise there would exist a point (y_1, z_1) with $u(y_1, z_1) \leqq c_0$ and a characteristic $y_1(t), z_1(t)$ with $y_1(t_0) = y_1, z_1(t_0) = z_1$, for which we would have

$$u(y_1(t_0 + T), z_1(t_0 + T)) > c_0,$$

and if we then let t^* denote the maximum in $t_0 \leqq t^* \leqq t_0 + T$ for which

$$u(y_1(t^*), z_1(t^*)) \leqq c_0,$$

it would follow from the mean value theorem that

$$\begin{aligned} &u(y_1(t_0 + T), z_1(t_0 + T)) - u(y_1(t^*), z_1(t^*)) \\ &\qquad = (t_0 + T - t^*) \frac{du}{dt}(y_1(\tau), z_1(\tau)) < 0 \qquad \text{with} \qquad t^* \leqq \tau \leqq t_0 + T, \end{aligned}$$

in contradiction to the assumption.

As an example to illustrate this proof we choose the differential equation

$$y'' + 3y^2y' + y^2 = \sin t. \tag{30}$$

The corresponding system is

$$\begin{aligned} y' + y^3 &= z, \\ z' + y^3 &= \sin t. \end{aligned} \tag{31}$$

The positive definite function

$$u(y, z) = y^2 + (y - z)^2$$

has as its level curves a family of ellipses centered on the origin, so that the conditions of the preceding paragraph are satisfied. From (31) it follows that along a characteristic we have

$$\begin{aligned} \frac{du}{dt} &= -2[(y^2 - 1)^2 + u + (y - z)\sin t - 1] \\ &\leqq -2[u - \sqrt{u} - 1] < 0 \qquad \text{for all} \qquad u \geqq c_0 = 3, \end{aligned}$$

which proves the existence of a periodic solution of (30).

Bibliography

[1] BIEBERBACH, L.: Differentialgleichungen. Springer, Berlin 1926.
[2] BIEBERBACH, L.: Einführung in die Theorie der Differentialgleichungen im reellen Gebiet. Springer, Berlin-Göttingen-Heidelberg 1956.

[3] HAHN, W.: Theorie und Anwendung der direkten Methode von Ljapunov. Ergebnisse der Mathematik, Heft 22, Springer, Berlin-Göttingen-Heidelberg 1959.

[4] KAMKE, E.: Differentialgleichunge, Vol. I., Akad. Verlagsges., Leipzig 1961.

[5] KAMKE, E.: Differentialgleichungen, Lösungsmethoden und Lösungen Vol. I., Akad. Verlagsges. Leipzig 1959.

[6] PERRON, O.: Über Differentialgleichungen erster Ordnung, die nicht nach der Ableitung aufgelöst sind. J. ber. DMV **22**, 356–368, 1913.

[7] RUNGE, C. and KÖNIG, H.: Numerisches Rechnen. Springer, Berlin 1924.

Bibliography added in Translation

BIRKHOFF G. and ROTA, G. C.: Ordinary differential equations. Ginn and Co., Boston 1962.

BOYCE, W. E.; DIPRIMA, R. C.: Elementary differential equations and boundary value problems, John Wiley & Sons, New York 1965.

CESARI, L.: Asymptotic behavior and stability problems in ordinary differential equations. 2nd ed., Academic Press, New York 1963.

CODDINGTON, E. A.: An introduction to ordinary differential equations Prentice-Hall, Englewood Cliffs, N.J. 1961.

COPPEL, W. A.: Stability & asymptotic behavior of differential equations. D. C. Heath & Co., Boston, Mass. 1965.

DAVIES, T. V.; JAMES E. M.: Nonlinear differential equations. Addison-Wesley Publ. Co., Reading, Mass. 1966.

HARTMAN, P.: Ordinary differential equations. John Wiley & Sons, New York 1964.

HOCHSTADT, H.: Differential equations: A modern approach. Holt, Rinehart & Winston, New York 1964.

MARTIN, W. T.; REISSNER E.: Elementary differential equations. 2nd ed., Addison-Wesley Publ. Co., Reading, Mass. 1961.

STRUBLE, R. A.: Nonlinear differential equations. McGraw-Hill Book Co., New York 1961.

TRICOMI, F. G.: Differential equations (translated from the Italian by E. A. McHarg). Hafner Publ. Co., New York 1961.

CHAPTER 9

Partial Differential Equations

1. Basic Concepts and the Simplest Examples

1.1 *Introduction*

A relation of the form

$$(1) \qquad F(x_1, x_2, \ldots, x_n, u, u_{x_1}, u_{x_2}, \ldots, u_{x_n}, u_{x_1x_1}, u_{x_1x_2}, \ldots, u_{x_nx_n}) = 0$$

with $n > 1$ is called a *partial differential equation of second order*. Here the order is determined by the derivative of highest order. We seek functions $u = u(x_1, \ldots, x_n)$ that satisfy (1) identically; such functions are called *solutions*. In (1) the partial derivatives of u have been denoted by $\dfrac{\partial u}{\partial x_i} \equiv u_{x_i}$. If $n = 1$, then (1) is an *ordinary differential equation* of second order $F(x_1, u, u', u'') = 0$ with $u' = \dfrac{du}{dx_1}$. In general, even an ordinary differential equation will have infinitely many solutions, from which we can select a single solution by imposing additional conditions. We now examine this question for the equation (1) and first consider the simplest examples, which are in fact of great importance in mathematics and physics. The equation

$$(2) \qquad \Delta_n u = f(x_1, \ldots, x_n) \qquad \text{with} \qquad \Delta_n u \equiv \sum_{i=1}^{n} u_{x_ix_i}$$

is called the *nonhomogeneous* (or if $f \equiv 0$, the *homogeneous*) *potential equation*. This equation not only dominates wide areas of physics (those problems in which the time t does not play a role) but also, for $n = 2$, plays a considerable role in the theory of functions of a complex variable. The relation

$$(3) \qquad \Delta_n u - u_{tt} = 0$$

is called the *homogeneous wave equation* in n-dimensional space. The special role played here by the independent variable t is justified by its physical importance as the time. This equation occurs in electrodynamics as a special case of the Maxwell equations, and also in the theory of elasticity, where for $n = 1, 2, 3$ it describes the vibrations of a string, a membrane and an elastic body. Strictly speaking, it occurs in the form $\Delta_n u - \frac{1}{c^2} u_{tt} = 0$ with $c = \text{const}$, but the transformation $\bar{t} = ct$ brings it into the form (3). Finally, the relation

$$(4) \qquad \Delta_n u - u_t = 0$$

is called the *homogeneous equation of heat conduction*. Let us consider these examples in some detail.

1.2. *The Wave Equation*

For the case of the vibrating string ($n = 1$) this equation has the form

$$(5) \qquad u_{xx} - u_{tt} = 0 \quad \text{for} \quad u_{\bar{x}\bar{t}} = 0,$$

the latter equation being obtained by the transformation of coordinates $\bar{x} = x + t$, $\bar{t} = x - t$. *All* the solutions of (5) are found at once in the form

$$(6) \qquad u(x, t) = w_1(\bar{x}) + w_2(\bar{t}) = w_1(x + t) + w_2(x - t)$$

with arbitrary functions w_ι. The function $w_1(x + t)$ is referred to as a wave advancing to the left with unit velocity; the reason for this terminology is as follows: if $w_1(x)$ is given for $t = 0$ by Figure 1, then Figure 2

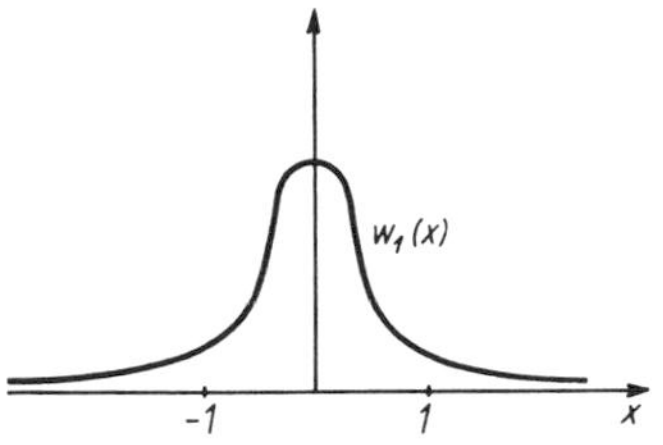

Fig. 1

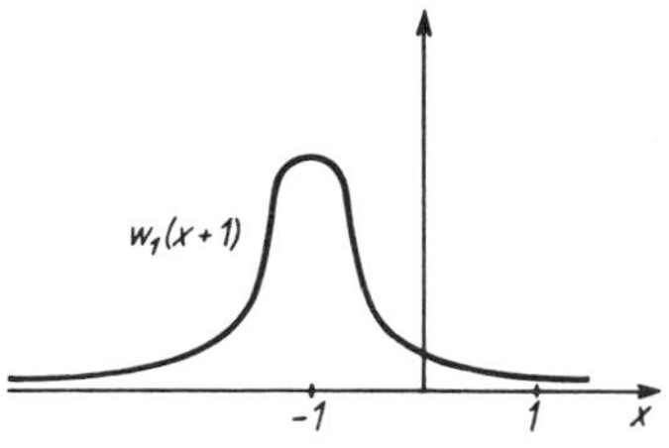

Fig. 2

describes the process at time $t = 1$. In order to determine suitable additional conditions for (5) we consider a string, with fixed ends, which in its position of equilibrium occupies the continuum $0 \leqq x \leqq l$. At time $t = 0$ the string is displaced and then released with a suitable velocity, so that it begins to vibrate. If we denote by $u(x, t)$ the displacement of a point x at the time t, then u satisfies the equation (5) with the *initial conditions*

$$(7) \qquad u(x, 0) = u_0(x), \qquad u_t(x, 0) = u_1(x), \qquad 0 \leqq x \leqq l,$$

where the functions $u_i(x)$ are sufficiently smooth but otherwise arbitrary, and u also satisfies the *boundary conditions* (fixed ends)

$$(8) \qquad u(0, t) = 0, \qquad u(l, t) = 0, \qquad 0 \leqq t < \infty.$$

In order that (5), (7), (8) may represent a significant *mathematical* (*physical*) *problem*, the following *two* (*three*) conditions must be satisfied: 1. *Uniqueness condition.* There exists at most one $u(x, t)$ satisfying all the conditions. 2. *Existence condition.* There exists at least one such u. 3. *Physical condition of continuous dependence on the initial data.* A slight change in $u_i(x)$, and possibly also in its derivatives, produces only a slight change in the solution. (In physical practice the $u_i(x)$ are determined by measurements, with the inevitable limits of error. If x ranges over an infinite interval, the change is to be made only in a finite subinterval.)

For the pure initial value problem with no boundary conditions on the infinitely long string ($-\infty < x < \infty$), we prove the following theorem.

THEOREM 1. *If* $u_0(x) \in \mathfrak{C}^2$, $u_1(x) \in \mathfrak{C}^1$ *for* $-\infty < x < \infty$, *then*

$$(9) \quad u(x, t) = M(t)u_1 + \frac{\partial}{\partial t}(M(t)u_0), \qquad \textit{with} \qquad M(t)u_i \equiv \frac{1}{2}\int_{x-t}^{x+t} u_i(\tau)d\tau$$

in $\mathfrak{C}^2$, *is a solution for* $-\infty < x, t < \infty$ *of the problem*

$$(10) \qquad u_{xx} - u_{tt} = 0; \qquad u(x, 0) = u_0(x), \qquad u_t(x, 0) = u_1(x),$$

and this problem satisfies the three conditions. By $u \in \mathfrak{C}^n$ *we mean that* u *is* n*-times continuously differentiable for any combination of mixed derivatives.*

From (6) it follows that

$$(11) \qquad \begin{aligned} u(x, 0) &= w_1(x) + w_2(x) = u_0(x), \\ u_t(x, 0) &= w_1'(x) - w_2'(x) = u_1(x). \end{aligned}$$

Integration of the last equation gives

$$w_1(x) - w_2(x) = \int_c^x u_1(\tau)d\tau$$

and thus

$$(12) \quad 2w_1(x) = u_0(x) + \int_c^x u_1(\tau)d\tau, \qquad 2w_2(x) = u_0(x) - \int_c^x u_1(\tau)d\tau.$$

From (6) we see finally that

$$(13) \qquad \begin{aligned} u(x, t) &= w_1(x + t) + w_2(x - t) \\ &= \tfrac{1}{2}\left\{u_0(x + t) + u_0(x - t) + \int_{x-t}^{x+t} u_1(\tau)d\tau\right\}, \end{aligned}$$

which coincides with (9). Here the conditions 2 and 1 are satisfied, the latter because of the "accidental circumstance" that in (6) we have all the solutions at our disposal. If u_i is slightly changed in the interval $x_0 \leqq x \leqq x_1$ of length L, then $\bar{u}_i(x) = u_i(x) + v_i(x)$ with $|v_i(x)| \leqq \epsilon$, and

$$(14) \qquad |\bar{u}(x, t) - u(x, t)| \leqq \epsilon(1 + L),$$

where $\bar{u}$ denotes the solution with initial values $\bar{u}_i$; thus condition 3 is satisfied.

We note that the solution u (13) at the point x_0, t_0 does not depend on the entire range of the initial values u_i, but only on the values $u_i(x)$ in the interval $x_0 - t_0 \leqq x \leqq x_0 + t_0$. This set of points is called the *domain of dependence* $\mathfrak{A}$ of the point x_0, t_0. If we change $u_i(x)$ outside $\mathfrak{A}$, the solution u is not thereby influenced at x_0, t_0. Now let us assign the initial values $u_i(x)$ only in the interval $a \leqq x \leqq b$ of the x-axis and ask at which points of the x, t-plane the solution can then be determined; the set of such points is called the *range of influence* $\bar{\mathfrak{B}}$ of the given initial values and from (13) we see that in the present case $\bar{\mathfrak{B}}$ is the square in Figure 3. The

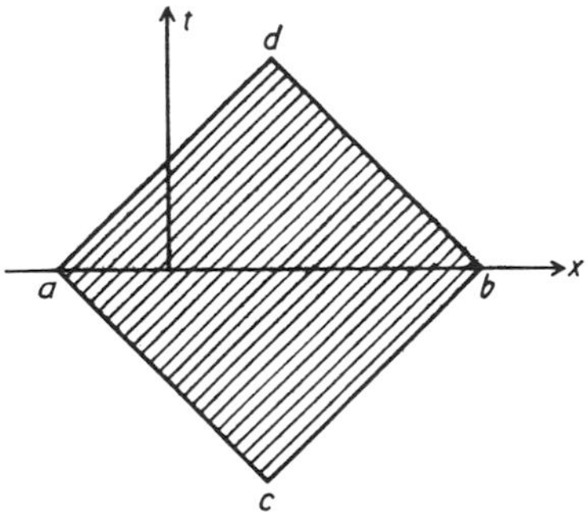

Fig. 3

boundaries $\dot{\mathfrak{B}}$ of $\bar{\mathfrak{B}}$ are called *characteristic manifolds* $\dot{\mathfrak{C}}$. For arbitrary a, b they are in this case $x \pm t =$ const. It follows readily from theorem 1 that: *if* $u_i(x) = 0$ *in* $a \leqq x \leqq b$, *then* $u \equiv 0$ *in* $\bar{\mathfrak{B}}$ (uniqueness theorem).

Another important problem, whose solution likewise satisfies the foregoing three conditions, is the problem with values given on the characteristics. Here u is arbitrarily preassigned on $\widehat{acb}$ (or $\widehat{cbd}$, etc.) and the solution is uniquely determined in $\bar{\mathfrak{B}}$ (Fig. 3).

The solution for the vibrating string with both initial and boundary conditions can also be found in the form (13). The $u_i(x)$ in (7) are now preassigned on $0 \leqq x \leqq l$. If we extend them for all x in such a way that the boundary conditions (8) are automatically satisfied by (13), then (13) is also a solution of the problem, where we assume that $u_i(x)$ are odd periodic functions with period $2l$:

$$(15) \quad u_i(-x) = -u_i(x), \quad u_i(x + 2l) = u_i(x), \quad u_i(0) = u_i(l) = 0, \quad i = 0, 1.$$

For then from (13) we have

$$(16)\quad \begin{aligned} u(0, t) &= \tfrac{1}{2}\left\{u_0(t) + u_0(-t) + \int_0^t u_1(\tau)d\tau + \int_{-t}^0 u_1(\tau)d\tau\right\} \\ &= \tfrac{1}{2}\int_0^t \{u_1(\tau) + u_1(-\tau)\}d\tau = 0. \end{aligned}$$

1.3. *The Wave Equation in* $\mathfrak{R}_2$, $\mathfrak{R}_3$

Here we adopt the vector notation $\mathfrak{x} = (x_1, x_2, x_3)$ with length

$$|\mathfrak{x}| = \sqrt{\sum_{i=1}^{3} x_i^2}$$

and write $u(x_1, x_2, x_3, t) = u(\mathfrak{x}, t)$. Also, we let $d\omega$ be the element of surface area on the unit cube and let $\nu = (\nu_1, \nu_2, \nu_3)$ be the outward normal unit vector with $|\nu| = 1$. Let us examine the initial value problem (3) in $\mathfrak{R}_3$, i.e., with $n = 3$. We have the following theorem.

Theorem 2. *If* $u_0(\mathfrak{x}) \in \mathfrak{C}^3$, $u_1(\mathfrak{x}) \in \mathfrak{C}^2$ *in* $-\infty < x_1, x_2, x_3 < \infty$, *then*

$$(17)\quad \begin{aligned} u(\mathfrak{x}, t) &= tM(t)u_1 + \frac{\partial}{\partial t}(tM(t)u_0), \\ \textit{with}\qquad M(t)u_i &\equiv \frac{1}{4\pi}\int_{|\nu|=1} u_i(\mathfrak{x} + \nu t)d\omega \end{aligned}$$

in $\mathfrak{C}^2$ *for* $-\infty < x_1, x_2, x_3, t < \infty$, *is a solution of the problem*

$$(18)\qquad \Delta_3 u - u_{tt} = 0; \qquad u(\mathfrak{x}, 0) = u_0(\mathfrak{x}), \qquad u_t(\mathfrak{x}, 0) = u_1(\mathfrak{x}),$$

and this problem satisfies the three conditions.

By way of explanation let us remark that in (17) the integration is taken over the surface of the unit sphere, so that the vector $\mathfrak{y} = \mathfrak{x} + \nu t$ describes the surface of the sphere $|\mathfrak{y} - \mathfrak{x}| = |t|$ with center $\mathfrak{x}$ and radius $|t|$. From (17) we obtain

$$(19)\quad \begin{aligned} u(\mathfrak{x}, 0) &= \lim_{t\to 0}\left\{M(t)u_0 + t\left(M(t)u_1 + \frac{\partial}{\partial t}M(t)u_0\right)\right\} = M(0)u_0 \\ &= \frac{1}{4\pi}u_0(\mathfrak{x})\int_{|\nu|=1} d\omega = u_0(\mathfrak{x}). \end{aligned}$$

The third condition is obviously satisfied and the first will be dealt with below in §3.4. For a complete proof we refer the reader to the readily available accounts in [5], [10], [16], [20]. Now the above theorem also provides the solution of the initial value problem in $\mathfrak{R}_2$ ($n = 2$ in (3), $\mathfrak{x} = (x_1, x_2)$, $u(\mathfrak{x}, t) = u(x_1, x_2, t)$), as is clear from the *Hadamard method of descent*: if the $u_i(\mathfrak{x})$ in theorem 2 depend only on x_1, x_2, then obviously the same is true of the solution $u(\mathfrak{x}, t)$ (17). But then it satisfies the wave equation in $\mathfrak{R}_2$, since the term $u_{x_3x_3} = 0$. After some simple recasting this solution appears, under the same hypotheses as in theorem 2, in the form

$$u(\mathfrak{x}, t) = M(t)u_1 + \frac{\partial}{\partial t}(M(t)u_0)$$

(20)

$$\text{with} \qquad M(t)u_i \equiv \frac{1}{2\pi}\int_{|\mathfrak{y}-\mathfrak{x}|\leqq|t|} \frac{u_i(\mathfrak{y})}{\sqrt{t^2 - |\mathfrak{y} - \mathfrak{x}|^2}}\,d\mathfrak{y}.$$

Here $\mathfrak{x}$ stands for (x_1, x_2) and $\mathfrak{y}$ for (y_1, y_2), and the integration is taken over the circle $|\mathfrak{y} - \mathfrak{x}| \leqq |t|$ with center $\mathfrak{x}$ and radius $|t|$. In this case $d\mathfrak{y} = dy_1 dy_2$ denotes the surface-element and thus is not a vector. The *domain of dependence* $\mathfrak{A}$ of $\mathfrak{x}^0, t^0$ is obviously, in the case (20), the circular disk $|\mathfrak{x} - \mathfrak{x}^0| \leqq |t^0|$, whereas in one dimension higher, namely for (17), it was merely the surface of the sphere $|\mathfrak{x} - \mathfrak{x}^0| = |t^0|$.[1] This phenomenon, depending on the number of dimensions, has many important consequences, one of which we now proceed to study. If for (20) the initial values $u_i(\mathfrak{x})$ are prescribed in the disk $|\mathfrak{x} - \mathfrak{x}^0| \leqq |t^0|$, we find that the corresponding *range of influence* $\mathfrak{B}$ consists of the interior and surface of the double cone in Figure 4 and the *characteristic manifolds* $\dot{\mathfrak{C}}$ are the

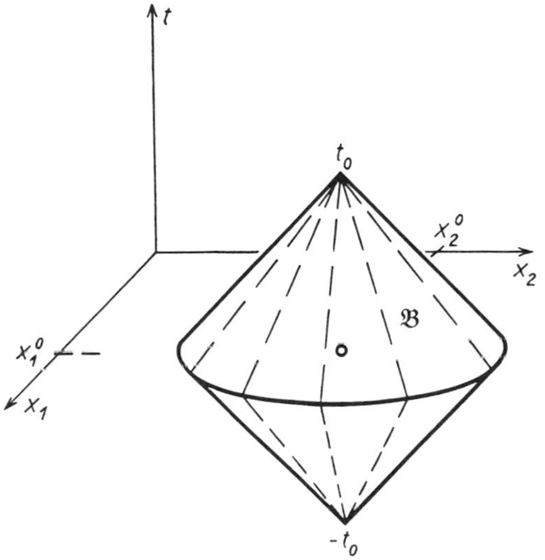

Fig. 4

[1] For let us consider (17) at the point $\mathfrak{x}^0, t^0$. It is obvious that $u(\mathfrak{x}^0, t^0)$ does not depend on the entire set of values of the u_i but only on the values assumed by u_i on the surface on the sphere with center $\mathfrak{x}^0$ and radius $|t^0|$. But in (20) the solution at the point $\mathfrak{x}^0, t^0$ obviously depends on the values assumed by u_i over the whole circular disk with center $\mathfrak{x}^0$ and radius $|t^0|$.

corresponding generators of the cone. The same remarks hold for (17) in a space with one more dimension. Condition 1 now reads: from $u_i(x) = 0$ in $\mathfrak{A}$ it follows that $u \equiv 0$ in $\mathfrak{B}$ (proof in §3.4).

Now for (17) and (20) let us consider the following problem: let the initial values $u_i(\mathfrak{x})$ be $\equiv 0$ except for a small "perturbation domain" $\mathfrak{S}$. Then, as t increases, how does the solution ("perturbation") behave at a fixed point $\mathfrak{x}^0 = (x_1^0, x_2^0)$ or (x_1^0, x_2^0, x_3^0)? The schematic Figure 5, in which

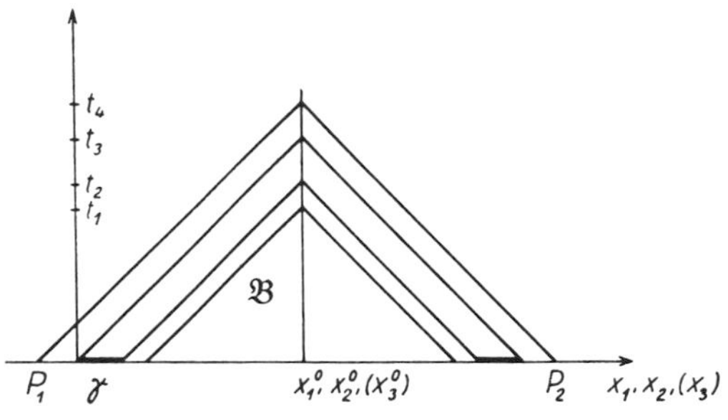

Fig. 5

we have compressed the spatial dimensions down to one, will serve to illustrate. At time t_1 there is equilibrium ($u \equiv 0$) (uniqueness theorem) at the point $\mathfrak{x}^0$. At time t_2 the perturbation begins sharply ($u \not\equiv 0$). But at time t_4 the situation is not the same in $\mathfrak{R}_2$ as it is in $\mathfrak{R}_3$. In $\mathfrak{R}_3$ (17) there is again equilibrium ($u \equiv 0$), since the solution there depends only on the initial values at P_1, P_2, where already $u_i = 0$. But in $\mathfrak{R}_2$ we have $u \not\equiv 0$, since the solution there depends on the values of u_i between $\widehat{P_1P_2}$ (the domain of dependence $\mathfrak{A}$ is of a different kind). We say that in $\mathfrak{R}_3$ the perturbation begins and ceases sharply, but in $\mathfrak{R}_2$ it only begins sharply and then decreases gradually but never again completely disappears. The situation in $\mathfrak{R}_n$ of higher dimensions is analogous to $\mathfrak{R}_3$ (or $\mathfrak{R}_2$) depending on whether n is odd > 1 (or n is even > 0).

For further investigation we now change (17) into a new form. Introducing the vector functions

$$(21) \qquad \mathfrak{u}_0(\mathfrak{x}) = (u_0(\mathfrak{x}), u_1(\mathfrak{x})), \qquad \mathfrak{u}(\mathfrak{x}, t) = (u(\mathfrak{x}, t), u_t(\mathfrak{x}, t))$$

we have from (17)

$$(22) \qquad \mathfrak{u}(\mathfrak{x}, t) = \mathfrak{T}(t)\mathfrak{u}_0(\mathfrak{x}) \qquad \text{with} \qquad \mathfrak{T}(t) = \begin{pmatrix} \dfrac{\partial}{\partial t} N(t), & N(t) \\[2ex] \dfrac{\partial^2}{\partial t^2} N(t), & \dfrac{\partial}{\partial t} N(t) \end{pmatrix}$$

$$\text{and} \qquad N(t) = tM(t).$$

Here we think of the matrix $\mathfrak{T}(t)$ as a *solution-operator* which allows us to determine the "state" $\mathfrak{u}$ at time t if we know the "state" $\mathfrak{u}$ at time $t = 0$. Obviously $\mathfrak{T}(0) = \mathfrak{E}$ is the unit matrix. This interpretation implies that we can obtain the state at time t by means of intermediate steps, namely, by first obtaining the state at time $t_1 < t$ and then, after n steps, obtaining it at time t (requirement of J. Hadamard, Determinism in Nature). This requirement demands that

$$\mathfrak{T}(t_1 + t_2) = \mathfrak{T}(t_2)\mathfrak{T}(t_1) \qquad \text{with} \qquad \mathfrak{T}(-t)\mathfrak{T}(t) = \mathfrak{E}, \tag{23}$$

i.e., the solution operators must form a *group* with respect to the group-operation indicated in (23). In fact, the state $\mathfrak{T}(t_1)\mathfrak{u}_0$ at time t_1 advances, in an interval of time t_2, to the state $\mathfrak{T}(t_2)\mathfrak{T}(t_1)\mathfrak{u}_0$. But this state must also be obtained if we view the system as proceeding in one step through the time $t_1 + t_2$. However, the *Hadamard requirement cannot be satisfied* in theorem 2. After the first step we have $u \in \mathfrak{C}^2$, $u_t \in \mathfrak{C}^1$, and these values can no longer be chosen as initial values, since it would otherwise be necessary to require that $u \in \mathfrak{C}^3$, $u_t \in \mathfrak{C}^2$. We say that *the initial values cannot be propagated* (the properties of the solution become less satisfactory with every step in time). *Initial values that can be propagated* have been given by K. O. Friedrichs and H. Lewy,[2] but they involve the concept of Lebesgue integration. For $n = 1$ the initial values given in theorem 1 can be propagated.

If for $n = 3$ we seek solutions of (3) depending only on $|\mathfrak{x}| = r$ and t, we see that since

$$\Delta_3 u = u_{rr} + \frac{2}{r} u_r = \frac{1}{r}(ru)_{rr}, \tag{24}$$

such solutions must satisfy the equation

$$(ru)_{rr} - (ru)_{tt} = 0. \tag{25}$$

From (6) we obtain the solutions $u(r, t) = \dfrac{w_1(r + t)}{r} + \dfrac{w_2(r - t)}{r}$, which may naturally be called advancing, spatially fading *spherical waves*. For $n = 2$ we have

$$\Delta_2 u = u_{rr} + \frac{1}{r} u_r = u_{tt}, \tag{26}$$

and thus by setting $u(r, t) = e^{i\omega t} f(r)$ we obtain for $f(r)$ the *Bessel differential equation* with the solution $J_0(\omega r)$. We are dealing here with rotation-symmetric stationary waves, the so-called *cylinder waves*.

[2] Über fortsetzbare Anfangsbedingungen . . ., Nachr. Wiss. Ges. Göttingen 26, pp. 135–143 (1932).

1.4. *The Potential Equation*

If we set up an initial value for equation (2), then in general condition 3 cannot be satisfied, and condition 2 can be satisfied only if the initial conditions are analytic, as is shown by the following examples.

Example 1 (J. Hadamard).

$$\Delta_2 u = 0, \qquad u(x_1, 0) = 0, \qquad u_{x_2}(x_1, 0) = \frac{1}{n} \sin nx_1$$

with the solution $u(x_1, x_2) = \dfrac{1}{n^2} \sin nx_1 \,\mathfrak{Sin}\, nx_2$. For large n the initial values differ arbitrarily little from zero, but the solution does not lie in the neighborhood of $u \equiv 0$, since $\mathfrak{Sin}\, nx_2$ behaves like e^{nx_2}.

Example 2. $\Delta_2 u = 0$; $u(x_1, 0) = 0$, $u_{x_2}(x_1, 0) = u_1(x_1)$. From the elementary theory of functions of a complex variable we know that the solution $u(x_1, x_2)$ can be regarded as the real part of a holomorphic (= complex-analytic) function $f(z) = u + iv$ with $z = x_1 + ix_2$. By the *Schwarz reflection principle*, which we shall not discuss here, $f(z)$ can be continued across the x_1-axis as a holomorphic function. For our function $u_1(x_1)$ this means that we must assume $u_1(x_1)$ to be real-analytic on the x_1-axis.

Thus we consider (2) in a simply connected open domain $\mathfrak{G}$ of $\mathfrak{R}_n$ (here $\mathfrak{x}$ again stands for the vector $(x_1, x_2, \ldots, x_n)$ and $|\mathfrak{x}|$ stands for $\sqrt{\sum_{i=1}^n) x_i^2}$ with sufficiently "well-behaved" boundary $\dot{\mathfrak{G}}$, and then we attempt to prescribe $u(\mathfrak{x}) = \varphi(\mathfrak{x})$ arbitrarily on $\dot{\mathfrak{G}}$ (*boundary value problem*). Let the boundary $\dot{\mathfrak{G}}$ be such that for the domain $\mathfrak{G}$ the Gauss theorem

$$\int_{\dot{\mathfrak{G}}} u(\mathfrak{x})\nu_i do = \int_{\mathfrak{G}} \frac{\partial u(\mathfrak{x})}{\partial x_i} d\mathfrak{x} \tag{27}$$

holds for functions $u(\mathfrak{x}) \in \mathfrak{C}^1$ in $\overline{\mathfrak{G}} = \mathfrak{G} + \dot{\mathfrak{G}}$. Here do is the surface-element of $\dot{\mathfrak{G}}$, and $d\mathfrak{x}$ is the volume-element $d\mathfrak{x} = dx_1 dx_2 \cdots dx_n$ (not a vector), whereas $\nu = (\nu_1, \nu_2, \ldots, \nu_n)$ is the "outward" directed unit normal vector (our powers of geometric visualization will be adequate in $\mathfrak{R}_2$ and $\mathfrak{R}_3$). We also need the well-known *Green's formulas*

$$\int_{\dot{\mathfrak{G}}} \left(v \frac{\partial u}{\partial \nu} - u \frac{\partial v}{\partial \nu}\right) do = \int_{\mathfrak{G}} (v\Delta_n u - u\Delta_n v) d\mathfrak{x}, \tag{28}$$

$$\int_{\dot{\mathfrak{G}}} v \frac{\partial u}{\partial \nu} do = \int_{\mathfrak{G}} v\Delta_n u d\mathfrak{x} + \sum_{i=1}^n \int_{\mathfrak{G}} u_{x_i} v_{x_i} d\mathfrak{x} \quad \text{with} \quad \frac{\partial}{\partial \nu} = \sum_{i=1}^n \nu_i \frac{\partial}{\partial x_i}, \tag{29}$$

for the first of which we make the assumptions: $u(\mathfrak{x}), v(\mathfrak{x}) \in \mathfrak{C}^1$ in $\overline{\mathfrak{G}}$, $\in \mathfrak{C}^2$ in $\mathfrak{G}$ and $\int_{\mathfrak{G}} \{\cdots\} d\mathfrak{x}$ exists.

We first determine the solutions $u(\mathfrak{x})$ of $\Delta_n u = 0$ that depend only on $r = |\mathfrak{a} - \mathfrak{x}| = \sqrt{\sum_{i=1}^n (a_i - x_i)^2}$, with a fixed vector $\mathfrak{a} \in \mathfrak{G}$. Setting $u(\mathfrak{x}) = f(r)$, we find

$$\Delta_n u = f''(r) + \frac{n-1}{r} f'(r) = 0 \tag{30}$$

with the solutions $f(r) = C_1 + C_2 r^{2-n}$ for $n > 2$ and

$$f(r) = C_1 + C_2 \log r \qquad \text{for} \qquad n = 2.$$

The expressions

$$\begin{aligned} s(\mathfrak{a}, \mathfrak{x}) &= \frac{1}{(n-2)\omega_n} |\mathfrak{a} - \mathfrak{x}|^{2-n} \qquad &(\text{for } n > 2), \\ &= -\frac{1}{2\pi} \log |\mathfrak{a} - \mathfrak{x}| \qquad &(\text{for } n = 2) \end{aligned} \tag{31}$$

are called *singularity functions* for $\Delta_n u = 0$, where ω_n is the area of the surface of the unit sphere ($\omega_2 = 2\pi$, $\omega_3 = 4\pi$, etc.). The function

$$\gamma(\mathfrak{a}, \mathfrak{x}) = s(\mathfrak{a}, \mathfrak{x}) + \varphi(\mathfrak{x}) \tag{32}$$

with

$$\varphi \in \mathfrak{C}^1 \text{ in } \overline{\mathfrak{G}}, \qquad \varphi \in \mathfrak{C}^2 \text{ in } \mathfrak{G}, \qquad \Delta_n \varphi = 0$$

is called a *fundamental solution*. For this solution we have the following theorem.

Theorem 3. *If $u(\mathfrak{x}) \in \mathfrak{C}^2$ in $\overline{\mathfrak{G}}$ is a solution of* (2), *then*

$$u(\mathfrak{a}) = \int_{\dot{\mathfrak{G}}} \left\{ \gamma(\mathfrak{a}, \mathfrak{x}) \frac{\partial u(\mathfrak{x})}{\partial \nu} - u(\mathfrak{x}) \frac{\partial \gamma(\mathfrak{a}, \mathfrak{x})}{\partial \nu} \right\} do - \int_{\mathfrak{G}} \gamma(\mathfrak{a}, \mathfrak{x}) f(\mathfrak{x}) d\mathfrak{x}. \tag{33}$$

For the proof we remove from $\mathfrak{G}$ the sphere $\mathfrak{K}$; $|\mathfrak{x} - \mathfrak{a}| \leqq \rho$, since $\gamma(\mathfrak{a}, \mathfrak{x})$ becomes singular for $\mathfrak{a} = \mathfrak{x}$. On $\mathfrak{G} - \mathfrak{K}$ we apply (28) and find

$$\begin{aligned} &\int_{\mathfrak{G} - \mathfrak{K}} (\gamma \Delta_n u - u \Delta_n \gamma) d\mathfrak{x} \\ &= \int_{\dot{\mathfrak{G}}} \left(\gamma \frac{\partial u}{\partial \nu} - u \frac{\partial \gamma}{\partial \nu} \right) do - \int_{|\mathfrak{x} - \mathfrak{a}| = \rho} \left(\gamma \frac{\partial u}{\partial \nu} - u \frac{\partial \gamma}{\partial \nu} \right) do. \end{aligned} \tag{34}$$

Substituting (32) into the last integral we have

$$\lim_{\rho \to 0} \int_{|\mathfrak{x} - \mathfrak{a}| = \rho} \left(\varphi \frac{\partial u}{\partial \nu} - u \frac{\partial \varphi}{\partial \nu} \right) do = 0, \tag{35}$$

so that we can estimate the integral

$$\int_{|\mathfrak{x} - \mathfrak{a}| = \rho} \left(s \frac{\partial u}{\partial \nu} - u \frac{\partial s}{\partial \nu} \right) do.$$

For $n > 2$ it follows that on $|\mathfrak{x} - \mathfrak{a}| = \rho$

$$\left| s \frac{\partial u}{\partial \nu} do \right| \leqq \left| \frac{\rho^{2-n}}{(n-2)\omega_n} \frac{\partial u}{\partial \nu} \rho^{n-1} d\omega \right| \leqq M\rho \to 0 \quad \text{for} \quad \rho \to 0,$$

$$-\int_{|\mathfrak{x}-\mathfrak{a}|=\rho} u(\mathfrak{x}) \frac{\partial s}{\partial \nu} do = \frac{\rho^{1-n}}{\omega_n} \rho^{n-1} \int_{|\nu|=1} u(\mathfrak{a} + \rho\nu) d\omega \to u(\mathfrak{a}) \quad \text{for} \quad \rho \to 0.$$

Thus for $\rho \to 0$ the equation (34) does in fact lead to (33). For $n = 2$ the argument is completely analogous.

If for fixed $\mathfrak{x} \in \mathfrak{G}$ the function $\gamma(\mathfrak{x}, \mathfrak{y})$ is a *fundamental solution* with respect to $\mathfrak{y}$ such that $\gamma(\mathfrak{x}, \mathfrak{y}) = 0$ for $\mathfrak{y} \in \dot{\mathfrak{G}}$, it is called a *Green's function* of the first kind and is denoted by $g(\mathfrak{x}, \mathfrak{y})$. Then theorem 3 gives us the representation

$$u(\mathfrak{x}) = -\int_{\dot{\mathfrak{G}}} \frac{\partial g(\mathfrak{x}, \mathfrak{y})}{\partial \nu} u(\mathfrak{y}) do - \int_{\mathfrak{G}} g(\mathfrak{x}, \mathfrak{y}) f(\mathfrak{y}) d\mathfrak{y}. \tag{36}$$

If for u on $\dot{\mathfrak{G}}$ we insert the arbitrary boundary values $\varphi(\mathfrak{x})$, a knowledge of g determines the right side and (36) appears to be the solution of the boundary value problem. But of course it still remains to prove that the Green's function g exists and then to show that the function $u(\mathfrak{x})$ in (36) actually has the desired properties. We shall first take up this question for the case of the sphere $|\mathfrak{y}| \leqq R$ with $\mathfrak{y} = (y_1, y_2, \ldots, y_n)$ and begin by assuming $n > 2$. We set[3]

$$g(\mathfrak{x}, \mathfrak{y}) = s(\mathfrak{x}, \mathfrak{y}) + \varphi(\mathfrak{x}, \mathfrak{y}) \quad \text{with} \quad \varphi(\mathfrak{x}, \mathfrak{y}) = -\frac{k}{(n-2)\omega_n} |\lambda\mathfrak{x} - \mathfrak{y}|^{2-n} \tag{37}$$

with suitable numbers k, λ and $\lambda\mathfrak{x} \neq \mathfrak{y}$. Then by (31) we have $\Delta_n \varphi = 0$ with respect to $\mathfrak{y}$. For $|\mathfrak{y}| = R$ we must have $g = 0$, and thus from (31) for $|\mathfrak{y}| = R$

$$|\mathfrak{x} - \mathfrak{y}|^{2-n} = k|\lambda\mathfrak{x} - \mathfrak{y}|^{2-n} \qquad \text{or} \qquad |\lambda\mathfrak{x} - \mathfrak{y}|^2 = k^{(2/n-2)}|\mathfrak{x} - \mathfrak{y}|^2. \tag{38}$$

If we write (38) out in full, noting that $|\mathfrak{y}| = R$ and $(\mathfrak{x}, \mathfrak{y}) = \sum_{i=1}^{n} x_i y_i$, we obtain

$$(1 - k^{(2/n-2)})R^2 = (k^{(2/n-2)} - \lambda^2)|\mathfrak{x}|^2 + 2(\lambda - k^{(2/n-2)})(\mathfrak{x}, \mathfrak{y}). \tag{39}$$

Setting $\lambda = k^{(2/n-2)}$, we see that λ is equal either to 1 or to $\dfrac{R^2}{|\mathfrak{x}^2|}$. The first of these values cannot be used (since it violates the restriction $\lambda\mathfrak{x} \neq \mathfrak{y}$), but the second gives us $k = \left(\dfrac{R}{|\mathfrak{x}|}\right)^{n-2}$. We also have $\lambda\mathfrak{x} \neq \mathfrak{y}$, since $R^2 > |\mathfrak{x}|^2$, and for spheres the function $g(\mathfrak{x}, \mathfrak{y})$ is explicitly determined. However, for $n = 2$ there is a simpler procedure which we shall discuss

[3] At the suggestion of E. Heinz.

just below. Finally, if we calculate $\frac{\partial g}{\partial \nu}$ for $|\mathfrak{y}| = R$, after some computation we find for $|\mathfrak{y}| = R$

$$\frac{\partial g}{\partial \nu} = \sum_{i=1}^{n} \nu_i \frac{\partial g}{\partial y_i} = \sum_{i=1}^{n} \frac{y_i}{R} \frac{\partial g}{\partial y_i} = -\frac{1}{R\omega_n} \frac{1}{|\mathfrak{x} - \mathfrak{y}|^n} (R^2 - |\mathfrak{x}|^2). \tag{40}$$

Then for $\Delta_n u = 0$ and $u = \varphi$ on $|\mathfrak{y}| = R$ the representation (36) becomes

$$u(\mathfrak{x}) = \frac{1}{R\omega_n} \int_{|\mathfrak{y}|=R} \frac{R^2 - |\mathfrak{x}|^2}{|\mathfrak{x} - \mathfrak{y}|^n} \varphi(\mathfrak{y}) do. \tag{41}$$

This is the familiar *Poisson formula*, which is also valid for $n = 2$, so that for the potential equation there is no essential dependence on the number of dimensions. We have the following theorem.

Theorem 4. *If $\varphi(\mathfrak{x}) \in \mathfrak{C}^0$ on $|\mathfrak{x}| = R$, then for $n \leqq 2$ the function*

$$u(\mathfrak{x}) = \begin{cases} \dfrac{R^2 - |\mathfrak{x}|^2}{R\omega_n} \displaystyle\int_{|\mathfrak{y}|=R} \frac{\varphi(\mathfrak{y})}{|\mathfrak{x} - \mathfrak{y}|^n} do & \text{for} \quad |\mathfrak{x}| < R \\ \varphi(\mathfrak{x}) & \text{for} \quad |\mathfrak{x}| = R \end{cases}$$

is in $\mathfrak{C}^0$ for $|\mathfrak{x}| \leqq R$ and in $\mathfrak{C}^2$ for $|\mathfrak{x}| < R$ and is a solution of the problem

$$\Delta_n u = 0 \quad \text{for} \quad |\mathfrak{x}| < R, \qquad u = \varphi(\mathfrak{x}) \quad \text{for} \quad |\mathfrak{x}| = R,$$

and this problem satisfies the three conditions.

If we attempt to use formula (36) to obtain a similar result for the general boundary value problem (2) with $u = \varphi$ on $\dot{\mathfrak{G}}$, our first difficulty is to prove the *existence of the Green's function.* Now it is a highly remarkable fact that this deep-lying difficulty can easily be dealt with (though only for the case $n = 2$) by means of the *Riemann mapping theorem in the theory of functions of a complex variable.* The theorem states that every simply connected domain $\mathfrak{G}$ (except for the complete plane or the plane with one deleted point) can be conformally mapped one-to-one by a holomorphic function onto the interior of the unit circle in such a way that an arbitrary point in $\mathfrak{G}$ and a direction at that point can be made to correspond to an arbitrary point and direction in the unit circle, e.g., the origin and the direction of the positive real axis. If $\mathfrak{G}$ lies in the ($z = y_1 + iy_2$)-plane and $w = f(z)$ is such a mapping onto the unit circle, taking the point $z_1 = x_1 + ix_2 \in \mathfrak{G}$ onto the point $w = 0$, then

$$g(\mathfrak{x}, \mathfrak{y}) = -\frac{1}{2\pi} \log |f(z)| \tag{42}$$

is the desired *Green's function.* For in fact $f(z)$ can be represented in the form

$$f(z) = (z - z_1)e^{h(z_1, z)}, \tag{43}$$

since $z = z_1$ must be a simple zero of $f(z)$ (conformality of the mapping).

We find

$$g(\mathfrak{x}, \mathfrak{y}) = -\frac{1}{2\pi} \log |z - z_1| - \frac{1}{2\pi} \operatorname{Re} h(z_1, z). \tag{44}$$

This is a fundamental solution, since the last term is the real part of a holomorphic function and thus satisfies the equation $\Delta_2 \operatorname{Re} h = 0$. Moreover, $|f(z)| = 1$ on $\dot{\mathfrak{G}}$ and thus $g(\mathfrak{x}, \mathfrak{y}) = 0$ for $\mathfrak{y} \in \dot{\mathfrak{G}}$. The corresponding differentiability properties on $\dot{\mathfrak{G}}$ are ensured by confining attention to "sufficiently well-behaved" domains $\mathfrak{G}$.

The Poisson formula has become the starting point for numerous further investigations. We say that $u(\mathfrak{x}) \in \mathfrak{C}^0$ has the *first* (*second*) *mean value property* in $\mathfrak{G}$ if

$$u(\mathfrak{x}) = \frac{1}{\omega_n R^{n-1}} \int_{|\mathfrak{x}-\mathfrak{y}|=R} u(\mathfrak{y})do, \tag{45a}$$

$$\left(u(\mathfrak{x}) = \frac{n}{\omega_n R^n} \int_{|\mathfrak{x}-\mathfrak{y}|\leqq R} u(\mathfrak{y})d\mathfrak{y}\right) \tag{45b}$$

for every sphere $|\mathfrak{x} - \mathfrak{y}| \leqq R \subset \mathfrak{G}$. It is obvious that the two definitions are equivalent. We also say that $u(\mathfrak{x}) \in \mathfrak{C}^0$ has the *strong maximum* (*minimum*) *property* in $\mathfrak{G}$ if the assumption that $u(\mathfrak{x})$ has a maximum (minimum) in $\mathfrak{G}$ implies that $u \equiv$ const in $\mathfrak{G}$.

Theorem 5. *If $u(\mathfrak{x})$ has the mean value property, then $u(\mathfrak{x})$ has the strong maximum* (*minimum*) *property.*

For the proof we let $\mathfrak{x}^0 \in \mathfrak{G}$ be the point at which $u : u(\mathfrak{x}^0) = M$ attains its maximum. Then it follows first of all that $u(\mathfrak{x}) \equiv M$ in every sphere $|\mathfrak{x}^0 - \mathfrak{y}| \leqq R \subset \mathfrak{G}$. For by the definition of the maximum we have $u < M$ everywhere in the sphere except at its center. From (45b) it follows that

$$M = u(\mathfrak{x}^0) = \frac{n}{\omega_n R^n} \int_{|\mathfrak{x}^0-\mathfrak{y}|\leqq R} u(\mathfrak{y})d\mathfrak{y} < M, \tag{46}$$

which is a contradiction. The set $\mathfrak{M}$ of points $\mathfrak{x} \in \mathfrak{G}$ at which $u(\mathfrak{x}) = M$ is therefore open. But it is also closed: for every sequence of points $\mathfrak{x}^\nu$ with $\lim_{\nu\to\infty} \mathfrak{x}^\nu = \mathfrak{x}$ and $u(\mathfrak{x}^\nu) = M$ shows that $u(\mathfrak{x}) = M$. Thus $\mathfrak{M} \equiv \mathfrak{G}$.

The important connection between the *mean value property* and the *harmonic functions* ($u(\mathfrak{x}) \in \mathfrak{C}^2$ satisfies $\Delta_n u = 0$ in $\mathfrak{G}$) is shown by the following theorem.

Theorem 6. *$u(\mathfrak{x}) \in \mathfrak{C}^0$ in $\mathfrak{G}$ is harmonic if and only if $u(\mathfrak{x})$ has the mean value property.*

For if $u(\mathfrak{x})$ is harmonic in $\mathfrak{G}$, then for every sphere centered on the origin the equation (41) implies

$$\begin{aligned} u(\mathfrak{x}) &= \frac{1}{R\omega_n} \int_{|\mathfrak{y}|=R} \frac{R^2 - |\mathfrak{x}|^2}{|\mathfrak{x} - \mathfrak{y}|^n} u(\mathfrak{y})do, \\ u(0) &= \frac{1}{\omega_n R^{n-1}} \int_{|\mathfrak{y}|=R} u(\mathfrak{y})do, \end{aligned} \tag{47}$$

from which (45a) follows by a simple shift of coordinate axes. On the other hand, let us assume that $u(\mathfrak{x})$ has the mean value property, and let $\mathfrak{K}$ be an arbitrary sphere $\subset \mathfrak{G}$. Let $v(\mathfrak{x})$ be the solution of the boundary value problem: $\Delta_n v = 0$ in $\mathfrak{K}$, $v = u$ on $\dot{\mathfrak{K}}$. Then $w = u - v$ has the mean value property in $\mathfrak{K}$ and thus, by theorem 5, the strong maximum property. It follows that $w \equiv 0$, since $w = 0$ on $\dot{\mathfrak{K}}$. Consequently we have shown that u is harmonic in every sphere $\subset \mathfrak{G}$ and thus in the whole of $\mathfrak{G}$.

1.5. *The Heat Equation*

We first consider the equation (4) in one spatial dimension x, so that no use is made of vector notation. The function

$$s(x, t) = \frac{1}{\sqrt{4\pi t}} e^{-x^2/4t} \tag{48}$$

satisfies the equation $u_{xx} - u_t = 0$ for $t > 0$, $-\infty < x < \infty$. It is called the *singularity function.* The following theorem will be familiar to the reader.

Theorem 7. *If $\varphi(x) \in \mathfrak{C}^0$ in $-\infty < x < \infty$ and $|\varphi(x)| \leqq M$, then*

$$u(x, t) = \int_{-\infty}^{+\infty} s(x - \xi, t)\varphi(\xi)d\xi \tag{49}$$

is a solution of the initial value problem. Also $u \in \mathfrak{C}^0$ in $0 \leqq t < \infty$, $-\infty < x < \infty$, and $u \in \mathfrak{C}^2$ in $0 < t < \infty$, $-\infty < x < \infty$, and this function $u(x, t)$ satisfies the equation (4) *for these values of x and t and $u(x, 0) = \varphi(x)$.*

The only advantage of this theorem is that it is easy to prove; it does not provide us with any deep insight into the situation. For that purpose we must proceed as follows.

Theorem 8. *If $\varphi(x) \in \mathfrak{C}^0$ in $-\infty < x < \infty$ and $|\varphi(x)| \leqq Me^{Ax^2}$, then* (49) *is a solution of the initial value problem for $0 \leqq t \leqq T$ with $T < \dfrac{1}{4A}$. Also, $u \in \mathfrak{C}^0$ in $-\infty < x < \infty, 0 \leqq t \leqq T$, and in $-\infty < x < \infty, 0 < t \leqq T$ the function u satisfies the equation* (4) *and $u(x, 0) = \varphi(x)$. Moreover, $u(x, t)$ satisfies the inequality $|u(x, t)| \leqq M_1 e^{A_1 x^2}$ in $-\infty < x < \infty$, $0 \leqq t \leqq T$.*

Thus condition 2 is satisfied. Condition 1 is seen to be satisfied as a result of the following theorem.

Theorem 9. *Let $u(x, t)$ satisfy the equation* (4) *with $n = 1$ in $-\infty < x < \infty$, $0 < t \leqq T$. Also let $u(x, t) \in \mathfrak{C}^0$ in $-\infty < x < \infty$, $0 \leqq t \leqq T$ with $u(x, 0) = 0$. Then if $u(x, t)$ satisfies an inequality of the form*

$$|u(x, t)| \leqq Me^{Ax^2} \quad in \quad 0 \leqq t \leqq T \tag{50}$$

with constants M and A, then $u(x, t) \equiv 0$.

We will now show by examples that none of these assumptions can be omitted.

Example 3. $u(x, t) = s_x(x, t)$ is a solution $\not\equiv 0$ of (4) with $n = 1$. Also, $\lim_{t\to\infty} u(x, t) = 0$ for fixed x. But $u(x, t)$ is not continuous on the x-axis. For if we approach the point (0, 0) along the curve $x^2 = t$, we see that

$$(51)\quad s_x = -\frac{x}{4\sqrt{\pi}t^{3/2}}e^{-x^2/4t} \qquad \text{with} \qquad s_x(\sqrt{t}, t) = -\frac{1}{4\sqrt{\pi t}}e^{-1/4}$$

is not even bounded for $t \to 0$.

Example 4. A solution of (4) for $n = 1$ is given by the function

$$(52)\qquad u(x, t) = \sum_{k=0}^{\infty} \frac{d^k f(t)}{dt^k}\frac{x^{2k}}{(2k)!},$$

as can be verified at once, provided that $f(t) \in \mathfrak{C}^\infty$ and the convergence is sufficiently rapid. Tihonov set $f(t) = e^{-1/t^2}$ for $t \neq 0$ and $f(0) = 0$ and showed that then $u \in \mathfrak{C}^0$ in $0 \leqq t$ and

$$(53)\qquad |u(x, t)| \leqq e^{-1/t(4/9t - x^2)}.$$

The latter statement follows at once from the inequality

$$\left|\frac{d^k f(t)}{dt^k}\right| \leqq \frac{2^k k!}{t^k}e^{-4/9t^2},$$

which is not altogether trivial (cf. [10], p. 53). From (53) it follows that $\lim_{t\to\infty} u(x, t) = 0$ uniformly for all x in every finite interval. But the condition (50) cannot be satisfied here. However, the consequences of this example go considerably further. In the case of the wave equation we could determine the solution of the initial value problem for the past ($t < 0$) as well as for the future, but for the solution (49) such a project is impossible, at least as matters stand, since (48) is not defined for $t < 0$. But in the Tihonov example t may be replaced by $-t$, which shows that the solution cannot be uniquely determined, either for the future $t > 0$ or for the past $t < 0$, without an inequality of the form (50). Here too, as in §1.2, the following initial-boundary problem is physically more meaningful:

$$(50)\qquad \begin{gathered} u_{xx} - u_t = 0, \qquad u(x, 0) = \varphi(x), \qquad u(0, t) = 0, \\ u(l, t) = 0, \qquad 0 \leqq x \leqq l, \end{gathered}$$

which, again as in §1.2, can be solved by a suitable continuation of $\varphi(x)$. If, as in §1.3, we regard the state of the process at the time t as being determined from its state at time $t = 0$ by the solution operator $T(t)$, the solution (49) becomes

$$(51)\quad u(x, t) = T(t)\varphi, \qquad T(t) = \int_{-\infty}^{+\infty} s(x - \xi, t)\{\cdots\}d\xi, \qquad T(0) = 1.$$

The *Hadamard requirement* shows here as in §1.3 that

$$(52) \qquad T(t_1 + t_2) = T(t_2)T(t_1),$$

but on the other hand $T(-t)$ is meaningless, i.e., the solution operators form a *semigroup*[4] with respect to the group-operation (52). Somewhat superficially we may say that the solution operators of reversible processes in nature form groups, and for irreversible processes they form semigroups. Condition 3 is obviously satisfied here, for if we change $\varphi(x)$ slightly in an interval $x_0 \leqq x \leqq x_1$ of length L: $\bar{\varphi}(x) = \varphi(x) + \psi(x)$ with $|\psi(x)| \leq \epsilon$, it follows from (49) that

$$(53) \quad |\bar{u}(x, t) - u(x, t)| \leqq \epsilon \int_{x_0}^{x_1} s(x - \xi, t)d\xi \leqq \epsilon \int_{-\infty}^{+\infty} s(x - \xi, t)d\xi = \epsilon,$$

since the last integral represents the solution with the initial values 1, or in other words $\equiv 1$.

2. Classification of Partial Differential Equations into Types; Normal Forms

2.1. *Characteristic Manifolds*

We consider the special case of (1)

$$(54) \qquad Du \equiv Au + f = 0 \qquad \text{with} \qquad Au \equiv \sum_{i,k=1}^{n} a_{ik} \frac{\partial^2 u}{\partial x_i \partial x_k}.$$

The equation (54) is said to be α) *quasilinear* if a_{ik}, f are functions of the $2n + 1$ variables $x_1, \ldots, x_n, u, \dfrac{\partial u}{\partial x_1}, \ldots, \dfrac{\partial u}{\partial x_n}$, and β) *almost linear if* a_{ik} are functions of $x_1, \ldots, x_n$ only and f is as in α). We consider (54) in the $(2n + 1)$-dimensional domain: $x_1, \ldots, x_n \in \mathfrak{G}$; $|u|, \left|\dfrac{\partial u}{\partial x_i}\right| \leq C$. To the equation (54) we assign the *quadratic form*

$$(55) \qquad Q(\xi_1, \ldots, \xi_n) = \sum_{i,k=1}^{n} a_{ik}\xi_i\xi_k.$$

In the case α) let us assume for the moment that a solution u of $Du = 0$, obtained in some manner, has been substituted into the a_{ik}, so that the coefficients in α) are also functions of $x_1, \ldots, x_n$ only. Then for a fixed

[4] The definition of a *semigroup* differs from that of a group solely in lacking the postulate of division.

point $\mathfrak{x}$: $x_1, \ldots, x_n \in \mathfrak{G}$, it is possible, by a transformation of coordinates with nonvanishing functional determinant

$$(56) \qquad \eta_1 = \eta_1(\xi_1, \ldots, \xi_n), \ldots, \qquad \eta_n = \eta_n(\xi_1, \ldots, \xi_n),$$

to refer the quadratic form

$$(57) \qquad \tilde{Q}(\eta_1, \ldots, \eta_n) = \sum_{i=1}^{n} \kappa_i \eta_i^2$$

to *principal axes.*

As is well known, the number of negative κ_i is called the *index of inertia* $\mathfrak{T}$, and the number of vanishing κ_i is called the *defect* $\mathfrak{D}$ of the quadratic form. The numbers $\mathfrak{T}$, $\mathfrak{D}$ are invariant under coordinate transformations, so that the equations in (54) can be *classified* with respect to these invariants.

Classification into types: at the point $\mathfrak{x}$: $x_1, \ldots, x_n$ the equation (54) is said to be

of *elliptic type*, if $\mathfrak{D} = 0$, $\mathfrak{T} = 0$ or n,
of *hyperbolic type*, if $\mathfrak{D} = 0$, $\mathfrak{T} = 1$ or $n - 1$,
of *parabolic type*, if $\mathfrak{D} > 0$,
of *ultrahyperbolic type*, if $\mathfrak{D} = 0$, $1 < \mathfrak{T} < n - 1$.

For a given point $\mathfrak{x} \in \mathfrak{G}$ this is a complete division into cases, the last case being possible only for $n > 3$, but for equations of type α) the classification is still indirectly dependent on our choice of the solution u. The expression Du is said to be of *elliptic type* in $\mathfrak{G}$ if Du is of elliptic type for every $\mathfrak{x} \in \mathfrak{G}$. For the first three types the most important representatives are equations (2), (3), (4), and for the last type say

$$u_{x_1x_1} + u_{x_2x_2} - u_{x_3x_3} - u_{x_4x_4} = 0.$$

To the quadratic form (55) we also assign a partial differential equation of the first order

$$(58) \qquad \sum_{i,k=1}^{n} a_{ik} \frac{\partial \Phi}{\partial x_i} \frac{\partial \Phi}{\partial x_k} = 0,$$

and attempt to find (of course, real) solutions $\Phi = \Phi(x_1, \ldots, x_n) = 0$, which we call *characteristic manifolds* $\dot{\mathfrak{C}}$. It is clear that for elliptic differential equations such $\dot{\mathfrak{C}}$ cannot exist. For the equation (54) of hyperbolic type they have many of the same properties as for the wave equation: they make up the boundary of the *range of influence* $\mathfrak{B}$ and in general consist of conoids, i.e., cone-like surfaces. For the wave equation in $\mathfrak{R}_2$ we obtain for (58), in an obvious notation, the readily verified solutions

$$(59) \qquad \begin{gathered} \Phi_{x_1}^2 + \Phi_{x_2}^2 - \Phi_t^2 = 0, \\ \text{with} \quad \Phi(x_1, x_2, t) = \sum_{i=1}^{2} (x_i - \gamma_i)^2 - (t - \delta)^2 = 0 \end{gathered}$$

with arbitrary constants γ_i, δ. If we set $\gamma_i = x_i^0$, $\delta = t^0$, we obtain the upper half of the surface of the double cone in Figure 4 (§1.3), and if we set $\gamma_i = x_i^0$, $\delta = -t^0$, we obtain the lower half. For the equation (4) we obtain for $\dot{\mathfrak{C}}$ the planes $t = \text{const}$. It is easy to show that the equation (58) for the characteristic manifolds $\dot{\mathfrak{C}}$ is invariant under coordinate transformations, so that the manifolds $\dot{\mathfrak{C}}$ have invariant significance, as must naturally be the case if they are to constitute the boundary of the range of influence $\mathfrak{B}$ in the hyperbolic case.

2.2. *Normal Forms*

The almost linear equation (54) can be considerably simplified, for $n = 2$, if we introduce the characteristic manifolds $\dot{\mathfrak{C}}$ as new coordinates. In the coordinates x, y, equation (54), written without the sign of summation, appears as

$$(60) \qquad Du \equiv au_{xx} + 2bu_{xy} + cu_{yy} + f(x, y, u, u_x, u_y) = 0,$$

which we shall examine in a domain $\mathfrak{G}$ with $a^2 + b^2 + c^2 > 0$. Furthermore, we may always assume that $a > 0$, if necessary after transformation to new coordinates. Then (55) becomes

$$(61) \qquad Q(\xi, \eta) = a\left[\left(\xi + \frac{b}{a}\eta\right)^2 + \frac{ac - b^2}{a^2}\eta^2\right],$$

and we see that (60) is of *elliptic*, *hyperbolic*, or *parabolic type* in $\mathfrak{G}$ as $ac - b^2 > 0, < 0, = 0$ in $\mathfrak{G}$. We shall assume below that $a, b, c \in \mathfrak{C}^1$ in $\mathfrak{G}$.

a) $ac - b^2 < 0$ in $\mathfrak{G}$; the equation (58) becomes

$$(62) \qquad a\varphi_x^2 + 2b\varphi_x\varphi_y + c\varphi_y^2 = 0,$$

and its solutions provide two families of curves $\varphi(x, y) = 0$, $\psi(x, y) = 0$. We employ these curves as new coordinates: $\xi = \varphi(x, y)$, $\eta = \psi(x, y)$. From (60) we then have

$$(63) \qquad \tilde{D}u \equiv \tilde{a}u_{\xi\xi} + 2\tilde{b}u_{\xi\eta} + \tilde{c}u_{\eta\eta} + \tilde{f}(\xi, \eta, u, u_\xi, u_\eta) = 0,$$

$$(64a) \qquad \begin{aligned} \tilde{a} &= a\varphi_x^2 + 2b\varphi_x\varphi_y + c\varphi_y^2, \\ \tilde{b} &= a\varphi_x\psi_x + b(\varphi_x\psi_y + \varphi_y\psi_x) + c\varphi_y\psi_y, \end{aligned}$$

$$(64b) \qquad \begin{aligned} \tilde{c} &= a\psi_x^2 + 2b\psi_x\psi_y + c\psi_y^2, \\ \tilde{a}\tilde{c} - \tilde{b}^2 &= (ac - b^2)(\varphi_x\psi_y - \varphi_y\psi_x)^2. \end{aligned}$$

Obviously $\tilde{a} = \tilde{c} = 0$, and $\tilde{b} \neq 0$, as follows from the last equation since $(\varphi_x\psi_y - \varphi_y\psi_x) \neq 0$. If we now make the transformation $\bar{\xi} = \frac{1}{2}(\xi + \eta)$, $\bar{\eta} = \frac{1}{2}(\eta - \xi)$, we obtain the two *normal forms* for *hyperbolic equations*

$$(65) \qquad u_{\xi\eta} + \tilde{f}(\xi, \eta, u, u_\xi, u_\eta) = 0, \qquad u_{\bar{\xi}\bar{\xi}} - u_{\bar{\eta}\bar{\eta}} + \hat{f}(\bar{\xi}, \bar{\eta}, u, u_{\bar{\xi}}, u_{\bar{\eta}}) = 0.$$

b) $ac - b^2 = 0$ in $\mathfrak{G}$. Then (62) provides only one family of curves $\varphi(x, y)$. However, we introduce the new coordinates as above except that now $\eta = \psi(x, y)$ is an arbitrary family of curves with the property that $(\varphi_x\psi_y - \varphi_y\psi_x) \neq 0$ and $\tilde{c} \neq 0$. Then we have $\tilde{a} = \tilde{b} = 0$, as follows from the last equation in (64b), and the *normal form* for *parabolic equations* becomes:

$$u_{\eta\eta} + \tilde{f}(\xi, \eta, u, u_\xi, u_\eta) = 0. \tag{66}$$

If in (60) the function f is linear in u, u_x, u_y, we can make further simplifications. For then we obtain (66) in the form

$$u_{\eta\eta} + \alpha(\xi, \eta)u_\xi + \beta(\xi, \eta)u_\eta + \gamma(\xi, \eta)u + \delta(\xi, \eta) = 0 \tag{67}$$

and may assume $\alpha \neq 0$. By the transformation $\bar{\eta} = h(\xi, \eta)$, $\bar{\xi} = \xi$ with $h = \int^\eta \sqrt{\pm\alpha(\xi, \tau)}d\tau$ (+ if $\alpha > 0$, $-$ if $\alpha < 0$) we obtain an equation of the form (67) in which, however, the term in $u_{\bar{\xi}}$ has the coefficient ± 1. Thus in (67) we may assume $\alpha = \pm 1$. By setting $u = vw$ with

$$\log w = -\int^\eta \frac{\beta(\xi, \tau)}{2}\, d\tau$$

we now introduce a new unknown function v. Then for v we have an equation (67) with $\alpha = \pm 1$ and $\beta = 0$, so that we can reduce the *normal form* for *linear parabolic equations* with $\alpha \neq 0$ to

$$v_{\eta\eta} \pm v_\xi + \bar{\gamma}(\xi, \eta)v + \bar{\delta}(\xi, \eta) = 0. \tag{68}$$

c) $ac - b^2 > 0$ in $\mathfrak{G}$. In this case we could begin by attempting to repeat the argument in a), except that now the coordinates ξ, η become complex conjugates. If then by means of

$$\bar{\xi} = \tfrac{1}{2}(\xi + \eta), \qquad \bar{\eta} = \frac{1}{2i}(\eta - \xi) \tag{69}$$

we return again to real coordinates, we have the *normal form* for *elliptic equations*

$$u_{\bar{\xi}\bar{\xi}} + u_{\bar{\eta}\bar{\eta}} + \tilde{f}(\bar{\xi}, \bar{\eta}, u, u_{\bar{\xi}}, u_{\bar{\eta}}) = 0. \tag{70}$$

But this path "through the complex plane" requires that the coefficients a, b, c be analytic in $\mathfrak{G}$, an assumption that is not acceptable in the theory of partial differential equations. Far preferable, though more difficult, is the following procedure. In the new coordinates $\xi = \varphi(x, y)$, $\eta = \psi(x, y)$ we wish to ensure that $\tilde{a} = \tilde{c}$, $\tilde{b} = 0$. Thus we must have

$$a\varphi_x^2 + 2b\varphi_x\varphi_y + c\varphi_y^2 = a\psi_x^2 + 2b\psi_x\psi_y + c\psi_y^2, \tag{71}$$

$$a\varphi_x\psi_x + b(\varphi_x\psi_y + \varphi_y\psi_x) + c\varphi_y\psi_y = 0. \tag{72}$$

From (72) it follows that $\varphi_x = \rho(b\psi_x + c\psi_y)$, $\varphi_y = -\rho(a\psi_x + b\psi_y)$, and from (71) that $\rho = \dfrac{1}{\sqrt{ac - b^2}}$. Finally, by differentiation with respect to x and y we obtain

$$\{\rho(a\psi_x + b\psi_y)\}_x + \{\rho(b\psi_x + c\psi_y)\}_y = 0. \tag{73}$$

Equation (73), which is also satisfied by φ, is called the *Beltrami differential equation of second order*. If we have found a solution $\psi = \psi(x, y) \not\equiv$ const valid in $\mathfrak{G}$, then by means of the above *Beltrami system* we at once obtain $\varphi(x, y)$ and with it the desired real transformation of coordinates. But this existence problem is by no means simple, even though we are only seeking an arbitrary solution of (73). The same problem arises in many important situations in mathematics, for example, in differential geometry in the question of *conformal mapping* of a *nonanalytic segment of a surface* $\mathfrak{F}$ onto the plane. Let $\mathfrak{F}$ be described by the *line-element*

$$(ds)^2 = c(dx)^2 - 2bdxdy + a(dy)^2 \qquad \text{with} \qquad ac - b^2 > 0. \tag{74}$$

Then we seek a mapping $\xi = \varphi(x, y)$, $\eta = \psi(x, y)$ such that (74) is transformed into

$$(ds)^2 = \lambda(\xi, \eta)((d\xi)^2 + (d\eta)^2) \qquad \text{with} \qquad \lambda > 0. \tag{75}$$

Here again φ, ψ must satisfy (73); in fact, (73) is the generalization of $\Delta_2\psi$ for the curved surface described by (74).

2.3. *Differential Equations of Mixed Type*

We confine our attention to the case $n = 2$ and retain the above notation. Even for $n = 2$ the *classification into types* in §2.1 is complete for a point $x, y \in \mathfrak{G}$, but not for a domain $\mathfrak{G}$. On the contrary, in $\mathfrak{G}$ we can find *differential equations* of *elliptic-parabolic*, of *hyperbolic-parabolic* and of *hyperbolic-parabolic-elliptic types*. The simplest example is

$$yu_{xx} + u_{yy} = 0 \tag{76}$$

for $-\infty < x < \infty$ and $y \geqq 0$ or $y \leqq 0$ or $-\infty < y < \infty$. An equation in $\mathfrak{G}$ that is not of the same type throughout is called an *equation of mixed type*. Let (60) be such an equation. Then we encounter new concepts. The set of points $\mathfrak{K} \subset \mathfrak{G}$ on which (60) is of *parabolic type* ($ac - b^2 = 0$) will be called *parabolic curves*. Here it is assumed that $\mathfrak{K}$ consists solely of sufficiently smooth arcs of curves $X_i(x, y) = 0$. For $\mathfrak{K}$ exactly *three cases* are to be distinguished: (I) $\mathfrak{K}$ is *characteristic*, i.e., $X_i(x, y) = 0$ satisfies (62); (II) $\mathfrak{K}$ is *not characteristic*, i.e., $X_i(x, y) = 0$ does not satisfy (62); or (III) $\mathfrak{K}$ is *partly characteristic* and *partly not characteristic*.

Example 5. (76) with $0 \leqq x, y < \infty$. $\mathfrak{K}$: $y = 0$ is of type (II). $u_{xx} + yu_{yy} = 0$, $0 \leqq x, y < \infty$. $\mathfrak{K}$: $y = 0$ is of type (I).

2.4. *Characteristic Manifolds for Systems*

We confine our attention to the simplest case and consider a *quasilinear system* of *two differential equations of first order in two independent variables* x, y and *two unknown functions* $u(x, y)$ $v(x, y)$. Such a system is of the form

$$\text{(77)}\qquad \begin{cases} a^1u_x + a^2u_y + b^1v_x + b^2v_y + f = 0, \\ \bar{a}^1u_x + \bar{a}^2u_y + \bar{b}^1v_x + \bar{b}^2v_y + \bar{f} = 0 \end{cases}$$

with functions $a^1, \ldots, \bar{f}$ of x, y, u, v. Here again we can assign to (77) a *quadratic form*

$$\text{(78)}\qquad Q(\xi, \eta) = \begin{vmatrix} a^1 & b^1 \\ \bar{a}^1 & \bar{b}^1 \end{vmatrix}\xi^2 + \left\{\begin{vmatrix} a^1 & b^2 \\ \bar{a}^1 & \bar{b}^2 \end{vmatrix} + \begin{vmatrix} a^2 & b^1 \\ \bar{a}^2 & \bar{b}^1 \end{vmatrix}\right\}\xi\eta + \begin{vmatrix} a^2 & b^2 \\ \bar{a}^2 & \bar{b}^2 \end{vmatrix}\eta^2$$

and make a *classification into types* in an analogous way.

Example 6 (O. Perron).

$$\text{(79)}\qquad u_x - u_y - v_y = 0, \qquad au_y - v_x + v_y + f(x + y) = 0$$

with constant a. We obtain

$$\text{(80)}\qquad Q(\xi, \eta) = -\xi^2 + 2\xi\eta + (a - 1)\eta^2 = -(\xi - \eta)^2 + a\eta^2.$$

Thus (79) is *hyperbolic* for $a > 0$, *parabolic* for $a = 0$, *elliptic* for $a < 0$. We have the following theorem.

Theorem 10. *For the initial value problem* (79) *with* $u(0, y) = 0$, $v(0, y) = 0$, *it is necessary and sufficient for the existence of a pair of solution functions* $u, v \in \mathfrak{C}^1$ *that* f *be continuous, twice continuously differentiable, or real-analytic in* $x + y$ *respectively, according as* (79) *is hyperbolic* ($a > 0$), *parabolic* ($a = 0$) *or elliptic* ($a < 0$). *The pair of functions forming the solution is in each case uniquely determined.*

Below we will be interested only in the first two statements (the hyperbolic and parabolic cases), which can be verified directly except for the uniqueness. For $a > 0$ we have

$$\text{(81)}\qquad \begin{aligned} u(x, y) &= \frac{1}{2a}\left\{\int_{x+y}^{x+y+\sqrt{a}x} f(t)dt - \int_{x+y-\sqrt{a}x}^{x+y} f(t)dt\right\}, \\ v(x, y) &= \frac{1}{2\sqrt{a}}\int_{x+y-\sqrt{a}x}^{x+y+\sqrt{a}x} f(t)dt, \end{aligned}$$

as may be verified by direct substitution. For $a = 0$ we find

$$\text{(82)}\qquad u(x, y) = \frac{x^2}{2}f'(x + y), \quad v(x, y) = xf(x + y) \quad \text{with} \quad ' \equiv \frac{d}{d(x + y)}.$$

3. Uniqueness Questions

It was not without reason that the uniqueness requirement in §1.2 was stated as condition 1, since in general it can be dealt with before the existence of a solution has been proved. This fact is important, because the second condition must be regarded as much more basic; moreover, the knowledge we shall gain in discussing the problem of uniqueness will give us the necessary insight to attack the problem of existence. For dealing with condition 1, there are at present two general methods at our disposal: 1. the *maximum-minimum principle* and 2. the *energy integral method.* Recent investigations have shown that in general both methods are available for all types of equations.

3.1. *The Elliptic Type*

We consider the linear equation (54)

$$Du \equiv \sum_{i,k=1}^{n} a_{ik} \frac{\partial^2 u}{\partial x_i \partial x_k} + \sum_{i=1}^{n} a_i \frac{\partial u}{\partial x_i} + au = f \tag{84}$$

with coefficients a_{ik}, a_i, a, f that are functions of $\mathfrak{x} = (x_1, x_2, \ldots, x_n)$ alone and are continuous in a domain $\overline{\mathfrak{G}}$. In $\mathfrak{G}$ let (84) be of elliptic type. By means of the definition in §2.1 this property of (84) may also be stated thus: the quadratic form $\sum_{i,k=1}^{n} a_{ik}\xi_i\xi_k$ is $\neq 0$, if $|\xi| \neq 0$. We may always assume (if necessary after multiplication of (84) by -1) that the form is positive. Let us assume that $u(\mathfrak{x}) \in \mathfrak{C}^0$ in $\overline{\mathfrak{G}}$ and $\in \mathfrak{C}^2$ in $\mathfrak{G}$, with $Du = f$. Then we have the following theorem.

Theorem 11.[5] *If* $a \leqq 0, f \leqq 0$ $(f \geqq 0)$ *in* $\mathfrak{G}$, *then u assumes a negative minimum* (*positive maximum*) *on* $\dot{\mathfrak{G}}$. ($\dot{\mathfrak{G}}$ *is the boundary of* $\mathfrak{G}$.)

Let us give the proof for $a \leqq 0$. We assume that $u(\mathfrak{x})$ has a negative minimum at $\mathfrak{x}^0 \in \mathfrak{G}$. Then

$$\left.\frac{\partial u}{\partial x_i}\right|_{\mathfrak{x}^0} = 0 \quad \text{and} \quad \sum_{i,k=1}^{n} \left.\frac{\partial^2 u}{\partial x_i \partial x_k}\right|_{\mathfrak{x}^0} \xi_i\xi_k \geqq 0.$$

Thus at $\mathfrak{x}^0$ we have two positive definite matrices

$$((a_{ik}(\mathfrak{x}^0))), \qquad \left(\left(\left.\frac{\partial^2 u}{\partial x_i \partial x_k}\right|_{\mathfrak{x}^0}\right)\right).$$

An elementary theorem of algebra[6] then shows that

$$\sum_{i,k=1}^{n} a_{ik} \left.\frac{\partial^2 u}{\partial x_i \partial x_k}\right|_{\mathfrak{x}^0} \geqq 0. \tag{85}$$

[5] E. Hopf, Elementare Bemerkungen über die Lösungen partieller Differentialgleichungen zweiter Ordnung vom elliptischen Typus. Sitzungsberichte Preuss. Akad. Wiss. Berlin, Band 19, 147–152 (1927).

[6] For every vector $\mathfrak{y} = (y_1, \ldots, y_n)$ let

$$\sum_{i,k=1}^{n} \alpha_{ik} y_i y_k \geqq 0 \quad \text{with} \quad \alpha_{ik} = \alpha_{ki}, \qquad \sum_{i,k=1}^{n} \beta_{ik} y_i y_k \geqq 0 \quad \text{with} \quad \beta_{ik} = \beta_{ki}.$$

Here α_{ik}, β_{ik} are suitably chosen real numbers. Then $\sum_{i,k=1}^{n} \alpha_{ik}\beta_{ik} \geqq 0$.

Since $au|_{\mathfrak{x}^0} > 0$, it follows from (84) that (85) is negative, which is a contradiction. Let us note that if $a \equiv 0$, $f \equiv 0$, then $u \equiv$ const is a solution, and the theorem remains valid in the above formulation, since every point is now a minimum (maximum) point.

In the uniqueness theorem for the boundary value problem with $a \leqq 0$, we assume that (84) has two solutions u_1, u_2 with $u_i = \varphi$ on $\dot{\mathfrak{G}}$. Then the difference $u = u_1 - u_2$, satisfies (84) with $f \equiv 0$, and $u = 0$ on $\dot{\mathfrak{G}}$.

Uniqueness theorem 12. *If $f \equiv 0$, $u = 0$ on $\dot{\mathfrak{G}}$, then $u \equiv 0$ in $\overline{\mathfrak{G}}$.*

For if $u(\mathfrak{x})$ were positive at any point in $\mathfrak{G}$, it would also be positive on $\dot{\mathfrak{G}}$, and the corresponding remark holds if $u(\mathfrak{x})$ is anywhere negative in $\mathfrak{G}$. Let us note that this proof provides us with the uniqueness theorem for equations of mixed type, since (84) can be of parabolic type on $\dot{\mathfrak{G}}$. For $a > 0$ these theorems are false.

Example 7. $\Delta_2 u + 2u = 0$ has the solution $u(\mathfrak{x}) = \sin x_1 \sin x_2$. On the boundary $\dot{\mathfrak{G}}$ of the rectangle $0 \leqq x_1, x_2 \leqq \pi$ we see that $u = 0$.

Theorem 13. *If $u_1(\mathfrak{x})$, $u_2(\mathfrak{x})$ are solutions of* (84) *with $u_1 = \varphi_1$, $u_2 = \varphi_2$ on $\dot{\mathfrak{G}}$ and if $|\varphi_1 - \varphi_2| \leqq \epsilon$ on $\dot{\mathfrak{G}}$, then*

$$|u_1 - u_2| \leqq \epsilon \quad \text{in} \quad \overline{\mathfrak{G}}.$$

For the proof we again consider the difference $u = u_1 - u_2$. If we had $u > \epsilon$ in $\mathfrak{G}$, then by theorem 11 the same inequality would also follow on $\dot{\mathfrak{G}}$, and so forth. Theorem 13 already satisfies condition 3, and this *maximum-minimum principle* provides us with possibilities for the proof of the *existence problem.*

Let us assume for the moment that the existence problem has been solved and that $u(\mathfrak{x})$ is a solution of (84): $Du = 0$ with $u = \varphi$ on $\dot{\mathfrak{G}}$. Then it is easy, as we shall see below, to construct functions $v_j(\mathfrak{x})$ such that $Dv_j \geqq 0$ and $v_j \leqq \varphi$ on $\dot{\mathfrak{G}}$. Then $D(u - v_j) \leqq 0$ and $u - v_j \geqq 0$ on $\dot{\mathfrak{G}}$. Thus it already follows from theorem 11 that $u(\mathfrak{x}) \geqq v_j(\mathfrak{x})$ in $\overline{\mathfrak{G}}$. Correspondingly, we can construct functions $V_k(\mathfrak{x})$ with $DV_k \leqq 0$ and $V_k \geqq \varphi$ on $\dot{\mathfrak{G}}$. Then $D(u - V_k) \geqq 0$ and $u - V_k \leqq 0$ on $\dot{\mathfrak{G}}$. Thus it follows again that $u(\mathfrak{x}) \leqq V_k(\mathfrak{x})$ in $\overline{\mathfrak{G}}$ and finally that

$$(86) \qquad v_j(\mathfrak{x}) \leqq u(\mathfrak{x}) \leqq V_k(\mathfrak{x}) \quad \text{in} \quad \overline{\mathfrak{G}}.$$

The v_j, (V_k) are called minorant (majorant) functions.[7] Inequalities (86) show that the solution is bounded above and below. If now, by means of suitable smoothing processes, we choose the sequences v_j, V_k carefully enough that $U(\mathfrak{x}) = \lim_{j\to\infty} v_j(\mathfrak{x}) = \lim_{k\to\infty} V_k(\mathfrak{x})$ in $\mathfrak{G}$, we may expect the limit function $U(\mathfrak{x})$ to be a solution $u(\mathfrak{x})$ of the boundary value problem, and in fact Perron[8] and Remak have solved the existence problem in this

[7] In the case $a \equiv 0$ the constants min φ, max φ, for example, are already minorant (majorant) functions on $\dot{\mathfrak{G}}$.

[8] Mathematische Zeitschrift 18 (1923).

elegant manner for $\Delta_n u = 0$ with $u = \varphi$ on $\dot{\mathfrak{G}}$. At the same time the proof provides us with the very natural conditions to be imposed on $\dot{\mathfrak{G}}$ and φ. This method of proof was later carried out successfully for the boundary value problem (84) also. However, it must be assumed here that (84) is of elliptic type in $\overline{\mathfrak{G}}$, since it is known today that arbitrarily prescribed values of $u = \varphi$ on $\dot{\mathfrak{G}}$ are not always permissible if (84) is of elliptic type on $\mathfrak{G}$ but of parabolic type on $\dot{\mathfrak{G}}$.

Below we shall discuss another method of proof for the existence problem, more general but more difficult. It makes use of *a priori* estimates for the solutions. The simplest such estimate is as follows. We consider (84) in $\mathfrak{G}$ with $u = \varphi$ on $\dot{\mathfrak{G}}$. In $\overline{\mathfrak{G}}$ let K be a bound for the coefficients: $|a_{ik}|, |a_i|, |a| \leqq K$. Moreover, let (84) be *uniformly elliptic* in $\overline{\mathfrak{G}}$;

$$\sum_{i,k=1}^{n} a_{ik}(\mathfrak{x})\xi_i\xi_k \geqq m \sum_{i=1}^{n} \xi_i^2 \qquad \text{with fixed} \qquad m > 0. \tag{87}$$

Theorem 14. *If $u(\mathfrak{x})$ is a solution of $Du = f$ with $u = \varphi$ on $\dot{\mathfrak{G}}$, then*

$$|u(\mathfrak{x})| \leqq \max|\varphi| + M \max|f| \qquad \text{in} \qquad \overline{\mathfrak{G}}, \tag{88}$$

where M is a function of K and m only. This inequality is called an a priori estimate since the solution $u(\mathfrak{x})$ is estimated by magnitudes that do not depend on $u(\mathfrak{x})$. From (88) it follows again that u is uniquely determined and depends continuously on φ and even continuously on f.

For the proof we construct a function $w(\mathfrak{x})$ with $Dw \leqq -\max|f|$, $w \geqq \max|\varphi|$ on $\dot{\mathfrak{G}}$. We then set

$$w(\mathfrak{x}) = \max|f|(e^{\alpha\xi} - e^{\alpha x_1}) + \max|\varphi| \tag{89}$$

and assume without loss of generality that $x_1 \geqq 0$ for $\overline{\mathfrak{G}}$. Then we choose $\xi > \max x_1$ with $x_1 \in \overline{\mathfrak{G}}$. Now $\alpha > 0$ is determined. We find that $w \geqq \max|\varphi|$on $\dot{\mathfrak{G}}$ and further

$$\begin{aligned} -Dw &= \max|f|\{-ae^{\alpha\xi} + e^{\alpha x_1}(a_{11}\alpha^2 + a_1\alpha + a)\} - a\max|\varphi| \\ &\geqq \max|f|\{m\alpha^2 - K(\alpha + 1)\}, \end{aligned} \tag{90}$$

where we have taken into account the inequalities (87) and $a \leqq 0$. For sufficiently large $\alpha(\alpha = \alpha(K, m))$ we have $\{\cdots\} > 1$ and $-Dw \geqq \max|f|$. Then, as we shall show at once, $|u| \leqq w$ in $\overline{\mathfrak{G}}$, so that the proof of (88) is complete. We consider $U = u - w$, so that $U \leqq 0$ on $\dot{\mathfrak{G}}$ and $DU = Du - Dw \geqq f + \max|f| \geqq 0$. From theorem 11 it follows that $u \leqq w$ in $\overline{\mathfrak{G}}$. Replacing w by the function $-w$, we see that $-w \leqq u$ in $\overline{\mathfrak{G}}$.

The possibilities offered by theorem 11 are by no means exhausted thereby. For example, let us consider the *Green's function* $g(\mathfrak{x}, \mathfrak{y})$ defined in §1.4. In the notation of that section we have $\varphi(\mathfrak{x}, \mathfrak{y}) = -s(\mathfrak{x}, \mathfrak{y})$ for $\mathfrak{y} \in \dot{\mathfrak{G}}$. Thus $\varphi < 0$ on $\dot{\mathfrak{G}}$ for $n \geqq 3$ and from theorem 11 we have $\varphi < 0$ in $\mathfrak{G}$. It follows that

$$g(\mathfrak{x}, \mathfrak{y}) = s(\mathfrak{x}, \mathfrak{y}) + \varphi(\mathfrak{x}, \mathfrak{y}) < s(\mathfrak{x}, \mathfrak{y}). \tag{91}$$

On the other hand, $g(\mathfrak{x}, \mathfrak{y}) = 0$ for $\mathfrak{y} \in \dot{\mathfrak{G}}$ and $g \to +\infty$ for $\mathfrak{x} \to \mathfrak{y}$, so that again theorem 11 provides us with the extremely useful estimate

$$0 \leqq g(\mathfrak{x}, \mathfrak{y}) \leqq s(\mathfrak{x}, \mathfrak{y}) \tag{92}$$

for $n \geqq 3$.

3.2. *The Parabolic Type*

We consider only the simplest case in two independent variables x, t and write the equation, in accordance with (68), in the form

$$Du \equiv u_{xx} - u_t + a(x, t)u = f(x, t). \tag{93}$$

For the domain $\mathfrak{F}$ we again consider a domain of the form shown in Figures 6 and 7, which is open for positive t. We close $\mathfrak{F}$ by means of a

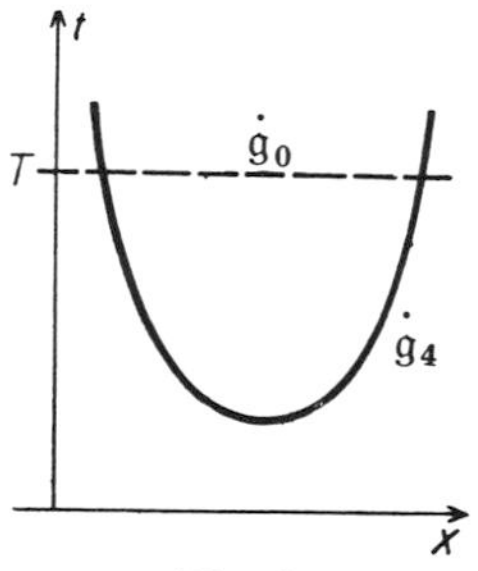

Fig. 6

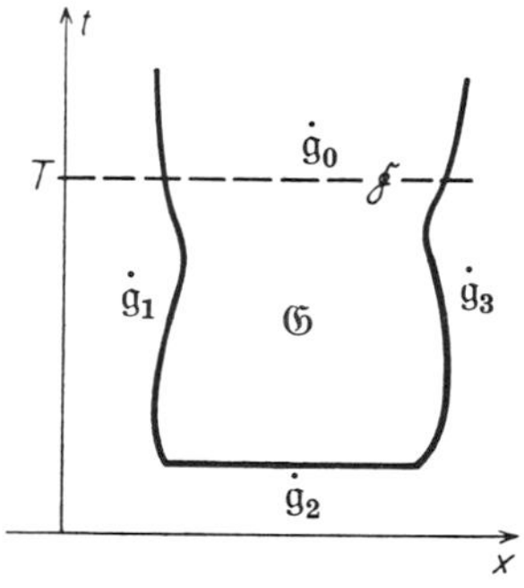

Fig. 7

line $t = T$, set $\dot{\mathfrak{g}}_1 + \dot{\mathfrak{g}}_2 + \dot{\mathfrak{g}}_3 = \dot{\mathfrak{g}}_4$ and denote the closed domain by $\overline{\mathfrak{G}} = \mathfrak{G} + \dot{\mathfrak{G}}$ with $\dot{\mathfrak{G}} = \dot{\mathfrak{g}}_0 + \dot{\mathfrak{g}}_4$.[9] For the function $u(x, t)$ we assume: $u \in \mathfrak{C}^0$ in $\overline{\mathfrak{G}}$; $u_{xx}, u_t \in \mathfrak{C}^0$ in $\mathfrak{G}$; and $Du = f$ in $\mathfrak{G}$. We also assume $a, f \in \mathfrak{C}^0$ in $\overline{\mathfrak{G}}$ and then prove the following theorem.

Theorem 15. *If $a \leqq 0$, $f \leqq 0$ ($f \geqq 0$) in $\overline{\mathfrak{G}}$, then u has a negative minimum (positive maximum) on $\dot{\mathfrak{g}}_4$.*

The proof will be given for $a < 0$. If P is the minimum point in $\mathfrak{G}$, we have $u(P) < 0$, $u_x(P) = u_t(P) = 0$ and $u_{xx}(P) \geqq 0$. But then (93), considered at the point P, provides a contradiction. If P lies on $\dot{\mathfrak{g}}_0$ (open), then $u(P) < 0$, $u_x(P) = 0$, $u_t(P) \leqq 0$, $u_{xx}(P) \geqq 0$, which is again in contradiction with (93) at the point P. The corresponding remarks hold for the maximum point.

[9] We take $\dot{\mathfrak{g}}_4$ as a closed arc and thus $\dot{\mathfrak{g}}_0$ as an open arc.

Then the uniqueness theorem follows if we consider (93) in $\mathfrak{F}$, close $\mathfrak{F}$ with an arbitrary $t = T$ and assign $u = \varphi$ on $\dot{\mathfrak{g}}_4$. If we assume the existence of two solutions u_1 and u_2, the difference $u = u_1 - u_2$ satisfies the problem $Du = 0$ with $u = 0$ on $\dot{\mathfrak{g}}_4$.

Uniqueness theorem 16. *If $f \equiv 0$, $u = 0$ on $\dot{\mathfrak{g}}_4$, it follows that $u \equiv 0$ in $\overline{\mathfrak{G}}$ for arbitrary a.*

For $a < 0$ the result follows from the proof of theorem 15. If the condition $a < 0$ is not satisfied, we determine a constant α such that $\alpha > a$ in $\overline{\mathfrak{G}}$, and set $u = ve^{\alpha t}$. Then the function v satisfies the equation

$$v_{xx} - v_t + b(x, t)v = 0 \qquad \text{with} \qquad b = a - \alpha < 0 \tag{94}$$

and $v = 0$ on $\dot{\mathfrak{g}}_4$. Thus theorem 15 is applicable and shows that $v \equiv 0$.

3.3. These theorems are false if in (93) the term u_t is taken with the positive sign or, in other words, if we replace t by $-t$, which again means that we are attempting to investigate the solutions not for the future but for the past. This result could of course be foreseen, since the solution operator (51) had the semigroup property. Let us consider the following illustration.

Example 8 (L. Bieberbach).

$$u_{xx} - u_t = 0$$

has the solution

$$u(x, t) = e^{-t} \cos x - \tfrac{1}{2}e^{-4t} \cos 2x. \tag{95}$$

Here $u = 0$ on the curve $t = \dfrac{1}{3} \log \dfrac{1}{2} \dfrac{\cos 2x}{\cos x}$. For $-\dfrac{\pi}{4} < x < \dfrac{\pi}{4}$ this curve $\dot{\mathfrak{g}}_4$ is represented in Figure 8. Thus the problem $u_{xx} - u_t = 0$ with $u = 0$ on $\dot{\mathfrak{g}}_4$ has the solution (95) and also the solution $u(x, t) \equiv 0$.

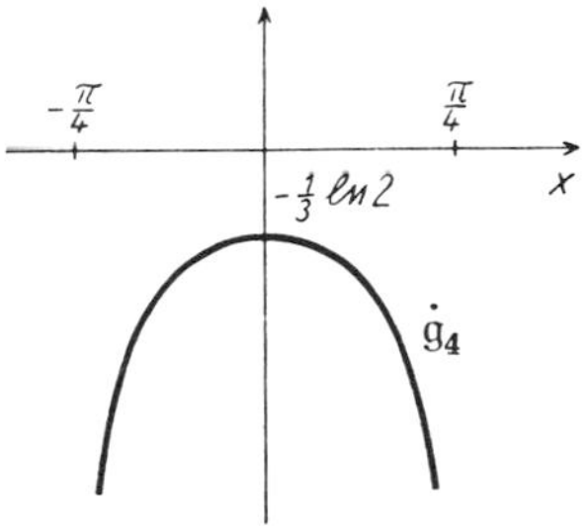

Fig. 8

3.4. *The Hyperbolic Case*

Here we shall illustrate the *energy integral method* only for the initial value problem of the wave equation in $\mathfrak{R}_2$ (the dependence on dimension does not arise for condition 1). We prove the following theorem.

Uniqueness theorem 17. (Without vector notation.) *If* $u(x, y, t)$ *is a solution of* $Du \equiv u_{xx} + u_{yy} + u_{tt} = 0$ *with* $u(x, y, 0)$, $u_t(x, y, 0) = 0$ *in* $-\infty < x, y, t < \infty$, *it follows that* $u \equiv 0$.

By making use of Figure 5 (§1.3) and similar notation we can formulate theorem 17 more conveniently: *from* $u(x, y, 0) = u_t(x, y, 0) = 0$ *in* $\mathfrak{A}$ *it follows that* $u \equiv 0$ *in* $\mathfrak{B}$. Since the wave equation is not changed if we replace t by $-t$, it is sufficient to give the proof for $t \geqq 0$. We first note (Figure 5) that

$$(96)\qquad \begin{aligned} 0 &= 2\int_{\mathfrak{B}^*} u_t Du\,dxdydt \\ &= \int_{\mathfrak{B}^*} \{2(u_t u_x)_x + 2(u_t u_y)_y - (u_x^2)_t - (u_y^2)_t - (u_t^2)_t\}dxdydt. \end{aligned}$$

Here $\mathfrak{B}^*$ denotes the cone (Figure 5) in $t \geqq 0$, which we have cut off at $t = t_1 < t_0$ (frustum of a cone). Then to (96) we apply the *Gauss integral theorem* (cf. III 4, §11). We find

$$(97)\qquad \begin{aligned} 0 = &\int_{\dot{\mathfrak{B}}_1^*} \{2(u_t u_x)\nu_1 + 2(u_t u_y)\nu_2 - (u_x^2 + u_y^2 + u_t^2)\nu_3\}do \\ &- \int_{\dot{\mathfrak{B}}_2^*} (u_x^2 + u_y^2 + u_t^2)do, \end{aligned}$$

where $\dot{\mathfrak{B}}_1^*$ denotes the slant surface of the frustum of the cone, $\nu = (\nu_1, \nu_2, \nu_3)$ is the normal vector and $\dot{\mathfrak{B}}_2^*$ is the horizontal top surface with $\nu = (0, 0, 1)$. The integral is not taken over the base, since u and its first derivatives vanish there. A short calculation shows that

$$(98)\qquad \begin{aligned} &\int_{\dot{\mathfrak{B}}_1^*} \{\cdots\}do \\ &= -\int_{\dot{\mathfrak{B}}_1^*} \frac{1}{\nu_3}\{(u_x\nu_3 - u_t\nu_1)^2 + (u_y\nu_3 - u_t\nu_2)^2 + u_t^2(\nu_3^2 - \nu_1^2 - \nu_2^2)\}do. \end{aligned}$$

Now we have $\nu_3^2 = \frac{1}{2}$ and $\nu_1^2 + \nu_2^2 = 1 - \nu_3^2 = \frac{1}{2}$. Thus the right side of (97) $\leqq 0$, so that

$$\int_{\dot{\mathfrak{B}}_2^*} (u_x^2 + u_y^2 + u_t^2)do = 0,$$

and thus $u \equiv$ const on $\dot{\mathfrak{B}}_2^*$. But $t_1 < t_0$ was chosen arbitrarily, so that $u \equiv$ const in $\mathfrak{B}$, and in view of the initial condition this constant can only be equal to zero.

The foregoing method is called the *energy integral method* because the important expression $(u_x^2 + u_y^2 + u_t^2)dxdy$ in it is precisely twice the entire energy of oscillation in $dxdy$, where the physical constants have been set equal to unity. Nowadays the name is taken to refer to any method in which, in order to prove uniqueness theorems for differential equations $Du = 0$, we begin with an expression of the form

$$0 = \int \alpha(\mathfrak{x}, u, u_{x_i}, \ldots) Du \, d\mathfrak{x}$$

and then by various transformations of the integral, which are often quite ingenious in making full use of the assigned conditions, we arrive at definite expressions which imply that $u \equiv 0$. As an example let us consider the problem of elliptic type

$$(99) \quad Du \equiv \Delta_n u + \sum_{i=1}^{n} a_i(\mathfrak{x}) \frac{\partial u}{\partial x_i} + a(\mathfrak{x})u = 0 \qquad \text{with} \qquad u = 0 \quad \text{on} \quad \dot{\mathfrak{G}}.$$

If we consider $0 = \int_{\mathfrak{G}} uDu\,d\mathfrak{x}$, we can easily show that for

$$a - \tfrac{1}{2} \sum_{i=1}^{n} a_{ix_i} \leqq 0 \qquad \text{in} \qquad \mathfrak{G}$$

the only solution is $u \equiv 0$. This condition and the other condition $a \leqq 0$ are not comparable with each other (they are both sufficient conditions). The *energy integral method* is also successful with parabolic equations, and for hyperbolic equations a *maximum-minimum principle* has been successfully employed in recent articles.[10]

3.5. *The Hyperbolic-Parabolic-Elliptic Case*

Here our initial position is considerably more unfavorable than in the preceding cases, since we do not yet know what sort of problems will turn out to be meaningful. But a uniqueness theorem that employs the smallest possible number of prescribed conditions will in itself contribute greatly to the classification of these questions. We consider the simplest equation in the variables x, y

$$(100) \qquad Du \equiv yu_{xx} + u_{yy} = f(x, y),$$

which is elliptic for $y > 0$ and hyperbolic for $y < 0$. The characteristic curves $\mathfrak{C}$ are given by

$$(101) \qquad \text{a)} \quad \frac{dx}{dy} = -\sqrt{-y}, \qquad \text{b)} \quad \frac{dx}{dy} = \sqrt{-y} \qquad \text{for} \qquad y \leqq 0.$$

We consider the domain $\mathfrak{G}$ in Figure 9. Here $\mathfrak{c}_1, \mathfrak{c}_3$ are characteristics in the family (101a), and $\mathfrak{c}_2, \mathfrak{c}_4$ are in the family (101b). Also, $\mathfrak{g}_1, \mathfrak{g}_2$ are monotone curves lying in the "wedges" bounded by the characteristics:

$$(102) \qquad \mathfrak{g}_1 : \frac{dx}{dy} \leqq -\sqrt{-y}, \qquad \mathfrak{g}_2 : \frac{dx}{dy} \geqq \sqrt{-y}.$$

[10] P. Germain and R. Bader (1952). See the generalization of S. Agmon, L. Nirenberg, and M. H. Protter. Commun. Pure Appl. Math. 6, p. 456 (1953).

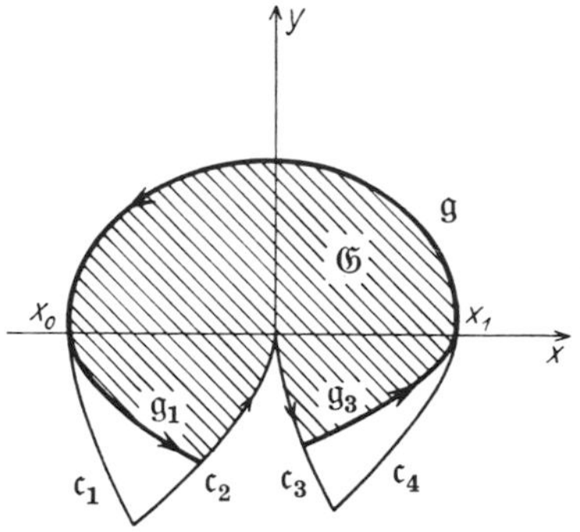

Fig. 9

Let $\mathfrak{g}$ be an arbitrary curve through the points x_0, x_1, with $x_0 < x_1$, that lies in $y \geqq 0$ and is starlike: $xdy - ydx \geqq 0$. We consider (100) with $u = \varphi$ on $\mathfrak{g} + \mathfrak{g}_1 + \mathfrak{g}_2$ (*Frankl's problem*) and assume that there exist two solutions $u_1, u_2 \in \mathfrak{C}^1$ in $\overline{\mathfrak{G}}$, where $\overline{\mathfrak{G}}$ is bounded by $\mathfrak{g}$, $\mathfrak{g}_1$, $\mathfrak{g}_2$, $\mathfrak{c}_2$, $\mathfrak{c}_3$. Then the difference $u = u_1 - u_2$ satisfies (100) with $f \equiv 0$ and $\varphi = 0$.

Uniqueness theorem 18.[11] *If* $Du = 0$ *and* $u = 0$ *on* $\mathfrak{g} + \mathfrak{g}_1 + \mathfrak{g}_2$ *then* $u \equiv 0$ *in* $\overline{\mathfrak{G}}$.

The proof depends on an ingenious application of the *energy integral method*; in fact, in working through the proof for the first time the reader is urged to give his attention to merely verifying the correctness of the formulas. With $\alpha = y$ for $y \geqq 0$, $\alpha = 0$ for $y \leqq 0$, it follows from the *Gauss theorem* in the form

$$\int_{\dot{\mathfrak{G}}} fdx + gdy = \iint_{\mathfrak{G}} (g_x - f_y)dxdy \tag{103}$$

that

$$\left\{\begin{aligned} 0 &= \iint_{\mathfrak{G}} (xu_x + \alpha u_y)(yu_{xx} + u_{yy})dxdy = \iint_{\mathfrak{G}} \{\tfrac{1}{2}y(xu_x^2)_x + x(u_xu_y)_y \\ &\quad - \tfrac{1}{2}(xu_y^2)_x + \alpha y(u_xu_y)_x - \tfrac{1}{2}(\alpha yu_x^2)_y + \tfrac{1}{2}(\alpha u_y^2)_y \\ &\quad + \tfrac{1}{2}u_x^2(-y + (\alpha y)_y) + \tfrac{1}{2}u_y^2(1 - \alpha_y)\}dxdy \\ &= \oint_{\dot{\mathfrak{G}}} \{(\tfrac{1}{2}yxu_x^2 - \tfrac{1}{2}xu_y^2 + \alpha yu_xu_y)dy + (-xu_xu_y + \tfrac{1}{2}\alpha yu_x^2 \\ &\quad - \tfrac{1}{2}\alpha u_y^2)dx\} + \iint_{\mathfrak{G}} \{\tfrac{1}{2}u_x^2(-y + (\alpha y)_y) + \tfrac{1}{2}u_y^2(1 - \alpha_y)\}dxdy. \end{aligned}\right. \tag{104}$$

[11] C. S. Morawetz, Commun. Pure Appl. Math. 7, p. 697 (1954).

On $\mathfrak{g} + \mathfrak{g}_1 + \mathfrak{g}_2$ we have $du = u_x dx + u_y dy = 0$; on $\mathfrak{c}_2$, $dx = \sqrt{-y}dy$, and on $\mathfrak{c}_3$, $dx = -\sqrt{-y}dy$. Thus from (104) we obtain

$$0 = \int_{\mathfrak{g}+\mathfrak{g}_1+\mathfrak{g}_2} \frac{1}{2}\left(y + \left(\frac{dx}{dy}\right)^2\right)u_x^2(xdy - \alpha dx)$$

$$(105) \qquad + \int_{\mathfrak{c}_2+\mathfrak{c}_3} \frac{1}{2}\left(\frac{du}{dy}\right)^2(-xdy - \alpha dx)$$

$$+ \iint_{\mathfrak{G}} \{\tfrac{1}{2}u_x^2(-y + (\alpha y)_y) + \tfrac{1}{2}u_y^2(1 - \alpha_y)\}dxdy.$$

Taking into account the direction of the circuit, we find in each case that the integral is $\geqq 0$. Thus the integral over $\mathfrak{G}$ must vanish. Consequently

$$(106) \quad \iint_{\mathfrak{G} \text{ with } y \geqq 0} \tfrac{1}{2}yu_x^2 dxdy + \iint_{\mathfrak{G} \text{ with } y \leqq 0} \tfrac{1}{2}(-yu_x^2 + u_y^2)dxdy = 0$$

and thus $u_x \equiv 0$ in $\overline{\mathfrak{G}}$. Now if (x, y) is an arbitrary point in $\mathfrak{G}$, there exists a corresponding point (x^0, y) on $\mathfrak{g} + \mathfrak{g}_1 + \mathfrak{g}_2$. Integration over the straight line through the two points then shows that $u \equiv 0$ in $\overline{\mathfrak{G}}$, in view of the assigned conditions on $\mathfrak{g} + \mathfrak{g}_1 + \mathfrak{g}_2$.

If in Figure 10 we let $\mathfrak{g}_1$ coincide with $\mathfrak{c}_1$, and $\mathfrak{g}_2$ with $\mathfrak{c}_4$, we obtain the *Gellerstedt problem*. Finally, if we put x_0 at the origin, we have the *Tricomi*

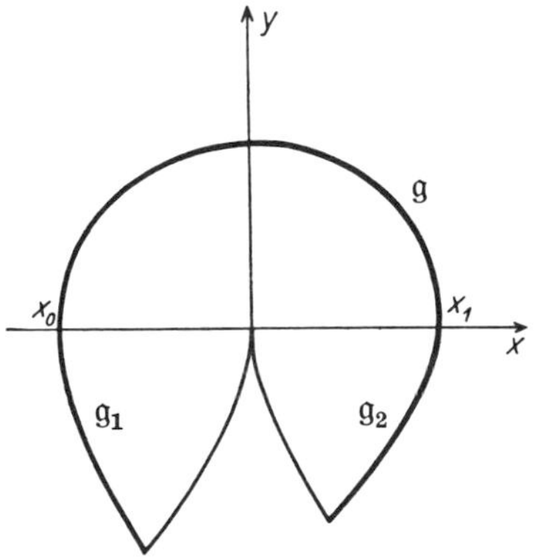

Fig. 10

problem[12] (Figure 11) from which the whole development began. In each case the boldface arcs carry the prescribed conditions on u. Heuristically it is easy to see that these conditions cannot be prescribed on the closed curve in Figure 11. For $y \leqq 0$ we would in this case have a *characteristic initial value problem*, as described for the wave equation in §1.2. If we

[12] Atti della Accad. Naz. dei Lin. (5) 14, p. 133 (1923).

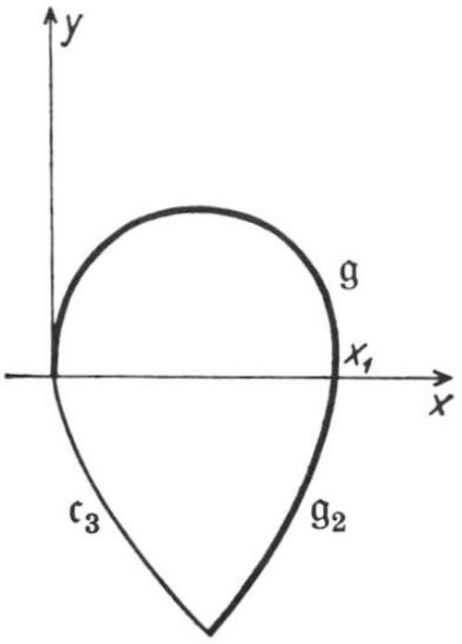

Fig. 11

assume that the same argument is valid here, then not only u, but in particular also the first derivatives of u, would be known on the x-axis in $0 \leqq x \leqq x_1$. For $y \geqq 0$ we would then have a *boundary value problem* for an equation which is *elliptic* for $y > 0$ and *parabolic* for $y = 0$. If the argument in §1.4 holds for this case (as in fact it does) we would be able to prescribe only the values of u on the closed boundary, and on $0 \leqq x \leqq x_1$ we would be forced to prescribe the derivatives of u also, which leads to a meaningless problem. Equations of mixed type govern the flow of gases that are in transition from subsonic to supersonic velocity. However, the mathematical results obtained to date are far from satisfying the demands of technology. Even the very simple question whether for the equation $yu_{xx} + u_{yy} + \lambda u = 0$ with $\lambda < 0$ theorem 18 remains valid for any of the problems in Figures 9, 10, 11 is still open. For $\lambda \geqq 0$ the answer is in the affirmative.

4. Questions of Existence

When existence questions for partial differential equations are discussed with the necessary generality, they become extremely difficult, and their solution may be regarded as one of the most important accomplishments of mathematical analysis. Every instrument that can be used to solve them, from whatever field of mathematics, is to be welcomed, but of course, we will give preference to proofs that remain entirely in the domain of partial differential equations (e.g., the *Perron proof* in §3.1).

4.1. *The Cauchy-Kovalevski Theorem*

Since the following arguments do not depend in any way on the number of dimensions, we shall confine our attention to two variables x, y. Thus we consider equation (1) in the form

(107) $$u_{yy} = f(x, y, u, u_x, u_y, u_{xx}, u_{xy})$$

with the initial conditions $u(x, y_0) = u_0(x)$, $u_y(x, y_0) = u_1(x)$. As long as we do not undertake any classification into types for (107), we know (examples 1, 2, 6) that we must assume the *analyticity* of all the functions and that condition 3 is not always satisfied. Thus the above formulation of the problem is important from the mathematical point of view but not for applications in physics.

Theorem 19. *Let $f(x, y, u, p, q, r, s)$ be a function which in a neighborhood of $x_0, y_0, u_0, p_0, q_0, r_0, s_0$ is analytic in the seven variables (i.e., can be expanded in a power series). Also let $u_0(x)$, $u_1(x)$ be analytic in a neighborhood of x_0. Here we have used the following notation: $u_0(x_0) = u_0$, $u_0'(x_0) = p_0$, $u_1(x_0) = q_0$, $u'(x_0) = s_0$, $u_0''(x_0) = r_0$. Then there exists exactly one function $u(x, y)$ which is analytic in a neighborhood of x_0, y_0 and is a solution of the initial value problem.*

For the proof we may assume $x_0 = y_0 = 0$ and set

$$u(x, y) = \sum_{i,k=0}^{\infty} \alpha_{i,k} x^i y^k. \tag{108}$$

Then the $\alpha_{i,k}$ are uniquely determined by (107) and the prescribed conditions. We find that $\alpha_{0,0} = u_0(0)$, $\alpha_{1,0} = u_0'(0)$, $\alpha_{0,1} = u_1(0)$, $\alpha_{2,0} = u_0''(0)$, $\alpha_{1,1} = u_1'(0)$, $\alpha_{0,2} = f(0, 0, u_0(0), u_0'(0), u_1(0), u_0''(0), u_1'(0))$, etc. Thus (108) will represent the solution, provided it can be proved to converge. Such a proof can be given, but we shall omit it here, though of course it is the essential part of the problem. The only remaining question is whether every equation (1) with $n = 2$ can be brought into the form (107). The answer is in the negative. In (60), for example, we must have $c \neq 0$. Also, if we allow arbitrary transformations of coordinates, we have as condition (64b)

$$a\psi_x^2 + 2b\psi_x\psi_y + c\psi_y^2 \neq 0. \tag{109}$$

But the exceptional cases obtained by setting (109) equal to zero lead precisely to the *characteristic manifolds* (which for $n = 2$ are curves). In the elliptic case such manifolds do not exist, so that the form (107) can be set up, but for the other types this is no longer true.

4.2. *The Boundary Value Problem for Elliptic Differential Equations*

In this section it will be desirable for the reader to have looked at Chapter 11 on functional analysis. Here again we employ the vector notation, so that $\mathfrak{x} = (x_1, \ldots, x_n)$ and $|\mathfrak{x}|$ is the norm of the vector. A function $u(\mathfrak{x})$ is said to be *Hölder continuous in* $\overline{\mathfrak{G}}$ if for all $\mathfrak{x}, \mathfrak{y} \in \overline{\mathfrak{G}}$

$$|u(\mathfrak{x}) - u(\mathfrak{y})| \leqq H|\mathfrak{x} - \mathfrak{y}|^\alpha \tag{110}$$

with constants H and α, $0 < \alpha < 1$, whereupon we write $u \in \mathfrak{C}^\alpha$ in $\overline{\mathfrak{G}}$. An obviously equivalent requirement is the existence of

$$H_\alpha[u] \equiv \underset{\substack{\mathfrak{x},\mathfrak{y} \in \overline{\mathfrak{G}} \\ \mathfrak{x} \neq \mathfrak{y}}}{\text{l.u.b.}} \frac{|u(\mathfrak{x}) - u(\mathfrak{y})|}{|\mathfrak{x} - \mathfrak{y}|^\alpha}. \tag{111}$$

For $u \in \mathfrak{C}^2$ we let D^2u denote *any* second derivative and then set

$$\|u\|_2 = \max_{x \in \overline{\mathfrak{G}}} |u| + \max_{x \in \overline{\mathfrak{G}}} |D^1 u| + \max_{x \in \overline{\mathfrak{G}}} |D^2 u|. \tag{112}$$

Finally, we must consider all functions $u \in \mathfrak{C}^{2+\alpha}$ in $\overline{\mathfrak{G}}$, i.e., functions u for which $D^2u \in \mathfrak{C}^\alpha$ with constants H, α independent of u. For such u we write

$$\|u\|_{2+\alpha} = \|u\|_2 + H_\alpha[D^2u]. \tag{113}$$

For (113) we see in particular that $\|u\|_{2+\alpha} \geqq 0$ and that the equality holds if and only if $u \equiv 0$. Furthermore, $\|u + v\|_{2+\alpha} \leqq \|u\|_{2+\alpha} + \|v\|_{2+\alpha}$, $\|\lambda u\|_{2+\alpha} = |\lambda|\ \|u\|_{2+\alpha}$, where λ is a number. In the terminology of functional analysis, this means that $\rho(u, v) = \|u - v\|_{2+\alpha}$ is a meaningful *definition of distance* for any two elements of the given set of functions. With this definition of distance the set of functions $u \in \mathfrak{C}^{2+\alpha}$ forms a *Banach space* $\mathfrak{B}$, which can be shown to be complete; i.e., if the sequence $\{u_n\} \in \mathfrak{B}$ satisfies the convergence condition $\|u_n - u_m\|_{2+\alpha} < \epsilon$ for $n, m > N(\epsilon)$, then there exists a $u \in \mathfrak{B}$ with $\|u - u_n\|_{2+\alpha} < \epsilon$ for $n > N(\epsilon)$.

We now describe the procedure for the proof of existence theorems.[13] We consider the equation

$$Du \equiv \sum_{i,k=1}^{n} a_{ik}(\mathfrak{x}) \frac{\partial^2 u}{\partial x_i \partial x_k} + \sum_{i=1}^{n} a_i(\mathfrak{x}) \frac{\partial u}{\partial x_i} + a(\mathfrak{x})u = f(\mathfrak{x}) \quad \text{in} \quad \mathfrak{G} \tag{114}$$

with $u = \varphi$ on $\dot{\mathfrak{G}}$ and assume in particular that a_{ik}, a_i, a, $f \in \mathfrak{C}^\alpha$ in $\overline{\mathfrak{G}}$ with $|a_{ik}|$, $|a_i|$, $|a| \leqq K$ and $a \leqq 0$. We also assume that Du is uniformly elliptic (87) in $\overline{\mathfrak{G}}$:

$$\sum_{i,k=1}^{n} a_{ik}\xi_i\xi_k \geqq m \sum_{i=1}^{n} \xi_i^2 \qquad \text{with fixed } m > 0 \text{ for all } \mathfrak{x} \in \overline{\mathfrak{G}}. \tag{115}$$

Furthermore, let $\varphi \in \mathfrak{C}^{2+\alpha}$ on $\dot{\mathfrak{G}}$. Here $\dot{\mathfrak{G}}$ is a "sufficiently well-behaved" boundary. Under these assumptions we have the following theorem.

Theorem 20. *The equation $Du = f$ with $u = \varphi$ on $\dot{\mathfrak{G}}$ has a solution $u \in \mathfrak{C}^{2+\alpha}$ in $\overline{\mathfrak{G}}$ which satisfies the three conditions.*

Our sketch of the proof falls into several parts:

1. The conditions 1 and 3 in §3.1 have already been dealt with.
2. We consider $\Delta_n v = 0$ with $v = \varphi$ on $\dot{\mathfrak{G}}$. By the classical *potential theory* or the *Perron method* in §3.1 we may consider that this problem has

[13] Following J. Schauder, Math. Zeitschrift 38, p. 257 (1934).

already been dealt with. For "well-behaved" $\dot{\mathfrak{G}}$ we have $v \in \mathfrak{C}^{2+\alpha}$ in $\overline{\mathfrak{G}}$ (*Kellogg theorem*). Setting $u = v + w$, we find

$$(116)\qquad Du = Dv + Dw = f \quad \text{or} \quad Dw = f - Dv = \tilde{f}$$
$$\text{with} \quad \tilde{f} \in \mathfrak{C}^{\alpha}, \quad w = 0 \quad \text{on} \quad \dot{\mathfrak{G}}.$$

Here we may assume that $\varphi = 0$ on $\dot{\mathfrak{G}}$.

3. The problem $Du = f$ with $u = 0$ on $\dot{\mathfrak{G}}$ is now embedded in a family of boundary value problems. We set

$$(117)\qquad D_t u \equiv tDu + (1 - t)\Delta_n u = f, \quad u = 0 \quad \text{on} \quad \dot{\mathfrak{G}}$$
$$\text{with} \quad 0 \leqq t \leqq 1.$$

For $t = 1$ we have the problem to be solved, and for $t = 0$ the "simple" problem of potential theory $\Delta_n u = f$ with $u = 0$ on $\dot{\mathfrak{G}}$. We introduce the set $\mathfrak{T}$ of values of the parameter t for which for every $f \in \mathfrak{C}^{\alpha}$ there exists a $u \in \mathfrak{C}^{2+\alpha}$ such that $D_t u = f$ with $u = 0$ on $\dot{\mathfrak{G}}$. If we let $\mathfrak{R}$ denote the set of all real numbers $0 \leqq t \leqq 1$, then obviously $\mathfrak{T} \subset \mathfrak{R}$. Thus the problem of existence is settled if we can show that $\mathfrak{T} \equiv \mathfrak{R}$. The proof consists of three steps:

α) $\mathfrak{T}$ is not the empty set; β) $\mathfrak{T}$ is an open set, i.e., from $t_0 \in \mathfrak{T}$ it follows that all t-values with $|t - t_0| < \epsilon(t_0)$ and $0 \leqq t \leqq 1$ from $\mathfrak{T}$ are also elements of $\mathfrak{T}$; γ) $\mathfrak{T}$ is a closed set, i.e., if $\{t_n\}$ is a sequence from $\mathfrak{T}$ converging to t, then t also belongs to $\mathfrak{T}$. From α, β, γ it then follows that $\mathfrak{T} \equiv \mathfrak{R}$, as desired.

Proof of α). We show that $t = 0$ belongs to $\mathfrak{T}$. For then (117) becomes

$$(118)\qquad \Delta_n u = f \quad \text{with} \quad u = 0 \quad \text{on} \quad \dot{\mathfrak{G}}.$$

But a theorem in potential theory tells us that for "well-behaved" $\dot{\mathfrak{G}}$ there exists, for every $f \in \mathfrak{C}^{\alpha}$, a function $u \in \mathfrak{C}^{2+\alpha}$ satisfying (118).

Proof of β). For (114) we have the *a priori* estimate (88), which in view of (112) can also be written as $\|u\|_0 \leqq M\|f\|_0$. Here $M = M(K, m)$. Thus for (117) it is easy to prove the inequality

$$(119)\qquad \|u\|_0 \leqq \tilde{M}\|f\|_0 \quad \text{with} \quad \tilde{M} = \tilde{M}(\tilde{K}, \tilde{m}) \quad \text{and}$$
$$\tilde{K} = \max(K, 1), \qquad \tilde{m} = \min(m, 1).$$

Now it was the great achievement of Schauder that he gave much sharper *a priori* estimates of this sort, such as the following for (117):

$$(120)\qquad \|u\|_{2+\alpha} \leqq C\|f\|_{\alpha} \quad \text{with} \quad C = C(\tilde{K}, \tilde{m}),$$

whereby for the first time the whole method of proof became possible. If we let $t_0 \in \mathfrak{T}$, we find from (117) that

$$(121)\quad D_{t_0} u = D_{t_0} u - D_t u + f = g \quad \text{and} \quad g = (t - t_0)\{\Delta_n u - Du\} + f$$
$$\text{with} \quad u = 0 \quad \text{on} \quad \dot{\mathfrak{G}}.$$

If $u \in \mathfrak{C}^{2+\alpha}$, then $g \in \mathfrak{C}^{\alpha}$. Moreover, since $t_0 \in \mathfrak{T}$, there exists a function $v \in \mathfrak{C}^{2+\alpha}$ such that $D_{t_0}v = g$ and $v = 0$ on $\dot{\mathfrak{G}}$. Since g is a function of u: $g = g(u)$, the v thus defined is also a function of $u : v = v(u)$. We now consider this v as produced from the u by an operator A in the form $v = Au$, where the operator A is not further known. This operator maps a $u \in \mathfrak{B}$ into a $v \in \mathfrak{B}$. For this mapping we seek a fixed point, i.e., a function $\Phi \in \mathfrak{B}$ for which $\Phi = A\Phi$. Then from (121) this Φ satisfies

$$(122) \quad D_{t_0}\Phi = g(\Phi) \quad \text{or} \quad D_{t_0}\Phi = D_{t_0}\Phi - D_t\Phi + f \quad \text{or} \quad D_t\Phi = f$$
$$\text{with} \qquad \Phi = 0 \quad \text{on} \quad \dot{\mathfrak{G}}.$$

Thus we have shown that $t \in T$. For $|t - t_0| < \epsilon(t_0)$ we prove the existence of Φ by means of a theorem from functional analysis.

Theorem 21. *In the complete Banach space $\mathfrak{B}$ with elements $u, v, w, \ldots$ and the distance function $\rho(u, v)$, let there be given an operator A which maps the elements of $\mathfrak{B}$ into $\mathfrak{B}$ in such a way that*

$$(123) \qquad \rho(Au, Av) \leqq p\rho(u, v) \qquad \text{with} \qquad p < 1.$$

Then there exists exactly one element $\Phi \in \mathfrak{B}$ with $A\Phi = \Phi$.

This theorem will be proved in III 11, §7.

For the Banach space $\mathfrak{B}$ we take our functions $u \in \mathfrak{C}^{2+\alpha}$ with $\rho(u, v) = \|u - v\|_{2+\alpha}$. If we set $v_1 = Au_1$, $v_2 = Au_2$, then from (121) we have

$$(124) \qquad D_{t_0}(v_1 - v_2) = (t - t_0)(\Delta_n - D)(u_1 - u_2) = g_1 - g_2.$$

But from (120) it follows in any case that

$$(125) \qquad \|v_1 - v_2\|_{2+\alpha} \leqq C\|g_1 - g_2\|_{\alpha}.$$

From (124) it is now easy to prove the inequality

$$\|g_1 - g_2\|_{\alpha} \leqq |t - t_0|C_1\|u_1 - u_2\|_{2+\alpha},$$

with suitably chosen C_1, so that finally

$$(126) \qquad \|Au_1 - Au_2\|_{2+\alpha} \leqq |t - t_0|CC_1\|u_1 - u_2\|_{2+\alpha}.$$

If we choose $|t - t_0| < \dfrac{1}{2CC_1}$, then we have (123) with $p = \frac{1}{2}$.

Proof of γ). Let $\{t_n\} \in \mathfrak{T}$ be a sequence of numbers converging to t. The corresponding $u^n \in \mathfrak{C}^{2+\alpha}$ are then solutions of

$$(127) \qquad D_{tn}u^n = f \qquad \text{with} \qquad u^n = 0 \quad \text{on} \quad \dot{\mathfrak{G}}.$$

From (120) $\|u^n\|_{2+\alpha} \leqq C\|f\|_{\alpha}$, it then follows that the sequence $u^1, u^2, \ldots$, together with the sequences of derivatives up to the second order, is uniformly bounded. Then it is possible, though we omit the proof here, to choose a subsequence u^{n_i} such that the u^{n_i}, together with their second derivatives, converge uniformly to a function $u \in \mathfrak{C}^{2+\alpha}$, together with its second derivatives. It follows that in (127) the passage to the limit $n \to \infty$ under the D-operator is valid. We obtain $D_t u = f$ with $u = 0$ on $\dot{\mathfrak{G}}$ and thus $t \in \mathfrak{T}$.

4.3. *The Eigenvalue Problem for Elliptic Differential Equations*

We consider the problem

$$(128)\quad Au = \lambda u \qquad \text{with} \qquad Au \equiv \frac{1}{k(\mathfrak{x})}\{-\Delta_n u + q(\mathfrak{x})u\}, \quad u = 0 \text{ on } \dot{\mathfrak{G}},$$

noting that again $\mathfrak{x}$ stands for the vector $(x_1, \ldots, x_n)$. It is obvious that $u \equiv 0$ is always a solution for every value of the parameter λ. But here we seek solutions $u \in \mathfrak{C}^0$ in $\overline{\mathfrak{G}}$, $u \in \mathfrak{C}^2$ in $\mathfrak{G}$ with $u \not\equiv 0$ and $u = 0$ on $\dot{\mathfrak{G}}$. Such u will exist only for suitable λ-values. These values are called *eigenvalues* and the corresponding functions u are *eigenfunctions*. Thus $\lambda = 2$ is an *eigenvalue* in example 7. We assume that $q, k \in \mathfrak{C}^0$ in $\overline{\mathfrak{G}}$ and $\in \mathfrak{C}^1$ in $\mathfrak{G}$ with $k > 0$ in $\overline{\mathfrak{G}}$, and we then call (128) a *regular eigenvalue problem*. On the other hand, if α) the assumptions on q, k hold only in $\mathfrak{G}$ with the possible exception of finitely many points, or β) if $\mathfrak{G}$ is not bounded but extends to infinity, or γ) if α) and β) both hold, then we call (128) a *singular eigenvalue problem*. In the singular problems (every eigenvalue problem of atomic physics is singular) the requirement $u = 0$ on $\dot{\mathfrak{G}}$ must in some cases be weakened, or even omitted altogether. Here we deal only with regular problems and assume $q \geqq 0$. Then as a consequence of theorem 12 we have at once that no number $\lambda \leqq 0$ can be an eigenvalue, since from (84) we see that

$$(129)\qquad a(\mathfrak{x}) = \lambda k(\mathfrak{x}) - q(\mathfrak{x}) \leqq 0.$$

In order to simplify the exposition, we now consider only the simplest case $q(\mathfrak{x}) \equiv 0$, $k(\mathfrak{x}) \equiv 1$ in (128). We first transform (128) into an *equivalent integral equation problem*. For this purpose we shall need the formula (36) for a solution. If in this formula we set $f = -\lambda u$, it is certainly reasonable to conjecture that problem (128) and

$$(130)\quad Ku = \mu u \qquad \text{with} \qquad \mu = \frac{1}{\lambda} \qquad \text{and} \qquad Ku = \int_{\mathfrak{G}} g(\mathfrak{x}, \mathfrak{y})u(\mathfrak{y})d\mathfrak{y}$$

are *equivalent problems*. The great advantage in this transformation consists in the fact that the operator K does not contain any differentiation. Thus we shall first try to prove that K is *symmetric* and *completely continuous* (III 11, §4), since for such an operator the *eigenvalue problem* is completely solved. As a suitable domain of definition for K in (130) we choose the *Hilbert space*

$$\text{(131)}\quad \mathfrak{H}: \begin{cases} u \in \mathfrak{C}^0 \quad \text{in } \bar{\mathfrak{G}} & \text{with the } \textit{inner product} \\ (u, v) = \int_{\mathfrak{G}} u(\mathfrak{x})v(\mathfrak{x})d\mathfrak{x} & \text{and the } \textit{norm} \quad \|u\| = (u, u)^{1/2}, \end{cases}$$

although this space as it stands is not complete (III 11, §2). The fact that K is symmetric in $\mathfrak{H}$ follows at once from the symmetry of the *Green's function* $g(\mathfrak{x}, \mathfrak{y}) = g(\mathfrak{y}, \mathfrak{x})$, but the proof of complete continuity is difficult, since $g(\mathfrak{x}, \mathfrak{y})$ becomes discontinuous as $\mathfrak{x} \to \mathfrak{y}$. We now borrow from III 11, §4 the fact that if the kernel of such an integral operator is continuous, then the operator itself is *completely continuous*. Thus we can approximate K with arbitrary accuracy by means of completely continuous operators and then make use of the following simple theorem 22.

Theorem 22. *The operator K is completely continuous in $\mathfrak{H}$ if for every $\epsilon > 0$ there exists a completely continuous operator K such that*

$$\text{(132)}\qquad \|(K - K_\epsilon)u\| \leqq \epsilon\|u\|$$

for every $u \in \mathfrak{H}$.

In order to apply this theorem to (130) we introduce the smoothing function $h_\epsilon(\xi)$ with the properties: $h_\epsilon(\xi)$ is continuous in $0 \leqq \xi < \infty$,

$$\text{(133)}\qquad \begin{cases} h_\epsilon(\xi) = 0 & \text{for} \quad 0 \leqq \xi \leqq \dfrac{1}{2\epsilon}, \\ 0 \leqq h_\epsilon(\xi) \leqq 1 & \text{for} \quad \dfrac{1}{2\epsilon} \leqq \xi \leqq \dfrac{1}{\epsilon}, \\ h_\epsilon(\xi) = 1 & \text{for} \quad \dfrac{1}{\epsilon} \leqq \xi < \infty. \end{cases}$$

If we set $g_\epsilon(\mathfrak{x}, \mathfrak{y}) = g(\mathfrak{x}, \mathfrak{y})h_\epsilon(|\mathfrak{x} - \mathfrak{y}|)$, then g_ϵ is continuous for every $\mathfrak{x}, \mathfrak{y}$ in $\mathfrak{G}$ and thus the operators

$$\text{(134)}\qquad K_\epsilon u = \int_{\mathfrak{G}} g_\epsilon(\mathfrak{x}, \mathfrak{y})u(\mathfrak{y})d\mathfrak{y}$$

are completely continuous. By the *Schwarz inequality* we obtain

$$\begin{aligned} \|(K - K_\epsilon)u\|^2 &= \int_{\mathfrak{G}} \left(\int_{\mathfrak{G}} \{g(\mathfrak{x}, \mathfrak{y}) - g_\epsilon(\mathfrak{x}, \mathfrak{y})\}u(\mathfrak{y})d\mathfrak{y}\right)^2 d\mathfrak{x} \\ \text{(135)}\qquad &\leqq \int_{\mathfrak{G}} \left(\int_{\mathfrak{G}} \sqrt{|g(\mathfrak{x}, \mathfrak{y}) - g_\epsilon(\mathfrak{x}, \mathfrak{y})|}\sqrt{|g(\mathfrak{x}, \mathfrak{y}) - g_\epsilon(\mathfrak{x}, \mathfrak{y})|}\ |u(\mathfrak{y})|d\mathfrak{y}\right)^2 d\mathfrak{x} \\ &\leqq \int_{\mathfrak{G}} \left(\int_{\mathfrak{G}} |g(\mathfrak{x}, \mathfrak{y}) - g_\epsilon(\mathfrak{x}, \mathfrak{y})|d\mathfrak{y} \int_{\mathfrak{G}} |g(\mathfrak{x}, \mathfrak{y}) - g_\epsilon(\mathfrak{x}, \mathfrak{y})|u^2(\mathfrak{y})d\mathfrak{y}\right)d\mathfrak{x}. \end{aligned}$$

Now we have

$$\begin{aligned}(136)\qquad \int_{\mathfrak{G}} |g(\mathfrak{x}, \mathfrak{y}) - g_\epsilon(\mathfrak{x}, \mathfrak{y})|d\mathfrak{y} &= \int_{|\mathfrak{y}-\mathfrak{x}|\leqq\epsilon} |g(\mathfrak{x}, \mathfrak{y}) - g_\epsilon(\mathfrak{x}, \mathfrak{y})|d\mathfrak{y} \\ &\leqq 2\int_{|\mathfrak{y}-\mathfrak{x}|\leqq\epsilon} |g(\mathfrak{x}, \mathfrak{y})|d\mathfrak{y}.\end{aligned}$$

Employing the inequality (91), we find for $n \geqq 3$

$$\int_{|\mathfrak{y}-\mathfrak{x}|\leqq\epsilon} |g(\mathfrak{x}, \mathfrak{y})|d\mathfrak{y} \leqq \epsilon^2.$$

Thus by interchange of the order of integration

$$\begin{aligned}\|(K - K_\epsilon)u\|^2 &\leqq \text{const } \epsilon^2 \int_{\mathfrak{G}} \left(\int_{\mathfrak{G}} |g(\mathfrak{x}, \mathfrak{y}) - g_\epsilon(\mathfrak{x}, \mathfrak{y})|d\mathfrak{x}\right)u^2(\mathfrak{y})d\mathfrak{y} \\ (137)\qquad &\leqq \text{const } \epsilon^4 \int_{\mathfrak{G}} u^2(\mathfrak{y})d\mathfrak{y} = \text{const } \epsilon^4\|u\|^2.\end{aligned}$$

In this way we have proved that K is completely continuous in $\mathfrak{H}$; thus by chapter 11 the equation (130) has infinitely many positive eigenvalues $\mu_1, \mu_2, \mu_3, \ldots$, with a limit point only at the origin, and the corresponding *eigenfunctions* $\varphi_1, \varphi_2, \ldots$, satisfy the orthogonality conditions

$$(138)\qquad (\varphi_i, \varphi_k) = \delta_{ik} = \begin{cases} 1, & \text{if } i = k, \\ 0, & \text{if } i \neq k. \end{cases}$$

Moreover, we have the *expansion theorem*, which states that for every function $u \in \mathfrak{H}$ the following expansion is valid:

$$(139)\qquad Ku = \sum_{i=1}^{\infty} (Ku, \varphi_i)\varphi_i = \sum_{i=1}^{\infty} \mu_i(u, \varphi_i)\varphi_i,$$

where the second of these two equations follows from $K\varphi_i = \mu_i\varphi_i$. In order to make use of these results for (128) we must first verify that the φ_i have the necessary differentiability. Now $\varphi_i \in \mathfrak{C}^0$, but $K\varphi_i \in \mathfrak{C}^1$, which shows at once that $\varphi_i \in \mathfrak{C}^1$, and finally, for (128) also, we can state the expansion theorem (139) with $\mu_i = \frac{1}{\lambda_i}$, where the eigenvalues λ_i of (128) have no limit point except ∞. The expansion theorem whose proof has been outlined here is one of the most important achievements of analysis, in view of the fact that the eigenvalues λ_i and the corresponding eigenfunctions φ_i can be calculated explicitly only for the simplest domains $\mathfrak{G}$.

4.4. *The Initial Value Problem for Hyperbolic Differential Equations*

We confine our attention to the simplest problem for a *linear differential equation of second order in two independent variables x, y* and thus make no use of vector notation. By (65) we can assume that the equation is in the form

$$u_{xy} + a(x, y)u_x + b(x, y)u_y + c(x, y)u + d(x, y) = 0. \tag{140}$$

The *characteristic manifolds* $\dot{\mathfrak{C}}$ for (140) are the lines $x = \text{const}$ and $y = \text{const}$. As initial curve $\mathfrak{K}$ we choose $\mathfrak{K}$: $x(t)$, $y(t)$ with $x(t), y(t) \in \mathfrak{C}^1$ in $t_0 \leqq t \leqq t_1$, which we assume to be nowhere coincident with a characteristic curve. This latter assumption is equivalent to the condition that $\mathfrak{K}$ is expressible both in the form $y = f(x)$ and also as $x = g(y)$. On $\mathfrak{K}$ the values of u, u_x, u_y, are arbitrarily prescribed in such a way that

$$\frac{du}{dt} = u_x \frac{dx}{dt} + u_y \frac{dy}{dt}. \tag{141}$$

Thus we can prescribe two functions, which we may do by writing

$$u(x, y) = \varphi(x) + \psi(y) \qquad \text{with} \qquad u_x = \frac{d\varphi}{dx}, \qquad u_y = \frac{d\psi}{dy} \quad \text{on} \quad \mathfrak{K}. \tag{142}$$

The rest of the argument refers to the rectangle $\mathfrak{R}$ (Figure 12). If P_2 has the coordinates (x, y), then P_1, P_3 have the coordinates $(g(y), y)$, $(x, f(x))$. Here $\mathfrak{B}$ is the *range of influence*, bounded by $\mathfrak{K}$ and the characteristics $\dot{\mathfrak{C}}$.

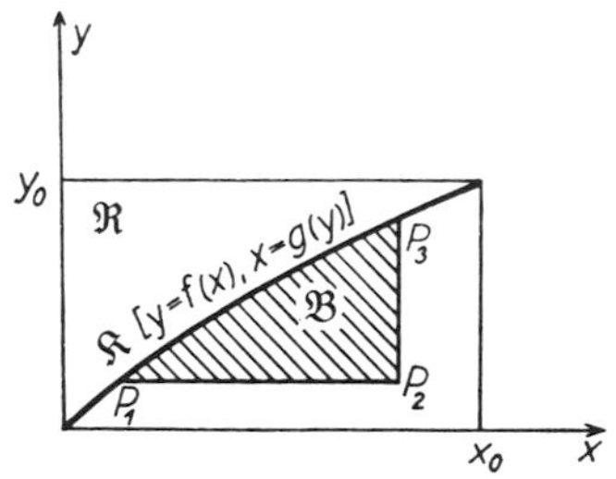

Fig. 12

Theorem 23. *Let the coefficients a, b, c, d belong to* $\mathfrak{C}^0$ *in* $\mathfrak{R}$, *and let* $\varphi(x)$, $\psi(y)$ *belong to* $\mathfrak{C}^1$ *in* $0 \leqq x \leqq x_0$, $0 \leqq y \leqq y_0$. *Then there exists exactly one solution* $u(x, y)$ *of* (140) *with* $u_{xy}, u_x, u_y \in \mathfrak{C}^0$ *in* $\mathfrak{R}$, *for which* u, u_x, u_y *coincide on* $\mathfrak{K}$ *with* $\varphi(x) + \psi(y)$, $\varphi'(x)$, $\psi'(y)$.

For the proof we first consider the function

$$u_0(x, y) = \varphi(x) + \psi(y) - \iint_{\mathfrak{B}} d(\xi, \eta)\, d\xi\, d\eta \tag{142}$$

$$\text{with} \qquad \iint_{\mathfrak{B}} = \int_{g(y)}^{x} \left(\int_{y}^{f(\xi)} \cdots d\eta \right) d\xi,$$

which is the desired solution for $u_{xy} + d = 0$ with corresponding initial values on $\mathfrak{K}$. Then $u_n(x, y)$ with $n > 0$ is defined recursively as the solution of

$$u_{n,xy} + au_{n-1,x} + bu_{n-1,y} + cu_{n-1} = 0 \tag{143}$$

with $u_n = 0$ on $\mathfrak{K}$.

As above, we obtain the representation

$$u_n(x, y) = -\iint_{\mathfrak{B}} (au_{n-1,\xi} + bu_{n-1,\eta} + cu_{n-1})d\xi d\eta. \tag{144}$$

We now prove that the series $u = \sum_{\nu=0}^{\infty} u_\nu$ converges uniformly. For this purpose we let M be an upper bound for $|a|, |b|, |c|, |d|, |u_0|, |u_{0,x}|, |u_{0,y}|$. We also let

$$\mu = x + y_0 - y \qquad \text{and} \qquad N = \max\{3M, 3M(x_0 + y_0)\}. \tag{145}$$

Then in $\mathfrak{R}$ we have the inequality

$$|u_n|, |u_{n,x}|, |u_{n,y}| \leqq \frac{MN^n\mu^n}{n!}, \tag{146}$$

which holds for $n = 0$. If these inequalities have already been proved for the index $n - 1$, then it follows from (144), (146) that

$$\begin{aligned} |u_n(x, y)| &\leqq 3M\frac{MN^{n-1}}{(n-1)!}\iint_{\mathfrak{B}} (\xi + y_0 - \eta)^{n-1}d\eta d\xi \\ &\leqq 3M\frac{MN^{n-1}}{(n-1)!}\int_0^x\int_y^{y_0} (\xi + y_0 - \eta)^{n-1}d\eta d\xi \\ &\leqq 3M\frac{MN^{n-1}}{(n-1)!}\frac{(x_0 + y_0 - y)^{n+1}}{n(n+1)} \\ &\leqq \frac{MN^{n-1}\mu^n}{n!}3M(x_0 + y_0). \end{aligned} \tag{147}$$

Correspondingly we obtain

$$\begin{aligned} |u_{n,x}(x, y)| &\leqq 3M\frac{MN^{n-1}}{(n-1)!}\int_y^{y_0} (x + y_0 - \eta)^{n-1}d\eta \\ &\leqq 3M\frac{MN^{n-1}}{(n-1)!}\mu^n \leqq \frac{MN^n\mu^n}{n!}, \end{aligned} \tag{148}$$

and analogous remarks hold for $|u_{n,y}|$. Thus $\sum_{\nu=0}^{\infty} u_\nu$ and $\sum_{\nu=0}^{\infty} u_{\nu,x}$, $\sum_{\nu=0}^{\infty} u_{\nu,y}$ are uniformly convergent and the limit function u satisfies the relation

$$u(x, y) = -\iint_{\mathfrak{B}} (au_\xi + bu_\eta + cu + d)d\xi d\eta, \tag{149}$$

which is equivalent to (140), as can be shown at once by differentiation with respect to x and y. Moreover, (149) satisfies the prescribed initial conditions on $\mathfrak{K}$. The proof of uniqueness and of continuous dependence on the initial data can be carried out in the usual way.

4.5. *Initial Value Problems and Boundary Value Problems for Hyperbolic and Parabolic Differential Equations*

In the applications it is ordinarily not the pure initial value problems that are of central interest; usually it is necessary to satisfy suitably chosen boundary conditions as well, and these can occur in a great variety of forms. The simplest example has been discussed already in §1.2. Moreover, it is desirable that the solutions should not only be proved to exist but should be given in explicit form, at least for sufficiently simple problems. As methods to such an end let us mention the method of *separation of variables*, which will be familiar to the reader in simple cases, and also the method of the *Laplace transform*, which depends to a great extent on the *semigroup properties* of the *solution operator*. Let us explain this method by a simple example without actually calculating the solution. We consider the partial differential equation in two independent variables x, t

$$Du \equiv Bu + k(x)\{r_1u_{tt} + r_2u_t\} = k(x)f(x, t) \tag{150}$$

with $Bu \equiv -(p(x)u_x)_x + q(x)u$, under the following assumptions: the variable t runs over the interval $0 \leqq t < \infty$, and in the interval $l \leqq x \leqq m$ the functions $p(x)$, $p'(x)$, $q(x)$, $k(x)$ are continuous, with $p > 0$, $k > 0$. The numbers r_1, r_2 are arbitrary with $r_1 > 0$, or if $r_1 = 0$, then $r_2 > 0$. Under these assumptions we see at once from §2.2 that in $l \leqq x \leqq m$, $0 \leqq t < \infty$ the equation (150) is of *hyperbolic type* for $r_1 > 0$ and of *parabolic type* for $r_1 = 0$, $r_2 > 0$. In addition to (150) we have suitable *initial conditions* (151) and *boundary conditions* (152):

$$u(x, 0) = u_0(x), \qquad u_t(x, 0) = u_1(x), \tag{151}$$

where the latter condition in (151) can be imposed only in the *hyperbolic case*. We shall choose the very simple boundary conditions:

$$\begin{cases} u(l, t)\ \cos\delta + u_x(l, t)\ \sin\delta = 0 \\ u(m, t)\cos\vartheta + u_x(m, t)\sin\vartheta = 0 \end{cases}, \qquad 0 \leqq \delta, \vartheta < \pi, \tag{152}$$

although it is precisely for complicated boundary conditions that the Laplace transform is superior to separation of variables.

We now first of all imagine that the solution $u(x, t)$ of the problem (150–152) is given, with "sufficiently good" properties. In particular, we assume that

$$\int_0^\infty e^{-st}u(x, t)dt = v(x, s) \tag{153}$$

is convergent for the value $s = s_0$ of the complex parameter $s = \xi + i\eta$. Then the integral also converges for every s with $\operatorname{Re} s > \operatorname{Re} s_0$. We say that v has been obtained from u by application of the *Laplace operator* $\int_0^\infty e^{-st}\{\ \}dt$, and we write $Lu = v$. Under these assumptions it already follows that $v(x, s)$ is analytic in s in the right-hand halfplane $\xi > \xi_0$ (ξ_0 is the real part of s_0). If u_t, u_{tt} become small very rapidly for large t, then by partial integration we find the *formal rules of calculation*

(154) $$Lu_t = sv - u(x, 0), \qquad Lu_{tt} = s^2 v - su(x, 0) - u_t(x, 0).$$

Finally we require the reciprocal operator L^{-1}, which is found, at least formally, from the *Fourier integral theorem*

(155) $$u(x, t) = L^{-1}v \equiv \frac{1}{2\pi i}\,\text{P.V.}\int_{\xi_1 - i\infty}^{\xi_1 + i\infty} e^{st}v(x, s)ds^{1}), \qquad \xi_1 > \xi_0.$$[14]

Here $v(x, s)$ must be analytic in s at least in the right-hand halfplane $\xi > \xi_0 > 0$. Incidentally we remark that under suitable assumptions the operator L^{-1} has the (not immediately obvious) *semigroup property* $L^{-1}(t_1 + t_2) = L^{-1}(t_1)L^{-1}(t_2)$ with respect to t, a fact which, in addition to the simple rules (154), forms the principal reason why the Laplace transform method is particularly well-adapted to our present problems. If we apply the *L-operator* to (150–152) and formally undertake interchange of the L-operator with suitable differentiations and limiting processes, we obtain, setting $Lf = j$,

(156)
$$Bv + k(x)\{r_1 s^2 + r_2 s\}v = k(x)\{(r_1 s + r_2)u_0(x) + r_1 u_1(x)\} + k(x)j(x, s),$$

(157) $$\begin{cases} R_1 v \equiv v(l, s)\cos\delta + v'(l, s)\sin\delta = 0, \\ R_2 v \equiv v(m, s)\cos\vartheta + v'(m, s)\sin\vartheta = 0, \end{cases}$$

with $v' \equiv \dfrac{dv}{dx}$. Introducing the new abbreviations

(157a)
$$-(r_1 s^2 + r_2 s) = \lambda, \qquad \lambda = \varphi + i\mu, \qquad h = (r_1 s + r_2)u_0(x) + r_2 u_1(x)$$

and the new operator

(158) $$Av \equiv \frac{1}{k(x)}\,Bv = \frac{1}{k(x)}\{-(pv')' + qv\},$$

we finally obtain

(159) $$Av = \lambda v + h + j, \qquad R_1 v = R_2 v = 0.$$

[14] Here P.V. denotes the Cauchy principal value.

This is a boundary value problem for ordinary differential equations and can be solved for every h, j, provided λ is not equal to one of the eigenvalues of the eigenvalue problem

$$Av = \lambda v, \qquad R_1 v = R_2 v = 0 \tag{160}$$

(cf. the discussion in III 8, §4). The eigenvalues $\lambda_1, \lambda_2, \lambda_3, \ldots$ are all real and are bounded from below: $\lambda_i \geqq a$. For (157) it follows that in the $(\lambda = \varphi + i\mu)$-plane these eigenvalues satisfy the condition $\varphi \geqq a$, $\mu = 0$, which certainly cannot be satisfied by the λ in (157a), since for this λ we have

$$\varphi = -(r_1\xi^2 - r_1\eta^2 + r_2\xi), \qquad \mu = -\eta(r_2 + 2\xi r_1). \tag{161}$$

Thus, $\mu = 0$ for $\eta = 0$ or $\xi = -\dfrac{r_2}{2r_1}$. By suitable choice of ξ_0 (any ξ with $\xi > \xi_0$ could also be chosen) the second case can be excluded. The first case gives

$$\varphi = -(r_1\xi^2 + r_2\xi), \tag{162}$$

from which for $\xi > \xi_0$ and sufficiently large ξ_0 it follows that $\varphi < a$. The solution of (159) can then be obtained in either of two ways: 1. by expansion in eigenfunctions and 2. by the "*Green's resolvent.*" In the second way, we obtain the solution in the form

$$v(x, s) = \int_l^m g(x, y, s)\{h(y, s) + j(y, s)\}k(y)dy. \tag{163}$$

The whole procedure can now be reversed. In other words, it is natural to conjecture that the solution $u(x, t)$ of (150–152) will appear in the form

$$\begin{aligned} u(x, t) &= L^{-1}v \\ &\equiv \frac{1}{2\pi i}\,\text{P.V.}\int_{\xi_1 - i\infty}^{\xi_1 + i\infty} e^{st}\left(\int_l^m g(x, y, s)(h(y, s) + j(y, s))k(y)dy\right)ds, \end{aligned} \tag{164}$$

where we have the additional advantage that under suitable circumstances the complex integral can be evaluated by the calculus of residues. It will then only remain to prove that (164) actually solves the problem (150–152), which will justify the whole procedure. This proof is a delicate task which, however, can be carried out successfully even for more general problems (cf. [10]).

Bibliography

The very limited size of the present chapter has meant that certain aspects of partial differential equations could not even be mentioned. This remark includes, for example, partial differential equations of first order, whose mathematical theory can largely be reduced to ordinary differential equations, the elegant relationship between partial differential equations and the calculus of variations, dealt with particularly in Courant-Hilbert, the whole theory of systems of partial differential equation, and their connection with the function theoretic method of kernel functions. So it may be useful to give a few references to the literature, including only text books and making no claim to completeness.

[1] BATEMAN, H.: Partial differential equations of mathematical physics. Dover Publications, New York 1944.

[2] BERGMAN, S. and SCHIFFER, M.: Kernel functions and elliptic differential equations in mathematical physics. Academic Press, New York 1953.

[3] BERNSTEIN, D. L.: Existence theorems in partial differential equations. University Press, Princeton.

[4] BIEBERBACH, L.: Theorie der Differentialgleichungen. Springer-Verlag, Berlin 1930.

[5] COURANT, R. and HILBERT, D.: Methods of mathematical physics, Interscience, New York-London. Vol. I, 1953; Vol. II, 1962.

[6] DOETSCH, G.: Handbuch der Laplace-Transformation, Vol. III. Birkhäuser-Verlag, Basel 1956.

[7] FRANK, P. and v. MISES, R.: Die Differential- und Integralgleichungen der Mechanik and Physik, Vols. I, II. Vieweg, Braunschweig 1930, 1935, 1961.

[8] Gunther, N. M.: La théorie du potentiel et ses applications aux problèmes fondamentaux de la physique mathématique. Gauthier-Villars, Paris, 1934.

[9] HADAMARD, J.: Lectures on Cauchy's problem. Dover, New Haven 1923.

[10] Hellwig, G.: Partial differential equations; an introduction. (Translated from the German.) Blaisdell, New York 1964.

[11] HORN, J.: Partielle Differentialgleichungen. de Gruyter, Berlin 1949.

[12] JOHN, F.: Plane waves and spherical means applied to partial differential equations. Interscience, New York 1955.

[13] KAMKE, E.: Differentialgleichungen, II. Teil. Akad. Verlagsgesellschaft, Leipzig 1961.

[14] KELLOGG, O. D.: Foundations of potential theory. Springer-Verlag, Berlin 1929.

[15] LICHTENSTEIN, L.: Neuere Entwicklungen der Theorie partieller Differentialgleichungen zweiter Ordnung vom elliptischen Typus. Enzyklopädie der math. Wiss., Vol. II/3_2.—Neuere Entwicklung der Potentialtheorie. Vol. II/3_1, as above.

[16] Petrovskii, I. G.: Lectures on partial differential equations. (Translation from the Russian.) Interscience (Wiley), New York 1955.

[17] Sauer, R.: Anfangswertprobleme bei partiellen Differentialgleichungen. 2nd revised edition. Springer-Verlag, Berlin 1958.

[18] Sommerfeld, A.: Partial differential equations in physics. (Translated from the German.) Academic Press, New York 1949.

[19] Sternberg, W. and Smith, T. L.: The theory of potential and spherical harmonics. Toronto 1944.

[20] Webster, A. and Szegö, G.: Partielle Differentialgleichungen der mathematischen Physik. Teubner, Leipzig 1930.

Bibliography added in Translation

Ames, W. F.: Nonlinear partial differential equations in engineering. Academic Press, New York 1965.

Carathéodory, C.: Calculus of variations and partial differential equations of the first order. Part I: Partial differential equations of the first order. (Translated from the German by R. B. Dean and J. J. Brandstatter.) Holden-Day, San Francisco 1965.

Duff, G. F. D.; Naylor, D.: Differential equations of applied mathematics. John Wiley & Sons, New York 1966.

Epstein, B.: Partial Differential equations: An introduction. McGraw-Hill, New York 1962.

Weinberger, H. F.: A first course in partial differential equations with complex variables and transform methods. Blaisdell Publ. Co., New York 1965.

Linear equations of mathematical physics. Edited by S. G. Mihlin, English translation edited by H. Hochstadt. Holt, Rinehart & Winston, Inc., New York 1967.

CHAPTER 10

Difference Equations and Definite Integrals

1. Introduction

1.1. *Recursive Sequences*

Various problems in pure and applied mathematics lead to sequences of numbers $\{u_n\}$ that are not explicitly defined by a known function $u_n = u(n)$ $(n = 1, 2, 3, \ldots)$ but by a "recursive law" for the terms of the sequence. Let us consider some simple examples.

1. In calculating the circumference of a circle from inscribed regular polygons of 2^n sides $(n = 2, 3, 4, \ldots)$ we find that the length of the side of an inscribed 2^{n+1}-gon (in the unit circle) is

$$s_{2^{n+1}} = \sqrt{2 - \sqrt{4 - s_{2^n}^2}}. \tag{1}$$

If we now set

$$4 - s_{2^n}^2 = u_{n+1}^2,$$

it follows that

$$s_{2^{n+1}} = \sqrt{2 - u_{n+1}},$$

and thus in view of (1) the sequence u_n satisfies the recursive relation

$$u_{n+1}^2 - u_n = 2. \tag{2}$$

From $s_{2^2} = \sqrt{2}$ we have the *initial condition* $u_2 = 0$, so that from (2) it is easy to calculate[1] the further terms of the sequence.

2. The book "Liber abbaci" written in 1202 by the Italian mathematician Fibonacci (Leonardo of Pisa) contains the famous *guinea-pig*

[1] For the number π it follows that

$$\pi = \lim_{n \to \infty} 2^n \sqrt{2 - u_{n+1}}.$$

It is easy to show that the sequence u_n is monotone increasing and converges to 2.

problem, which we shall formulate as follows. A pair of guinea pigs born at time 0 creates a further pair in every month of its existence, beginning with the second month, and its descendants behave in the same way. How many pairs will there be after n months?

We can set up a table and enter the number u_n of pairs (the *Fibonacci numbers*) according to the above rule:

Month	0	1	2	3	4	5	6	...
Number of Pairs	1	1	2	3	5	8	13	...
	u_1	u_2	u_3	u_4	u_5	u_6	u_7	...

The law for the formation of the *Fibonacci sequence* is obviously

(3) $$u_{n+2} - u_{n+1} - u_n = 0.$$

The *initial conditions* are $u_1 = u_2 = 1$.

3. The function

$$f(x) = \frac{1}{a_0 + a_1 x + \cdots + a_l x^l}$$

is to be expanded in a power series in the neighborhood of the origin. We adopt the method of undetermined coefficients:

$$f(x) = v_0 + v_1 x + v_2 x^2 + \cdots.$$

Then we must have

$$1 = (a_0 + a_1 x + \cdots + a_l x^l)(v_0 + v_1 x + v_2 x^2 + \cdots).$$

After the multiplication, the coefficients of x^r for $r \geqq 1$ must vanish. Thus

(4) $$a_0 v_0 = 1, \qquad a_0 v_1 + a_1 v_0 = 0, \qquad \text{etc.}$$

In general, for $m \geqq 1$,

(4′) $$a_0 v_m + a_1 v_{m-1} + \cdots + a_l v_{m-l} = 0.$$

The equation (4′) also holds for $0 < m < l$ if we agree to set the $v_{-\nu}$ equal to 0. Then the successive coefficients v_n of the power series for $f(x)$ can be calculated from (4) and (4′).

In order that (4′) may be better adapted in a purely formal way to the relations (2) and (3), we introduce a new notation for the subsequence

$$v_l, v_{l+1}, v_{l+2}, \ldots.$$

We set $v_{m-1} = u_n$. Then for $n \geqq 0$ we have from (4′):

(5) $$a_0 u_{n+l} + \cdots + a_l u_n = 0.$$

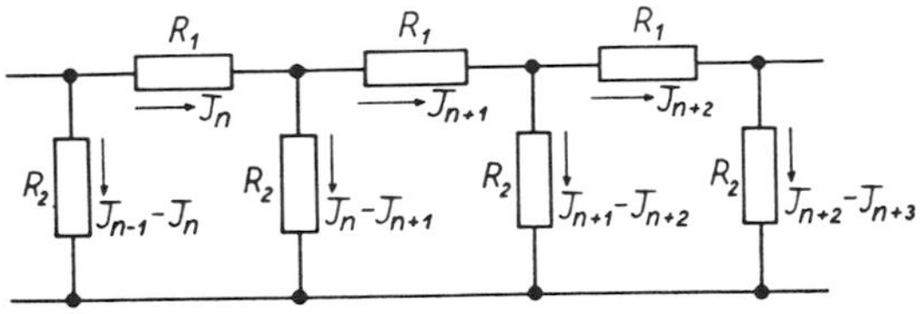

Fig. 1

4. Finally we give an example from physics. For a *chain conductor* with resistances R_1 and R_2 and amperages J_m in the *cell* with index m (see Figure 1) the Kirchhoff law states that the sum of the voltage differences in each cell is equal to 0. Thus in the cell with index $n + 1$ we have

$$R_1 J_{n+1} + R_2(J_{n+1} - J_{n+2}) + R_2(J_{n+1} - J_n) = 0$$

or

(6) $$-R_2 J_n + (R_1 + 2R_2)J_{n+1} - R_2 J_{n+2} = 0).$$[2]

1.2. *Definition of Difference Equations*

We now consider the equations (2), (3), (5), and (6) without regard to the particular problems that gave rise to them. If we look on the sequences u_n as functions $u(n)$ for positive (or nonnegative) integral arguments, we may regard (2), (3), (5), and (6) as *functional equations*. Such functional equations of the form

(7) $$\Phi(u_n, u_{n+1}, \ldots, u_{n+l}, n) = 0$$

are called *difference equations* of *order* 1. Of particular interest are the *linear difference equations*

(8) $$a_l(n)u_{n+l} + \cdots + a_0(n)u_n = b_n.$$

They are called *homogeneous* if $b_n = 0$, and *nonhomogeneous* if $b_n \neq 0$. Thus (3), (5), and (6) are linear homogeneous difference equations with constant coefficients. The equation (5) is of order l, and (3) and (6) are of order 2. The equation (2) is also a difference equation, but it is not linear.

In many cases it is very easy, under suitable *initial conditions*, to determine the *solution* u_n by a process of *recursion*. But the answer to the following question is not quite so simple.

How can we represent the solution u_n of a difference equation with given initial conditions as an explicit function of n?

For the *Fibonacci sequence* (example 2), for example, it is quite easy to

[2] Further examples from technology are to be found in [2], [8], [9].

determine the successive terms recursively, but the explicit representation of the sequence u_n as a function $u(n)$ is not trivial (see §4.1).

Another natural problem is the following one.

How can we find the totality of all solutions of a given difference equation?

Before taking up these questions, let us make a natural generalization of the problem. We have interpreted the recursive sequences as functions $u(n)$ for integral arguments. But instead we could ask for functions which for arbitrary real x, or also for complex values z in a suitable domain of the Gauss plane, satisfy a prescribed difference equation. Naturally we will then admit functions of x or z as coefficients of the difference equation. The linear difference equations, for example, will have the form

$$a_l(z)u(z + l) + \cdots + a_0(z)u(z) = b(z). \tag{8'}$$

These difference equations have a formal similarity with linear differential equations. In fact, when discussing the solution of (8′) for constant coefficients we arrive at statements completely analogous to the theorems about linear differential equations with constant coefficients. However, the theory of difference equations occupies a special position in analysis. For it has turned out that there exist relatively simple difference equations such that the functions representing their solutions do not satisfy any algebraic differential equation[3] (theorem of Hölder; see §3.5). This theorem is significant for the following reason.

By means of difference equations it is possible to define functions that cannot be obtained from the theory of differential equations. Thus the generalization of the original statement of the problem for difference equations is of particular interest for analysis.

In most of the applications of difference equations to problems in statics and electrotechnology, the solution is required only for integral arguments; but of course, every solution of say (8′) for a suitable domain of the complex plane is also a solution for the corresponding problem (8). Thus the function-theoretic methods are also important for the practical engineer. On the other hand, certain methods of solution for the difference equation (8) (in particular, with constant coefficients on the left side) are specially adapted to the case of integral arguments.[4]

In the present chapter we discuss the general function-theoretic approach, but only for the linear difference equation (8′). Thus we seek analytic solutions $u(z)$ of (8′) that are holomorphic in a certain domain $\mathfrak{B}$ of the complex plane.

In this domain $\mathfrak{B}$ the coefficients $a_\lambda(z)$ are also assumed to be single-valued holomorphic functions. Since $u(z)$ must always be defined for

[3] The definition of an "algebraic differential equation" is given in §3.5, p. 375.

[4] See Zypkin [9], and D. F. Lawden, Linear Difference Equations, The Mathematical Gazette 36, pp. 193–196, London 1952.

$z+1, z+2, \ldots, z+n$, the domain $\mathfrak{B}$ must consist of a "right half-strip," by which we mean a set of points in the Gauss plane for which

$$\operatorname{Re} z \geqq a, \qquad b_1 \leqq \operatorname{Im} z \leqq b_2$$

(see Figure 2).

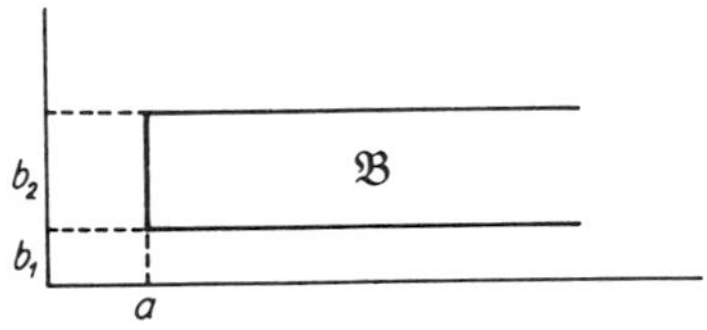

Fig. 2

We now discuss certain special cases of (8′); among their solutions we will encounter functions of particular interest for the whole of analysis; for example, the *Euler Γ-function*, whose properties gain greatly in clearness when it is regarded as the solution of a difference equation. At the end of the chapter we give some general methods for the solution of (8′) (with constant and with variable coefficients).

2. Simple Difference Equations

2.1. *The Summation Method*

We begin with the special case

(9) $$u(z+1) - u(z) = b(z),$$

which is important for certain practical applications and also as an auxiliary equation in the treatment of difference equations of higher order. For given $b(z)$ it is often possible to determine $u(z)$ in the following way.

From (9),

(10) $$u(z+h+1) - u(z+h) = b(z+h),$$

and thus by summation over h:

(10′) $$u(z) = -\sum_{h=0}^{n-1} b(z+h) + u(z+n).$$

For $z = 1$, we have

(10″) $$u(1) = -\sum_{h=0}^{n-1} b(1+h) + u(1+n).$$

From (10′) and (10″) it follows by subtraction that

(11)

$$u(z) - u(1) = -\sum_{h=0}^{n-1} [b(z + h) - b(1 + h)] + [u(z + n) - u(1 + n)].$$

Let us now assume that the sum in (11) is uniformly convergent in every interior subdomain of $\mathfrak{B}$. The limit of the series in (11) then defines a function $G_1(z)$ holomorphic for all $z \in \mathfrak{B}$:

(12) $$G_1(z) = -\sum_{h=0}^{\infty} [b(z + h) - b(1 + h)].$$

This function $G_1(z)$ is already a solution of (9), since we have

$$G_1(z + 1) = -\sum_{h=0}^{\infty} [b(z + 1 + h) - b(1 + h)] = G_1(z) + b(z).$$

We shall call $G_1(z)$ the *fundamental solution* of (9). If the sum in (11) converges, then for all $z \in \mathfrak{B}$ the limit

$$E_1(z) = \lim_{n \to \infty} [u(z + n) - u(1 + n)]$$

also exists, and from (11)

$$u(z) - u(1) = G_1(z) + E_1(z).$$

Since $u(z)$ and $G_1(z)$ satisfy the difference equation (9), it follows that $E_1(z)$ is periodic with the period 1:

(13) $$E_1(z + 1) - E_1(z) = u(z + 1) - u(z) - [G_1(z + 1) - G_1(z)] = 0.$$

On the other hand, if $E^*(z)$ is an *arbitrary* periodic function of period 1, satisfying the regularity conditions in $\mathfrak{B}$ (e.g., $\sin \pi z$), then $G_1(z) + E^*(z)$ is also a solution of (9).

In certain cases the summation procedure can be further simplified. If the sum

(14) $$\sum_{h=0}^{n-1} b(z + h)$$

in (10′) converges, it is not necessary to subtract $u(1)$. Then we may choose

(14′) $$G(z) = -\sum_{h=0}^{\infty} b(z + h)$$

as the "fundamental solution."

For this function $G(z)$ it is also obvious that

$$G(z+1) - G(z) = b(z),$$

and thus we obtain the general solution (9) in the form

$$u(z) = G(z) + E(z).$$

Here again $E(z)$ is a periodic function (with period 1), since every function of the form

$$G(z) + \text{periodic function}$$

is a solution of (9), and conversely the difference of two solutions of (9) is always periodic.

Thus we have obtained the following theorem.

Theorem 1. *If the sum*

$$\sum_{h=0}^{n-1} [b(z+h) - b(1+h)]$$

converges in a "right halfstrip"[5] $\mathfrak{B}$, *then the general solution of* (9) *in* $\mathfrak{B}$ *can be represented in the form*

$$u(z) = G_1(z) + E_1(z).$$

Here $G_1(z)$ *is the function defined by* (12) *and* $E_1(z)$ *is a periodic function* (*with period* 1).

If

$$\sum_{h=0}^{n-1} b(z+h)$$

is convergent, then the general solution can also be represented in the following way:

$$u(z) = G(z) + E(z),$$

where $G(z)$ *is defined by* (14′) *and* $E(z)$ *is again periodic.*

Also, from $E_1(z) = \lim_{n\to\infty} [u(z+n) - u(1+n)]$ (see p. 356) it follows, in view of (9), that:

$$\begin{aligned} 0 = E_1(z+1) - E_1(z) &= \lim_{n\to\infty} [u(z+1+n) - u(z+n)] \\ &= \lim_{n\to\infty} b(z+n). \end{aligned}$$

Thus we have a *necessary criterion* for applicability of the "summation method" to the solution of (9):

$$\lim_{n\to\infty} b(z+n) = 0.$$

This condition is satisfied, for example, by $b(z) = z^{-r}$, if $r \geqq 1$.

[5] See p. 355. Of course the halfplane $\operatorname{Re} z > a$ is also an admissible domain.

For $r > 1$ it is even true that $\sum_{h=0}^{\infty} \dfrac{1}{(z+h)^r}$ is convergent and we can employ the second, simplified summation method (theorem 1). On the other hand, for $r = 1$

$$\sum_{h=0}^{\infty} \frac{1}{(z+h)}$$

is divergent, but the necessary criterion $\lim_{n\to\infty} \dfrac{1}{z+n} = 0$ for the "first" method is nevertheless satisfied.

2.2. *Difference Equation for the ψ-Function*

This procedure leads in fact to the solution of the difference equation

$$u(z+1) - u(z) = \frac{1}{z}. \tag{15}$$

For if Re $z > 0$, the series

$$\sum_{h=0}^{\infty} \left(\frac{1}{z+h} - \frac{1}{1+h}\right) = \sum_{h=0}^{\infty} \frac{1-z}{(z+h)(1+h)} \tag{16}$$

is convergent. The convergence is uniform in every compact, i.e., closed and bounded, subdomain of the halfplane. Let us assume that

$$\operatorname{Re} z > M > 0, \qquad |z| < N.$$

Then

$$\left|\frac{1-z}{(z+h)(1+h)}\right| < \frac{1+N}{(M+h)(1+h)} < \frac{1+N}{h^2},$$

from which it follows at once that (16) is convergent.

The "fundamental solution" of the difference equation (15) will be denoted by $\psi_1(z)$:

$$\psi_1(z) = -\sum_{h=0}^{\infty} \left(\frac{1}{z+h} - \frac{1}{1+h}\right). \tag{17}$$

For the time being it is defined for all z with Re $z > 0$. However, it can be continued analytically into the halfplane Re $z < 0$, although it has poles of the first order $z = 0, -1, -2, \ldots$.

2.3. "*Summation*" *of Polynomials*

Another function for which the summation method is successful is

$$b(z) = we^{wz},$$

if

$$\operatorname{Re} w < 0.$$

In this case $\sum_{h=0}^{\infty} b(z+h)$ is convergent and we can employ the "simplified" summation procedure. As a solution of the difference equation

$$u(z+1) - u(z) = we^{wz} \tag{18}$$

we then obtain the following geometric series (which is convergent since $\operatorname{Re} w < 0$):

$$G(z, w) = -\sum_{h=0}^{\infty} b(z+h) = -we^{wz} \sum_{h=0}^{\infty} e^{wh} = \frac{we^{wz}}{e^{w}-1}. \tag{19}$$

As can be seen at once from the power series for ew^{wz} and $e^{w} - 1$, the function $G(z, w)$ with fixed z is everywhere holomorphic in w. We have $G(z, 0) = 1$, and we can expand $G(z, w)$ in a series of powers of w, with coefficients that are functions of z:

$$G(z, w) = \sum_{k=0}^{\infty} \frac{B_k(z)}{k!} w^k. \tag{20}$$

From the difference equation (18) it then follows that

$$\sum_{k=0}^{\infty} \left[\frac{B_k(z+1)}{k!} - \frac{B_k(z)}{k!}\right] w^k = \sum_{k=0}^{\infty} \frac{z^{k-1}k}{k!} w^k.$$

Comparison of coefficients gives

$$B_k(z+1) - B_k(z) = kz^{k-1}. \tag{21}$$

Thus the functions $B_k(z)$ are solutions of the difference equation (21). This fact is of particular importance, since (21) certainly cannot be solved by the summation method, the necessary condition

$$\lim_{n \to \infty} b(z+n) = 0$$

failing to hold for $b(z) = kz^{k-1}$. Let us therefore try to determine these functions $B_k(z)$ from relations between power series. To this end we write

$$G(z, w) = e^{zw} h(w) \tag{22}$$

with

$$h(w) = \frac{w}{e^{w}-1} = \sum_{k=0}^{\infty} \frac{B_k}{k!} w^k. \tag{23}$$

Then the coefficients B_k are to be determined subsequently by comparison of power series.

A series expansion for $h(w)$ must be possible, since $h(w)$ is regular at $w = 0$ and we have $h(0) = 1$, as can be seen at once if the denominator of $h(w)$ is written as a series.

We first multiply together the power series for e^{zw} and $h(w)$; thus

$$G(z, w) = \sum_{h=0}^{\infty} \frac{z^h w^h}{h!} \sum_{k=0}^{\infty} \frac{B_k}{k!} w^k.$$

If we set $h + k = m$, it follows that

$$G(z, w) = \sum_{m=0}^{\infty} \left\{ \sum_{k=0}^{m} \binom{m}{k} B_k z^{m-k} \right\} \frac{w^m}{m!},$$

and by (20)

$$B_m(z) = \sum_{k=0}^{m} \binom{m}{k} B_k z^{m-k}. \tag{24}$$

Consequently, our functions $B_m(z)$ are seen to be polynomials of degree m. It remains only to determine the numbers B_k defined by (23). From (23) it follows that

$$w = h(w)(e^w - 1) = \sum_{k=0}^{\infty} \frac{B_k}{k!} w^k \sum_{h=1}^{\infty} \frac{w^h}{h!}$$

or

$$w = \left(B_0 + \frac{B_1}{1!} w + \frac{B_2}{2!} w^2 + \cdots\right)\left(w + \frac{w^2}{2!} + \frac{w^3}{3!} + \cdots\right). \tag{25}$$

The product on the right side of (25) must therefore produce a power series in w in which the coefficient of w is equal to 1, whereas all the other coefficients vanish:

$$B_0 = 1, \qquad \sum_{k=0}^{m-1} B_k \binom{m}{k} \frac{1}{m!} = 0 \qquad \text{for} \qquad m > 1. \tag{26}$$

The numbers defined by (26) are called *Bernoulli numbers* (after Jacob Bernoulli, 1654–1705), and the polynomials $B_k(z)$ are *Bernoulli polynomials*.[6] Summing up, we have the following theorem.

Theorem 2. *The ("simplified") summation procedure leads to the special solution* (19), *of the difference equation* (18). *The coefficients (multiplied by k!) of the expansion of $G(z, w)$ in powers of w are the Bernoulli polynomials $B_k(z)$. These polynomials are in themselves solutions of the difference equation* (21).

[6] The definition of *Bernoulli numbers* and *Bernoulli polynomials* in the literature is not uniform. Sometimes the coefficients of the expansion of $\frac{1}{w}[G(z, w) - G(1, w)]$ or other similarly constructed functions, are called *Bernoulli polynomials*.

We are now in a position to give the solution of the difference equation (9) in a number of special cases.

1. If

$$\sum_{h=0}^{\infty} b(z+h) \qquad \text{or} \qquad \sum_{h=0}^{\infty} [b(z+h) - b(1+h)]$$

converges, the "summation procedure" can be applied.

2. If $b(z)$ is a polynomial, the solution of (9) can be built up as a linear combination of Bernoulli polynomials.

For it is easy to see from (21) that

$$u(z) = \sum_{k=1}^{l} \alpha_k \frac{B_k(z)}{k}$$

is a solution of the difference equation

$$u(z+1) - u(z) = \sum_{k=1}^{l} \alpha_k z^{k-1}.$$

For the time being we shall not investigate the general methods of solution any further, but shall turn our attention to the functions, important in many applications, that can be computed by the simple methods discussed up to this point. We begin with some remarks about Bernoulli polynomials.

2.4. *Properties of the Bernoulli Polynomials*

From (24) and (26) we can calculate the Bernoulli numbers and the Bernoulli polynomials. For example,

$$B_1 = -\tfrac{1}{2}, \qquad B_2 = \tfrac{1}{6}, \qquad B_3 = 0,$$
$$B_4 = -\tfrac{1}{30}, \qquad B_5 = 0, \qquad B_6 = \tfrac{1}{42}, \ldots.$$

The first few polynomials are:

$$B_0(x) = 1, \qquad B_1(x) = x - \tfrac{1}{2},$$
$$B_2(x) = x^2 - x + \tfrac{1}{6},$$
$$B_3(x) = x^3 - \tfrac{3}{2}x^2 + \tfrac{1}{2}x,$$
$$B_4(x) = x^4 - 2x^3 + x^2 - \tfrac{1}{30}.$$

The corresponding curves for the interval [0; 1] are given in Figure 3.

In this interval the Bernoulli polynomials have remarkable properties of symmetry. For it follows from (21) that

$$B_m(0) = B_m(1), \qquad m > 1,$$

and for $m \geqq 1$ we have

$$\int_0^1 B_m(t)\,dt = 0. \tag{27}$$

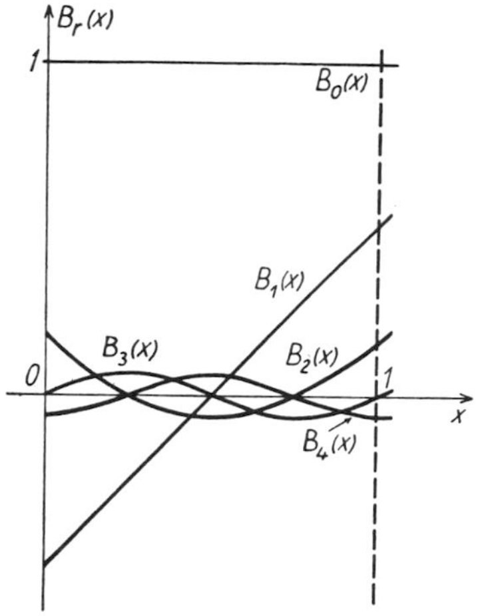

Fig. 3

In order to prove (27) we differentiate (24):

$$B'_m(x) = \sum_{k=0}^{m} (m - k)\binom{m}{k} B_k x^{m-k-1}.$$

Since

$$(m - k)\binom{m}{k} = m\binom{m-1}{k},$$

we thus have

(27′) $$B'_m(x) = mB_{m-1}(x).$$

By integration we obtain

$$B_{m+1}(x + 1) - B_{m+1}(x) = (m + 1)\int_x^{x+1} B_m(t)dt.$$

But from the difference equation (21) it follows that

$$B_{m+1}(x + 1) - B_{m+1}(x) = (m + 1)x^m.$$

The comparison of these two equations gives

$$\int_x^{x+1} B_m(t)dt = x^m.$$

For $x = 0$ the equation (27) follows immediately. Furthermore,

$$\int_0^n B_m(t)dt = \sum_{k=1}^{n-1} k^m = 1^m + 2^m + \cdots + (n - 1)^m.$$

It was this relation that led Bernoulli to investigate the polynomials subsequently named after him.[7]

[7] For this reason we should use the term *Bernoulli polynomials* to refer only to the polynomials defined by (24); cf. footnote 6.

The Bernoulli numbers have the remarkable property that all B_k with odd $k > 1$ are equal to 0, as can be seen, for example, in the following way. For the function $h(w)$ defined by (23) we have

$$h(w) - h(-w) = \frac{w}{e^w - 1} - \frac{w}{1 - e^{-w}} = -w = \sum_{k=0}^{\infty} \frac{B_k}{k!}(w^k - (-w)^k).$$

Thus $B_1 = -\frac{1}{2}$ (as we already knew) and

$$B_3 = B_5 = B_7 = \cdots = 0.$$

2.5. *The Euler Summation Formula*

In order to derive the summation formula named after Euler we define for real x the following function, discontinuous for all integers $x = n$,

$$B_n^*(x) = B_n(x - [x])^8 \tag{28}$$

Thus in every interval $[n; n + 1]$ the function $B_n^*(x)$ has the same values as $B_n(x)$ in the interval $[0; 1]$. (See Figure 4 for $n = 1$.) Consequently, $B_1^{*\prime}(x)$ is equal to 1 at all points where it is continuous.

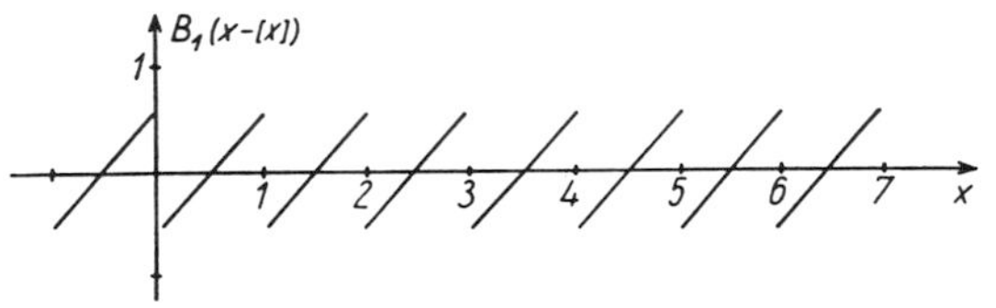

Fig. 4

Now let $f(x)$ be a function which for $x \geqq 0$ is continuous, together with all the derivatives that occur in the following discussion. Then by integration by parts

$$\int_k^{k+1} f(x)dx = \int_k^{k+1} f(x)B_1^{*\prime}(x)dx$$

$$= \tfrac{1}{2}[f(k) + f(k + 1)] - \int_k^{k+1} B_1^*(x)f'(x)dx,$$

and by summation from 0 to m

$$\int_0^m f(x)dx = \tfrac{1}{2}[f(0) + f(m)] + \sum_{k=1}^{m-1} f(k) - \int_0^m B_1^*(x)f'(x)dx.$$

Adding $\frac{1}{2}(f(0) + f(m))$ to both sides of the equation, we obtain

$$\sum_{k=0}^{m} f(k) = \int_0^m f(x)dx + \tfrac{1}{2}(f(0) + f(m)) + \int_0^m B_1^*(x)f'(x)dx. \tag{29}$$

[8] For $n \leqq x < n + 1$ we have $[x] = n$.

Let us apply the *Euler summation formula* (29) to the function

$$f(x) = \frac{1}{1 + x}.$$

Then from (29)

$$(30)\qquad \sum_{k=0}^{m} \frac{1}{1 + k} = \log(1 + m) + \frac{1}{2}\left(1 + \frac{1}{m + 1}\right) - \int_0^m \frac{B_1^*(x)}{(1 + x)^2}\,dx.$$

If we now write $m + 1 = n$, $1 + k = \nu$, we have

$$(31)\qquad \sum_{\nu=1}^{n} \frac{1}{\nu} - \log n = \frac{1}{2}\left(1 + \frac{1}{n}\right) - \int_0^{n-1} \frac{B_1^*(x)}{(1 + x)^2}\,dx.$$

But by the definition of $B_1^*(x)$ it follows that $|B_1^*(x)| \leqq \frac{1}{2}$, so that (for $N > n$);

$$\left|\int_n^N \frac{B_1^*(x)}{(1 + x)^2}\,dx\right| < \frac{1}{2}\int_n^N \frac{dx}{(1 + x)^2} = -\frac{1}{2}\left(\frac{1}{1 + N} - \frac{1}{1 + n}\right).$$

Thus the limit

$$\lim_{n \to \infty} \int_0^{n-1} \frac{B_1^*(x)}{(1 + x)^2}\,dx$$

exists, and (31) implies convergence of the sequence

$$a_n = \sum_{\nu=1}^{n} \frac{1}{\nu} - \log n.$$

The limit

$$(32)\qquad C = \lim_{n \to \infty} \left(\sum_{\nu=1}^{n} \frac{1}{\nu} - \log n\right)$$

is called the *Euler constant*. It is still an open question whether this number is rational or not. Its decimal expansion begins with $C = 0.5772156649\ldots$.

For many purposes we may wish to transform the integral on the right side of (29) by integration by parts. Taking into account the properties of the Bernoulli polynomials derived in §2.4, we obtain

$$(33)\qquad \sum_{k=0}^{m} f(k) = \int_0^m f(x)dx + \tfrac{1}{2}(f(0) + f(m))$$

$$+ \sum_{k=1}^{n} \frac{B_{2k}}{(2k)!}\left(f^{(2k-1)}(m) - f^{(2k-1)}(0)\right)$$

$$+ \frac{mB_{2n-2}}{(2n - 2)!} f^{(2n+2)}(\vartheta m), \qquad 0 < \vartheta < 1.$$

3. The Γ-Function

3.1. *The Product Representation*

The linear homogeneous difference equation

$$(34)\qquad v(z+1) - zv(z) = 0 \qquad \text{or} \qquad v(z)z = v(z+1)$$

is satisfied for positive integers by the function $a(n-1)!$ (for arbitrary a). But in accordance with the program developed in §1.2, we now seek holomorphic solutions of such a difference equation.

Of particular interest for many applications are the holomorphic solutions of (34) which for all positive integers n assume the value $(n-1)!$. In order to solve this "interpolation problem," we reduce (34), by means of logarithmic differentiation, to a difference equation already familiar to us:

$$(35)\qquad \frac{v'(z+1)}{v(z+1)} - \frac{v'(z)}{v(z)} = \frac{1}{z}.$$

Setting $\dfrac{v'(z)}{v(z)} = u(z)$, we have the difference equation (15). In addition to the "fundamental solution"

$$\psi_1(z) = -\sum_{h=0}^{\infty}\left(\frac{1}{z+h} - \frac{1}{1+h}\right),$$

every function $\psi(z) + E(z)$ (with periodic $E(z)$) is also a solution of (15). Since the constants are also periodic functions, we may choose as a special solution:

$$\psi(z) = \psi_1(z) - C.$$

Here C is the *Euler constant* (32) introduced in §2.5. This special choice of the additive constant has the advantage, as will be seen below, that the solution of (34) obtained by integration from

$$(36)\qquad \frac{v'(z)}{v(z)} = \psi(z)$$

has a particularly simple form. For $v(z)$ it is clear that we still have a multiplicative constant at our disposal, since with $v(z)$ every function $bv(z)$ also satisfies the condition (36). We determine this constant by the condition that $v(1) = 1$. Then from (34) it follows that $v(n) = (n-1)!$ for all positive integers n. This special solution of (34), first investigated by Euler, is called the Γ-function. Thus we now write $\Gamma(z)$ instead of $v(z)$.

Integration of (36) between the limits 1 and z gives:

$$\log \Gamma(z) - 0$$
$$= \sum_{h=1}^{\infty}\left(\frac{z}{h} - \frac{1}{h} - \log(z+h) + \log(1+h)\right) - \log z - Cz + C.$$

Let us add and subtract $\log h$ under the summation sign:

$$\log \Gamma(z) = \sum_{h=1}^{\infty} \left[\frac{z}{h} - \log\left(1 + \frac{z}{h}\right)\right] + \sum_{h=1}^{\infty} \left[\log\left(1 + \frac{1}{h}\right) - \frac{1}{h}\right]$$
$$+ C - \log z - Cz.$$

The second sum is now precisely equal to $-C$, since (cf. (32).)

$$\lim_{N\to\infty} \sum_{h=1}^{N} \left[\log\left(1 + \frac{1}{h}\right) - \frac{1}{h}\right] = \lim_{N\to\infty} \sum_{h=1}^{N} \left[\log(1 + h) - \log h - \frac{1}{h}\right]$$
$$= \lim_{N\to\infty} \sum_{h=1}^{N} \left(\log(1 + N) - \frac{1}{h}\right)$$
$$= -\lim_{N\to\infty} \left[\left(\sum_{h=1}^{N} \frac{1}{h} - \log N\right) - \log\frac{N+1}{N}\right]$$
$$= -C.$$

It is now clear why C was taken above for the additive constant. For the function $\Gamma(z)$ we thus have the representation

$$(37) \qquad \Gamma(z) = \frac{1}{z} e^{-Cz} \prod_{h=1}^{\infty} e^{z/h}\left(1 + \frac{z}{h}\right)^{-1}.$$

Obviously $\Gamma(z)$ is a meromorphic function with simple poles at $z = 0, -1, -2, -3, \ldots$, so that for the transcendental entire function $\Gamma(z)^{-1}$ we have the *Weierstrass product representation*

$$(38) \qquad \frac{1}{\Gamma(z)} = ze^{Cz} \prod_{h=1}^{\infty} e^{-z/h}\left(1 + \frac{z}{h}\right).$$

If for C we substitute (32) in this formula, we obtain a well-known formula, due to Gauss, for the Γ-function:

$$\frac{1}{\Gamma(z)} = \lim_{n\to\infty} \left[z \prod_{h=1}^{n} \left(1 + \frac{z}{h}\right) \exp\left(-\sum_{h=1}^{n} \frac{z}{h}\right) \exp\left(+\sum_{h=1}^{n} \frac{z}{h}\right) \exp(-z \log n)\right]$$
$$= \lim_{n\to\infty} \left[\frac{z(z+1)\cdots(z+n)}{1\cdot 2\cdots n} n^{-z}\right]$$

or

$$(39) \qquad \Gamma(z) = \lim_{n\to\infty} \frac{n!n^z}{z(z+1)\cdots(z+n)}.$$

3.2. *Functional Equations*

The following argument leads to an important functional equation for the Euler Γ-function. The product

$$\frac{1}{\Gamma(z)}\frac{1}{\Gamma(-z)}$$

is a transcendental entire function which vanishes for all integers $z = n$. This function will presumably have a close connection with the function $\sin \pi z$, which has the same property. From (38) it follows that

$$\frac{1}{\Gamma(z)}\frac{1}{\Gamma(-z)} = \prod_{\nu=1}^{\infty}\left(1 + \frac{z}{\nu}\right)\left(1 - \frac{z}{\nu}\right)z(-z),$$

and since it is well known that[9]

$$\sin \pi z = \pi z \prod_{\nu=1}^{\infty}\left(1 - \frac{z^2}{\nu^2}\right),$$

we have in fact

$$\text{(40)} \qquad \sin \pi z = \frac{\pi}{(-z)\Gamma(z)\Gamma(-z)}.$$

Since $\Gamma(z)$ is a solution of (34), it follows for the argument $-z$ that

$$(-z)\Gamma(-z) = \Gamma(1 - z).$$

Thus we obtain the *Euler functional equation* for the Γ-function:

$$\text{(40')} \qquad \frac{\pi}{\sin \pi z} = \Gamma(z)\Gamma(1 - z).$$

It follows at once that $\Gamma(\frac{1}{2}) = \sqrt{\pi}$, and by logarithmic differentiation we obtain the following "supplementary theorem" for the function $\psi(z)$. By differentiating

$$\log \pi - \log \sin \pi z = \log \Gamma(z) + \log \Gamma(1 - z)$$

we see from (36) (for $\Gamma(z) = v(z)$) that

$$\text{(41)} \qquad -\pi \operatorname{ctg} \pi z = \psi(z) - \psi(1 - z).$$

Summing up our results on the Γ-function we have the following theorem.[10]

[9] See, e.g., H. Meschkowski [5].

[10] A detailed treatment of the Γ-function is given in Chapter 5 of H. Meschkowski.

Theorem 3. The function $\Gamma(z)$ represented by (37) or (39) is a special solution of the difference equation (34). This function $\Gamma(z)$ is meromorphic with simple poles at $z = 0, -1, -2, \ldots$, and may be regarded as obtained by "interpolation" from the function $\Gamma(n) = (n-1)!$ defined for positive integers n. The Γ-function satisfies the functional equation (40).

3.3. *The Stirling Formula*

We now make use of the Euler summation formula (29) in order to determine the asymptotic behavior of the Γ-function for large values of z (*Stirling formula*). For this purpose we set

$$f(x) = \log(z+x)$$

in (29):

$$\begin{aligned}\sum_{k=0}^{m} \log(z+k) &= \int_0^m \log(z+x)dx + \frac{1}{2}[\log(z+m) + \log z] \\ &\quad + \int_0^m \frac{B_1^*(x)}{z+x}\,dx \\ &= \left(z+m+\frac{1}{2}\right)\log(z+m) - m - \left(z-\frac{1}{2}\right)\log z \\ &\quad + \int_0^m \frac{B_1^*(x)}{z+x}\,dx.\end{aligned}$$

For $z = 1$ we obtain

$$\sum_{k=0}^{m} \log(1+k) = \left(\frac{3}{2}+m\right)\log(1+m) - m + \int_0^m \frac{B_1^*(x)}{1+x}\,dx.$$

Subtracting one of these formulas from the other and adding

$$-(z-1)\log m$$

to both sides we obtain, after a slight simplification,

$$\begin{aligned}&\sum_{k=0}^{m} \log\frac{z+k}{1+k} - (z-1)\log m \\ &\qquad = (z-1)\log\frac{z+m}{m} + \left(m+\frac{3}{2}\right)\log\frac{z+m}{1+m} \\ &\qquad\qquad -\left(z-\frac{1}{2}\right)\log z + \int_0^m \frac{B_1^*(x)}{z+x}\,dx - \int_0^m \frac{B_1^*(x)}{1+x}\,dx\end{aligned}$$

or

$$\log\left[\frac{z(z+1)\cdots(z+m)}{(m+1)!}m^{1-z}\right]$$
$$= (z-1)\log\left(1+\frac{z}{m}\right) + (1+m)\log\left(1+\frac{z-1}{1+m}\right)$$
$$+\frac{1}{2}\log\left(1+\frac{z-1}{1+m}\right) - \left(z-\frac{1}{2}\right)\log z$$
$$+\int_0^m \frac{B_1^*(x)}{z+x}\,dx - \int_0^m \frac{B_1^*(x)}{1+x}\,dx.$$

Noting that

$$\lim_{m\to\infty}\log\left(1+\frac{z}{m}\right) = \lim_{m\to\infty}\log\left(1+\frac{z-1}{1+m}\right) = 0$$

and

$$\lim_{m\to\infty}(1+m)\log\left(1+\frac{z-1}{1+m}\right) = \lim_{m\to\infty}\log\left(1+\frac{z-1}{1+m}\right)^{1+m}$$
$$= \log e^{z-1} = z-1,$$

we thus obtain from (39):

(42) $$\log\Gamma(z) = (z-\tfrac{1}{2})\log z - z + 1 - J(z) + J(1),$$

where

$$J(z) = \int_0^\infty \frac{B_1^*(x)}{z+x}\,dx.$$

The improper integrals $J(z)$ and $J(1)$ converge, as can easily be shown from (28) and (27′) by integration by parts of

$$\int_0^m \frac{B_1^*(x)}{z+x}\,dx \qquad \text{and} \qquad \int_0^m \frac{B_1^*(x)}{1+x}\,dx.$$

We shall omit this proof but shall show by similar methods that

(43) $$\lim_{z\to\infty} J(z) = 0$$

for pure imaginary $z = iy$. From (43) we shall then be able to determine the constant $J(1)$ by a very elegant method. For the proof of (43) we integrate

$$\int_0^n B_1^*(x)\frac{dx}{z+x}$$

by parts and make use of (27′) and (28):

$$\int_0^n \frac{B_1^*(x)}{z+x}\,dx = \left[\frac{1}{2}\frac{B_2^*(x)}{z+x}\right]_0^n + \frac{1}{2}\int_0^n \frac{B_2^*(x)}{(z+x)^2}\,dx.$$

Thus

$$\int_0^\infty \frac{B_1^*(x)}{z+x}\,dx = -\frac{1}{12}\frac{1}{z} + \frac{1}{2}\int_0^\infty \frac{B_2^*(x)}{(z+x)^2}\,dx;$$

for we have $B_2^*(x) = (x - [x])^2 - (x - [x]) + \frac{1}{6}$ and, since $n - [n] = 0$, also $B_2^*(n) = \frac{1}{6}$.

From the definition of $B_2^*(x)$ it is easy to see that $|B_2^*(x)| < |\frac{1}{6}|$ for all x. Thus for pure imaginary $z = ir$ we obtain the following inequality, by substituting $x = r\tau$:

$$\left|\int_0^\infty \frac{B_2^*(x)}{(z+x)^2}\,dx\right| \leqq \frac{1}{6}\int_0^\infty \frac{dx}{|z+x|^2}$$

$$= \frac{1}{6r}\int_0^\infty \frac{d\tau}{|i+\tau|^2} < \frac{1}{6r}\left[\int_0^1 \frac{d\tau}{1} + \int_1^\infty \frac{d\tau}{\tau^2}\right] = \frac{1}{3r}.$$

Then (43) follows at once from $r \to \infty$.

We now proceed to determine the constant $J(1)$, as can be done most simply by taking into account the fact that

$$|\Gamma(ir)|$$

can easily be calculated. For we see at once from (37) that $\Gamma(\bar{z}) = \overline{\Gamma(z)}$ and thus from (40)

$$\Gamma(ir)\Gamma(-ir) = \Gamma(ir)\Gamma(\overline{ir}) = \Gamma(ir)\overline{\Gamma(ir)} = |\Gamma(ir)|^2$$

$$= -\frac{\pi}{ir\sin i\pi r} = \frac{2\pi}{r(e^{\pi r} - e^{-\pi r})}.$$

Consequently

$$(44) \qquad |\Gamma(ir)| = \sqrt{\frac{2\pi}{r(e^{\pi r} - e^{-\pi r})}}.$$

From (42) we now conclude that

$$(45) \qquad \operatorname{Re}\log\Gamma(ir) = \operatorname{Re}(ir - \tfrac{1}{2})\log(ir) + 1 - \operatorname{Re}J(ir) + J(1).$$

But since

$$\operatorname{Re}(ir\log(ir)) = -\operatorname{Im} r\log ir = -\frac{\pi}{2}r; \qquad \operatorname{Re}\log(ir) = \log r,$$

it follows from (45) that

$$-J(1) = -\operatorname{Re}(J(ir)) + 1 + \left(-\frac{\pi}{2}r - \frac{1}{2}\log r - \log|\Gamma(ir)|\right),$$

and from (44) that

$$J(1) = +\operatorname{Re}(J(ir)) - 1 + \log\sqrt{\frac{2\pi r e^{\pi r}}{r(e^{\pi r} - e^{-\pi r})}}.$$

Then by passage to the limit $r \to \infty$ we have from (43):

$$J(1) = -1 + \log \sqrt{2\pi}.$$

Thus we have determined the constant $J(1)$ in (42) and from (42) we obtain the *Stirling formula*

$$\Gamma(z) = z^{z-1/2}e^{-z}\sqrt{2\pi}e^{-J(z)}.$$

From (43)[11] we now have the following theorem.

Theorem 4. *The asymptotic behavior of the function $\Gamma(z)$ for large values of $|z|$ is expressed by the formula*

$$\lim_{z\to\infty} \frac{\Gamma(z)}{z^{z-1/2}e^{-z}\sqrt{2\pi}} = 1.$$

For integers $n > 0$ we thus have

$$(n-1)! \sim n^{n-1/2}e^{-n}\sqrt{2\pi}$$

or, in other words,

$$n! \sim n^n e^{-n}\sqrt{2\pi n}.$$

3.4. *The Euler Integral Representation of the Γ-Function*

For $\operatorname{Re} z > 0$ we can also represent the function $\Gamma(z)$ by an improper integral

$$\Gamma(z) = \int_0^\infty t^{z-1}e^{-t}dt,$$

as was done by Euler. We first show that the limit

$$(46)\qquad \lim_{n\to\infty} \int_{1/n}^{n} t^{z-1}e^{-t}dt = \lim_{n\to\infty} \left[\int_{1/n}^{n} t^{z-1}e^{-t}dt + \int_1^n t^{z-1}e^{-t}dt\right]$$
$$= \lim_{n\to\infty} (J_{1n} + J_{2n})$$

exists if $\operatorname{Re} z > 0$. Let $z = x + iy$, $x > r > 0$. Then $|t^{iy}| = 1$ and we obtain

$$|J_{1n}| \leqq \left|\int_{1/n}^{1} t^{x-1}\cdot 1\cdot e^{-t}dt\right| < \int_{1/n}^{1} t^{x-1}dt = \frac{1}{x}[t^x]_{1/n}^{1} < \frac{1}{r}.$$

Consequently, $\lim J_{1n}$ exists. We now write the second integral in (46) in the form

$$J_{2n} = \int_1^N t^{z-1}e^{-t}dt + \int_N^n t^{z-1}e^{-t}dt.$$

[11] We have proved (43) only for pure imaginary z. But it is easy to show, by a slight modification in the proof, that $\lim J(z) = 0$ holds for arbitrary z: see, e.g., Bieberbach [1], p. 312.

Since, as is well known,

$$t^{x-1}e^{-t/2}$$

converges to 0 (for $t \to \infty$), we have:

$$t^{x-1}e^{-t} < e^{-t/2} \tag{47}$$

for sufficiently large t. If we choose N so large that (47) is satisfied for $t > N$, then

$$\left| \int_N^n t^{z-1}e^{-t}dt \right| \leqq \int_N^n t^{x-1}e^{-t}dt < \int_N^n e^{-t/2}dt < 2e^{-N/2}.$$

Thus the convergence of the second integral J_{2n} is also proved. Now it is easy to show, as follows, that the analytic function defined by the improper integral

$$\Gamma_1(z) = \int_0^\infty t^{z-1}e^{-t}dt \tag{48}$$

for $\operatorname{Re} z > 0$ satisfies the difference equation (34). Integration by parts gives

$$\Gamma_1(z+1) = \int_0^\infty t^z e^{-t}dt = [-e^{-t}t^z]_0^\infty + z\int_0^\infty t^{z-1}e^{-t}dt,$$

$$\Gamma_1(z+1) = 0 + z\Gamma_1(z).$$

Thus

$$\Gamma_1(n+z+1) = \Gamma_1(z)z(z+1)\cdots(z+n),$$

and since $\Gamma_1(1) = 1$ we have

$$\Gamma_1(n+1) = n!.$$

Thus the functions $\Gamma(z)$ and $\Gamma_1(z)$ coincide for $z = n$ ($n > 0$, an integer), and the quotient $\dfrac{\Gamma(z)}{\Gamma_1(z)}$ is a periodic function

$$\frac{\Gamma(z+1)}{\Gamma_1(z+1)} = \frac{z\Gamma(z)}{z\Gamma_1(z)} = E(z).$$

We now show that this function $E(z)$ is identically equal to 1. By the "*identity theorem*" *of the theory of functions of a complex variable*[12] it is sufficient to prove this equality for the interval $0 < x \leqq 1$. For the proof we note that for $0 < x \leqq 1$ and $t \leqq n$ we have

$$t^x \leqq n^x, \qquad t^{x-1} \geqq n^{x-1}, \tag{49a}$$

and for $0 < x \leqq 1$, $n \leqq t$

$$t^x \geqq n^x, \qquad t^{x-1} \leqq n^{x-1}. \tag{49b}$$

[12] See III 6, §1.10.

Thus

$$\text{(50a)}\qquad \Gamma_1(x+n+1) \leqq n^x \int_0^n t^n e^{-t}dt + n^{x-1}\int_n^\infty t^{n+1}e^{-t}dt,$$

$$\text{(50b)}\qquad \Gamma_1(x+n+1) \geqq n^{x-1} \int_0^n t^{n+1} e^{-t}dt + n^{x}\int_n^\infty t^{n}e^{-t}dt.$$

The integrals involving the factor t^{n+1} can be transformed by integration by parts:

$$\int_n^\infty t^{n+1}e^{-t}dt = [-t^{n+1}e^{-t}]_n^\infty + (n+1)\int_n^\infty t^n e^{-t}dt$$

$$= n^{n+1}e^{-n} + (n+1)\int_n^\infty t^n e^{-t}dt.$$

Thus

$$n^{x-1}\int_n^\infty t^{n+1}e^{-t}dt = n^{n+x}e^{-n} + n^{x-1}\int_n^\infty t^n e^{-t}dt + n^x\int_n^\infty t^n e^{-t}dt.$$

If we substitute this result into (50a), inserting the last summand into the first integral in (50a), we obtain

$$\text{(51a)}\quad \Gamma_1(x+n+1) \leqq n^x\int_0^\infty t^n e^{-t}dt + e^{-n}n^{n+x} + n^{x-1}\int_n^\infty t^n e^{-t}dt.$$

Similarly, integration by parts of the first integral in (50b) leads to the inequality

$$\text{(51b)}\quad \Gamma_1(x+n+1) \geqq n^x\int_0^\infty t^n e^{-t}dt - e^{-n}n^{n+x} + n^{x-1}\int_0^n t^n e^{-t}dt.$$

In (51a) and (51b) the factors of n^x are equal to the integral $\Gamma_1(n+1)$ and thus are equal to $n!$; on the other hand, the factors of n^{x-1} are smaller than $n!$, since the integrals do not extend from 0 to ∞. Thus it follows from the two inequalities (51a) and (51b) that

$$\text{(52)}\qquad 1 - \frac{n^n}{e^n n!} < \frac{\Gamma_1(n+x+1)}{n^x n!} < 1 + \frac{1}{n} + \frac{n^n}{e^n n!}.$$

Also, as we shall show just below,

$$\text{(53)}\qquad \lim_{n\to\infty}\frac{n^n}{e^n n!} = 0.$$

Thus from (52) and (53) it follows, in view of (39), that:

$$\lim_{n\to\infty}\frac{\Gamma_1(n+x+1)}{n^x n!} = \lim_{n\to\infty}\frac{\Gamma_1(x)x(x+1)\cdots(x+n)}{n^x n!} = \frac{\Gamma_1(x)}{\Gamma(x)} = 1.$$

So it only remains to prove (53). But this result follows[13] from the power series for e^x:

$$e^x = 1 + x + \frac{x^2}{2!} + \cdots + \frac{x^n}{n!} + \cdots,$$

since then

$$e^n \geqq \frac{n^n}{n!}\left(1 + \frac{n}{n+1} + \frac{n^2}{(n+1)(n+2)} + \cdots\right)$$

or

$$\frac{e^n n!}{n^n} \geqq \varphi(n) = 1 + \frac{n}{n+1} + \frac{n^2}{(n+1)(n+2)} + \cdots.$$

For fixed n the series on the right-hand side is convergent, and on the other hand

$$\lim_{n\to\infty} \varphi(n) = \infty. \tag{54}$$

In order to prove this last relation, let us consider the $(k + 1)$-th summand in the series for $\varphi(n)$:

$$a(k+1, n) = \frac{n^k}{(n+1)\cdots(n+k)}.$$

Since $\lim_{n\to\infty} a(k + 1, n) = 1$, it follows that $a(k + 1, n)$ is greater than $\frac{1}{2}$ for sufficiently large n. But obviously

$$a(\kappa + 1, n) > a(k + 1, n) \qquad \text{for} \qquad \kappa < k.$$

Consequently $a(\kappa + 1, n) > \frac{1}{2}$ for all values of $a(\kappa + 1, n)$. Thus $\varphi(n) > \frac{1}{2}k$ for correspondingly large n. Since k can be chosen arbitrarily large, we have therefore proved (54) and (53). Thus we have the following theorem.

Theorem 5. *For* Re $z > 0$ *the function* $\Gamma(z)$ *can be represented by the improper integral* (48)

$$\Gamma(z) = \int_0^\infty t^{z-1}e^{-t}dt.$$

Our first representations for the function $\Gamma(z)$ were obtained from the "summation" of a difference equation. On the other hand, the Euler integral formula presents itself in "ready-made" form without any recognizable relationship to the above problems. Thus we might be inclined to wonder how, for Re $z > 0$, this formula can represent the function $\Gamma(z)$ so familiar to us from difference equations. In §4 we shall see that the Euler formula can also be subsumed under a more general procedure for the solution of certain types of difference equations.

[13] The equality (53) can also be derived from the Stirling formula (theorem 4).

3.5. *The Hölder Theorem*[14]

The properties of the *Euler Γ-function* have long been familiar to mathematicians. In addition to being a solution of a difference equation, it is involved in many functional relations with other important functions of analysis. But in the 19th century the attention of mathematicians was attracted by the fact that, in contrast to many other well-known functions, no differential equation was known for $\Gamma(z)$. In particular, Hölder spent a great deal of time trying to find such a differential function. After many unsuccessful attempts, he finally came to the conclusion that an algebraic differential equation for the solution of the difference equation (34) *cannot exist* and eventually he was able to prove this conjecture.[15] He first showed that the function which we have called $\psi_1(z)$ cannot satisfy any algebraic differential equation and then used that fact to show that the corresponding statement holds for $\Gamma(z)$. As indicated above in §1.2, the important role played in analysis by the theory of difference equations is largely due to this *Hölder theorem.*

We now proceed to prove a generalization of the first part of the Hölder theorem.

Theorem 6. *The solutions of the difference equation*

$$u(z+1) - u(z) = az^{-k} \qquad (k > 0, \text{ an integer}) \tag{55}$$

do not satisfy any algebraic differential equation.

For let us assume that $u(z)$ satisfies an algebraic differential equation of the form

$$G(z, u(z), u'(z), \ldots, u^{(n-1)}(z)) = 0. \tag{56}$$

In other words, G is a rational entire function, not vanishing identically,

$$G(z, y_1, \ldots, y_n) = \sum_{\nu_1 = \nu_2 = \cdots = 0}^{\nu_1 + \nu_2 + \cdots + \nu_n = m} a_{\nu_1 \cdots \nu_n} y_1^{\nu_1} \cdots y_n^{\nu_n}, \tag{56'}$$

with coefficients that are rational (but not necessarily entire) functions of z. If for y_ρ we substitute $u^{(\rho-1)}(z)$, then (56) is to be satisfied identically. We will show that such a function G cannot exist.

The function $G(z, y_1, y_2, \ldots, y_n)$, as a function of the y_ρ, is of degree m, and by division we can always arrange that (at least) one term of degree m has the coefficient 1.

From (55) it now follows that

$$u^{(\rho)}(z+1) = u^{(\rho)}(z) + a(-1)^\rho \frac{(k+\rho-1)!}{(k-1)!} \frac{1}{z^{k+\rho}}, \tag{57}$$

[14] This section is somewhat more difficult than the rest of the chapter but has been included because of the fundamental importance of the Hölder theorem. At first reading it can be omitted without prejudice to the rest of the chapter.

[15] Math. Annalen 28, 1887, pp. 1–13.

and from (56) and (56′) that

$$(58)\quad G(z+1, u(z+1), \ldots, u^{(n-1)}(z+1)) - G(z, u(z), \ldots, u^{(n-1)}(z)) = 0,$$

since from (56) it also follows that

$$G(z+1, u(z+1), \ldots, u^{(n-1)}(z+1)) = 0.$$

If we substitute (57) into (58), we obtain a differential equation of the form

$$(59)\qquad H(z, u(z), u'(z), \ldots, u^{(n-1)}(z)) = 0.$$

The left side of (59) is a sum whose general term can be written in the form

$$R_\sigma^*(z)(u(z)^{\nu_1}u'(z)^{\nu_2}\cdots u^{(n-1)}(z)^{\nu_n}) = R_\sigma^*(z)P_{\sigma m}^*(z).$$

The $R_\sigma^*(z)$ are rational functions of z. The sum $\nu_1 + \nu_2 + \cdots + \nu_n$ (of course, some of the ν_i may be equal to 0) is called the *dimension* of the term. We now ask whether (59) contains terms of dimension m. Such terms could only arise from terms of dimension m in (56). Let the aggregate of these terms be

$$(60)\qquad P_{1m}(z) + R_2(z)P_{2m}(z) + \cdots + R_{sm}(z)P_{sm}(z).$$

Here the $P_{\sigma m}(z)$ are products of the form

$$(60')\quad P_{\sigma m}(z) = u(z)^{\nu_1}u'(z)^{\nu_2}\cdots u^{(n-1)}(z)^{\nu_n}, \qquad \nu_1 + \nu_2 + \cdots + \nu_n = m,$$

where $P_{1m}(z)$ has the coefficient 1.

If now in $P_{\sigma m}(z+1)$ we replace $u(z+1), u'(z+1), \ldots,$ in accordance with (55) and (57), by

$$u^{(\rho)}(z) + b_\rho \frac{1}{z^{k+\rho}} \left(b_\rho = (-1)^\rho a \frac{(k+\rho-1)!}{(k-1)!}\right),$$

we obtain for $P_{\sigma m}(z+1)$ a product of the form

$$\left(u(z) + \frac{b_0}{z^k}\right)^{\nu_1} \cdots \left(u^{(n-1)}(z) + \frac{b_{n-1}}{z^{k+n-1}}\right)^{\nu_n}.$$

Multiplying out, we obtain *only one* summand of dimension m, which by (60′) is precisely $P_{\sigma m}(z)$. Thus the aggregate of terms of dimension m in (59) is

$$(61)\quad (R_2(z+1) - R_2(z))P_{2m}(z) + \cdots + (R_s(z+1) - R_s(z))P_{sm}(z).$$

But here the number of summands is one smaller than in (60), since

$$R_1(z+1) = R_1(z) = 1.$$

The fundamental idea of the proof is now the following.

This process of "simplification" can be continued. From (59) *we infer the existence of a differential equation of degree smaller than m and in this way finally arrive at a very simple equation, which is immediately seen to be self-contradictory.*

The chief difficulty lies in proving that the algebraic functions obtained in this process of "simplification" have coefficients different from 0.

We first ask, can a factor of $P_{\sigma m}(z)$ in (61) be equal to 0? If $R_\nu(z + 1) = R_\nu(z)$ for every value of z, then

$$R_\nu(z) = R_\nu(z + 1) = R_\nu(z + 2) = \cdots$$

and, since $R_\nu(z)$ is *rational*, it would follow that $R_\nu(z)$ is a constant. Thus if one of the functions $R_\nu(z)$ is not a constant, (59) is not an identity but is an algebraic differential equation of degree m. However, the number of terms of degree m in (59) is smaller by one than in (56). Thus we can proceed until either

a) all coefficients of terms of dimension m are constant, or
b) there is only one nonconstant term in the aggregate of dimension m.

But when we have reached this stage, we can divide by the coefficient of this one term, and then we have only one term of m-th dimension with the constant coefficient 1. This is a special case of a), to which we must now turn our attention. Let

$$G_0(z, u(z), u'(z), \ldots, u^{(n-1)}(z)) = 0 \tag{62}$$

be an equation of degree m in which all the terms of dimension m have constant coefficients. Then

$$\begin{aligned} H_0(z, u(z), \ldots, u^{(n-1)}(z)) = {} & G_0(z + 1, u(z + 1), \ldots, u^{(n-1)}(z + 1)) \\ & - G_0(z, u(z), \ldots, u^{(n-1)}(z)) = 0 \end{aligned} \tag{63}$$

is an equation of degree at most $(m - 1)$. It remains to show that not all the coefficients of (63) can be equal to 0. For this purpose we consider the terms of degree $m - 1$ occurring on the left side of (63).

These terms can arise from terms of dimension $(m - 1)$ in (62) or from the aggregate

$$P_{1m}(z) + c_2 P_{2m}(z) + \cdots + c_s P_{sm}(z), \tag{64}$$

which is taken to represent the terms of dimension m in (62). The coefficients of (62) are constant, and again c_1 is equal to 1. Let us now write $P_{1m}(z)$ in the following form:

$$P_{1m}(z) = (u^{(k_1)}(z))^{\alpha_1}(u^{(k_2)}(z))^{\alpha_2} \cdots (u^{(k_r)}(z))^{\alpha_r}.$$

Here we must have $\alpha_1 + \alpha_2 + \cdots + \alpha_r = m$, and $\alpha_\rho \neq 0$. *One* of the numbers k_r can be equal to 0; in this case let $u^{(0)}(z)$ denote $u(z)$. Then the difference $P_{1m}(z + 1) - P_{1m}(z)$ contains, among others, the term

$$a(-1)^k \frac{(k + k_1 - 1)!}{(k - 1)!} \frac{1}{z^{k+k_1}} (u^{(k_1)}(z))^{\alpha_1 - 1}(u^{(k_2)}(z))^{\alpha_2} \cdots (u^{(k_r)}(z))^{\alpha_r}. \tag{64'}$$

This term is different from 0, but can under certain circumstances be combined with other terms of (63). We first ask, can (64) give rise to a further term of the form

$$R(z)(u^{(k_1)}(z))^{\alpha_1-1}(u^{(k_2)}(z))^{\alpha_2}\cdots(u^{(k_r)}(z))^{\alpha_r}? \tag{65}$$

This is possible only if at least one of the terms $P_{\nu m}(z)$ ($\nu = 2, 3, \ldots$) is of the form

$$P_{\nu m}(z) = u^{(l)}(z)(u^{(k_1)}(z))^{\alpha_1-1}(u^{(k_2)}(z))^{\alpha_2}\cdots(u^{(k_r)}(z))^{\alpha_r}.$$

Here l is one of the numbers $0, 1, 2, \ldots, n$. Obviously, l is different from k_1, since $P_{\nu m}$ is different from P_{1m}. But of course l can be equal to $k_2, k_3, \ldots, k_n$. In the expansion of

$$c_\nu P_{\nu m}(z + 1) - c_\nu P_{\nu m}(z)$$

there occurs, by (57), a term of the form

$$\frac{c}{z^{k+l}}(u^{(k_1)}(z))^{\alpha_1-1}(u^{(k_2)}(z))^{\alpha_2}\cdots(u^{(k_r)}(z))^{\alpha_r}; \tag{65'}$$

for by (57) we certainly have

$$u^{(l)}(z + 1) = u^{(l)}(z) + a(-1)^l\frac{(k + l - 1)!}{(k - 1)!}\frac{1}{z^{k+l}}.$$

It is important that the exponent $k + l$ in (65′) is different from $k + k_1$. A term obtained in this way cannot compensate for the term (64′).

We can now combine into *one* term all the terms, arising from the aggregate (64), which make a contribution of the form (65); this single term will then have the factor $R = \sum_{\mu=0}^{n} A_\mu z^{-\mu-k}$. Since $k_1 \neq l$, at least one of these coefficients must be different from 0, namely, the one corresponding to the denominator $k + k_1$:

$$A_{k_1} = a(-1)^{k_1}\frac{(k + k_1 - 1)!}{(k - 1)!}.$$

We must now investigate whether terms of $(m - 1)$-th dimension in (62) can also make a contribution to the equation (63) of the form (65). This will be the case if and only if one of the terms of dimension $(m - 1)$ in (62) is itself of the form (65), since then we have the contribution

$$(R(z + 1) - R(z))(u^{(k_1)}(z))^{\alpha_1-1}(u^{(k_2)}(z))^{\alpha_2}\cdots(u^{(k_r)}(z))^{\alpha_r}.$$

To sum up: the product (65) occurs in the equation (63) with the coefficient

$$\sum_{\mu=0}^{n}\frac{A_\mu}{z^{\mu+k}} + R(z + 1) - R(z). \tag{67}$$

Here $R(z)$ is a rational function of z. We assert that (67) cannot vanish identically. Since at least the coefficient $A_\mu \neq 0$, the sum in (67) will become infinite for $z = 0$. Thus, if (67) vanished identically, $R(z + 1) - R(z)$ would also become infinite for $z = 0$. But then the function $R(z)$ would itself have poles and, since $R(z)$ is rational, there would be finitely many of them. We now arrange these poles in the order of magnitude of their *real parts*:

$$z_1, z_2, \ldots, z_t, \qquad \operatorname{Re} z_\lambda \leqq \operatorname{Re} z_\mu, \qquad \text{if} \qquad \lambda \leqq \mu.$$

In particular $\operatorname{Re} z_1 \leqq \operatorname{Re} z_t$. The difference $R(z + 1) - R(z)$ would then certainly have poles for $z = z_1 - 1$ and $z = z_t$, and thus for two *distinct* values z. Since the sum in (67) can have only one pole, namely $z = 0$, the whole coefficient (67) cannot vanish identically.

So we have the following result: the "reduction procedure," by means of which we pass from a differential equation of degree m to one of degree $m - 1$, cannot lead to trivial equations, i.e., to equations in which all the coefficients are equal to 0. Thus we can carry out the procedure until we arrive at an equation of degree 1

$$\sum R_\mu(z)u^{(\mu)}(z) + R(z) = 0.$$

Then, exactly as on page 376, we obtain a corresponding equation with constant coefficients,

$$\sum c_\mu u^{(\mu)}(z) + R^*(z) = 0,$$

from which it again follows that

$$\sum c_\mu(u^{(\mu)}(z + 1) - u^{(\mu)}(z)) + R^*(z + 1) - R^*(z) = 0.$$

From (57) we now have a relation of the form

$$\sum \frac{d_\mu}{z^{\mu + k}} + R^*(z + 1) - R^*(z) = 0. \tag{68}$$

Here (68) must be satisfied identically in z. But this is impossible, as can be seen by repeating the arguments that led to (67). So we have completed the proof that a differential equation of the form (56) cannot exist for the solution of (55).

4. Methods of Solution for Linear Difference Equations

4.1. *Homogeneous Difference Equations with Constant Coefficients*

The *Hölder theorem* shows that certain difference equations have solutions which cannot satisfy any algebraic differential equation. However, there exists a far-reaching formal analogy between the methods for solution of many types of linear difference equations and of the corresponding differential equations.

We begin with equations with constant coefficients, with some prefatory remarks on linear difference equations in general (i.e., with variable coefficients as well).

If $u_1(z)$ and $u_2(z)$ are solutions of the homogeneous difference equation

$$a_0(z)u(z+n) + \cdots + a_n(z)u(z) = 0, \tag{69}$$

then

$$\pi_1(z)u_1(z) + \pi_2(z)u_2(z)$$

is also a solution, where $\pi_1(z)$ and $\pi_2(z)$ are arbitrary periodic functions with period 1. In particular, we may choose constants for these periodic functions. The difference between two solutions of the nonhomogeneous equation

$$a_0(z)u(z+n) + \cdots + a_n(z)u(z) = b(z) \tag{69'}$$

is always a solution of the corresponding homogeneous equation.

Thus if we have already found the *general solution* of the homogeneous equation, we obtain the solution of the corresponding nonhomogeneous difference equation by adding a *special solution* of the nonhomogeneous equation to the *general solution* of the homogeneous equation.

Let us now seek solutions for the homogeneous equations with constant coefficients, for which we have already had many examples in §1. It is natural to look for the solution of the difference equation[16]

$$a_0u(z+n) + a_1u(z+n-1) + \cdots + a_nu(z) = 0 \tag{70}$$

in the form

$$u(z) = \alpha^z.$$

We then obtain

$$\alpha^z(a_0\alpha^n + a_1\alpha^{n-1} + \cdots + a_n) = 0.$$

If this equation is to be satisfied identically in z, we must have

$$a_0\alpha^n + a_1\alpha^{n-1} + \cdots + a_n = 0. \tag{71}$$

Equation (71) is called the *characteristic equation* of the difference equation (70). Without loss of generality we may assume that

$$a_0 = 1, \qquad a_n \neq 0. \tag{71'}$$

[16] The difference equation

$$u(z+n) + a_1u(z+n-1) + \cdots + a_{n-\rho}u(z+\rho) = 0 \qquad (a_{n-\rho} \neq 0)$$

can be reduced to the *normal case* by the substitution $v(z) = u(z+\rho)$.

For the time being, let us also assume that all the roots of (71) are *distinct*. Then the functions

$$\alpha_1^z, \alpha_2^z, \ldots, \alpha_n^z$$

are solutions of (70). But every linear combination

$$\pi_1(z)\alpha_1^z + \pi_2(z)\alpha_2^z + \cdots + \pi_n(z)\alpha_n^z \tag{72}$$

with periodic function $\pi(z)$ (having period 1) is also a solution of our difference equation. It is easy to see that (72) represents the *general solution* of (70): for if $u(z)$ is an arbitrary solution of the given difference equation, we have

$$\sum_{\nu=0}^{n} a_\nu u(z + n - \nu) = 0. \tag{70'}$$

But the powers α_N^z are also solutions:

$$\sum_{\nu=0}^{n} a_\nu \alpha_N^{z+n-\nu} = 0, \qquad N = 1, 2, \ldots, n. \tag{70''}$$

The two equations (70') and (70"), taken together, can be interpreted as a homogeneous system of linear equations in the *unknowns* $a_0, a_1, \ldots, a_n$. Since this system has a nontrivial solution, its determinant must vanish for all values of z:

$$D(z) = \begin{vmatrix} u(z+n) & \cdots & u(z) \\ \alpha_1^{z+n} & \cdots & \alpha_1^z \\ \vdots & & \vdots \\ \alpha_n^{z+n} & \cdots & \alpha_n^z \end{vmatrix} = 0.$$

Expansion by the elements of the last column gives

$$u(z)D_0(z) + \alpha_1^z D_1(z) + \cdots + \alpha_n^z D_n(z) = 0. \tag{73}$$

But, by well-known theorems from the theory of determinants, the sub-determinant $D_0(z)$ is different from 0, since

$$D_0(z) = \alpha_1^{z+1}\alpha_2^{z+1}\cdots\alpha_n^{z+1}\begin{vmatrix} \alpha_1^{n-1} & \cdots & 1 \\ \alpha_2^{n-1} & \cdots & 1 \\ \vdots & & \vdots \\ \alpha_n^{n-1} & \cdots & 1 \end{vmatrix}$$

$$= \alpha_1^{z+1}\alpha_2^{z+1}\cdots\alpha_n^{z+1}\prod_{i<j}(\alpha_i - \alpha_j) \neq 0.$$

Thus it follows from (73) that

$$u(z) = -\frac{D_1(z)}{D_0(z)}\alpha_1^z - \cdots - \frac{D_n(z)}{D_0(z)}\alpha_n^z. \tag{74}$$

In this way we have found a representation for $u(z)$ of the form (72). It remains only to prove that the functions

$$\frac{D_1(z)}{D_0(z)}, \frac{D_2(z)}{D_0(z)}, \ldots, \frac{D_n(z)}{D_0(z)}$$

are periodic. For this purpose we first consider

$$D_1(z) = -\begin{vmatrix} u(z+n) & \cdots & u(z+1) \\ \alpha_2^{z+n} & \cdots & \alpha_2^{z+1} \\ \vdots & & \vdots \\ \alpha_n^{z+n} & \cdots & \alpha_n^{z+1} \end{vmatrix}.$$

In view of (70), the functions in the first column of $D_1(z)$ can now be replaced by the following sums:

$$u(z+n) = -[u(z+n-1)a_1 + \cdots + u(z+1)a_{n-1} + u(z)a_n],$$
$$\alpha_N^{z+n} = -[\alpha_N^{z+n-1}a_1 + \cdots + \alpha_N^{z+1}a_{n-1} + \alpha_N^z a_n] \qquad (N = 2, 3, \ldots, n).$$

Omitting the sum of multiples of columns two through n, we obtain

$$D_1(z) = +a_n\begin{vmatrix} u(z) & u(z+n-1) & \cdots & u(z+1) \\ \alpha_2^{z} & \alpha_2^{z+n-1} & \cdots & \alpha_2^{z+1} \\ \vdots & & & \vdots \\ \alpha_n^{z} & \alpha_n^{z+n-1} & & \alpha_n^{z+1} \end{vmatrix}.$$

Rearrangement of columns gives

$$D_1(z+1) = a_n(-1)^n D_1(z). \tag{75}$$

But for $D_0(z+1)$ we obviously have

$$D_0(z+1) = \alpha_1\alpha_2\cdots\alpha_n D_0(z). \tag{76}$$

Since by the *Vieta rule for the roots*

$$\alpha_1\alpha_2\cdots\alpha_n = (-1)^n a_n \qquad \text{(cf. (71))},$$

it follows from (75) and (76) that:

$$\frac{D_1(z+1)}{D_0(z+1)} = \frac{D_1(z)}{D_0(z)}.$$

Analogously, we can prove

$$\frac{D_\nu(z+1)}{D_0(z+1)} = \frac{D_\nu(z)}{D_0(z)} \qquad \text{for} \qquad \nu = 2, 3, \ldots, n.$$

Thus we have shown that $\dfrac{D_\nu(z)}{D_0(z)}$ is periodic. Consequently, the linear combination (72) provides *all* the solutions of (70).

Let us consider an example. The *Fibonacci sequence* (3) in §1.1 has the characteristic equation

$$\alpha^2 - \alpha - 1 = 0.$$

This equation has two distinct real roots:

$$\alpha_1 = \tfrac{1}{2} + \tfrac{1}{2}\sqrt{5}; \qquad \alpha_2 = \tfrac{1}{2} - \tfrac{1}{2}\sqrt{5}.$$

Thus the general solution of (3) is:

$$u(z) = \pi_1(z)\alpha_1^z + \pi_2(z)\alpha_2^z$$

with periodic functions $\pi_1(z)$ and $\pi_2(z)$. In special cases, of course, these periodic functions can also be constants. By suitable choice of A and B in

$$u_n = A\alpha_1^n + B\alpha_2^n,$$

we can satisfy the *initial conditions* $u_1 = u_2 = 1$ in the *Fibonacci problem.* Substituting these conditions, we obtain

$$A = \frac{1}{\sqrt{5}}, \qquad B = -\frac{1}{\sqrt{5}},$$

and thus

$$u_n = \frac{1}{\sqrt{5}}\left[\left(\frac{1+\sqrt{5}}{2}\right)^n - \left(\frac{1-\sqrt{5}}{2}\right)^n\right]$$

is the general expression for the *Fibonacci numbers.*

By means of this formula we can prove, among other results, the following interesting theorem of Lamé, which we state here without proof:

Let a and b be natural numbers with a < b. Then the number of divisions necessary for determining the greatest common factor of a and b by the Euclidean algorithm is at most five times as great as the number of digits in the expression for a in the decimal system.

If (71) has multiple roots, the general solution of (70) cannot be represented in the form (72), since then we do not have n distinct roots α_ν. But in this case also the general solution of (70) is a sum of multiples

$$\pi_1(z)\varphi_1(z) + \pi_2(z)\varphi_2(z) + \cdots + \pi_n(z)\varphi_n(z).$$

Here again the functions $\varphi_\nu(z)$ will be powers α_ν^z if the roots α_ν are simple. But if α_1 is a $(k + 1)$-fold root, then corresponding to this root we have $k+1$ solutions $\varphi_1(z), \ldots, \varphi_{k+1}(z)$, namely,

$$\alpha_1^z,\ z\alpha_1^z,\ z^2\alpha_1^z,\ \ldots,\ z^k\alpha_1^z. \tag{75}$$

We now show that under the given assumptions the functions (75) constitute all the solutions. If α_1 is a $(k + 1)$-fold root of (71), then for

$\alpha = \alpha_1$ the derivatives of (71) up to the k-th order also vanish:

$$(71')\qquad \begin{aligned} &\sum_{\nu=0}^{n} a_\nu(n-\nu)\alpha_1^{n-\nu-1} = 0,\\ &\qquad\vdots\\ &\sum_{\nu=0}^{n} a_\nu(n-\nu)(n-\nu-1)\cdots(n-\nu-k+1)\alpha_1^{n-\nu-k} = 0.\end{aligned}$$[17]

For the following computation we now note the identity

$$(76)\qquad \alpha^{k+1} = \alpha(\alpha-1)\cdots(\alpha-k) + \mathfrak{P}_k(\alpha).$$

Here $\mathfrak{P}_k(\alpha)$ is a polynomial of degree k in α, and (76) is verified at once by multiplying out the product

$$\alpha(\alpha-1)\cdots(\alpha-k).$$

For our purposes we do not need to compute $\mathfrak{P}_k(\alpha)$; it is sufficient that the degree of this polynomial is smaller than $k+1$. After these preliminary remarks we substitute

$$u(z) = \alpha_1^z z^l \qquad (l \leqq k)$$

into (70), obtaining:

$$(77)\qquad \begin{aligned} \sum_{\nu=0}^{n} a_\nu \alpha_1^{z+n-\nu}(z+(n-\nu))^l &= \alpha_1^z\left\{\sum_{\nu=0}^{n}\sum_{\lambda=0}^{l} a_\nu\alpha_1^{n-\nu}z^{l-\lambda}\binom{l}{\lambda}(n-\nu)^\lambda\right\}\\ &= \alpha_1^z\left\{\sum_{\lambda=0}^{l} z^{l-\lambda}\binom{l}{\lambda}\gamma_\lambda\right\},\end{aligned}$$

where

$$(78)\qquad \gamma_\lambda = \sum_{\nu=0}^{n} a_\nu\alpha_1^{n-\nu}(n-\nu)^\lambda.$$

We now show that all the coefficients γ_λ vanish. In the first place, $\gamma_0 = 0$, since α_1 is a root of (70).

We now proceed by induction. Let

$$\gamma_0 = \gamma_1 = \cdots = \gamma_\lambda = 0; \qquad \lambda < k.$$

We assert that then also $\gamma_{\lambda+1} = 0$. For by (76) we have

$$(n-\nu)^{\lambda+1} = (n-\nu)(n-\nu-1)\cdots(n-\nu-\lambda) + \mathfrak{P}_\lambda(n-\nu),$$

[17] For formal reasons we write all the sums in (71′) with upper limit n. Summands for which the exponent of α_1 becomes negative are equal to zero.

so that

$$(79)\quad \gamma_{\lambda+1} = \sum_{\nu=0}^{n} a_\nu \alpha_1^{n-\nu}(n-\nu)^{\lambda+1}$$

$$= \sum_{\nu=0}^{n} a_\nu \alpha_1^{n-\nu}(n-\nu)\cdots(n-\nu-\lambda) + \sum_{\nu=0}^{n} a_\nu \alpha_1^{n-\nu}\mathfrak{P}_\lambda(n-\nu).$$

The second sum in (79) vanishes by the induction assumption (since $\mathfrak{P}_\lambda$ is at most of degree λ), and for the first sum we can write

$$\sum_{\nu=0}^{n} a_\nu \alpha_1^{n-\nu}(n-\nu)\cdots(n-\nu-\lambda)$$

$$= \alpha_1^{\lambda+1} \sum_{\nu=0}^{n} a_\nu \alpha_1^{n-\nu+\lambda-1}(n-\nu)\cdots(n-\nu-\lambda).$$

But this sum vanishes, since for $\alpha = \alpha_1$ the $(\lambda + 1)$-th derivative of (71) is also equal to 0. Thus it is clear that all the coefficients γ_λ in (77) vanish, which means that $\alpha_1^z z^l$ is a solution of (70).

Up to now we have confined our attention to homogeneous equations. But the nonhomogeneous linear difference equations (with constant coefficients) are easily reduced to this case. For if in

$$(80)\qquad u(z+n) + a_1 u(z+n-1) + \cdots + a_n u(z) = b$$

we replace z by $z + 1$, we obtain

$$u(z+n+1) + a_1 u(z+n) + \cdots + a_n u(z+1) = b.$$

Subtraction of one of these equations from the other gives a homogeneous difference equation of $(n + 1)$-th order. Thus the solutions of (80) are to be sought among the solutions of this equation.

Let us now sum up our results about the homogeneous equations (with constant coefficients).

Theorem 7. *The general solution of the difference equation* (70) *is*

$$(81)\qquad u(z) = \sum_{\nu=1}^{n} \pi_\nu(z)\varphi_\nu(z).$$

Here the functions $\pi_\nu(z)$ are arbitrary periodic functions with period 1, *and the functions $\varphi_\nu(z)$ correspond to the roots of the characteristic equation* (70) *in the following way: to the simple root α_N there corresponds a function α_N^z, and to the multiple root $\alpha_M = \alpha_{M+1} = \cdots = \alpha_{M+\nu}$ correspond the functions*

$$\alpha_M^z,\ z\alpha_M^z,\ \ldots,\ z^\nu \alpha_M^z.$$[18]

[18] We have not given a complete proof of theorem 7. The fact that every solution of (70) can be represented in the form (81) was proved above only for the case of simple roots.

4.2. *Application of the Laplace Transform*

Let us now turn to linear homogeneous differential equations with linear coefficients. In order to solve

$$(82)\quad (a_0z + b_0)u(z+n) + (a_1z + b_1)u(z+n-1) + \cdots + (a_nz + b_n)u(z) = 0$$

let us introduce the *Laplace transform*

$$(83)\qquad u(z) = \int_p^q t^{z-1}f(t)dt,$$

which is also useful in many problems involving differential equations. Here p and q are numbers still at our disposal. Substituting (83) into (82) we have

$$(84)\qquad \int_p^q t^{z-1}f(t)\varphi(t)dt + \int_p^q zt^{z-1}f(t)\psi(t)dt = 0.$$

Here

$$\psi(t) = a_0t^n + a_1t^{n-1} + \cdots + a_n,$$

$$(85)\qquad \varphi(t) = b_0t^n + b_1t^{n-1} + \cdots + b_n.$$

By integrating the second integral by parts we obtain from (84)

$$(86)\qquad \int_p^q t^{z-1}\left[f(t)\varphi(t) - t\,\frac{df((t)\cdot\psi(t))}{dt}\right]dt + [t^zf(t)\psi(t)]_p^q = 0.$$

The limits of the integrals in (83) are now to be so chosen that

$$(87a)\qquad [t^zf(t)\psi(t)]_p^q = 0.$$

Then (86) is certainly satisfied if we arrange that the integrand in (86) vanishes

$$(87b)\qquad f(t)\varphi(t) = t\,\frac{d(f(t)\psi(t))}{dt}.$$

But this is a simple linear differential equation for the unknown function $f(t)$, which we can easily solve. We have

$$\frac{f'(t)}{f(t)} = \frac{\varphi(t) - t\psi'(t)}{t\psi(t)},$$

from which it follows at once that

$$(88)\qquad f(t) = \frac{1}{\psi(t)}\exp\left(\int_{t_0}^t \frac{\varphi(\tau)}{\tau\psi(\tau)}\,d\tau\right).$$

In order to calculate $f(t)$ from (88) we decompose $\dfrac{\varphi(t)}{t\psi(t)}$ into partial fractions. We shall give the result here only for the case that all the roots $\alpha_1, \ldots, \alpha_n$ of the polynomial $t\psi(t)$ defined by (85) are simple.

Then

$$t\psi(t) = t(t - \alpha_1)\cdots(t - \alpha_n),$$

and for $\dfrac{\varphi(t)}{t\psi(t)}$ we find a decomposition into partial fractions of the form

$$\frac{\varphi(t)}{t\psi(t)} = \frac{\beta_0}{t} + \frac{\beta_1}{t - \alpha_1} + \cdots + \frac{\beta_n}{t - \alpha_n} + P(t).$$

Here $P(t)$ is a polynomial in t that vanishes if the degree of $\varphi(t)$ is smaller than that of $t\psi(t)$. From (88) it now follows that

$$\text{(89)} \qquad f(t) = t^{\beta_0}(t - \alpha_1)^{\beta_1 - 1}\cdots(t - \alpha_n)^{\beta_n - 1}\cdot \mathrm{c} \qquad (\mathrm{c} = \text{const}).$$

Setting (89) in (83), we obtain a solution of the difference equation (82). It remains only to see how the condition (87a) can be satisfied.

We set (89) in (87a) and obtain

$$\text{(87c)} \qquad [K] \equiv [t^{z+\beta_0}(t - \alpha_1)^{\beta_1}\cdots(t - \alpha_n)^{\beta_n}]_p^q = 0.$$

If we now assume that all the real parts of β_ν ($\nu = 1, 2, \ldots, n$) are > 0, and that $\operatorname{Re} z > 0$, then K in (87c) has zeros for

$$t = 0; \qquad t = \alpha_1, \qquad t = \alpha_2, \ldots, \qquad t = \alpha_n.$$

Then we can choose $p = 0$ and for the upper limit q set

$$q = \alpha_1, \qquad q = \alpha_2, \ldots, \qquad q = \alpha_n.$$

In this way we obtain, for (83) and (89), n distinct solutions for the difference equation

$$\text{(90)} \quad u_\nu(z) = \int_0^{\alpha_\nu} t^{z+\beta_0-1}(t - \alpha_1)^{\beta_1 - 1}\cdots(t - \alpha_n)^{\beta_n - 1}dt, \quad \nu = 1, 2, \ldots, n.$$

On the other hand, if $\operatorname{Re} \beta_N < 0$ for some N, the factor $(t - \alpha_N)^{\beta_N}$ in K is singular. For integral $\beta_N(\beta_N = -m)$ the function $(t - \alpha_N)^{-m}$ has a pole at $t = \alpha_N$, and otherwise has a branch point there. Then we cannot simply choose α_N as the upper limit for the integral, since (87c) would not be satisfied. But in this case also, which we do not discuss here in detail, we obtain a solution of our difference equation if for the path of integration we choose a closed curve around the singularity α_N. Let us carry out the solution for another special case. Consider the difference equation

$$u(z + 1) - zu(z) = 0,$$

which is a special case of (82). From (85) we now have

$$\psi(t) = -1; \qquad \varphi(t) = t; \qquad \frac{\varphi(t)}{t\psi(t)} = -1.$$

Thus

$$f(t) = -e^{-t}, \qquad K = t^z e^{-t},$$

and the condition (87a) becomes:

$$[t^z e^{-t}]_p^q = 0.$$

Since

$$\lim_{t\to\infty} t^z e^{-t} = 0,$$

we may choose 0 and ∞ as the limits here, and then for the solution we obtain the integral representation (see above) of the Γ-function:

$$u(z) = \int_0^\infty t^{z-1}e^{-t}dt = \Gamma(z).$$

This discussion makes it clear that, like the product representation, the *Euler integral formula*, introduced in §3.4 without "motivation," can be obtained by a general procedure for certain types of difference equations.

4.3. *Existence Theorems for Linear Difference Equations*

In our study of linear difference equations with linear coefficients we arrived at a result which can be regarded as a generalization of theorem 7. Just as for constant coefficients, we have obtained n distinct solutions $u_\nu(z)$, $\nu = 1, \ldots, n$, (for $\operatorname{Re}\beta_\nu > 0$). It is clear that every linear combination

$$\pi_1(z)u_1(z) + \cdots + \pi_n(z)u_n(z)$$

with periodic functions $\pi_\nu(z)$ is also a solution of (82). Analogous statements hold in the case $\operatorname{Re}\beta_\nu < 0$ and are also valid for the general linear difference equation (69). In the present chapter we cannot give a proof for such existence theorems but we will state at least a few of the most important results concerning general linear difference equations.

For this purpose we must introduce a generalization, particularly adapted to the theory of difference equations, of the concept of the *linear independence* of functions. In general, n functions $\varphi_\nu(z)$ ($\nu = 1, \ldots, n$) are said to be linearly independent if from a relation

$$\sum_{\nu=1}^{n} c_\nu\varphi_\nu(z) = 0$$

with constants c_ν it follows that

$$c_1 = c_2 = \cdots = c_n = 0.$$

But for difference equations the periodic functions (with period 1) play a role similar to the role of constants in other parts of analysis. Thus it is desirable here to define the concept of *linear independence* in the following way.

A system of m solutions $\varphi_\mu(z)$ of the difference equation (69) with analytic coefficients is said to be linearly independent if there do not exist m periodic functions $\pi_\mu(z)$ for which

$$\sum_{\mu=1}^{m} \pi_\mu(z)\varphi_\mu(z)$$

vanishes identically. For the periodic functions $\pi_\mu(z)$ we do not even assume continuity. It is sufficient to require that for at least one value of z, not congruent to a singularity of (69),[19] these functions remain finite and do not all vanish.

Then we have the following theorem.

Theorem 8. *For the difference equation of n-th order* (69) *there exists a "fundamental system" of n linearly independent solutions* $\varphi_\nu(z)$, $(\nu = 1, 2, \ldots, n)$. *Every solution of* (69) *can be written in the form*

$$u(z) = \sum_{\nu=1}^{n} \varphi_\nu(z)\pi_\nu(z)$$

with periodic functions $\pi_\nu(z)$.

Finally (again without proof) we give a criterion for *linear independence* in the sense of our new definition.

Theorem 9. *The solutions* $\varphi_\nu(z)$, $(\nu = 1, 2, \ldots, n)$ *of* (69)[20] *are linearly independent if and only if the determinant*

$$D(z) = \begin{vmatrix} \varphi_1(z) & \cdots & \varphi_n(z) \\ \varphi_1(z+1) & \cdots & \varphi_n(z+1) \\ \vdots & & \vdots \\ \varphi_1(z+n-1) & \cdots & \varphi_n(z+n-1) \end{vmatrix}$$

is different from 0 *for all points a not congruent to singularities of* (69).

This determinant, a "counterpart" of the *Wronskian determinant* in the theory of linear differential equations, is already familiar to us in a special case: the determinant $D_0(z)$ in §4.1 corresponds (up to a constant factor) to the determinant $D(z)$ in theorem 9 for the "fundamental system" $\alpha_1^z, \alpha_2^z, \ldots, \alpha_n^z$.

Bibliography

[1] BIEBERBACH, L.: Funktionentheorie, Vol. 1, Teubner, Leipzig 1934.

[2] BLEICH and MELAN: Die gewöhnlichen und partiellen Differenzengleichungen der Baustatik, Springer, Berlin 1927.

[19] The *singularities* of (69) are the values of z for which the coefficients $a_\nu(z)$ become singular, or else $a_0(z)$ or $a_n(z)$ vanishes. Here a is said to be congruent to b if the difference $a - b$ is an integer.

[20] For the solution of the nonhomogeneous equation (69′) see §4.1.

[3] BÖHMER, P. E.: Differenzengleichungen und bestimmte Integrale. K. F. Koehler, Leipzig 1939.
[4] GEL'FOND, A. O.: Difference Calculation. Gittl, Moscow and Leningrad 1952 (in Russian).
French translation: Calcul des différences finies, Coll. Univ. de Math. XII, Dunod, Paris 1963.
[5] MESCHKOWSKI, H.: Differenzengleichungen. Vandenhoeck & Ruprecht, Göttingen 1959.
[6] MILNE-THOMSON, L. M.: The calculus of finite differences. Macmillan, London 1951.
[7] NÖRLUND, N. E.: Differenzenrechnung. Springer, Berlin 1924.
[8] SCHWANK, F.: Randwertprobleme. Teubner, Leipzig 1951.
[9] ZYPKIN, J. S.: Differenzengleichungen der Impuls- und Regeltechnik. Verlag Technik, Berlin 1956.

On the subject of difference equations, see the list of significant recent journal articles in the book by MESCHKOWSKI [5].

Bibliography added in translation

HILDEBRAND, F. B.: Introduction to numerical analysis. McGraw-Hill, New York 1956.
LEVY, H. and LESSMAN, F.: Finite difference equations. Macmillan Co., New York 1961.
MILLER, K. S.: An Introduction to the calculus of finite differences and difference equations. Dover Publications, New York 1966.
PINNEY, E.: Ordinary difference-differential equations. University of California Press, Berkeley 1958.
RICHTMYER, R. D.: Difference methods for initial-value problems. Interscience Publishers, New York 1957.

CHAPTER 11

Functional Analysis

A. Introduction

The subject of analysis deals with variables and with functions of variables, where the nature of the variables is quite at will. Usually they are variable numbers; for example, the real numbers on an interval of the real line, or the complex numbers in a given domain of the complex plane.

But there is an extensive field of mathematics, namely, *functional analysis*, in which the functions do not depend on individual numbers, but on the *entire range of a numerical function* or on a *vector* with several components; in short, on an "argument" which already has a complicated mathematical structure, though here it must be considered as a single whole, as a "point," so to speak, of the "space" in question.

We have already had many examples. An ordinary definite integral

$$\int_a^b y(s)ds, \qquad \int_{(\mathfrak{B})} \cdots \int y(t_1, \ldots, t_m)dt_1 \cdots dt_m$$

denotes a number whose value is determined by the entire range of the function y and changes if the range is changed. More generally, the situation is the same for an expression of the form

$$\int_a^b \int_a^b k(s, t)y(s)y(t)dsdt, \qquad \int_a^b \int_a^b \Phi(s, t, y(s), y(t))dsdt,$$

or again for the angle between two vectors $\mathfrak{a}$ and $\mathfrak{b}$ in n-dimensional space, which is defined by

$$\cos \varphi = \frac{\sum a_\alpha b_\alpha}{\sqrt{\sum a_\alpha^2 \sum b_\alpha^2}}$$

and thus depends on all the components of both vectors. The last example, a function of two vectors or points $\mathfrak{a}$, $\mathfrak{b}$, is not very different from an ordinary function of $2n$ variables, but from an abstract point of view the distinction is clear.

The essential feature of the situation is that there must first of all be defined a *fundamental domain*, over which the variables range. For example, this domain may be the totality of all continuous functions in the interval (0, 1), of the functions that are square-integrable in the *sense of Lebesgue*, or of all vectors or points in a certain space. In general, we speak of "elements" of the fundamental domain.

A correspondence which to each element of a given fundamental domain assigns a real or complex number is called a functional over the given fundamental domain, the latter being called the domain of definition of the functional.

A well-known special case occurs in the classical *calculus of variations*, in the problem, for example, of finding a function $y(s)$ that gives an extreme value to an integral of the form

$$\int_a^b F(s, y, y', \ldots, y^{(n)})ds.$$

Here the fundamental domain is the totality of all sufficiently differentiable functions y satisfying certain boundary conditions and the problem is to find an extreme value for the functional. Consequently, the calculus of variations is one of the oldest fields of functional analysis; we shall return to it below.

But instead of assigning *numbers* to the elements of the fundamental domain, it may happen that to these elements the correspondence assigns elements of the same domain or of a different one.

A correspondence in which to each element of a fundamental domain there is assigned an element of the same or another domain is called an operator over the given fundamental domain, which in turn is called the domain of definition of the operator.

As an illustration, let us consider the equation

$$z(s) = \int_a^b F(s, t, y(t))dt,$$

which to every function $y(t)$ assigns a function $z(s)$. The extent of the fundamental domain depends on the nature of the given function F; for example, if F is continuous in all its variables, then for $y(t)$ we may choose an arbitrary function continuous in $a \leqq t \leqq b$.

In view of the great number of possibilities here it is essential to inquire about the simplest functionals and operators. In the domain of functions of one or more ordinary variables, the simplest ones are certainly the linear functions, with easily stated laws and properties. Consequently, we now define and investigate (Part B) the concepts of *linear functional* and *linear operator*. Since it will be necessary to restrict the domains of definition, in order not to lose ourselves in excessive generality, we will also be involved in the study of suitable fundamental domains, with emphasis on *Hilbert* and *Banach spaces*.

For nonlinear functions in the ordinary sense an especially important role is played by the *derivative*, which becomes trivial in the linear domain. Thus it will be our task (Part C) to extend this concept in some meaningful way to functionals and operators; in particular, we shall see that the familiar *Euler equation* for the calculus of variations is very closely connected with the concept of a derivative. We shall present some of the consequences and applications of the new concepts, especially for integral equations and integral transforms, both linear and nonlinear. In the nonlinear case the so-called fixed-point theorems will be particularly important.

B. Linear Theory

I. Linear Functionals and Operators

Functionals will be denoted by Latin lower-case letters and operators by German capitals. Thus

$$f = f(y), \qquad \mathfrak{A} = \mathfrak{A}(y), \qquad y \in \mathfrak{B}$$

are a functional and an operator respectively, with the domain of definition $\mathfrak{B}$.[1] In order to arrive at the concept of a *linear* functional or operator, we first require that $\mathfrak{B}$ be a linear domain; i.e., with every two elements y_1, y_2 in $\mathfrak{B}$, every linear combination $c_1 y_1 + c_2 y_2$ with arbitrary scalars c_1, c_2 also belongs to $\mathfrak{B}$, where multiplication and addition obey the usual rules. The coefficients are real or complex numbers, depending on the individual case. We further require that

$$f(c_1 y_1 + c_2 y_2) = c_1 f(y_1) + c_2 f(y_2),$$
$$\mathfrak{A}(c_1 y_1 + c_2 y_2) = c_1 \mathfrak{A}(y_1) + c_2 \mathfrak{A}(y_2).$$

If these conditions are satisfied, we speak of a *linear functional or operator*.

For example, if $\mathfrak{B}$ is the real *n-dimensional Euclidean space* $\mathfrak{R}^n$, and its elements are the points (or equivalently, the vectors) of this space, then every element $y \in \mathfrak{B}$ can be written in the form

$$y = a_1 e_1 + \cdots + a_n e_n,$$

where $e_1, \ldots, e_n$ is a system of n mutually orthogonal basis elements of the space and $a_1, \ldots, a_n$ are the scalar components of the given element with respect to the basis. A linear *functional* f can obviously be written as

$$f(y) = a_1 f(e_1) + \cdots + a_n f(e_n),$$

and can thus be regarded as the scalar product of a vector y with the fixed vector $f(e_1)e_1 + \cdots + f(e_n)e_n$. If the scalars are complex numbers, we must

[1] Use of the same notation for operators and domains will not give rise to confusion.

adopt the usual definition of the scalar product as $\overline{f(e_1)}e_1 + \cdots + \overline{f(e_n)}e_n$. Conversely, it is obvious that every scalar product of a variable and a fixed vector is a linear functional of the variable vector.

As for linear *operators* in $\mathfrak{R}^n$, we assume that for every vector y in $\mathfrak{R}^n$ the symbol $\mathfrak{A}(y)$ again denotes a vector, say in a space $\mathfrak{R}^m$. If we let the latter space be represented by an orthogonal basis $g_1, \ldots, g_m$, we have, in particular,

$$\mathfrak{A}(e_1) = a_{11}g_1 + \cdots + a_{m1}g_m$$
$$\cdots\cdots\cdots\cdots$$
$$\mathfrak{A}(e_n) = a_{1n}g_1 + \cdots + a_{mn}g_m,$$

and, for the linear operator $\mathfrak{A}(y)$,

$$\mathfrak{A}(y) = (a_{11}a_1 + \cdots + a_{1n}a_n)g_1 + \cdots + (a_{m1}a_1 + \cdots + a_{mn}a_n)g_m.$$

Thus the components $b_1, \ldots, b_m$ of $\mathfrak{A}(y)$ are

$$b_\alpha = \sum_{\beta=1}^{n} a_{\alpha\beta}a_\beta \qquad (\alpha = 1, \ldots, m),$$

so that every linear operator is characterized by a linear transformation with the matrix $(a_{\alpha\beta})$. Conversely, it is clear that every such linear transformation defines a linear operator. The most important case is the transformation of a space $\mathfrak{R}^n$ onto itself ($m = n$).

From these simple remarks it is already clear that the characterization of a linear functional or operator depends in an essential way on the domain of definition admissible in each case. In the next section we take up the problem of deciding which domains of definition are particularly advantageous for given purposes.

2. Hilbert Space and Banach Space

Instead of immediately defining *Hilbert space* by setting up a suitable, purely abstract system of axioms, we shall begin by showing how we are led to such a space, almost inevitably, by the very concept of a linear functional.

Hilbert's original idea was to replace the n-dimensional Euclidean space $\mathfrak{R}^n$ by a space of the same sort with a *countable infinity of dimensions*, preserving as far as possible the character of a Euclidean space. Thus we begin with countably many unit vectors $e_1, e_2, \ldots$, assumed to be linearly independent, and describe an arbitrary vector in terms of the corresponding complex scalar components $a_1, a_2, \ldots$; in other words, we write

$$y = \sum_{\alpha=1}^{\infty} a_\alpha e_\alpha,$$

without paying any attention, at first, to the question of convergence. Thus every element is defined by the sequence of its components $a_1, a_2, \ldots$, for which reason we also speak of a *Hilbert sequence space*. For the time being we shall denote this space by our general symbol for a domain, namely $\mathfrak{B}$.

Now if $f(y)$ is a linear functional in $\mathfrak{B}$ and we form the "segment-elements" $\eta_n = \sum_{\alpha=1}^{n} a_\alpha e_\alpha$ of y, we have

$$f(\eta_n) = \sum_{\alpha=1}^{n} a_\alpha f(e_\alpha),$$

and the most obvious way of defining $f(y)$ itself is to set

$$f(y) = \lim_{n\to\infty} \sum_{\alpha=1}^{n} a_\alpha f(e_\alpha) = \sum_{\alpha=1}^{\infty} a_\alpha f(e_\alpha).$$

But here of course the sum must converge, and if the domain of definition is to be the whole space $\mathfrak{B}$, this convergence must hold for all admissible systems $a_1, a_2, \ldots$. Thus we must investigate the linear form $\sum_{\alpha=1}^{\infty} a_\alpha c_\alpha$ with $c_\alpha = f(e_\alpha)$, in order to determine the necessary restrictions on the coefficients c_α and the variables a_α.

The most natural answer to this still somewhat vaguely worded question is given by a well-known formula, namely, the *Schwarz inequality*

$$\left|\sum_{\alpha=1}^{m} a_\alpha c_\alpha\right| \leqq \sum_{\alpha=1}^{m} |a_\alpha|\,|c_\alpha| \leqq \sqrt{\sum_{\alpha=1}^{m} |a_\alpha|^2 \sum_{\alpha=1}^{m} |c_\alpha|^2},$$

which shows that if the a_α and also the c_α are required to be absolutely square-summable (i.e., such that the sum of their squares is absolutely convergent), then $\sum_{\alpha=1}^{\infty} a_\alpha c_\alpha$ will also be convergent in every case. Thus it is natural to admit only those sequences that are absolutely square-summable, and to make the same requirement for the coefficients of the linear form $c_\alpha = f(e_\alpha)$. This decision is based on the following important theorem, stated here without proof.

If $\sum_{\alpha=1}^{\infty} a_\alpha c_\alpha$ converges for every absolutely square-summable system a_α, then $\sum_{\alpha=1}^{\infty} |c_\alpha|^2 = M^2$ is also convergent, and in every case

$$\left|\sum_{\alpha=1}^{\infty} a_\alpha c_\alpha\right| \leqq M\sqrt{\sum_{\alpha=1}^{\infty} |a_\alpha|^2}.$$

Thus if we require convergence of $\sum_{\alpha=1}^{\infty} |a_\alpha|^2$, *every linear functional in the domain of definition $\mathfrak{B}$ can be represented in the form $\sum_{\alpha=1}^{\infty} a_\alpha f(e_\alpha)$, where $\sum_{\alpha=1}^{\infty} |f(e_\alpha)|^2$ is convergent.*

It is customary to call $\|y\| = \sqrt{\sum_{\alpha=1}^{\infty} |a_\alpha|^2}$ the *norm of y*. Then for every linear functional defined in $\mathfrak{B}$ there exists a positive constant M such that

$$|f(y)| \leqq M\|y\|.$$

Such a functional is said to be bounded, so that we have the following theorem.

Every linear functional with the domain of definition $\mathfrak{B}$ *is bounded.*

It is also natural to make the following definition of *continuity*. We say that a functional is continuous at a point y if $\|h\| \to 0$ implies

$$|f(y + h) - f(y)| \to 0.$$

Of course, for a linear functional this means that

$$|f(h)| \to 0, \qquad f(h) \to 0.$$

Then the Schwarz inequality shows at once that *every linear functional is continuous in* $\mathfrak{B}$, *for every* y *in* $\mathfrak{B}$.

If for two elements y and z with the components a_α and b_α we define the *scalar product* by $(y, z) = \sum_{\alpha=1}^{\infty} a_\alpha \bar{b}_\alpha$, every linear functional can be represented as the scalar product of a fixed element with a variable element.

A sequence of elements $y_1, y_2, \ldots$ is said to be *strongly convergent* to an element y, in symbols $y_n \to y$, if

$$\lim_{n\to\infty} \|y_n - y\| = 0.$$

The concept of *weak convergence* is also important. A sequence $y_1, y_2, \ldots$ is said to weakly convergent to y, in symbols $y_n \rightharpoonup y$, if for every linear functional $f(y)$ in $\mathfrak{B}$ the sequence $f(y_n)$ is convergent to $f(y)$ in the ordinary sense. It follows at once from the Schwarz inequality that every strongly convergent sequence is also weakly convergent but, as we shall see below, the converse does not hold.

Every strongly convergent sequence $y_1, y_2, \ldots$ satisfies the *Cauchy condition*:

$$\|y_n - y_m\| < \delta \qquad (\delta > 0 \text{ arbitrary}),$$

for all m and n greater than a suitably chosen $N(\delta)$. In this sense the sequence is said to be a *Cauchy sequence*. But conversely, every Cauchy sequence is also strongly convergent, i.e., there exists a limit element in $\mathfrak{B}$, as can be proved as follows.

In the first place, it is easy to prove that the scalar product is continuous, namely, that $x_n \to x$, $y_m \to y$ implies $(x_n, x_m) \to (x, y)$, so that, in particular, the norm is continuous: $x_n \to x$ implies $\|x_n\| \to \|x\|$. Now if x_n is a Cauchy sequence, then for every r we have $(x_n - x_m, e_r) \to 0$ as $n, m \to \infty$. Thus the limit $\lim_{n\to\infty} (x_n, e_r) = a_r$ exists for every r. Furthermore,

$$|(x_n, x_n) - (x_m, x_m)| \leqq (\|x_n\| + \|x_m\|)\|x_n - x_m\|,$$

and $\|x_n\|$ is bounded for every n. So by the Cauchy convergence criterion, $\|x_n\|^2 \to q \geqq 0$, as desired.

But now for $x_n = x_n = \sum_{r=1}^{\infty} a_r^{(n)} e_r$ we have

$$q - \sum_{r=1}^{k} |a_r|^2 = q - \lim_{n\to\infty} \sum_{r=1}^{k} |(x_n, e_r)|^2 = q - \lim_{n\to\infty} \sum_{r=1}^{k} |a_r^{(n)}|^2$$

$$= \lim_{n\to\infty} \sum_{r=1}^{\infty} |a_r^{(n)}|^2 - \lim_{n\to\infty} \sum_{r=1}^{k} |a_r^{(n)}|^2 = \lim_{n\to\infty} \sum_{r=k+1}^{\infty} |a_r^{(n)}|^2 \geqq 0,$$

and thus $\sum_{r=1}^{k} |a_r|^2 \leqq q$, which implies the desired convergence of $\sum_{r=1}^{\infty} |a_r|^2$. For this reason the Hilbert sequence space is said to be *complete*.

Abstractly formulated, we have hereby defined a space called the *Hilbert space* $\mathfrak{H}$, which can be characterized in the following way.

H 1. *$\mathfrak{H}$ is a linear space, i.e., the ordinary linear vector laws hold with respect to addition and to multiplication by scalar factors; these scalars, denoted by a, b, c, . . . , are complex, or often real, numbers.*

H 2. *For every two elements y and z there is defined a scalar product, denoted by (y, z), which obeys the following laws:*

$$(y, z) = \overline{(z, y)},$$
$$(cy, z) = c(y, z); \qquad (y_1 + y_2, z) = (y_1, z) + (y_2, z).$$

H 3. *The expression $\|y\| = \sqrt{(y, y)}$, called the norm of y, is in every case $\geqq 0$. From $\|y\| = 0$ it follows that $y = 0$, and conversely. The Schwarz inequality holds:*

$$|(y, z)| \leqq \|y\| \cdot \|z\|.$$

H 4. *The dimension is countable; i.e., there exists a countable set of elements $e_1, e_2, \ldots$, whose linear combinations $a_1, e_1, + \ldots, + a_n e_n$ approximate every element y with arbitrary accuracy, i.e., so that*

$$\|y - a_1 e_1 - \cdots - a_n e_n\| < \delta$$

for arbitrary $\delta > 0$ and sufficiently large n.

H 5. *The space is complete; i.e., every Cauchy sequence of elements $y_1, y_2, \ldots$ (for which $\|y_n - y_m\| \to 0$ with $n, m \to \infty$) is strongly convergent to an element belonging to the space.*

On the basis of these axioms, all of which are satisfied by the Hilbert sequence space, it can be shown that every *linear functional* in $\mathfrak{H}$ is representable as the scalar product (y, y_0) of a variable y with a fixed element y_0 and that such a functional is continuous and bounded in the sense of the norm. Consequently, these linear functionals, on the basis of their correspondence with the elements of the Hilbert space $\mathfrak{H}$, themselves form a Hilbert space, denoted by $\mathfrak{H}^*$ and called the *adjoint space* of $\mathfrak{H}$.

Two elements y and z in $\mathfrak{H}$ are said to be *orthogonal* if $(y, z) = 0$. In $\mathfrak{H}$ there exist *orthonormal systems* $\varphi_1, \varphi_2, \ldots$, namely, systems such that

$$(\varphi_\alpha, \varphi_\beta) = \delta_{\alpha\beta} \qquad (\alpha, \beta = 1, 2, \ldots);$$

in other words, the vectors are pairwise orthogonal and are also "*normalized*," i.e., $\|\varphi_\alpha\| = 1$. In the sequence space, for example, the unit vectors $e_1, e_2, \ldots$ with components

$$\begin{matrix} 1, 0, 0, \ldots \\ 0, 1, 0, \ldots \\ 0, 0, 1, \ldots \\ . \quad . \quad . \quad . \end{matrix}$$

form an orthonormal system. It can be proved that every orthonormal system contains at most countably many elements. If for an arbitrary element y we form the *Fourier coefficients* $a_\alpha = (y, \varphi_\alpha)$ with respect to the orthonormal system $\varphi_1, \varphi_2, \ldots$, we find the relation

$$\left\| y - \sum_{\alpha=1}^{m} a_\alpha \varphi_\alpha \right\|^2 = \|y\|^2 - \sum_{\alpha=1}^{m} |a_\alpha|^2.$$

It follows that $\sum_{\alpha=1}^{\infty} |a_\alpha|^2$ converges and at the same time satisfies the *Bessel inequality*

$$\sum_{\alpha=1}^{\infty} |a_\alpha|^2 \leqq \|y\|^2.$$

If the sign of equality holds in every case, we obtain the *completeness relation*

$$\sum_{\alpha=1}^{\infty} |a_\alpha|^2 = \|y\|^2.$$

Thus an orthonormal system is said to be *complete*[2] if this equation holds for every $y \in \mathfrak{H}$ and, in the sense of strong convergence,

$$\sum_{\alpha=1}^{m} a_\alpha \varphi_\alpha \rightarrow y = \sum_{\alpha=1}^{\infty} a_\alpha \varphi_\alpha.$$

Every complete orthonormal system can serve as a basis for the space $\mathfrak{H}$, and the element y is then uniquely determined by the sequence of its Fourier coefficients; conversely, for every sequence whose partial sums are absolutely square-summable, there exists a corresponding element y.

[2] The same word is used in two different senses: as a property of spaces and as a property of orthonormal systems.

More generally, it follows from the completeness relation that, for every pair of elements y and z,

$$(y, z) = \sum_{\alpha=1}^{\infty} (y, \varphi_\alpha)(\varphi_\alpha, z).$$

From the Bessel inequality it further follows that $a_\alpha = (y, \varphi_\alpha)$ approaches zero as $\alpha \to \infty$. Since this statement holds for every y in $\mathfrak{H}$, the sequence $\varphi_1, \varphi_2, \ldots$ converges weakly to zero. However, it does not converge strongly to zero, since then we would have $\|\varphi_\alpha\| \to 0$, whereas in fact $\|\varphi_\alpha\| = 1$. Thus there exist weakly convergent sequences that are not strongly convergent.

A particularly important realization of a *Hilbert space* in the domain of functions of one or more variables is provided by the functions which, together with their squares, are Lebesgue-integrable in a fixed domain (we shall say that such functions are square-integrable), where the functions may be either real-valued or complex-valued. By the scalar product of two functions we mean the integral

$$\int f(P)\overline{g(P)}d\tau_p,$$

with the norm $\sqrt{\int |f(P)|^2 d\tau_p}$. Two functions whose values differ only on a set of measure zero determine the same element in the Hilbert space and have the same Fourier coefficients with respect to a complete orthonormal system. It is easy to show that all the axioms for a Hilbert space are satisfied in the present case and that the resulting Hilbert space is isomorphic to the sequence space of Fourier coefficients with respect to any complete orthonormal system.[3] If we confine ourselves to real-valued functions, the domain of scalars for the corresponding Hilbert space is restricted to the totality of real constants.

Next to the Hilbert space the most important concept in functional analysis is that of a *Banach space*. This concept is more general, since now we require only the existence of a norm for the elements, but not of a scalar product. The system of axioms is as follows:

B 1. *A Banach space $\mathfrak{B}$ is a linear space.*

B 2. *For every element $y \in \mathfrak{B}$, there is defined a norm which is $\geqq 0$, in symbols $\|y\| \geqq 0$. From $\|y\| = 0$ it follows that $y = 0$, and conversely. Furthermore,*

$$\|cy\| = |c| \cdot \|y\|, \qquad \|y + z\| \leqq \|y\| + \|z\|.$$

B 3. *The space is complete.*

[3] This statement is valid for any Hilbert space, since the existence of a complete orthonormal system follows from the axioms.

The definitions of the various other concepts proceed in precisely the same way as for a Hilbert space. Every Hilbert space is a Banach space, since the norm in a Hilbert space is easily seen to satisfy axiom B2.

An important realization of a *Banach space* is provided by the totality of all functions $f(t)$ continuous in a closed interval, with $\|f\| = \max |f(t)|$. Strong convergence in $\mathfrak{B}$, defined by means of the norm in the same way as in any Banach space, is seen to be equivalent to uniform convergence in the given interval.

For every Banach space $\mathfrak{B}$, the bounded linear functionals $f(y)$, in other words, those functionals for which

$$|f(y)| \leqq M\|y\|,$$

form a Banach space with the norm

$$\|f\| = \inf M.$$

This space is called the *adjoint* (*or conjugate, or dual*) *Banach space* $\mathfrak{B}^*$.

In the foregoing example of a Banach space of continuous functions, the linear functionals are represented by the Stieltjes integrals $\int y d\varphi$ (cf. III 3, §1.6), where φ is a function of bounded variation in the given interval.[4]

Another important example of a Banach space is given by the totality of functions L-integrable up to their p-th powers ($p > 1$) in an interval $a \leqq t \leqq b$. In view of the fact that the integral

$$\int_a^b |y(t)|^p dt$$

exists, we may set $\|y\| = \sqrt[p]{\int_a^b |y(t)|^p dt}$, since this definition of a norm is easily seen to satisfy all the required conditions. The *adjoint* of this space L^p is L^q, with $\frac{1}{p} + \frac{1}{q} = 1$. The linear functionals in L^p are defined by

$$\int_a^b y(t)z(t)dt,$$

with y in L^p and z in L^q. For $p = 2$ we obtain the Hilbert space L^2, and for the convergence theorem on page 395 there is a corresponding generalization.[5]

3. Linear Operators in $\mathfrak{H}$

After this short discussion of linear *functionals* we now turn to linear *operators*, which for convenience we shall take to be defined on a Hilbert space. For the definition of a linear operator, see §1.

[4] Cf. [4], pp. 106–110.

[5] Cf., E. Landau, Über einen Konvergenzsatz. Göttinger Nachrichten (1907), pp. 25–27. See also [2].

Thus let $z = \mathfrak{A}(y)$ define a linear operator, where z is in a second Hilbert space $\mathfrak{H}^*$,[6] which may also be identical with $\mathfrak{H}$. For example, with certain requirements on $K(s, t)$, the integral

$$z(s) = \int_a^b K(s, t)y(t)dt \qquad (c \leqq s \leqq d)$$

defines a linear integral operator with the kernel $K(s, t)$, where $\mathfrak{H}$ is the space $L^2(a, b)$, and $\mathfrak{H}^*$ is $L^2(c, d)$.

Since $\mathfrak{H}^*$ has a scalar product, we may form the product $(A(y), u)$ for each u in $\mathfrak{H}^*$. In particular, if u runs through the elements $\psi_1, \psi_2, \ldots$ of a complete orthonormal system in $\mathfrak{H}^*$, we obtain (setting $\mathfrak{A}(y) = \mathfrak{A}y$):

$$(\mathfrak{A}y, \psi_\alpha) = (z, \psi_\alpha).$$

But for every α the scalar product on the left-hand side is now a linear functional in $\mathfrak{H}$. Thus by §2 there exists an element $h_\alpha \in \mathfrak{H}$ such that

$$(\mathfrak{A}y, \psi_\alpha) = (y, h_\alpha),$$

which, by the general form of the completeness relation, may be written

$$(y, h_\alpha) = \sum_{\beta=1}^{\infty} (y, \varphi_\beta)(\varphi_\beta, h_\alpha),$$

where $\varphi_1, \varphi_2, \ldots$ is a complete orthonormal system in $\mathfrak{H}$. If we also set

$$(y, \varphi_\beta) = y_\beta, \qquad (z, \psi_\alpha) = z_\alpha \qquad (\alpha, \beta = 1, 2, \ldots),$$

our operator equation $z = \mathfrak{A}y$ becomes the system of equations

$$z_\alpha = \sum_{\beta=1}^{\infty} (\varphi_\beta, h_\alpha)y_\beta = \sum_{\beta=1}^{\infty} (\mathfrak{A}\varphi_\beta, \psi_\alpha)y_\beta,$$

since $(\varphi_\beta, h_\alpha) = (\mathfrak{A}\varphi_\beta, \psi_\alpha)$.

This system of equations enables us to calculate the Fourier coefficients of z, for the system $\psi_1, \psi_2, \ldots$, in terms of the Fourier coefficients of the given element y for $\varphi_1, \varphi_2, \ldots$. Since every element of a Hilbert space is determined by its Fourier coefficients with respect to a complete orthonormal system, the above system of equations can be regarded as equivalent to the given operator equation. Consequently, just as in the n-dimensional space $\mathfrak{R}^n$, *an operator defined everywhere in* $\mathfrak{H}$ *can be characterized, in terms of two complete orthonormal systems, by a matrix*

$$(a_{\alpha\beta}) = (\mathfrak{A}\varphi_\beta, \psi_\alpha) \qquad (\alpha, \beta = 1, 2, \ldots)$$

called the kernel matrix.

[6] This space has nothing to do with the space $\mathfrak{H}^*$ of linear functionals in $\mathfrak{H}$ defined above.

Of course, in contrast to the case for $\mathfrak{R}^n$, this matrix has *infinitely many rows and columns.* But it also has other important properties, to which we now turn.

As before, for an *arbitrary* element u in $\mathfrak{H}^*$ we can find an $h \in \mathfrak{H}$ such that

$$(\mathfrak{A}y, u) = (y, h) = \sum_{\beta=1}^{\infty} (y, \varphi_\beta)(\varphi_\beta, h)$$

$$= \sum_{\beta=1}^{\infty} y_\beta \left[\sum_{\alpha=1}^{\infty} (\mathfrak{A}\varphi_\beta, \psi_\alpha)(\psi_\alpha, u)\right] = \sum_{\beta=1}^{\infty} y_\beta \left(\sum_{\alpha=1}^{\infty} a_{\alpha\beta}\bar{u}_\alpha\right).$$

We obtain a bilinear form in the y_β and $\bar{u}_\alpha = (\psi_\alpha, u)$, with coefficients determined by the kernel matrix. This bilinear form will be convergent if $\sum_{\beta=1}^{\infty} |y_\beta|^2$ and $\sum_{\alpha=1}^{\infty} |u_\alpha|^2$ are convergent. For such a bilinear form we have the following theorems:[7]

1. *The value of a bilinear form under "column summation"*

$$\sum_{\beta=1}^{\infty} y_\beta \left(\sum_{\alpha=1}^{\infty} a_{\alpha\beta}\bar{u}_\alpha\right)$$

is equal to its value under "row summation"

$$\sum_{\alpha=1}^{\infty} \bar{u}_\alpha \left(\sum_{\beta=1}^{\infty} a_{\alpha\beta} y_\beta\right)$$

and also equal to its value under "segment summation"

$$\lim_{n\to\infty} \sum_{\alpha,\beta=1}^{n} a_{\alpha\beta}\bar{u}_\alpha y_\beta.$$

2. *There exists a positive constant M such that for all n*

$$\left|\sum_{\alpha,\beta=1}^{n} a_{\alpha\beta}\bar{u}_\alpha y_\beta\right| \leqq M \sqrt{\sum_{\alpha=1}^{n} |u_\alpha|^2 \sum_{\beta=1}^{n} |y_\beta|^2}.$$

The bilinear form is then said to be bounded.

It follows at once that

$$\|\mathfrak{A}y\| \leqq M\|y\|,$$

so that the operator $\mathfrak{A}$ is also said to be *bounded.* Such an operator is obviously *continuous,* since it is immediately clear that

$$\|\mathfrak{A}y\| \to 0 \quad \text{for} \quad \|y\| \to 0.$$

[7] Cf. O. Toeplitz, Grundlagen einer Theorie der unendlichen Matrizen, Math. Annalen 69.

A linear operator defined everywhere in $\mathfrak{H}$ is necessarily continuous and bounded.

For every $u \in \mathfrak{H}^*$, the equation

$$(\mathfrak{A}y, u) = (y, h)$$

also defines an element $h - h(u)$ in $\mathfrak{H}$, and this element is unique. For if h_1 were a second element of the same kind, we would have $(y, h - h_1) = 0$ for all $y \in \mathfrak{H}$, so that in particular

$$(h - h_1, h - h_1) = 0, \qquad h - h_1 = 0, \qquad h = h_1$$

(cf. §2, axiom H3). The operator thus defined is linear, as can be seen at once, and is defined everywhere in $\mathfrak{H}^*$. We set

$$h = \mathfrak{A}^*u$$

and call $\mathfrak{A}^*$ the *adjoint operator* of $\mathfrak{A}$. From

$$(\mathfrak{A}y, u) = (y, h) = (y, \mathfrak{A}^*u)$$

it follows that the adjoint operator of $\mathfrak{A}^*$ is again $\mathfrak{A}$. The two operators have the same bound

$$\inf M = \|\mathfrak{A}\| = \|\mathfrak{A}^*\|.$$

As an example, let us consider the *Fourier transform*

$$z(s) = \frac{1}{\sqrt{2\pi}} \int_{-\infty}^{\infty} e^{ist} y(t)dt = \mathfrak{A}y.$$

Here the adjoint operator is

$$\mathfrak{A}^*u = \frac{1}{\sqrt{2\pi}} \int_{-\infty}^{\infty} e^{-ist} u(s)ds.$$

More generally, for

$$\mathfrak{A}y = \int_a^b K(s, t)y(t)dt \qquad (c \leqq s \leqq d),$$

the adjoint operator is

$$\mathfrak{A}^*u = \int_c^d \overline{K(s, t)}u(s)ds.$$

As an example for the calculation of the elements of the *kernel matrix* let us again consider the Fourier transform, where for the orthonormal system, in s and in t, we take the Hermite functions

$$\varphi_\alpha(s) = \frac{(-1)^\alpha}{\sqrt{2^\alpha \cdot \alpha! \sqrt{\pi}}} e^{s^2/2} \frac{d^\alpha e^{-s^2}}{ds^\alpha} \qquad (\alpha = 0, 1, 2, \ldots).$$

The result is

$$a_{\alpha\beta} = \frac{1}{\sqrt{2\pi}} \int_{-\infty}^{\infty} \int_{-\infty}^{\infty} e^{ist} \varphi_\beta(t) \varphi_\alpha(s) dt ds = \delta_{\alpha\beta} i^\alpha$$

$$= \begin{pmatrix} 1 & 0 & 0 & 0 & \cdots \\ 0 & i & 0 & 0 & \cdots \\ 0 & 0 & -1 & 0 & \cdots \\ 0 & 0 & 0 & -i & \cdots \\ \cdot & \cdot & \cdot & \cdot & \cdot \end{pmatrix}.$$

So in this case the system of equations is not only easy to set up, but also easy to solve; we see that

$$z(s) = \frac{1}{\sqrt{2\pi}} \int_{-\infty}^{\infty} e^{ist} y(t) dt$$

leads conversely to

$$y(t) = \frac{1}{\sqrt{2\pi}} \int_{-\infty}^{\infty} e^{-ist} z(s) ds,$$

since the solution follows at once by interchange of i and $-i$. It is also obvious that the kernel matrix is bounded, with the bound 1. Further properties of the Fourier transform are given by the *Parseval equation*

$$\int_{-\infty}^{\infty} |z(s)|^2 ds = \int_{-\infty}^{\infty} |y(t)|^2 dt$$

and the *convolution theorem*[8]

$$z(s) \cdot z_1(s) = \frac{1}{2\pi} \int_{-\infty}^{\infty} e^{ist} \left[\int_{-\infty}^{\infty} y(t - x) y_1(x) dx \right] dt,$$

where

$$z(s) = \frac{1}{\sqrt{2\pi}} \int_{-\infty}^{\infty} e^{ist} y(t) dt \quad \text{and} \quad z_1(s) = \frac{1}{\sqrt{2\pi}} \int_{-\infty}^{\infty} e^{ist} y_1(t) dt.$$

Applications of the Fourier transform in various branches of analysis are extraordinarily numerous.

As a further example, important both in theory and for its applications, let us mention the *Laplace transform*

$$z(s) = \int_0^{\infty} e^{-st} y(t) dt,$$

[8] The integral $\int_{-\infty}^{\infty} y(t - x) y_1(x) dx = \int_{-\infty}^{\infty} y_1(t - x) y(x) dx$ is called the "convolution" of $y(x)$ and $y_1(x)$.

which is closely related, as can be seen at once, to the Fourier transform. The solution of this equation is generally expressed by means of the *Riemann-Mellin inversion formula*

$$y(t) = \frac{1}{2\pi}\int_{\eta-i\infty}^{\eta+i\infty} e^{st}z(s)ds,$$

where the convergence requires a special investigation. There is a voluminous literature on this topic, as a result of the great importance of the *Laplace transform.*[9] From the formulas

$$sz(s) - y(0) = \int_0^\infty e^{-st}y'(t)dt,$$

$$\frac{z(s)}{s} = \int_0^\infty e^{-st}\left(\int_0^t y(\tau)d\tau\right)dt$$

it follows that differentiation and integration of the function $y(t)$ correspond to simpler, algebraic operations on $z(s)$, a fact which explains the wide range of applications of the transformation. Thus for the differential equation

$$y''(t) + \omega^2 y(t) = f(t)$$

multiplication with e^{-st} and integration from 0 to ∞ with respect to t leads to the algebraic equation

$$s^2z(s) - sy(0) - y'(0) + \omega^2 z(s) = F(s) = \int_0^\infty e^{-st}f(t)dt$$

for the corresponding $z(s)$, so that

$$z(s) = \frac{F(s) + sy(0) + y'(0)}{s^2 + \omega^2}.$$

The summands on the right-hand side are easily recognizable Laplace transforms, so that we have the solution

$$y(t) = \frac{1}{w}\int_0^t \sin\omega(t-\tau)f(\tau)d\tau + y(0)\cos\omega t + \frac{y'(0)}{\omega}\sin\omega t$$

of the given differential equation.

The example of the Laplace transform clearly shows that our choice of the Hilbert space of square-integrable functions as the domain of definition of an operator is not always well-adapted to the requirements of the problems that arise, since, e.g., the square-integrable functions are not all differentiable. However, let us again turn our attention to this space, since

[9] Cf. G. Doetsch, Einführung in Theorie und Anwendung der Laplace-Transformation. Birkhauser, Basel, Stuttgart 1958.

the various possibilities in it, in particular for solving the above (kernel-matrix) system of equations, are especially simple and clear-cut.

For the given operator equation

$$z = \mathfrak{A}y,$$

we have the system of equations, constructed by means of the kernel matrix,

$$z_\alpha = \sum_{\beta=1}^{\infty} a_{\alpha\beta} y_\beta \qquad A = (a_{\alpha\beta}),$$

whose solution we now seek in the form of a system

$$y_\beta = \sum_{\gamma=1}^{\infty} r_{\beta\gamma} z_\gamma \qquad R = (r_{\beta\gamma})$$

such that for all z_α:

$$z_\alpha = \sum_{\beta=1}^{\infty} a_{\alpha\beta}\left(\sum_{\gamma=1}^{\infty} r_{\beta\gamma} z_\gamma\right).$$

It is clear that we are dealing here with multiplication of infinite matrices (cf. the corresponding discussion of finite matrices in IB3, §2.2) and that, in particular, we must have

$$AR = E = \delta_{\beta\gamma}.$$

In the first place, we can easily show that for two bounded matrices A, B the product matrix

$$AB = \left(\sum_{\beta=1}^{\infty} a_{\alpha\beta} b_{\beta\gamma}\right) = (c_{\alpha\gamma}) = C$$

always exists, since the relevant sums are all convergent. Moreover, this matrix is again bounded, with the bound $\|\mathfrak{A}\| \cdot \|\mathfrak{B}\|$, where $\|\mathfrak{A}\|$ is a bound for $\mathfrak{A}$, and $\|\mathfrak{B}\|$ is a bound for $\mathfrak{B}$. Also, for three bounded matrices A, B, C we have the associative law:

$$A(BC) = (AB)C,$$

so that the whole formal calculus of matrices (IB3, §2.2) can be transferred from finite matrices to bounded infinite matrices.

Our problem now requires the determination of a "right reciprocal" R with $AR = E$. Here we have the following four possibilities:

I. A has a unique bounded reciprocal R, which is both a left and a right reciprocal, so that

$$AR = RA = E.$$

II. A has no left reciprocal but has infinitely many bounded right reciprocals $AR = E$.

III. A has no right reciprocal, but has infinitely many bounded left reciprocals $RA = E$.

IV. A has neither a left nor a right reciprocal.

An example of the first possibility is provided by the Fourier transform.

As an example for Case II, consider the integral transform

$$z(s) = \frac{1}{\pi}\int_0^\pi \frac{\sin s}{\cos t - \cos s}\, y(t)dt \qquad (0 \leqq s \leqq \pi),$$

where the integral is taken in the sense of the Cauchy principal value.[10] Using the two orthonormal systems

$$\varphi_\alpha(s) = \sqrt{\frac{2}{\pi}}\sin \alpha s \qquad (\alpha = 1, 2, \ldots),$$

$$\psi_0(t) = \frac{1}{\sqrt{\pi}}, \qquad \psi_\beta(t) = \sqrt{\frac{2}{\pi}}\cos \beta t \qquad (\beta = 1, 2, \ldots),$$

we find from the formula

$$\frac{1}{\pi}\int_0^\pi \frac{\cos nt}{\cos t - \cos s}\, dt = \frac{\sin ns}{\sin s} \qquad (n = 0, 1, 2, \ldots)$$

that the kernel matrix is

$$(a_{\alpha\beta}) = \begin{pmatrix} 0 & 1 & 0 & 0 & \cdots \\ 0 & 0 & 1 & 0 & \cdots \\ 0 & 0 & 0 & 1 & \cdots \\ . & . & . & . & . \end{pmatrix},$$

which proves the desired assertion.

As an example for Case III we take

$$z(s) = \frac{1}{\pi}\int_0^\pi \frac{\sin t}{\cos t - \cos s}\, y(t)dt \qquad (0 \leqq s \leqq \pi)$$

with the same orthonormal systems as before, except for interchange of s and t, and find the kernel matrix

$$(a_{\alpha\beta}) = \begin{pmatrix} 0 & 0 & 0 & 0 & \cdots \\ -1 & 0 & 0 & 0 & \cdots \\ 0 & -1 & 0 & 0 & \cdots \\ 0 & 0 & -1 & 0 & \cdots \\ . & . & . & . & . \end{pmatrix}.$$

[10] $z(s) = \lim_{\epsilon \to 0} \frac{1}{\pi}\left(\int_0^{s-\epsilon} + \int_{s+\epsilon}^{\pi}\right).$

A kernel matrix of the fourth type is

$$(a_{\alpha\beta}) = \begin{pmatrix} 1 & 0 & 0 & 0 & \cdots \\ 0 & \frac{1}{4} & 0 & 0 & \cdots \\ 0 & 0 & \frac{1}{9} & 0 & \cdots \\ 0 & 0 & 0 & \frac{1}{16} & \cdots \\ \cdot & \cdot & \cdot & \cdot & \cdot \end{pmatrix},$$

with a left and right reciprocal which is unique but not bounded. Below we shall find this reciprocal as the kernel matrix of an integral transform with continuous kernel between finite limits.

4. Symmetric and Completely Continuous Linear Operators

In the present section we consider the Hilbert space $\mathfrak{H}$ with real scalars and a linear operator $\mathfrak{A}y$ for which

$$\mathfrak{A} = \mathfrak{A}^*.$$

Such an operator is said to be *self-adjoint*. An example is the integral operator

$$z(s) = \int_a^b K(s, t)y(t)dt \qquad (a \leqq s \leqq b)$$

with $K(s, t) = K(t, s)$,[11] where $K(s, t)$ is real and continuous in (s, t), and the domain $a \leqq s, t \leqq b$ is finite.

We now examine the double integral

$$\mu = \int_a^b \int_a^b K(s, t)y(s)y(t)dsdt,$$

regarding it as a (nonlinear) functional of y, where y is assumed to be "normalized" by the side condition

$$\|y\| = \sqrt{\int_a^b y^2(s)ds} = 1 \qquad (y(s) \text{ real}).$$

Since by the Schwarz inequality for double integrals[12] the values of this functional are bounded by

$$|\mu| = \left|\int_a^b \int_a^b K(s, t)y(s)y(t)dsdt\right| \leqq \sqrt{\int_a^b \int_a^b K^2(s, t)dsdt \int_a^b y^2(s)ds}$$

$$= \sqrt{\int_a^b \int_a^b K^2(s, t)dsdt},$$

[11] From this equation it follows that $\mathfrak{A}$ is self-adjoint.

[12] $|\iint f(s, t)g(s, t)dsdt|^2 \leqq \iint f^2(s, t)dsdt \iint g^2(s, t)dsdt.$

the least upper bound μ_1 of all these values must exist. Consequently there exists a sequence $y_1(s), y_2(s), \ldots$ of normalized functions for which

$$\int_a^b \int_a^b K(s, t)y_m(s)y_m(t)dsdt \to \mu_1 \qquad \text{for} \qquad m \to \infty.$$

In particular, for a fixed complete orthonormal system of continuous functions $w_1, w_2, \ldots$ we can choose the sequence $y_1, y_2, \ldots$ with

$$y_m(s) = x_1^{(m)}w_1(s) + \cdots + x_m^{(m)}w_m(s)$$

and such that the double integral

$$\int_a^b \int_a^b K(s, t)y_m(s)y_m(t)dsdt$$

$$= \sum_{\alpha,\beta=1}^{m} \int_a^b \int_a^b K(s, t)w_\alpha(s)w_\beta(t)dsdt x_\alpha^{(m)}x_\beta^{(m)} = \mu^{(m)},$$

where m is fixed, attains its maximum under the side condition

$$\sum_{\alpha=1}^{m} x_\alpha^{(m)2} = 1.$$

Then the differential calculus, as applied to $x_1^{(m)}, \ldots, x_m^{(m)}$, shows that

$$\int_a^b w_\alpha(s) \sum_{\beta=1}^{m} x_\beta^{(m)} \int_a^b K(s, t)w_\beta(t)dtds - \lambda x_\alpha^{(m)} = 0$$

and therefore $\lambda = \mu^{(m)}$, as we see by multiplication with $x_\alpha^{(m)}$ and addition of these equations.

The equations for the $x_\alpha^{(m)}$ can be put into a more concise form if we replace the kernel $K(s, t)$ by an approximation

$$K^{(m)}(s, t) = \sum_{\alpha,\beta=1}^{m} a_{\alpha\beta}^{(m)}w_\alpha(s)w_\beta(t)$$

so chosen that in the whole domain

$$|K^{(m)}(s, t) - K(s, t)| < \delta$$

uniformly for sufficiently large m. For such an m an approximation of this sort is always possible. If we now again set

$$\int_a^b \int_a^b K^{(m)}(s, t)y_m(s)y_m(t)dsdt = \sum_{\alpha,\beta=1}^{m} a_{\alpha\beta}^{(m)}x_\alpha^{(m)}x_\beta^{(m)} = \mu^{(m)},$$

the system of conditions on the $x_\alpha^{(m)}$ becomes

$$\sum_{\beta=1}^{m} a_{\alpha\beta}^{(m)} x_\beta^{(m)} - \mu^{(m)} x_\alpha^{(m)} = 0 \qquad (\alpha = 1, \ldots, m)$$

or

$$\int_a^b K^{(m)}(s, t) y_m(t) dt - \mu^{(m)} y_m(s) = 0.$$

Letting δ approach zero, we see that for the corresponding $\mu^{(m)}$ we again have $\mu^{(m)} \to \mu_1$. If $\mu_1 > 0$, as may be assumed without loss of generality, it follows from the last equation that the $y_m(s)$ are uniformly bounded for all m, since $K^{(m)}(s, t)$ approaches $K(s, t)$ uniformly in s, t. Moreover, the $y_m(s)$ are uniformly continuous[13] for all sufficiently large m. So by the well-known theorem of Arzelà we can select from the sequence of the $y_m(s)$ a subsequence uniformly convergent to a continuous limit function $y(s)$, for which the above equation shows that

$$\int_a^b K(s, t) y(t) dt - \mu_1 y(s) = 0, \qquad \|y\| = 1.$$

Thus we have the following theorem:

Every continuous symmetric real kernel in the finite interval (a, b) *has an "eigenvalue"* μ_1 *and a corresponding normalized "eigenfunction"* $y(s)$ *such that the above equation holds.*

Integral equations of this type are called *integral equations of the second kind.* The unknown function occurs not only under the integral sign but also outside of it. In its present form the equation is said to be *homogeneous,* but with a given nonzero function $f(s)$ on the right-hand side it is *inhomogeneous.*

An example is provided by the *theory of vibrations* of a stretched string or a beam. If we denote by $Y(s, \tau)$ the displacement at the point s $(0 \leqq s \leqq l)$ for time τ, by $\rho(s)$ the density, i.e., the mass per unit length, and by $E(s, t)$ the influence function (or Green function), i.e., the (stationary) displacement at the point s produced by a unit force applied at the point t, then the total displacement at the point s for time τ will consist, in the case of a free vibration, of the superposition of all the displacements produced at any point t by the d'Alembert inertia $-\dfrac{\partial^2 Y(t, \tau)}{\partial \tau^2} \rho(t) dt$. Thus

$$Y(s, \tau) = -\int_0^l \frac{\partial^2 Y(t, \tau)}{\partial \tau^2} \rho(t) dt.$$

[13] For every $\epsilon > 0$ there exists a number $\delta > 0$, independent of m (but not in general of s) such that $|y_m(s + h) - y_m(s)| < \epsilon$ for all $|h| < \delta$.

Setting $Y(s, \tau) = y(s) \cos \omega\tau$, or in other words, assuming that the vibration is harmonic with frequency ω per 2π seconds, we obtain the following condition:

$$y(s) = \omega^2 \int_0^l E(s, t)\rho(t)y(t)dt.$$

Since from the definition of the influence function it follows that $E(s, t) = E(t, s)$, we thus obtain the symmetric kernel

$$\sqrt{\rho(s)\rho(t)}E(s, t) = K(s, t)$$

and, for $\sqrt{\rho(s)}y(s) = y_1(s)$, the homogeneous integral equation of the second kind

$$\mu_1 y_1(s) = \int_0^l K(s, t)y_1(t)dt,$$

where we have set $\dfrac{1}{\omega^2} = \mu_1$. The difference between the string and the beam consists simply in the choice of the influence function, which we cannot discuss here for lack of space.

Since μ_1 is the greatest value of the quadratic integral form (see p. 409), we thus obtain the smallest possible frequency ω, or in other words, the fundamental vibration. But it is easy to see that this is not the only way to satisfy the equation

$$\mu y(s) = \int_0^l K(s, t)y(t)dt$$

by a normalized function $y(s)$. Further eigenvalues and corresponding eigenfunctions are obtained as follows.

If the normalized function $y(s)$ is not completely arbitrary but is so chosen as to be orthogonal to a fixed function, then among the values of the functional

$$\mu = \int_a^b \int_a^b K(s, t)y(s)y(t)dsdt$$

attained for such $y(s)$ there will again be a greatest, which will be assumed for a definite $y(s)$. Of course, the value of μ depends on the choice of the fixed function, and there will be a choice for which the corresponding value of μ will be as small as possible; again, there will exist a $y(s)$ for which this *minimax* (minimum of all the maxima) will actually be attained. If we call this function $y_2(s)$ and write

$$\mu_2 = \int_a^b \int_a^b K(s, t)y_2(s)y_2(t)dsdt,$$

we can show as before that

$$\mu_2 y_2(s) = \int_a^b K(s, t) y_2(t) dt, \qquad \| y_2 \| = 1, \qquad \mu_2 \leqq \mu_1.$$

In this way we have obtained the second eigenvalue. The further procedure is analogous; for the k-th eigenvalue we begin with $k - 1$ fixed functions and determine the maximum of the functional for all normalized $y(s)$ orthogonal to these functions; then we find the minimum (as the fixed functions are varied) of all these maxima (as $y(s)$ is varied), which will actually be attained for a certain function $y_k(s)$. We again have

$$\mu_k y_k(s) = \int_a^b K(s, t) y_k(t) dt, \qquad \| y_k \| = 1, \qquad \mu_k \leqq \cdots \leqq \mu_2 \leqq \mu_1.$$

Proceeding in the same way for the negative eigenvalues, we obtain an at most countable set of eigenvalues, which can have a limit point only at zero, with corresponding normalized eigenfunctions, which must be orthogonal to one another. Every eigenvalue can have only finitely many corresponding linearly independent eigenfunctions.

For the vibrating string, the higher eigenfrequencies are integral multiples of the fundamental frequency, so that the eigenvalues are proportional to the squares of the reciprocals of the integers, and the kernel with respect to the orthonormal system of eigenfunctions, which is in this case complete, is essentially the matrix given at the end of §3.

In the general case, the eigenfunctions of a continuous symmetric kernel do not necessarily form a complete orthonormal system. But then every element orthogonal to all the eigenfunctions is also orthogonal to the kernel, i.e., satisfies the equation $\int_a^b K(s, t) z(t) dt = 0$, and conversely. Moreover, we have the important Expansion theorem: *Every function*

$$z(s) = \int_a^b K(s, t) y(t) dt$$

representable by means of a continuous kernel can be expanded in a uniformly convergent series of eigenfunctions

$$z(s) = \sum_{k=1}^{\infty} y_k(s) \int_a^b z(t) y_k(t) dt = \sum_{k=1}^{\infty} \mu_k y_k(s) \int_a^b y(t) y_k(t) dt.$$

Since the complete orthonormal systems occurring most frequently in mathematics can be represented as the eigenfunctions of a continuous kernel, this theorem is of particular importance in the applications.

The most important theorems in the theory remain valid for an operator $\mathfrak{K}$ which in the Hilbert space $\mathfrak{H}^* = \mathfrak{H}$ is

a) *completely continuous, i.e., transforms every weakly convergent sequence* $y_n \rightharpoonup y$ *into a strongly convergent sequence* $\mathfrak{K} y_n \to \mathfrak{K} y$.

b) *self-adjoint, i.e., satisfies the condition* $\mathfrak{K}^* = \mathfrak{K}$.

If condition b) is not required, we nevertheless have the following theorems of Fredholm for *completely continuous* operators in $\mathfrak{H}$:

1. The alternative theorem. *Either the equation*

$$y - \mathfrak{K}y = f$$

has a unique solution for every $f \in \mathfrak{H}$, *or else*

$$y = \mathfrak{K}y = 0$$

has a nonzero solution.

2. *The number of linearly independent solutions of each of the two equations*

$$y - \mathfrak{K}y = 0, \qquad z - \mathfrak{K}^*z = 0$$

is finite and is the same for both equations.

3. *A necessary and sufficient condition for the existence of a solution of*

$$y - \mathfrak{K}y = f$$

is that

$$(f, z) = 0,$$

where z *is an arbitrary solution of* $z - \mathfrak{K}^*z = 0$.

5. Spectral Theory of Selfadjoint Operators

In the present section we assume only that the linear operator K is self-adjoint and bounded; the scalars are the complex numbers and the domain of definition is $\mathfrak{H} = \mathfrak{H}^*$. Neither the theory of eigenvalues sketched in the preceding sections, which depends on the complete continuity of the operator, nor the Fredholm theorems necessarily hold in the present case.

We examine the equation

$$\mu y - \mathfrak{K}y = f$$

and note that, depending on the choice of μ, the following two cases may occur.

1. For every $f \in H$ there exists exactly one solution in $\mathfrak{H}$.
2. There exist elements $f \in \mathfrak{H}$ for which it is not true that the equation has a unique solution in $\mathfrak{H}$.

Those values of μ for which the second case occurs form the so-called *spectrum* of the operator $\mathfrak{K}$. In the case of a completely continuous operator the spectrum consists of the set of eigenvalues with a limit point at zero (*the point spectrum*). In our general case it turns out that all the numbers in the spectrum are real and lie in a bounded interval of the real μ-axis, where they form a closed set of points, which may include the entire interval (*the continuous spectrum*). Thus for all complex values of μ ($\text{Im}\,\mu \neq 0$) the first case occurs, i.e., the equation is always solvable.

The expansion theorem, i.e., the representation of an arbitrary function of the form $z = \mathfrak{K}y$, now involves not only an expansion, in an infinite *series* of eigenfunctions, corresponding to the point spectrum, but also an *integral*, corresponding to the continuous spectrum. In order to give at least some idea of the situation, let us consider an example.

Taking

$$\mathfrak{K}y = \int_{-\infty}^{\infty} e^{-(s-t)^2} y(t)dt,$$

we see, on the basis of the easily proved formula

$$\begin{aligned} e^{-(s-t)^2} &= \frac{1}{2\sqrt{\pi}} \int_{-\infty}^{\infty} e^{-w^2/4} \cos w(s - t)dw \\ (*) \qquad &= \frac{1}{2\sqrt{\pi}} \int_{-\infty}^{\infty} e^{-w^2/4} (\cos ws \cos wt + \sin ws \sin wt)dw, \end{aligned}$$

that

$$\begin{aligned} z = \mathfrak{K}y &= \frac{1}{2\sqrt{\pi}} \int_{-\infty}^{\infty} \int_{-\infty}^{\infty} e^{-w^2/4} (\cos ws \cos wt + \sin wx \sin wt)y(t)dwdt \\ &= \int_{-\infty}^{\infty} \left(\frac{\cos ws}{\sqrt{2\pi}} \int_{-\infty}^{\infty} \frac{y(t) \cos wt}{\sqrt{2\pi}} dt + \frac{\sin ws}{\sqrt{2\pi}} \int_{-\infty}^{\infty} \frac{y(t) \sin wt}{\sqrt{2\pi}} dt \right) \\ &\quad \times e^{-w^2/4} \sqrt{\pi} dw. \end{aligned}$$

Comparing this result with the expansion on page 412, we find the following analogy.

The earlier sum $k = 1, 2, \ldots$ corresponds here to the outer integral with respect to w; the functions $y_k(s)$, in the earlier expansion, to the functions $\dfrac{\cos ws}{\sqrt{2\pi}}$ and $\dfrac{\sin ws}{\sqrt{2\pi}}$; the Fourier coefficients $\int_a^b y(t)y_k(t)dt$, to the integrals

$$\int_{-\infty}^{\infty} \frac{y(t) \cos wt}{\sqrt{2\pi}} dt \qquad \text{and} \qquad \int_{-\infty}^{\infty} \frac{y(t) \sin wt}{\sqrt{2\pi}} dt;$$

and the earlier μ_k, to the values $\mu = \sqrt{\pi} e^{-w^2/4}$. To each value μ with $0 \leqq \mu \leqq \sqrt{\pi}$, there correspond in this way "eigenfunctions" $\dfrac{\cos ws}{\sqrt{2\pi}}$ and $\dfrac{\sin ws}{\sqrt{2\pi}}$, for which in fact

$$\int_{-\infty}^{\infty} e^{-(s-t)^2} \frac{\cos wt}{\sqrt{2\pi}} dt = \sqrt{\pi} e^{-w^2/4} \frac{\cos ws}{\sqrt{2\pi}},$$

$$\int_{-\infty}^{\infty} e^{-(s-t)^2} \frac{\sin wt}{\sqrt{2\pi}} dt = \sqrt{\pi} e^{-w^2/4} \frac{\sin ws}{\sqrt{2\pi}},$$

but these functions are *not* square-integrable in $(-\infty, \infty)$. The continuous spectrum consists of all values $0 \leqq \mu \leqq \sqrt{\pi}$. Outside this range, the equation

$$\mu y(s) - \int_{-\infty}^{\infty} e^{-(s-t)^2} y(t)dt = f(s)$$

has a square-integrable solution $y(s)$, provided $f(s)$ is square-integrable. There is no point spectrum in this case, since there is no value of μ for which the homogeneous equation has a nontrivial solution $y \in \mathfrak{H}$. Let us also remark that the integral representation (*) of the kernel, easily verified for the above example, is in general much more difficult to define and verify. But the whole spectral theory depends upon this integral representation of the operator.

The word "spectrum" is not used here by mere chance but indicates the close association of the theory with important branches of physics that make basic use of such operators and of their spectral representation.

The most important operators, namely, those arising in atomic theory, do not have the simple structure described up to now, since this structure depends on the fact that the given operator is defined for *all* elements of the underlying space $\mathfrak{H}$; the operators of atomic physics are defined only in a subspace of $\mathfrak{H}$ (though usually this subspace is everywhere dense in $\mathfrak{H}$) and, as a result, they are no longer bounded. Nevertheless many of the concepts described above for bounded operators can be extended to the general case, in particular the concept of self-adjointness, characterized by the equation $\mathfrak{A}^* = \mathfrak{A}$. For unbounded operators, it is also possible to develop an integral theory, and a corresponding spectral representation, of fundamental importance in physics. Here the spectrum is not necessarily confined to an *interval* of the real μ-axis, but can be a closed point set distributed over the *entire real* μ-axis.

A simple example of such an operator in the domain $-\infty < s < \infty$ is given by

$$\mathfrak{A}y = sy(s).$$

This operator is self-adjoint in the subspace of $\mathfrak{L}^2(-\infty, \infty)$ consisting of those elements that remain square-integrable in $\mathfrak{L}^2(-\infty, \infty)$ after multiplication by s; its domain is everywhere dense in $\mathfrak{L}^2(-\infty, \infty)$.[14] There is no point spectrum, since from

$$(\mu - s)y(s) = 0$$

[14] In other words, every function in $\mathfrak{L}^2(-\infty, +\infty)$ can be approximated by functions defined on these subdomains, with arbitrary accuracy in the sense of the norm defined by the quadratic integral.

it follows that $y(s) = 0$. On the other hand, every real value μ belongs to the spectrum, since the equation

$$(\mu - s)y(s) = f(s)$$

is not solvable for every $f(s)$ in $\mathfrak{L}^2(-\infty, \infty)$.

If we define an operator $\mathfrak{E}_\lambda$ by setting

$$\mathfrak{E}_\lambda y = \begin{cases} y(s) & \text{for } s \leqq \lambda, \\ 0 & \text{for } s > \lambda, \end{cases}$$

we find that if $\lambda_n < \lambda_{n-1} < \cdots < \lambda_1 < \lambda$, the expression

$$\lambda(\mathfrak{E}_\lambda - \mathfrak{E}_{\lambda_1}) + \lambda_1(\mathfrak{E}_{\lambda_1} - \mathfrak{E}_{\lambda_2}) + \cdots + \lambda_{n-1}(\mathfrak{E}_{\lambda_{n-1}} - \mathfrak{E}_{\lambda n})$$

applied to $y(s)$ defines an operator which produces the function $\lambda y(s)$ for $\lambda_1 < s \leqq \lambda$, the function $\lambda_1 y(s)$ for $\lambda_2 < s \leqq \lambda_1$, and so forth, and produces zero outside the interval (λ_n, λ). Thus we arrive at the "spectral representation"

$$\mathfrak{A}y = sy(s) = \int_{-\infty}^{\infty} \lambda d\mathfrak{E}_\lambda y,$$

where the integral on the right-hand side is defined in a suitable way.

Going a little further in our discussion of the *applications of spectral theory in physics*, and in particular in quantum mechanics, let us first remark that such a theory provides an important example of the way in which present-day physics sets up a completely abstract correspondence between a physical event and a mathematical theory, without any justification other than that the results of the mathematical theory are in sufficiently close agreement with observation.

Now let us note that if a mechanical system of $f + 1$ point masses is characterized by its position coordinates $x_0, y_0, z_0, \ldots, x_f, y_f, z_f$ and its corresponding momentum coordinates $p_{x\lambda}, p_{y\lambda}, p_{z\lambda}$, $\lambda = 0, 1, \ldots, f$, then in classical mechanics the energy is defined by

$$(*) \qquad T + U = \sum_{\lambda=0}^{f} \frac{1}{2\mu_\lambda}(p_{x\lambda}^2 + p_{y\lambda}^2 + p_{z\lambda}^2) + U(q),$$

where the kinetic energy is given by the first part and the potential energy by the second; we assume that the potential energy depends only on the position coordinates, which for brevity are now denoted by $q_0, \ldots, q_f$. The symbol μ_λ denotes the mass of the point λ.

The passage from classical mechanics to quantum theory consists in the introduction of a "wave function" Ψ, which for the time being has no physical significance but is simply a complex-valued function of $q_0, \ldots, q_f$ assumed to belong to the Hilbert space of square-integrable functions of

$q_0, \ldots, q_f$ over the fundamental domain of these variables. This function satisfies the *Schrödinger differential equation* (in nonrelativistic form):

$$H\Psi + \frac{h}{2\pi i}\frac{\partial \Psi}{\partial t} = 0,$$

and represents the state of the system in a way that is described more precisely below; here the constant h is the Planck quantum of action. We now introduce the "energy operator" H.

The significance of this operator is as follows: in the expression (*) for the energy, the momentum coordinates p_x, p_y, p_z are replaced by the operators $\frac{h}{2\pi i}\frac{\partial}{\partial x}, \frac{h}{2\pi i}\frac{\partial}{\partial y}, \frac{h}{2\pi i}\frac{\partial}{\partial z}$, with the result that H is now the operator

$$H = \sum_{\lambda=0}^{f'} -\frac{h^2}{8\pi^2\mu_\lambda}\left(\frac{\partial^2}{\partial x_\lambda^2} + \frac{\partial^2}{\partial y_\lambda^2} + \frac{\partial^2}{\partial z_\lambda^2}\right) + U(q),$$

to be applied to Ψ.

For example, the motion of an electron around a (fixed) nucleus is given by the Schrödinger equation with

$$U(q) = -\frac{e^2}{\sqrt{x_0^2 + y_0^2 + z_0^2}} = -\frac{e^2}{r_0} \qquad (e \text{ denotes the charge}),$$

or (omitting the index 0)

$$-\frac{h^2}{8\pi^2\mu}\Delta\Psi - \frac{e^2\Psi}{r} + \frac{h}{2\pi i}\frac{\partial\Psi}{\partial t} = 0.$$

The solutions of particular interest are the standing waves

$$\Psi = \psi(q)e^{i\omega t},$$

for which, in general, we obtain the operator equation

$$H\psi = -\frac{h^2}{8\pi^2\mu}\Delta\psi + U(q)\psi = -\frac{h\omega}{2\pi}\psi.$$

This equation for the "eigenfunction" ψ has the form of an eigenvalue problem for the operator H; the values $-\frac{h\omega}{2\pi}$ occurring here are the eigenvalues. In the present example, the eigenvalues are found to be the stationary (negative) values of the energy, each of which is characterized by a definite frequency of evolution ω. The values of ω calculated in this way are found to be in excellent agreement with observation; the frequencies $\nu = \frac{\omega}{2\pi}$ correspond to the observed spectral lines.

If the operator H, which in most cases turns out to be self-adjoint in Hilbert space, has not only a point spectrum consisting of the eigenvalues, but also a (real) continuous spectrum, a physical interpretation can be found for the latter spectrum as well, namely, the band spectra observed, for example, in the case of a molecule.

But quite apart from these special applications to quantum mechanics, and more generally to other branches of physics, the concepts of the general theory of operators play a fundamental and constantly increasing role in the applications of mathematics. Examples, particularly in practical analysis, will be found in Part C.

Bibliography

[1] AHIEZER, N. I., and GLAZMAN, I. M.: The Theory of linear operators in Hilbert space. (Translated from the Russian.) Ungar, New York 1961.

[2] BANACH, S.: Théorie des opérations linéaires. Monografie Matematyczne, Warsaw 1932. (Reprinted by Chelsea, New York 1957.)

[3] COURANT, R., and HILBERT, D.: Methods of mathematical physics, Interscience, New York-London, Vol. I, 1953.

[4] VON NEUMAN, J.: Mathematical foundations of quantum mechanics. (Translated from the German by R. T. Beyer.) Princeton University Press, Princeton 1955.

[5] RIESZ, F., and SZ.-NAGY, B.: Functional Analysis. (Translated from the second French edition by L. Boron.) Ungar, New York 1955.

[6] SCHMEIDLER, W.: Linear operators in Hilbert Space. (Translated from the German by J. Strum.) Academic Press, New York 1965.

[7] SNEDDON, J. N.: Functional analysis, Handbuch der Physik II. Springer, Berlin 1955.

[8] TAYLOR, ANGUS E.: Introduction to functional analysis. John Wiley & Sons, New York and London 1958.

C. The Nonlinear Theory

6. The Fréchet Differential

In the nonlinear theory, it is important to formulate a concept that corresponds to the ordinary derivative in the theory of functions. We base our discussion on the concepts of a functional and an operator as given in the introduction.

Thus let $f(y)$ be a functional with the domain of definition ϑ. We form the expression ($y \in \vartheta$, $\eta \in \vartheta$):

$$\frac{f(y + \epsilon\eta) - f(y)}{\epsilon}$$

and assume that $y + \epsilon\eta$ with $\epsilon > 0$ also belongs to ϑ. Now if there exists a decomposition

$$\frac{f(y + \epsilon\eta) - f(y)}{\epsilon} = df(y, \eta) + (\epsilon),$$

where the first term on the right is independent of ϵ and the second term approaches zero as $\epsilon \to 0$, we say that $df(y, \eta)$ is the *Fréchet differential* of $f(y)$.[15]

Our assumption of the existence of a Fréchet differential clearly implies that

$$\lim_{\epsilon \to 0} [f(y + \epsilon\eta) - f(y)] = 0.$$

Furthermore, the Fréchet differential is a linear functional in η, provided that ϑ is a linear space. For from

$$f(y + \epsilon\eta_1) - f(y) = \epsilon df(y, \eta_1) + \epsilon(\epsilon)_1$$
$$f(y + \epsilon\eta_2) - f(y) = \epsilon df(y, \eta_2) + \epsilon(\epsilon)_2$$

we have

$$f(y + \epsilon\eta_1) + f(y + \epsilon\eta_2) - 2f(y) = \epsilon[df(y, \eta_1) + df(y, \eta_2)] + \epsilon(\epsilon),$$

and, on the other hand,

$$f(y + \epsilon(\eta_1 + \eta_2)) - f(y) = \epsilon df(y, \eta_1 + \eta_2) + \epsilon(\epsilon)_3.$$

Subtraction then gives

$$f(y + \epsilon\eta_1) + f(y + \epsilon\eta_2) - f(y + \epsilon(\eta_1 + \eta_2)) - f(y)$$
$$= \epsilon[df(y, \eta_1) + df(y, \eta_2) - df(y, \eta_1 + \eta_2)] + \epsilon(\epsilon)_4,$$

and since after division by ϵ the left-hand side approaches zero as $\epsilon \to 0$, it must also be true that

$$df(y, \eta_1) + df(y, \eta_2) - df(y, \eta_1 + \eta_2) = 0.$$

Similarly we can show that

$$df(y, c\eta) = c df(y, \eta).$$

As a simple example, let us consider the double integral in §4

$$f(y) = \int_a^b \int_a^b K(s, t) y(s) y(t) ds dt, \qquad K(s, t) = K(t, s)$$

for which we find

$$df(y, \eta) = 2 \int_a^b \int_a^b K(s, t) \eta(s) y(t) ds dt.$$

[15] The definition given by Fréchet is somewhat different, but since he was the first to introduce a concept of this sort our terminology seems justified.

If we consider the functional

$$f(y) = \mu \int_a^b y^2(s)ds - \int_a^b \int_a^b K(s, t)y(s)y(t)dsdt,$$

we obtain

$$df(y, \eta) = 2 \int_a^b \eta(s)\Big[\mu y(s) - \int_a^b K(s, t)y(t)dt\Big]ds.$$

If for μ we insert the maximum value of the double integral under the side condition $\int_a^b y^2(s)ds = 1$, then $f(y)$ will be greater than or equal to zero for all not identically vanishing $y(s)$ and zero will be the minimum value of $f(y)$. If $y(s)$ is the function for which this minimum value is attained, then $f(y + \epsilon\eta)$, regarded as a function of ϵ, must have a minimum at $\epsilon = 0$; thus $df(y, \eta) = 0$ for every choice of η. But from the fact that

$$\int_a^b \eta(s)\Big[\mu y(s) - \int_a^b K(s, t)y(t)dt\Big]ds = 0$$

for all $\eta(s)$ it necessarily follows that the expression in square brackets must vanish, since otherwise $\eta(s)$ could be so chosen as to make the integral positive. Thus, as was shown above, $y(s)$ is a solution of the homogeneous integral equation

$$\mu y(s) - \int_a^b K(s, t)y(t)dt = 0.$$

As a somewhat more general setting of the question we may consider the functional

$$f(y) = \mu^n \int_a^b y^{n+1}(s)ds$$

$$- \sum_{\substack{\alpha+\alpha_1+\cdots+\alpha_\nu \\ = n+1}} \int_a^b \cdots \int_a^b K_{\alpha\alpha_1\cdots\alpha_\nu}(st_1\cdots t_\nu)y^{\alpha}(s)y^{\alpha_1}(t_1)\cdots y^{\alpha_\nu}(t_\nu)dsdt_1\cdots dt_\nu$$

and the maximum value of the sum under the side condition

$$\int_a^b y^{n+1}(s)ds = 1,$$

which is a natural one for odd values of n. The corresponding argument leads to an equation of the form

$$\mu^n y^n(s) - \sum_{\beta=0}^{n-1} y^{\beta}(s) \sum_{(\beta_1+\cdots+\beta_\nu=n-\beta)} \int_a^b \cdots \int_a^b L_{\beta\beta_1\cdots\beta_\nu}(st_1\cdots t_\nu)$$
$$\times\, y^{\beta_1}(t_1)\cdots y^{\beta_\nu}(t_\nu)dt_1\cdots dt_\nu = 0,$$

in which the kernels L and the exponents β are related in a simple way to the K and the α; this is an example of a homogeneous "*algebraic integral equation*" of the n-th degree (cf. also p. 435).

The argument outlined here remains valid in more general cases, for example for the functional

$$f(y) = \int_a^b F(s, y(s), y'(s))ds \qquad (F \text{ twice continuously differentiable}),$$

where for the domain of definition we take the set of all twice continuously differentiable functions $y(s)$ with the fixed values A and B at the points a and b. Insertion of $y + \epsilon\eta$ with $\eta(a) = \eta(b) = 0$ then gives

$$\frac{f(y + \epsilon\eta) - f(y)}{\epsilon} = \frac{1}{\epsilon}\int_a^b [F(s, y + \epsilon\eta, y' + \epsilon\eta') - F(s, y, y')]ds$$

$$= \int_a^b \left[\eta \frac{\partial F}{\partial y}(s, y, y') + \eta' \frac{\partial F}{\partial y'}(s, y, y')\right]ds + (\epsilon)$$

and, after integration by parts,

$$df(y, \eta) = \int_a^b \eta\left[\frac{\partial F}{\partial y} - \frac{d}{ds}\left(\frac{\partial F}{\partial y'}\right)\right]ds.$$

If the functional is to have an extreme value for $y(s)$, the Fréchet differential must vanish for all admissible $\eta(s)$, from which we obtain the equation, already known to Euler,

$$\frac{\partial F}{\partial y} - \frac{d}{ds}\left(\frac{\partial F}{\partial y'}\right) = 0,$$

namely, a differential equation of the second order with the side conditions $y(a) = A$, $y(b) = B$. In the same way, for

$$f(y) = \int_a^b F(s, y, y', \ldots, y^{(n)})ds$$

we obtain the Fréchet differential

$$df(y, \eta) = \int_a^b \eta\left[\frac{\partial F}{\partial y} - \frac{d}{ds}\left(\frac{\partial F}{\partial y'}\right) + \frac{d^2}{ds^2}\left(\frac{\partial F}{\partial y''}\right) - \cdots + (-1)^n \frac{d^n}{ds^n}\left(\frac{\partial F}{\partial y^{(n)}}\right)\right]ds$$

and from it the Euler differential equation

$$\frac{\partial F}{\partial y} - \frac{d}{ds}\left(\frac{\partial F}{\partial y'}\right) + \cdots + (-1)^n \frac{d^n}{ds^n}\left(\frac{\partial F}{\partial y^{(n)}}\right) = 0,$$

a differential equation of order $2n$ with fixed values for $y, y', \ldots, y^{(n-1)}$ at the points a and b.

In the textbooks on the calculus of variations one can find a more or less detailed discussion of numerous examples. Here we shall mention only a few of them briefly; for the details of the calculation in each case we refer the reader to the standard literature.

Smallest area for a *surface of rotation* of a curve between two fixed points: here $\int_a^b y\sqrt{1+y'^2}ds = \min$ leads to the Euler equation

$$\sqrt{1+y'^2} - \frac{d}{dx}\left(\frac{yy'}{\sqrt{1+y'^2}}\right) = 0.$$

The result is a catenary between the fixed points.

Curves of shortest length between two points of a fixed surface (geodesic lines): if $\mathfrak{r} = \mathfrak{r}(u, v)$ is the parametric representation of the surface,

$$\int_{t_a}^{t_b} \sqrt{E\dot{u}^2 + 2F\dot{u}\dot{v} + G\dot{v}^2}dt = \min$$

leads to the Euler equations $\mathfrak{r}_u\ddot{\mathfrak{r}} = 0$, $\mathfrak{r}_v\ddot{\mathfrak{r}} = 0$ (the principal normal of the curve coincides with the normal to the surface).

Isoperimetric problem in the plane: a closed curve of given length enclosing the maximum area,

$$\tfrac{1}{2}\int_0^{2\pi} r^2(\varphi)d\varphi = \max, \qquad \int_0^{2\pi} \sqrt{r^2 + r'^2}d\varphi - l = 0.$$

The Euler equation for $H = \frac{1}{2}r^2 + \lambda\sqrt{r^2 + r'^2}$ means that the curve is of constant curvature and is therefore a circle.

Isoperimetric problem in space: the volume of an oval around the origin $V = \frac{1}{3}\int_0^{2\pi}\int_0^{\pi} \mathfrak{r}(\mathfrak{r}_\vartheta \times \mathfrak{r}_\varphi)d\vartheta d\varphi$. The surface is $O = \int_0^{2\pi}\int_0^{\pi}\sqrt{EG - F^2}d\vartheta d\varphi$, and $H = V + \lambda O$ leads to an Euler equation signifying that the mean curvature must be constant (a sphere).

There remains the question whether the Euler equation is not only necessary but also sufficient for the functional to assume an extreme value. This question, which is the fundamental problem of the calculus of variations, cannot be discussed here. Let us only make the further remark that for functionals involving functions of several variables and their *partial* derivatives there are corresponding Euler differential equations, also involving partial derivatives. For example, the condition that the mean curvature of a segment of a surface is equal to zero is equivalent to the Euler equation for the variational problem of finding a surface of smallest area (minimal surface) bounded by a given closed curve.

The Fréchet differential can be constructed not only for functionals but also for operators. For the operator $\mathfrak{A}(y)$ ($y \in \vartheta$) we form the expression

$$\frac{\mathfrak{A}(y + \epsilon\eta) - \mathfrak{A}(y)}{\epsilon}$$

and require the existence of a decomposition of the form

$$\frac{\mathfrak{A}(y + \epsilon\eta) - \mathfrak{A}(y)}{\epsilon} = d\mathfrak{A}(y, \eta) + (\epsilon),$$

where the second part approaches zero as $\epsilon \to 0$ and the first part is independent of ϵ. Then $d\mathfrak{A}(y, \eta)$ is called the Fréchet differential of the operator $\mathfrak{A}(y)$. If ϑ is a linear space, as we shall assume, it can be shown as before that the operator $d\mathfrak{A}(y, \eta)$ is linear in η.

If $\vartheta = \mathfrak{R}^n$ is the n-dimensional Euclidean space and $\mathfrak{A}$ is a mapping of this space onto itself, the equation $z = \mathfrak{A}(y)$, written out in the components, becomes:

$$z_1 = f_1(y_1, \ldots, y_n),$$
$$\cdot \quad \cdot \quad \cdot \quad \cdot \quad \cdot \quad \cdot \quad \cdot$$
$$z_n = f_n(y_1, \ldots, y_n);$$

where we assume the existence and continuity in $\mathfrak{R}^n$ of the partial derivatives of $f_1, \ldots, f_n$ of first and second order. Then for $d\mathfrak{A}(y, \eta)$ we obtain the operator, linear in η, with the components

$$\sum_{\alpha=1}^{n} \frac{\partial f_1}{\partial y_\alpha} \eta_\alpha, \ldots, \sum_{\alpha=1}^{n} \frac{\partial f_n}{\partial y_\alpha} \eta_\alpha,$$

where the transformation matrix is the functional (Jacobian) matrix $\left(\dfrac{\partial f_\alpha}{\partial y_\beta}\right)$.

Of particular interest again are the *operator equations* arising from the concept of an operator. As suggested by the linear case, we direct our attention chiefly to the existence of a solution of

$$y = \mathfrak{A}(y).$$

If we interpret the operator equation $z = \mathfrak{A}(y)$ as a mapping of $\mathfrak{R}^n$ into itself, the existence of the desired solution is equivalent to the presence of a *fixed point of the mapping*. Thus in the following section we shall attempt to set up theorems on such fixed points, not only in $\mathfrak{R}^n$, but also in more general spaces.

7. Fixed Point Theorems

Fixed point theorem of Banach. *Let $z = \mathfrak{A}(y)$ be a contractive mapping of a complete metric space E onto itself. Then the mapping has exactly one fixed point.*

Here we need the following definitions. A *metric* space is defined by the property that to every pair of points (x, y) in the space there is assigned a distance, denoted by $\rho(x, y)$, which is equal to zero if and only if $x = y$ and is otherwise positive. Furthermore, the distance is symmetric, $\rho(x, y) = \rho(y, x)$, and satisfies the triangle inequality

$$\rho(x, y) \leqq \rho(x, z) + \rho(z, y) \qquad (x, y, z \in \mathfrak{E}).$$

A sequence of points is said to be convergent to a limit point $x_n \to x$ if

$$\lim_{n\to\infty} \rho(x_n, x) = 0.$$

Every convergent sequence is a Cauchy sequence in the sense that $\rho(x_n, x_m) \to 0$ for $n, m \to \infty$. If, conversely, every Cauchy sequence has a limit point in $\mathfrak{E}$, then the space is said to be *complete*. The mapping $z = \mathfrak{A}(y)$ is *contractive* if there exists a positive number $q < 1$ such that

$$\rho(\mathfrak{A}(x), \mathfrak{A}(y)) \leqq q\rho(x, y)$$

for all x, y. Such a mapping is obviously continuous, since $\rho(x, y) < \delta$ implies $\rho(\mathfrak{A}(x), \mathfrak{A}(y)) < \delta$.

Now to prove the existence of a fixed point we start from an arbitrary element y_0 and construct the sequence $y_{n+1} = \mathfrak{A}(y_n)$ $(n = 0, 1, 2, \ldots)$. Then

$$\rho(y_{n+1}, y_n) = \rho(\mathfrak{A}(y_n), \mathfrak{A}(y_{n-1})) \leqq q\rho(y_n, y_{n-1})$$

and thus

$$\rho(y_{n+1}, y_n) \leqq q^n\rho(y_1, y_0); \qquad \rho(y_n, y_{n+p}) \leqq \frac{q^n}{1-q}\rho(y_1, y_0).$$

Consequently, the sequence is a Cauchy sequence and is therefore convergent. If y is its limit point, then

$$\rho(y_{n+1}, \mathfrak{A}(y)) = \rho(\mathfrak{A}(y_n), \mathfrak{A}(y)) \leqq q\rho(y_n, y) \to 0$$

for $n \to \infty$. Thus

$$\rho(y, \mathfrak{A}(y)) \leqq \rho(y, y_{n+1}) + \rho(y_{n+1}, \mathfrak{A}(y)) \to 0,$$

i.e.,

$$\rho(y, \mathfrak{A}(y)) = 0, \qquad y = \mathfrak{A}(y).$$

This fixed point is unique. For if there were a second z, we would have

$$\rho(y, z) = \rho(\mathfrak{A}(y), \mathfrak{A}(z)) \leqq q\rho(y, z),$$

which is possible only for $\rho(y, z) = 0$. Thus the Banach fixed point theorem is completely proved. Since this theorem also provides a method for the actual construction of the desired fixed point, it is often employed in applied mathematics for the solution of linear or nonlinear systems of equations, after they have been brought into a suitable form.

As an important application let us consider a *system of n ordinary differential equations* (III 8, §2.4)

$$\frac{dy_\alpha}{ds} = f_\alpha(s, y_1, \ldots, y_n) \qquad (\alpha = 1, 2, \ldots, n) \quad (0 \leqq |s| \leqq a),$$

where the $f_\alpha(s, y_1, \ldots, y_n)$ are continuous functions and $y_\alpha(0) = a_\alpha$ are arbitrarily preassigned values. For continuous solutions we may write

$$y_\alpha(s) = a_\alpha + \int_0^s f_\alpha(t, y_1(t), \ldots, y_n(t))dt,$$

so that, by setting

$$z_\alpha(s) = a_\alpha + \int_0^s f_\alpha(t, y_1(t), \ldots, y_n(t))dt, \qquad z = \mathfrak{A}(y),$$

we define an operator mapping the space of vectors $y_1(t), \ldots, y_n(t)$ with continuous components $y_\alpha(t)$ onto the same space. Let us see whether the conditions of the Banach fixed point theorem can be satisfied.

If $|y_\alpha(t)| \leqq M_0$ $(0 \leqq |t| \leqq a)$ and if for all these $y_\alpha(t)$ the maximum of the absolute value of $|f_\alpha(t, y_1, \ldots, y_n)| < M$ is less than M, then $|z_\alpha(s)| \leqq \max |a_\alpha| + aM < M_0$, provided $M_0 > \max |a_\alpha|$ and a is sufficiently small. Thus, if for every admissible vector $y = \{y_1(t), \ldots, y_n(t)\}$ we set the distance

$$\rho(y, 0) = \max |y_\alpha(t)| \leqq M_0,$$

and if more generally for two vectors

$$x = \{x_1(t), \ldots, x_n(t)\} \qquad y = \{y_1(t), \ldots, y_n(t)\}$$

we define the distance by $\rho(x, y) = \max |x_\alpha(t) - y_\alpha(t)|$, then for the image vector we have

$$\rho(z, 0) = \max |z_\alpha(s)| \leqq M_0.$$

Then $\mathfrak{E}$ can be regarded as the set of all vectors y with $\rho(y, 0) \leqq M_0$. This space is bounded, metric and complete, since every Cauchy sequence $y^{(1)}, y^{(2)}, \ldots, (\rho(y^{(n)}, y^{(m)}) \to 0, m, n \to \infty)$ is uniformly convergent in all the components to a continuous limit vector in $\mathfrak{E}$. Thus it remains only to determine whether the mapping $z = \mathfrak{A}(y)$ is contractive. For this purpose we form

$$\rho(\mathfrak{A}(x), \mathfrak{A}(y)) = \max \left| \int_0^s [f_\alpha(t, x) - f_\alpha(t, y)] \right| dt, \qquad \begin{pmatrix} x \in \mathfrak{E} \\ y \in \mathfrak{E} \end{pmatrix}$$

and must then make an additional assumption on the functions $f_\alpha(t, x)$ in order to estimate the difference under the integral sign by means of the differences $|x_\alpha - y_\alpha|$. To this end we impose a *Lipschitz condition*, namely, that there exists a positive constant C such that for all s in $0 \leqq |s| \leqq a$:

$$|f_\alpha(s, x) - f_\alpha(s, y)| < C \max |x_\alpha - y_\alpha| \qquad (\alpha = 1, \ldots, n).$$

Under this condition we have

$$\rho(\mathfrak{A}(x), \mathfrak{A}(y)) < aC \max |x_\alpha - y_\alpha| = aC\rho(x, y),$$

and thus, if $aC = q < 1$, as can always be ensured by taking a sufficiently small, the mapping is in fact contractive. Consequently, by applying the Banach fixed point theorem to our system of differential equations, we guarantee the *existence of a unique continuous solution with arbitrary initial values, provided the right-hand sides are continuous and satisfy a Lipschitz condition; here the existence and uniqueness of the solution is guaranteed in a sufficiently small interval about the initial point.*

As a further application we consider the *Volterra integral equation*

$$y(x) = \int_0^x K(x, \xi, y(\xi))d\xi = \mathfrak{A}(y),$$

where the continuous kernel satisfies a Lipschitz condition

$$|K(x, \xi, y) - K(x, \xi, z)| \leqq L|y - z|;$$

a special case is the ordinary differential equation of first order

$$y' = f(x, y), \qquad y(0) = 0, \qquad y(x) = \int_0^x f(\xi, y(\xi))d\xi.$$

Here we take as the norm

$$\|y\| = \max_{0 \leqq x \leqq l} \left| \frac{y(x)}{e^{\mathfrak{L}x}} \right|,$$

which satisfies the axioms for a distance. Then

$$\begin{aligned}
\|\mathfrak{A}(y) - \mathfrak{A}(z)\| &\leqq \max e^{-\mathfrak{L}x} \int_0^x |K(x, \xi, y(\xi)) - K(x, \xi, z(\xi))| d\xi \\
&\leqq \max e^{-\mathfrak{L}x} \int_0^x L|y(\xi) - z(\xi)| d\xi \\
&\leqq \max e^{-\mathfrak{L}x} L \int_0^x e^{\mathfrak{L}\xi} \|y - z\| d\xi \\
&= \|y - z\| L \max e^{-\mathfrak{L}x} \frac{e^{\mathfrak{L}x} - 1}{\mathfrak{L}} \leqq \|y - z\| \frac{L}{\mathfrak{L}}.
\end{aligned}$$

If $\mathfrak{L}$ is large enough so that $L/\mathfrak{L} = q < 1$, the Banach fixed point is applicable, and we obtain the solution of the integral equation in the whole interval $0 \leqq x \leqq l$.

A further example for this iterative procedure is the well-known *method of alternation*, due to H. A. Schwarz, for the first boundary-value problem in potential theory. Let $\mathfrak{G}$ and $\mathfrak{G}'$ be two planar bounded domains for which the boundary-value problem is solvable, and let them be bounded by the curves $\mathfrak{C} = \mathfrak{C}_a + \mathfrak{C}_i$ and $\mathfrak{C}' = \mathfrak{C}'_a + \mathfrak{C}'_i$, where $\mathfrak{C}_i$ is that part of the boundary of $\mathfrak{G}$ which lies inside $\mathfrak{G}'$, and $\mathfrak{C}'_i$ is that part of the boundary of $\mathfrak{G}'$ which lies inside $\mathfrak{G}$ (see Figure 1); then we seek a function U which is harmonic in $\mathfrak{G} \cup \mathfrak{G}'$ and takes preassigned continuous values on the boundary $\mathfrak{C}_a + \mathfrak{C}'_a$.

Let $u, v, \ldots$ be harmonic in $\mathfrak{G}$ and continuous in $\mathfrak{G} + \mathfrak{C}$ and let them assume the preassigned boundary values on $\mathfrak{C}_a$; and also let the corresponding conditions hold for $u', v', \ldots$ in $\mathfrak{G}'$ and on $\mathfrak{G}' + \mathfrak{C}'$ and $\mathfrak{C}'_a$.

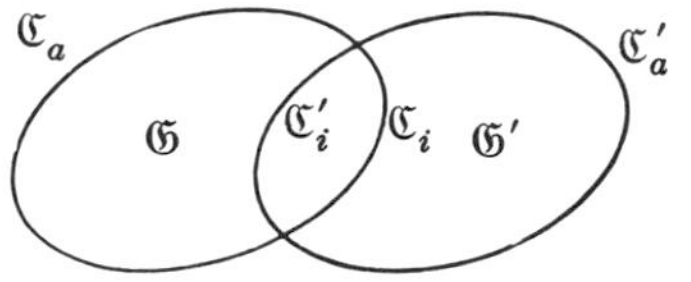

Fig. 1

We start from a u_0 with values on $\mathfrak{C}_i$ that are continuous extensions of the prescribed values on $\mathfrak{C}_a$ but are otherwise arbitrary; and similarly u'_0 is chosen with values on $\mathfrak{C}'_i$ that are arbitrary continuous extensions of the prescribed boundary values on $\mathfrak{C}'_a$. We then construct the sequences

$$u_\nu = u'_{\nu-1} \quad \text{on } \mathfrak{C}_i \quad \text{and} \quad u'_\nu = v_{\nu-1} \quad \text{on } \mathfrak{C}'_i.$$

The theorem of Schwarz asserts the convergence of $u_\nu \to w$ in $\mathfrak{G} + \mathfrak{C}$, $u'_\nu \to w'$ in $\mathfrak{G}' + \mathfrak{C}'$ and $w = w'$ in $\mathfrak{G} \cap \mathfrak{G}'$, and that

$$U = \begin{cases} w \text{ in } \mathfrak{G} \\ w' \text{ in } \mathfrak{G}' \end{cases}$$

is the solution of the boundary-value problem.

For the proof we consider the space $\mathfrak{E}$ of pairs $\xi = (u, u'), \eta = (v, v'), \ldots$ with the distance function (satisfying all the requirements)

$$\rho(\xi, \eta) = \max \{|u - v|_{\mathfrak{C}_i}; |u' - v'|_{\mathfrak{C}'_i}\}.$$

Then it is easy to prove that the space $\mathfrak{E}$ is complete.[16] Let our maping be

$$\mathfrak{A}\xi = (\bar{u}, \bar{u}'), \qquad \mathfrak{A}\eta = (\bar{v}, \bar{v}') \qquad \text{with} \quad \begin{Bmatrix} \bar{u} = u' \\ \bar{v} = v' \end{Bmatrix} \text{ on } \mathfrak{C}_i,$$

$$\begin{Bmatrix} \bar{u}' = u \\ \bar{v}' = v \end{Bmatrix} \text{ on } \mathfrak{C}'_i.$$

Then

$$\rho(\mathfrak{A}\xi, \mathfrak{A}\eta) = \max \{|u' - v'|_{\mathfrak{C}_i}; |u - v|_{\mathfrak{C}'_i}\}.$$

Since

$$|u' - v'|_{\mathfrak{C}_i} \leqq q|u' - v'|_{\mathfrak{C}'_i}; \qquad |u - v|_{\mathfrak{C}'_i} \leqq q|u - v|_{\mathfrak{C}_i}$$

with $q < 1$, we have

$$\rho(\mathfrak{A}\xi, \mathfrak{A}\eta) \leqq q\rho(\xi, \eta),$$

[16] Cf. Courant-Hilbert, Methoden der mathematischen Physik II, p. 266.

so that

$$(u_n, u_n') = \mathfrak{A}(u_{n-1}, u_{n-1}')$$

is convergent, and so forth.

It is sometimes possible to set up an iteration procedure not only by a suitable choice of the space and the corresponding distance function but also by a suitable transformation of the equation.

For the integral equation $y = \lambda\mathfrak{K}y + f$, it is well known that the Neumann series is convergent for all λ with $|\lambda| < |\lambda_1|$, where $\lambda_1, \lambda_2, \ldots$ is the sequence of eigenvalues arranged in the order of magnitude of their absolute values and it is assumed that the operator is completely continuous. It was shown by G. Wiarda [13] that the transformed equation

$$y = \Theta y + \lambda(1 - \Theta)\mathfrak{K}y + (1 - \Theta)f$$

enables us to make use of the iteration procedure

$$y_{n+1} = \Theta y_n + \lambda(1 - \Theta)\mathfrak{K}y_n + (1 - \Theta)f,$$

which is always convergent for $\left|\Theta + \frac{\lambda(1 - \Theta)}{\lambda_i}\right| < 1$ and $|\Theta| < 1$. This idea has been generalized by Bückner [2] to the effect that for a periodic sequence of constants $\Theta_{\nu+\rho} = \Theta_\nu$ ($\nu = 1, 2, \ldots, \rho$ fixed) the iteration procedure

$$y_{n+1} = \Theta_{n+1}y_n + \lambda(1 - \Theta_{n+1})\mathfrak{K}y_n + (1 - \Theta_{n+1})f,$$

or in other words

$$y_{(n+1)\rho} = \mathfrak{K}^{(\rho)}y_{n\rho} + f^{(\rho)} \qquad (n = 0, 1, \ldots),$$

always converges to the solution of the integral equation, provided λ is not an eigenvalue and the Θ's satisfy the conditions

$$\prod_{\nu=1}^{\rho} |\Theta_\nu| < 1 \qquad \text{and} \qquad \prod_{\nu=1}^{\rho} \left|\Theta_\nu + \frac{\lambda(1 - \Theta_\nu)}{\lambda_i}\right| < 1 \qquad (i = 1, 2, \ldots).$$

The existence of such constants Θ_ν is proved.

For the application of "inclusion theorems" in "problems of monotonic type," which are also based on the methods of functional analysis, see Collatz [3].

The theorem on contractive mappings makes use of a very general concept of space, but a very special type of mapping. Let us now attempt to introduce a simple notion of space, namely, the Euclidean $\mathfrak{R}^n$, but to weaken the requirements on the mapping. If we join each point to its image point, we have a vector field; we assume that this vector field is continuous, and for the space in question we choose a sphere in $\mathfrak{R}^n$, or any part of $\mathfrak{R}^n$ that is topologically equivalent to a sphere. The existence of a fixed point now means that the field vector vanishes at this point. If the field vector is nonzero on the whole surface of the sphere, we ask whether it perhaps vanishes at some point in the interior.

We begin with $n = 2$ and consider a square on whose boundary the field vector is everywhere different from zero. If we make one circuit of the boundary the field vector returns to its original position, and has therefore traced out an angle $\gamma \cdot 2\pi$; the integer γ is called the *revolution number*. If this number is different from zero, we assert the existence of a fixed point in the interior.

For the proof we divide the square into four equal squares. Then either there is a zero point of the vector field on the boundary of one of the smaller squares, or else for each of these squares we can again assign a revolution number, whereupon we see at once that the sum of the four revolution numbers is equal to the revolution number of the whole square. Consequently, at least one of the smaller squares must have a nonzero revolution number. We divide this square again into four equal squares and continue the process. In this way we obtain a sequence of infinitely many telescoped squares with area approaching zero, so that we have thereby defined a unique point situated in the interior of every square of the sequence, and this point has the property that any neighborhood of it, however small, contains one of the squares with nonzero revolution number and therefore contains vectors in every direction. But then by continuity the vector at this point itself must vanish, so that the proof of the existence of a fixed point is now complete.

For a continuous mapping $z = \mathfrak{A}y$ *of a domain into itself, where the domain is a one-to-one continuous image of the sphere* $y_1^2 + \cdots + y_n^2 \leqq 1$ *in* R^n *and has a nonzero "revolution number," there always exists at least one fixed point, i.e., a solution of the equation* $y = \mathfrak{A}(y)$.

We have formulated this theorem for general n but have proved it for $n = 2$. In order to complete the proof we must first give a general definition of the concept "revolution number" in $\mathfrak{R}^n$. For this purpose we make use of the *Gauss-Kronecker integral*

$$J = \frac{1}{V_{n-1}} \int \cdots \int_{(V_{n-1})} \begin{vmatrix} v_1 & \cdots & v_n \\ \dfrac{\partial v_1}{\partial u_1} & \cdots & \dfrac{\partial v_n}{\partial u_1} \\ \cdot & \cdot\ \cdot\ \cdot & \cdot \\ \dfrac{\partial v_1}{\partial u_{n-1}} & \cdots & \dfrac{\partial v_n}{\partial u_{n-1}} \end{vmatrix} \frac{d\omega}{\sqrt{v_1^2 + \cdots + v_n^2}^{\,n}}$$

(where the $v_1, \ldots, v_n$ components of the vector field).

In this case V_{n-1} is the $(n-1)$-dimensional surface of our domain, and $d\omega$ is the corresponding surface-element expressed in terms of the surface parameters $u_1, \ldots, u_{n-1}$. Here it is assumed that the parameters $u_1, \ldots, u_{n-1}$ represent the whole surface; however, a given parametric system may serve for only a part of the surface, provided the various parts cover the whole surface, since it is easy to verify that the integral is invariant under transformation of the parameters. For $n = 2$ we have

$$\frac{1}{2\pi R}\int_0^{2\pi} \frac{v_1 \dfrac{\partial v_2}{\partial \varphi} - v_2 \dfrac{\partial v_1}{\partial \varphi}}{v_1^2 + v_2^2} R d\varphi = \frac{1}{2\pi}\int_0^{2\pi} \frac{\partial}{\partial \varphi}\left(\frac{v_2}{v_1}\right) \frac{v_1^2}{v_1^2 + v_2^2} d\varphi$$

or with $v_1 = \sqrt{v_1^2 + v_2^2}\cos\alpha$, $v_2 = \sqrt{v_1^2 + v_2^2}\sin\alpha$:

$$= \frac{1}{2\pi}\int_0^{2\pi} \frac{\partial}{\partial \varphi}(\operatorname{tg}\alpha)\cos^2\alpha d\varphi = \frac{1}{2\pi}\int_0^{2\pi} \frac{d\alpha}{d\varphi} d\varphi = \frac{1}{2\pi}(\alpha(2\pi) - \alpha(0)) = \gamma.$$

In proving the fixed point theorem we confine our attention to the case $n = 3$, since the case for general n does not call for any basic changes. First of all, it is clear that the integral always exists if $v_1^2 + v_2^2 + v_3^2 \neq 0$ on the whole surface, as we shall assume.[17] We now wish to prove that under this assumption the value of the integral is independent of the form of the surface. For this purpose we introduce Cartesian coordinates and write J in the form

$$J = \iint A dydx + B dzdx + C dxdy,$$

where we have set

$$A = \frac{\begin{vmatrix} v_1 & v_2 & v_3 \\ v_{1y} & v_{2y} & v_{3y} \\ v_{1z} & v_{2z} & v_{3z} \end{vmatrix}}{\sqrt{v_1^2 + v_2^2 + v_3^2}^3}, \qquad B = \frac{\begin{vmatrix} v_1 & v_2 & v_3 \\ v_{1z} & v_{2z} & v_{3z} \\ v_{1x} & v_{2x} & v_{3x} \end{vmatrix}}{\sqrt{v_1^2 + v_2^2 + v_3^2}^3}, \qquad C = \frac{\begin{vmatrix} v_1 & v_2 & v_3 \\ v_{1x} & v_{2x} & v_{3x} \\ v_{1y} & v_{2y} & v_{3y} \end{vmatrix}}{\sqrt{v_1^2 + v_2^2 + v_3^2}^3},$$

$$\left(v_{1x} = \frac{\partial v_1}{\partial x} \text{ etc.}\right)$$

as is clearly allowable. We now consider a second surface, completely contained in the first one, such that on both surfaces, and in the space (Z) between them, we have $v_1^2 + v_2^2 + v_3^2 \neq 0$, with the result that the partial derivatives of A, B, C exist and are continuous in this whole intermediate domain. Application of the Gauss theorem (see its coordinate-free form III 4, §11, p. 181) then shows that the difference between the two surface integrals, each of them taken in the same sense, is

$$J_1 - J_2 = \iiint_{(z)} (A_x + B_y + C_z) dxdydz.$$

[17] For the time being we also assume the existence and continuity of all partial derivatives occurring in the formulas; see below.

Now we can show by a calculation, quite simple but rather inconvenient to write down, that the divergence (cf. III 4, §11, p. 180) on the right-hand side is always zero, so that

$$J_1 = J_2,$$

which shows that the integral is independent of the form of the surface. But if this integral is different from zero on the given surface and if we also assume that $v_1^2 + v_2^2 + v_3^2 \neq 0$ throughout the interior of the given domain, we are necessarily led to a contradiction. For instead of taking the integral over the given surface we can take it over an arbitrarily small surface around an arbitrary inner point and will then obtain the same value; but it follows from the continuity of v_1, v_2, v_3 that as we close down on this point the integral must converge to zero.

Thus if $J_1 \neq 0$, we necessarily have

$$v_1 = v_2 = v_3 = 0$$

somewhere in the interior, which completes the proof of the existence of a fixed point.

In contrast to the case $n = 2$, this argument fails to show that the value of the integral is a whole number. In the literature this fact is usually proved first, and if it is already proved, then the proof that the value of the integral remains unchanged under continuous deformation of the surface is very easy. But the proof of the fixed point theorem as a whole is easiest in the form in which we have given it. Moreover, the above argument enables us to show very easily that if the vector field has *finitely many* zeros in the given domain, then the value of the integral is in fact a whole number; for it is clear that every zero for which the functional determinant

$$\begin{vmatrix} v_{1x} & v_{2x} & v_{3x} \\ v_{1y} & v_{2y} & v_{3y} \\ v_{1z} & v_{2z} & v_{3z} \end{vmatrix}$$

is positive makes a contribution of $+1$, and every zero for which it is negative, makes the contribution -1. The case where the functional determinant vanishes at a zero of the vector field requires further investigation, which cannot be given here.

On the other hand it is important to extend the proof from a twice continuously differentiable vector field to a field in which we require only continuity of the components. But it is well known that a continuous vector field is uniformly continuous in a closed domain and that such a vector field can be uniformly approximated with arbitrary accuracy by a twice continuously differentiable vector field. The Gauss-Kronecker integral of such an approximating field remains unchanged if we extend the field a little beyond its domain of definition by a vector field for which the components are everywhere twice continuously differentiable and $v_1^2 + v_2^2 + v_3^2 \neq 0$. If on this enclosing surface we always choose the same boundary field, we see that the value of the integral is the same for all the approximating fields and can thus be taken as the "revolution number" for the merely continuous vector field. If this number is different from zero, each of the countably many approximating fields has a zero; and therefore a limit point of these zeros, which cannot lie on the boundary (since there we have assumed $v_1^2 + v_2^2 + v_3^2 \neq 0$ for all fields), must be the desired zero of the given continuous vector field.

For the sake of completeness we also make the following remark, usually deduced from the fact (not proved here for the general case) that the value in question is a whole number: the integral retains the same value under continuous deformations of the vector field (without change of the surface), provided we continue to have $v_1^2 + v_2^2 + v_3^2 \neq 0$. For here also we can extend the vector field slightly outwards and undertake a continuous deformation of the field such that the boundary values on the enclosing surface remain unchanged; but then the assertion follows at once.

The fixed point theorem can be generalized still further to the following Brouwer fixed point theorem.

If $\mathfrak{M}$ is a convex closed bounded subset of $\mathfrak{R}^n$ which is mapped continuously onto itself, then the mapping has a fixed point.

A set $\mathfrak{M}$ is said to be convex if for every two points $y^{(1)}$ and $y^{(2)}$ in $\mathfrak{M}$ every combination $\dfrac{c_1 y^{(1)} + c_2 y^{(2)}}{c_1 + c_2}$ with positive coefficients c_1 and c_2, i.e., every point of the line-segment joining $y^{(1)}$ and $y^{(2)}$, is also contained in $\mathfrak{M}$. Also, as is well known, a set $\mathfrak{M}$ is said to be closed if every limit point of points in $\mathfrak{M}$ again belongs to $\mathfrak{M}$. Such a point set can always be regarded as the topological image of an m-dimensional sphere (together with its surface) in $\mathfrak{R}^n$ ($m \leqq n$). The corresponding revolution number for such a mapping is necessarily positive, as can easily be seen by calculating the integral for the sphere. Thus the existence of a fixed point follows from the theorem proved above.

Of particular interest is the extension, due to Schauder, of the Brouwer theorem to Banach spaces. The theorem of Schauder reads:

Every continuous mapping $z = \mathfrak{A}(y)$ of a convex bounded closed set $\mathfrak{M}$ of a Banach space $\mathfrak{B}$ under which the image of $\mathfrak{M}$ is a sequentially compact set belonging to $\mathfrak{M}$, has at least one fixed point.

By a *sequentially compact* set we mean a set in which every sequence of elements that is bounded in the sense of the norm of the Banach space has a convergent subsequence.

For the proof we consider the image

$$\mathfrak{Q} = \mathfrak{A}(\mathfrak{M})$$

of the set $\mathfrak{M}$ and a so-called ϵ-covering on it, i.e., a finite number of points $\eta_1, \ldots, \eta_m$ such that the neighborhoods $\|y - \eta_\nu\| < \epsilon$ with $y \in \mathfrak{Q}$ cover the entire set $\mathfrak{Q}$. The existence of such a covering set for every compact set is guaranteed by the so-called Heine-Borel theorem.

We now construct the following mapping of $\mathfrak{Q}$:

$$H_\epsilon(y) = \frac{\sum_{\nu=1}^{m} \eta_\nu p_\nu(y)}{\sum_{\nu=1}^{m} p_\nu(y)}, \qquad p_\nu(y) = \max\left\{0, 1 - \frac{\|y - \eta_\nu\|}{\epsilon}\right\}.$$

Thus $p_\nu(y) = 0$ for all those values of ν for which $\|\varphi - \eta_\nu\| \geqq \epsilon$; but for least one ν we have $p_\nu(y) > 0$, namely, if the corresponding ϵ-covering of η_ν contains the point y. This mapping is obviously continuous in y. Since $\mathfrak{M}$ is convex, the element $H_\epsilon(y)$ is contained in $\mathfrak{M}$ and belongs, moreover, to the linear vector space L' generated by $\eta_1, \ldots, \eta_m$. Thus the set $H_\epsilon(y)$ is contained in the intersection $\mathfrak{M} \cap L' = L$. Also, it is clear that

$$\|H_\epsilon(y) - y\| \leqq \frac{\sum_{\nu=1}^{m} \|y - \eta_\nu\| p_\nu(y)}{\sum_{\nu=1}^{m} p_\nu(y)} \leqq \epsilon.$$

If we now form $H_\epsilon(\mathfrak{A}(y)) - \mathfrak{A}(y)$, then for every $y \in \mathfrak{M}$ we have

$$\|H_\epsilon(\mathfrak{A}(y)) - \mathfrak{A}(y)\| \leqq \epsilon.$$

If we choose y from L, then $H_\epsilon(y)$ is also contained in L and L is convex, since it is the intersection of two convex sets $\mathfrak{M}$ and L'. Since the space L' is finite-dimensional and the convex closed bounded subset L is mapped continuously onto itself, the Brouwer fixed point theorem guarantees the existence of a fixed point y for which $H_\epsilon(\mathfrak{A}(y_\epsilon)) = y_\epsilon$. Thus for this fixed point

$$\|y_\epsilon - A(y_\epsilon)\| \leqq \epsilon.$$

If we now choose a sequence of positive numbers $\epsilon_1, \epsilon_2, \ldots$ converging to zero, the corresponding elements y_{ϵ_ρ} and the elements generated by them $\mathfrak{A}(y_{\epsilon_\rho})$ satisfy the equation

$$\|y_{\epsilon_\rho} - \mathfrak{A}(y_{\epsilon_\rho})\| \to 0 \qquad \text{for} \qquad \rho \to \infty.$$

The sequence $\mathfrak{A}(y_{\epsilon_\rho})$ is uniformly bounded and belongs to the sequentially compact set $\mathfrak{Q}$; consequently it has a convergent subsequence, say $\mathfrak{A}(y_{\epsilon_\sigma}) \to z \in \mathfrak{M}$. Thus for all sufficiently large indices σ:

$$\|y_{\epsilon_\sigma} - z\| < \delta,$$

which means that the sequence y_{ϵ_σ} is also convergent, namely, to z. But then, since $\mathfrak{A}$ is continuous, we also have $\mathfrak{A}(y_{\epsilon_\sigma}) \to \mathfrak{A}(z)$. Thus $\mathfrak{A}(z) = z$, and the existence of a fixed point is proved.

8. Application of the Fixed Point Theorems to Nonlinear Operator Equations

Example 1. We seek a periodic solution for the equation of nonlinear vibration

$$\frac{d^2x}{dt^2} + g(x) = p(t),$$

where we assume that

1. $p(t + T) = p(t)$ is periodic and continuous for all t,
2. $g(x)$ is continuous and $|g(x)| \leqq M$ for all real x.

A well-known special case is the so-called "*Duffing equation*" with $g(x) = \alpha \sin x$.

We introduce the *Green's function* for the differential expression $\ddot{x}(t)$ with vanishing boundary values for $s = 0$ and $s = T$, namely,

$$K(s, t) = 2T \sum_{n=1}^{\infty} \frac{1}{\pi^2 n^2} \sin \frac{n\pi s}{T} \sin \frac{n\pi t}{T}.$$

It is easy to verify that the solution $x(t)$ of the equation

$$\ddot{x}(t) = f(t)$$

vanishing at the end points $t = 0$ and $t = T$ can be represented in the form

$$x(s) = -\int_0^T K(s, t) f(t) dt.$$

In the present case we see that every solution of our problem can be represented in the form

$$x(s) = \int_0^T K(s, t)(g(x(t)) - p(t))dt + as + b,$$

where for the time being a and b are arbitrary constants. If we also set

$$f(s) = -\int_0^T K(s, t)p(t)dt = f(s + 2T) \qquad (f(0) = f(T) = 0),$$

then

$$x(s + 2T) = f(s) + a(s + 2T) + b + \int_0^T K(s, t)g(x(t))dt;$$

thus the desired periodicity of the solution requires that $a = 0$, and for a periodic solution we have the integral equation

$$x(s) = f(s) + b + \int_0^T K(s, t)g(x(t))dt.$$

We must now ask whether this integral equation has a solution. We consider the mapping

$$y(s) + f(s) + b + \int_0^T K(s, t)g(x(t))dt = \mathfrak{A}(x)$$

for all continuous functions $x(t)$ with absolute value

$$|x(t)| \leqq \max |f(s) + b| + KTM = N,$$

where $K = \max |K(s, t)|$. For such $x(t)$ we then also have $|y(s)| \leqq N$.

The set of functions continuous in the interval $0 \leqq t \leqq 2T$ with $x(0) = x(2T)$ and absolute value $|x(t)| \leqq N$ now form a convex closed and bounded set $\mathfrak{M}$ in the Banach space with the norm

$$\|x\| = \max |x(t)|,$$

and the image functions $\mathfrak{A}(x)$ form a sequentially compact subset of $\mathfrak{M}$. For in fact every infinite sequence of image functions is uniformly bounded and uniformly continuous, so that by the theorem of Arzelà it contains a uniformly convergent subsequence, which is therefore convergent in the sense of the norm of the Banach space.

Consequently we can apply the fixed point theorem of Schauder and find a fixed point for the mapping, i.e., a function $x(t)$ with

$$x(s) = f(s) + b + \int_0^T K(s, t)g(x(t))dt.$$

Thus we have found a periodic solution of the differential equation for our vibration, and in fact such a solution exists for every fixed choice of b.

Example 2. We consider an "algebraic integral equation" of the form

$$\mu^n y^n(s) - \sum_{\beta=0}^{n-1} \mu^\beta y^\beta(s) \sum_{(\alpha_1 + \cdots + \alpha_\nu = n - \beta)} \int_0^1 \cdots \int_0^1 K_{\beta\alpha_1 \cdots \alpha_\nu}(st_1, \ldots, t_\nu) y^{\alpha_1}(t_1) \cdots y^{\alpha_\nu}(t_\nu) dt_1 \cdots dt_\nu = 0,$$

which for $n = 1$ becomes a homogeneous linear integral equation. We wish to generalize the theorem of Jentzsch guaranteeing, for a positive kernel, the existence of a positive eigenvalue μ and a positive eigenfunction.

From the homogeneity of the equation it follows that for every $y(s)$ the function $cy(s)$ is also a solution. Moreover, if we consider the equation

$$z^n(s) - \sum_{\beta=0}^{n-1} z^\beta(s)\mathfrak{K}_\beta(s; y)$$

$$\equiv z^n(s) - \sum_{\beta=0}^{n+1} z^\beta(s) \sum_{(\alpha_1+\cdots+\alpha_\nu=n-\beta)} \int_0^1 \cdots \int_0^1 K_{\beta\alpha_1\cdots\alpha_\nu}(st_1\cdots t_\nu)y^{\alpha_1}(t_1) \cdots y^{\alpha_\nu}(t_\nu)dt_1\cdots dt_\nu = 0$$

and if $z(s)$ is a solution of this algebraic equation for z, then $cz(s)$ is a solution of the equation arising from the given equation by substitution of cy in place of y.

The kernels are assumed to be continuous in all variables, and also, for the aggregate of all the integral terms ($\beta = 0$) free of $z(s)$ and for $y(s) \geqq 0$ and $\int_0^1 y(s)ds = 1$, we assume that

$$\mathfrak{K}_0(s; y) \geqq \delta > 0.$$

Then the left-hand side of the last equation is always negative for $z = 0$ but is positive for sufficiently large z, so that for every s there exists a positive root z. We choose the greatest positive root $z(s)$ and also assume that the discriminant of the equation for z is nonzero for all s, provided y is one of the functions described above. For example, these assumptions are satisfied if all the kernels vanish for $\beta > 0$, while the given assumptions hold for $\beta = 0$.

In view of this requirement on the discriminant it is easy to see that the solution $z(s)$ is continuous for all s. If we set

$$\mu = \mu(y) = \int_0^1 z(s)ds,$$

then

$$z^*(s) = \frac{z(s)}{\int_0^1 z(s)ds}$$

is normalized and is a solution of the equation

$$\mu^n z^*(s)^n - \sum_{\beta=0}^{n-1} \mu^\beta z^*(s)^\beta \mathfrak{K}_\beta(s; y) = 0.$$

Let us now examine the mapping $z^* = z^*(y)$ thus defined for the positive functions $y(s)$ normalized by

$$\|y\| = \int_0^1 y(s)ds = 1.$$

These functions form a convex, normwise bounded closed set M of the Banach space of all positive integrable normalized functions, the fact that the set is closed being easily proved as follows: let $y_n(s) \geqq 0$ be a sequence with $\int_0^1 y_n(s)ds = 1$ and let $\int_0^1 |y_n(s) - y(s)|ds \to 0$ for $n \to \infty$. If for the limit function we had $y(s) < 0$ on a set of positive measure $\mathfrak{N}$, it would follow that

$$\int_0^1 |y_n(s) - y(s)|ds \geqq \int_{(\mathfrak{N})} |y_n(s) - y(s)|ds = \int_{(\mathfrak{N})} (y_n(s) - y(s))ds$$
$$\geqq \int_{\mathfrak{N}} (-y(s))ds > 0 \qquad \text{for all } n,$$

which is a contradiction. Thus $y(s) \geqq 0$ except on a set of measure zero, and

$$\left|\int_0^1 (y_n(s) - y(s))ds\right| \leqq \int_0^1 |y_n(s) - y(s)|ds \to 0,$$

i.e., $\int_0^1 y(s)ds = 1$. The image of $\mathfrak{M}$ is a sequentially compact set belonging to $\mathfrak{M}$.

For if $y_1, y_2, \ldots$ is a sequence of elements of $\mathfrak{M}$, then the corresponding functions $z_1^*, z_2^*, \ldots$ form a sequence of positive equicontinuous uniformly bounded functions, since $\mathfrak{K}_0(s; y_m) \geqq \delta$ implies $z_m^* \geqq \epsilon(\delta) > 0$ for all m and thus $\mu(y_m) \geqq \epsilon$, so that the uniform boundedness of $z_m(s)$ implies the same property for $z_m^*(s)$. The uniform boundedness of $z_m(s)$ then follows from the fact that for $M > 0$ the function $\dfrac{z_m(s)}{M}$ satisfies an equation whose coefficients

$$\left|\frac{\mathfrak{K}_\beta(s; y_m)}{M^{n-\beta}}\right| \leqq \frac{K\int_0^1 y_m^{\alpha_1}ds \cdots \int_0^1 y_m^{\alpha_\nu}ds}{M^{n-\beta}} \leqq \frac{KM^{\alpha_1 - 1 + \cdots + \alpha_\nu - 1}}{M^{n-\beta}} \leqq KM^{-\nu}$$

become arbitrarily small if $y_m \leqq M$ and M increases beyond all bounds. Thus by the theorem of Arzelà there exists a uniformly convergent sequence $z_{\rho_1}^*, z_{\rho_2}^*, \ldots$, which is therefore convergent in the sense also of the norm of the Banach space; in other words, the space of the z^* is sequentially compact. Application of the Schauder fixed point theorem shows the existence of a normalized $y(s)$ which is the solution of the given equation; thus $\mu = \mu(y)$ is a positive eigenvalue.

9. More General Nonlinear Integral Equations

In our discussion of functional equations up to now we have begun in each case with an operator $z = \mathfrak{A}(y)$ and have looked for a fixed point $z = y$. But in the more general cases now to be investigated we consider an equation of the form

$$F(s; y, z) = 0,$$

which, when one function y is given, determines another function z, depending on y. If y and z are ordinary variables, the problem is to determine the function $z = z(y)$ implicitly defined by the equation, and our present task is to extend this problem to the domain of functional analysis. To avoid excessive abstraction we will deal with the special case, also of importance in ordinary analysis, that the problem can be solved by means of expansions in series. For this purpose we proceed as follows.

We assume that the functions $y(s)$ and $z(s)$ are continuous in the interval $0 \leqq s \leqq 1$ and that the left-hand side of the given equation

$$F(s; y(s), z(s)) = 0$$

can be represented by an integral power series:

$$F(s; y, z) = \sum_{\substack{\alpha+\alpha_1+\cdots+\alpha_\nu=m \\ \beta+\beta_1+\cdots+\beta_\nu=n \\ n+m=1,2\ldots}} \int_0^1 \cdots \int_0^1 K_{\substack{\alpha\alpha_1\cdots\alpha_\nu \\ \beta\beta_1\cdots\beta_\nu}}(s, t_1, \ldots, t_\nu) y^\alpha(s) y^{\alpha_1}(t_1) \cdots y^{\alpha_\nu}(t_\nu) z^\beta(s) z^{\beta_1}(t_1) \cdots z^{\beta_\nu}(t_\nu) dt_1 \cdots dt_\nu.$$

Here the kernel functions are continuous in all their variables, the indices are integers $\geqq 0$, and, if all the indices α_ρ and β_ρ vanish, the corresponding term is a continuous function of s multiplied by $y^\alpha(s)z^\beta(s)$. Moreover, the series is maximally convergent in the sense that there exist two positive numbers U and V such that

$$\sum_{m+n=1,2,\ldots} K_{mn} U^m V^n$$

converges, where

$$K_{mn} = \max \left(\sum_{\substack{\alpha+\cdots+\alpha_\nu=m \\ \beta+\cdots+\beta_\nu=n}} \left| K_{\substack{\alpha\alpha_1\cdots\alpha_\nu \\ \beta\beta_1\cdots\beta_\nu}} \right| \right).$$

Then the series converges uniformly for all s, if y and z are so chosen that $\max |y(s)| \leqq U$ and $\max |z(s)| \leqq V$.

Since $F(s; y, z)$ does not contain any term independent of y and z, it follows that

$$F(s; 0, 0) = 0.$$

Our problem now consists of finding, for a fixed preassigned y with $\max |y(s)| \leqq U$, a function $z(s)$ such that $\max |z(s)| \leqq V$, with consequent convergence of the integral power series, and such that for all s in $0 \leqq s \leqq 1$ we have

$$F(s; y(s), z(s)) = 0.$$

Before taking up this problem in greater detail let us discuss its meaning. If to the terms on the left-hand side we add a term $f(s)$ independent of

y and z and *assume* that the resulting equation has a solution $y_0(s)$, $z_0(s)$, and if we then rearrange the equation in powers of $y - y_0$ and $z - z_0$, we obtain an equation in the differences $y - y_0$ and $z - z_0$ which, like the equation from which we originally began, has no free term. Thus, our problem can be interpreted in the following way: *in the neighborhood of a preassigned system of solutions* $y_0(s)$, $z_0(s)$ (or in simplified notation, near the point 0, 0) we seek a further system of solutions. The situation is exactly the same as in the special case of an implicit function obtained by omitting all terms that involve an integral. Thus we are dealing with a *local theory*, whereas in the preceding section our statements were *global*. In this sense the present discussion represents a restriction of earlier results, although it is an extension of them in the sense that the equation in question is much more general. The reader should keep this distinction clearly in mind in what follows.

In order to solve the problem we first write the given equation more precisely:

$$a(s)z(s) - \int_0^1 K(s, t)z(t)dt = b(s)y(s) + \int_0^1 \mathfrak{L}(s, t)y(t)dt + \sum_{m+n\geqq 2} \cdots,$$

by putting the terms of first order at the front in a simplified notation. In order to avoid computation that has nothing to do with the essence of the problem we assume that $a(s) \neq 0$ in the whole interval and can therefore be set equal to unity without loss of generality.

In the special problem of calculating an implicit function $z(y)$ it is customary to assume that $\dfrac{\partial F}{\partial z}(0, 0) \neq 0$. For our general case the corresponding assumption is that the Fréchet differential

$$\left[\lim_{\epsilon\to 0} \frac{F(s; y, z + \epsilon\eta) - F(s; y, z)}{\epsilon}\right]_{y=z=0}$$

is nonzero for every choice of η. Then we obviously have

$$\eta(s) - \int_0^1 K(s, t)\eta(t)dt \neq 0$$

for every function $\eta(s)$ not identically equal to zero, which means that the equation

$$\eta(s) - \int_0^1 K(s, t)\eta(t)dt = 0$$

has the unique solution $\eta(s) \equiv 0$. In this case the nonhomogeneous linear equation

$$z(s) - \int_0^1 K(s, t)z(t)dt = f(s)$$

has a unique solution, which by means of the "resolvent kernel" $L(s, t)$ can be written in the form[18]

$$z(s) = f(s) + \int_0^1 L(s, t)f(t)dt.$$

We first examine the principal case. Applying the above remarks to our present equation we obtain

$$z(s) = b(s)y(s) + \int_0^1 \mathfrak{L}(s, t)y(t)dt + \sum_{m+n\geqq 2} \cdots$$
$$+ \int_0^1 L(s, t)\left[b(t)y(t) + \int_0^1 \mathfrak{L}(t, u)y(u)du + \sum_{m+n\geqq 2} \cdots\right]dt,$$

which, if we set

$$b(s)y(s) + \int_0^1 \mathfrak{L}(s, t)y(t)dt$$
$$+ \int_0^1 L(s, t)\left[b(t)y(t) + \int_0^1 \mathfrak{L}(t, u)y(u)du\right]dt = w(s),$$

becomes

$$z(s) = w(s) + \sum_{m+n\geqq 2} U_{mn}(s; y, z) = w(s) + \mathfrak{A}(y, z).$$

In the summand $U_{mn}(s; y, z)$ we have combined all those terms arising from the given equation which are of total order $m + n$ ($\geqq 2$) with respect to y and z.

In order to solve the last equation, we construct the iteration scheme

$$z_1(s) = w(s),$$
$$z_2(s) = w(s) + \sum_{m+n\geqq 2} U_{mn}(s; y, z_1),$$
$$\cdots\cdots\cdots\cdots\cdots$$
$$z_h(s) = w(s) + \sum_{m+n\geqq 2} U_{mn}(s; y, z_{h-1})$$

[18] The theory of linear integral equations is closely related to systems of linear equations, where the distinction must be made between the *principal case* (determinant $\neq 0$) and the *special case* (determinant $= 0$).

In the principal case the nonhomogeneous system of equations has a unique solution for an arbitrary right-hand side, while in the special case the homogeneous system has a nontrivial solution. The situation is exactly analogous in integral equations, where the two cases (now called the principal case and the *branching case*, are mutually exclusive in exactly the same way. The latter theory is named after Fredholm, who has also shown that in the principal case the solution can be obtained by means of the *resolvent kernel*.

and proceed to prove that all the sums in it are in fact convergent. For this purpose we let W denote the maximum of the absolute values of $y(s)$ and $w(s)$; it is clear that if max $|y(s)|$ is chosen sufficiently small, then W can also be made as small as desired. If we can so arrange matters that all the $|z_1|, |z_2|, \ldots, |z_h|$ are smaller than V, the convergence of the above sums is guaranteed.

We now consider the operator

$$w(s) + \mathfrak{A}(y, z)$$

and assert that we can find a constant $A > 0$ such that

$$|w(s) + \mathfrak{A}(y, z)| \leqq W + A(U^2 + UV + V^2).$$

For in fact, for the majorant series[19]

$$G(U, V) = \sum_{m+n \geqq 2} U_{mn}(U, V)$$

it follows from the mean value theorem, in view of the conditions

$$G(0, 0) = 0, \qquad \frac{\partial G}{\partial U}(0, 0) = 0, \qquad \frac{\partial G}{\partial V}(0, 0) = 0,$$

that

$$G(U, V) = \tfrac{1}{2}U^2 F_{UU}(\vartheta U, \vartheta V) + UVF_{UV}(\vartheta U, \vartheta V) + \tfrac{1}{2}V^2 F_{VV}(\vartheta U, \vartheta V),$$

which proves the desired assertion for a suitably chosen constant A.

We now require that for a suitable τ

$$W + A(\tau^2 + \tau W + W^2) = \tau,$$

whereupon with this τ we also have, for $V \leqq \tau$,

$$|w(s) + \mathfrak{A}(y, z)| \leqq W + A(\tau^2 + \tau W + W^2) = \tau.$$

But the desired condition on τ is easily satisfied; we need only set

$$\tau = \frac{1}{2A} - \frac{W}{2} - \sqrt{\left(\frac{1}{2A} - \frac{W}{2}\right)^2 - W\left(\frac{1}{A} + W\right)}.$$

The expansion of the right-hand side in powers of W begins with

$$\tau = W + AW^2 + \cdots,$$

and is therefore positive for small positive W and approaches zero for $W \to 0$. Thus if max $|y(s)|$ is sufficiently small, we can also make τ as small as desired. If we then choose $V \leqq \tau$, the convergence of

$$\sum U_{mn}(s; y, z_h)$$

is guaranteed for all $h = 1, 2, \ldots$.

[19] This series is formed from $U_{mn}(s; y, z)$ in the same way as before and is likewise convergent for U, V.

We must now ask whether the iterative sequence $z_1, z_2, \ldots$ converges to a limit function. To answer this question we form the expression

$$|z_{h+1} - z_h| = \left| \sum_{m+n \geqq 2} U_{mn}(s; y, z_h) - U_{mn}(s; y, z_{h-1}) \right|.$$

Then it is easy to see that there exists a positive constant B such that the absolute value of the right-hand side is $\leqq B(U + V) \max |z_h - z_{h-1}|$. If we choose W small enough so that $B(W + \tau) \leqq q < 1$, the desired uniform convergence is thereby guaranteed. The limit function $z(s)$ then satisfies the integral equation, and $|z(s)| \leqq \tau$.

In order to prove that the solution is unique we assume that for a sufficiently small $\max |y(s)|$ there exist two distinct solutions $z(s)$ and $z^*(s)$. Then it follows from the integral equation that

$$|z(s) - z^*(s)| = \left| \sum_{m+n \geqq 2} U_{mn}(s; y, z) - \sum_{m+n \geqq 2} U_{mn}(s; y, z^*) \right|$$

and thus

$$|z(s) - z^*(s)| \leqq q \max |z(s) - z^*(s)|,$$

which is possible only if $z(s) - z^*(s) = 0$, $z(s) = z^*(s)$.

It is also easy to show that this unique solution can be expanded in an integral power series in $y(s)$ which is convergent for sufficiently small values of $\max |y(s)|$.

The whole theory can be extended to the case of several parameter functions $y_1(s), \ldots, y_m(s)$, where the given equation is represented as an integral power series in $y_1, \ldots, y_m$ and the solution $z(s)$ can also be expanded in an integral power series in $y_1, \ldots, y_m$. In particular, the parameters may be constants, in which case the solution appears as a power series in these parameters.

In the *branching case*[20] the equation

$$\eta(s) - \int_0^1 K(s, t)\eta(t)dt = 0$$

has p linearly independent solutions $\varphi_\alpha(s)$ $(\alpha = 1, \ldots, p)$; we then let $\psi_\beta(t)$ $(\beta = 1, \ldots, p)$ be the linearly independent solutions of the transposed integral equation

$$\zeta(t) - \int_0^1 K(s, t)\zeta(s)ds = 0,$$

where we may assume that the systems are both orthonormal.

Following E. Schmidt (cf. [11] in the literature for Part C), we then obtain the kernel

$$K_1(s, t) = K(s, t) + \sum_{\beta=1}^{p} \overline{\varphi_\beta(t)}\overline{\psi_\beta(s)}$$

[20] Cf. footnote 18.

in the principal case, so that the earlier theory is applicable to it. Thus we write our equation in the form

$$z(s) - \int_0^1 K_1(s, t)z(t)dt$$

$$= b(s)y(s) + \int_0^1 \mathfrak{L}(s, t)y(t)dt - \sum_{\beta=1}^{p} \overline{\psi_\beta(s)} \int_0^1 z(t)\overline{\varphi_\beta(t)}dt + \sum_{m+n\geqq 2} U_{mn}(s; y, z).$$

Then if $\mathfrak{L}_1(s, t)$ is the resolvent kernel for $K_1(s, t)$ and if $\int_0^1 z(t)\overline{\varphi_\beta(t)}dt$ is denoted by x_β, and if we set

$$U(s) = b(s)y(s) + \int_0^1 \mathfrak{L}(s, t)y(t)dt,$$

we obtain the representation:

$$z(s) = U(s) - \sum_{\beta=1}^{p} x_\beta \overline{\psi_\beta(s)} + \sum_{m+n\geqq 2} U_{mn}(s; y, z)$$

$$+ \int_0^1 \mathfrak{L}_1(s, u)\left[U(u) - \sum_{\beta=1}^{p} x_\beta \overline{\psi_\beta(u)} + \sum_{m+n\geqq 2} U_{mn}(u; y, z)\right]du.$$

By multiplication with $\overline{\varphi_\alpha(s)}$ and integration, we reach the result

$$0 = -\int_0^1 U(s)\psi_\alpha(s)ds - \sum_{m+n\geqq 2} \int_0^1 U_{mn}(s; y, z)\psi_\alpha(s)ds \quad (\alpha = 1, \ldots, p).$$

If for z we substitute the solution of the integral equation with the modified kernel K_1, which still contains the parameters $x_1, \ldots, x_p$, this result leads to p conditions on the parameters. These conditions are called the *branching equations* of the problem.

Example.

$$\ddot{x}(t) + \alpha^2 \sin x(t) = -\beta f(t) \qquad (0 \leqq t \leqq \pi), \qquad x(0) = x(\pi) = 0.$$

Let $f(t)$ be an odd periodic function of t. The transformation to an integral equation leads to:

$$x(t) - \alpha^2 \int_0^\pi G(t, \tau) \sin x(\tau)dt = \beta h(t) = \beta \int_0^\pi G(t, \tau)f(\tau)dt.$$

$$G(t, \tau) = \frac{2}{\pi} \sum_{n=1}^{\infty} \frac{\sin nt \sin n\tau}{n^2} = \begin{cases} t\left(1 - \dfrac{\tau}{\pi}\right) & (t \leqq \tau), \\ \tau\left(1 - \dfrac{t}{\pi}\right) & (\tau \leqq t). \end{cases}$$

There can be a branching of the solution only if the equation

$$\ddot{\varphi}(t) = \alpha^2 \cos x(t)\varphi(t) = 0, \qquad \varphi(0) = \varphi(\pi) = 0$$

has a nontrivial solution, or if the integral equation

$$\varphi(t) - \alpha^2 \int_0^\pi G(t, \tau) \cos x(\tau)\varphi(\tau)d\tau = 0$$

has a nontrivial solution. For $x(t) = 0$ this is possible only for $\alpha^2 = 1, 4, 9, \ldots$. These points thus provide the possible branchings of the solutions in the neighborhood of the solution $x(t) = 0$ corresponding to $\beta = 0$. The branching possibilities in the neighborhood of the periodic solution $x(t)$ determined above have not yet been investigated. (Cf. R. Iglisch: Reelle Lösungsfelder der elliptischen Differentialgleichung $\Delta u = F(u)$. Math. Annalen 101 (1929), 98–119; Zur Theorie der Schwingungen, II. Mitteilung, Monatshefte für Mathematik und Physik 39 (1932), 173–220.)

Bibliography

[1] Alexandroff, P. and Hopf, H.: Topologie, Vol. I. Springer-Verlag, Berlin 1935.

[2] Bückner, H.: Die praktische Behandlung von Integralgleichungen. Springer-Verlag, Berlin-Göttingen-Heidelberg 1952.

[3] Collatz, L.: Einige Anwendungen funktionalanalytischer Methoden in der praktischen Analysis. ZAMM 4 (1953).

[4] Courant, R., and Hilbert, D.: Methods of mathematical physics. (Translated from the German.) Interscience, New York 1953.

[5] Fréchet, M.: La notion de différentielle dans l'analyse générale. Ann. Ec. Norm. Sup. 42 (1925).

[6] Grüss, G.: Variationsrechnung. Quelle & Meyer, Heidelberg 1955.

[7] Hadamard, J.: Sur quelques applications de l'indice de Kronecker. In: J. Tannery, Introduction à la théorie des fonctions d'une variable II, 2me. éd. (1910).

[8] Hukuhara: Sur l'existence des points invariants d'une transformation dans l'espace fonctionnel. Jap. Journ. of Math., Vol. XX (1950).

[9] Lichtenstein, L.: Vorlesungen über einige Klassen nichtlinearer Integralgleichungen und Integro-Differentialgleichungen. Springer-Verlag, Berlin 1931.

[10] Schauder, J.: Der Fixpunktsatz in Funktionalräumen. Studia Mathematica II (1930).

[11] Schmidt, E.: Auflösung der allgemeinen linearen Integralgleichung. Math. Ann. **64**, 161–174 (1907).

[12] Weissinger, J.: Zur Theorie und Anwendung des Iterationsverfahrens. Mathematische Nachrichten **8** (1952).

[13] Wiarda, G.: Integralgleichungen. Teubner-Verlag, Leipzig-Berlin 1930.

Bibliography added in Translation

Bachman, G., and Navici, L.: Functional analysis. Academic Press, New York 1966.

Berberian, S. K.: Introduction to Hilbert space. Oxford University Press, New York 1961.

Collatz, L.: Functional analysis and numerical mathematics. (Translated from the German by H. Oser.) Academic Press, New York 1966.

Dunford, N., and Schwartz, J.: Linear Operators I. General Theory; II. Spectral Theory. Self adjoint operators in Hilbert space. Interscience Publishers, New York 1958.

Goffman, C., and Pedrick, G.: First course in functional analysis. Prentice-Hall, Englewood Cliffs, N.J., 1965.

Goldberg, S.: Unbounded linear operators: Theory and applications. McGraw-Hill, New York 1966.

Goldstein, A. A.: Constructive real analysis. Harper and Row, New York 1967.

Gould, S. H.: Variational methods for eigenvalue problems. 2nd ed. University of Toronto Press, Toronto; Oxford University Press, London 1966.

Kato, T.: Perturbation theory for linear operators. Springer-Verlag, New York 1966.

Ljusternik, L. A., and Sobolev, V. I.: Elements of functional analysis. (Translated from the Russian by A. E. Labarre, H. Izbicki, and H. W. Crawley.), Ungar, New York 1961.

Riesz, F., and Sz.-Nagy, B.: Functional analysis (Translated from the French by L. F. Boron). Ungar, New York 1955.

Wilansky, A.: Functional analysis. Blaisdell, New York 1964.

Yosida, K.: Functional Analysis. Academic Press, New York; Springer-Verlag, Berlin 1965.

CHAPTER 12

Real Functions

Introduction

1. *Various Interpretations of "Larger" or "Richer"*

The statement that a given set A is "larger" or "richer" than another set B can be interpreted in various ways. Among the simplest are to define A as being "richer" than B if B is properly contained in A, in symbols $A \supset B$, i.e., if $A \neq B$ and A contains every point of B, or again if the power[1] of A is greater than that of B, in particular if A is infinite and B is finite. But we also have the two modern concepts at our disposal: a set may be of *measure* 0 or of *measure* $\neq 0$,[2] and secondly, it may be of *category* 1 or of *category* 2.[3] Any set whose measure (or category) $\neq 0$ (or $\neq 1$) can be regarded as "richer" than any other set whose measure (or category) is 0 (or 1). Here is an impressive example:

2. *A Nowhere Differentiable, Real Function[4] on the Set R of Real Numbers (Banach Theorem).*

We consider the set St of all real functions defined and continuous on the real numbers R; let M_1 be the subset of St consisting of those functions which are differentiable at one point at least. Many strenuous efforts were made to prove the equality $St = M_1$ or $M_2 = St - M_1 = \varnothing$.[5] But in reality $M_2 \neq \varnothing$, as has been known since Weierstrass produced a real continuous function on R which is not differentiable at any point of R. Now it is natural to ask which of the two sets M_1, M_2 is "richer"; yet

[1] On the concept of power see IA, §7.3.

[2] If A and B are, for example, sets of real numbers. For the concept of the measure of a set of real numbers see III 3, §2.

[3] For the concept of category, see Introduction 2, in particular §1.2.6.

[4] By a real function f we mean a mapping of a set S into the set R of the real numbers. In the present chapter, R will always denote this set.

[5] $\varnothing$ is the symbol for the empty set.

the number of elements[6] in each of them is $c = 2^{\aleph_0}$, i.e., the same number as in the whole of *St*. For if f_i is in M_i, $i = 1, 2$, then $f_i + k \in M_i$ for every real number k, and if $k \neq k'$ we have $f_i + k = f_i + k'$. Thus M_i is at least of the same power as the real numbers *R*. But since *St* has the power c, and since M_i, as a subset of *St*, can be of power c at most, the sets M_i have exactly c elements.[7]

But the size of the sets M_1, M_2 can be considered from another point of view. For this purpose we first define the following concept. In the set $[0, 1]_R$ of real numbers $0 \leqq x \leqq 1$, we consider the ternary Cantor set *T*, which contains $c = 2^{\aleph_0}$ points, like *R* itself. But although *T* is of measure 0, the set $[0, 1]_R$ is of measure 1. Moreover, *T* is nowhere dense,[8] and $[0, 1]_R$ is not only everywhere dense (and thus not, like *T*, nowhere dense) but it is not even the union of countably many nowhere dense sets. This last fact can also be expressed in the following way: $[0, 1]_R$ is not of "first category," but is of "second category" (with respect to *R*). *The set of the irrational numbers is also of second category with respect to R* (Baire). On the other hand, the ternary Cantor set *T* (although it contains c points), is of the first category with respect to *R*,[9] as are also the set *Q* of rational numbers and the set *A* of algebraic numbers.

In order to introduce the concept "of first (or second) category with respect to *R*" we have only needed to know (in addition to the idea of countability) what is meant by "nowhere dense in *R*." So we can define by analogy what it means to say that a subset of *St* is "nowhere dense in *St*" and then speak of the concepts "of first (or second) category with respect to *St*."

After these explanations the meaning of the following theorem of Banach will be clear.

Theorem 1 (Banach). *M_1 is of first category with respect to St; and M_2 is of second category with respect to St.*

In this sense M_2 is "richer" than M_1.

3. *Metric, Descriptive, and Topological Properties*

The foregoing theorems of Baire and Banach are stated in terms of a typical *descriptive*, nonmetric concept of the theory of functions: namely, first and second category. On the other hand, the *metric* theory of functions investigates such metric properties as measure, length, extreme values, etc. and the *topological* theory deals with topological properties, i.e., properties preserved under topological mappings, among them, for example, the concept of continuity, the *Bolzano theorem*, and the like. The descriptive

[6] See IA, §§7.3 and 7.5.
[7] See the theorem of Bernstein, IA, §7.3.
[8] For this concept see §1.2.5.
[9] See §1.2.6.

theory of functions, lying between topology and the metric theory, is concerned chiefly with questions of density and limit processes,[10] by means of which it is possible to build up complicated functions (sets) from relatively simple ones (*Baire classes of functions, Borel sets, A-sets, projective sets*, etc.[11]). For example, the following very interesting theorem of Baire is definitely of a descriptive character:

A real function f defined on R is the limit of continuous real functions on R if and only if every perfect set[12] $P \subseteqq R$ *contains a point of continuity of* $f|P$ (cf. §1.1).

Thus the Dirichlet function χ ($\chi(x) = 1$ for rational $x \in R$ and $\chi(x) = 0$ for irrational $x \in R$) is not such a function, since, e.g., the perfect set R has no point of continuity of χ (although χ is the limit of a sequence of functions of the first Baire class (cf. §4.3)); for example,

$$\chi(x) = \lim_{m\to\infty} \lim_{n\to\infty} \cos((m!\pi x)^{2n}).$$

4. *Purpose of the Chapter*

Beginning with simple descriptive definitions we present some of the characteristic results of the descriptive theory of functions, leading up to the questions of logical existence, constructibility, and the like, introduced by Luzin, Gödel and others and culminating in the continuum hypothesis.[13]

1. Basic Concepts of the Descriptive Theory of Sets and Functions

All sets occurring in the following section, unless specifically described otherwise, are subsets of real Euclidean spaces of numbers.[14] *All mappings* (see §1.1) *are therefore mappings of subsets of Euclidean spaces into Euclidean spaces.*

1.1. *Functions: Their Restrictions and Extensions*

Let f be a mapping of the set M into the set N, i.e., for every element $x \in M$ let there be assigned an element $fx \in N$; a mapping of the set M is also called a function on M. If B is a subset of M, then $f|B$ denotes the well-defined function whose domain of definition is B and whose values there coincide with those of f. Thus $f|B$ is defined by $(f|B)x = fx$ for $x \in B$. Then $f|B$ is said to be a *restriction* of f. If we wish to emphasize the domain of definition M, we can also write f itself in the form $f|M$.

[10] See §4 of this chapter.
[11] For these concepts see §§4, 5, 6, 7 of this chapter.
[12] For this concept see §1.2.4.
[13] See §7.
[14] The arguments apply to a wide class of topological spaces but for simplicity we confine our attention to subsets of Euclidean spaces of real numbers.

Let $X \subseteq M$; then fX is the union of all fx, $x \in X$. For example, for the set R of real numbers we have $\cos R = [-1, +1]_R$. In this way every mapping f of M induces a mapping of the power set $\mathbf{P}M$ into the power set $\mathbf{P}N$. (The power set $\mathbf{P}A$ of a set A consists of all the subsets of A.) We stipulate that the induced mapping takes the empty set into itself. The restrictions of this induced mapping of $\mathbf{P}M$ are also called extensions of the original mapping f of M, provided they are defined for every one-point set $\{x\} \subseteq M$.

It should be emphasized if B, B' are two distinct subsets of M, then the corresponding restrictions of f are to be considered as distinct functions. For example, if D is the set of integers and $D\pi$ is set $\{n\pi | n \in D\}$, then $\sin | D\pi$ is a function whose domain of definition is $D\pi$ and whose range is $\{0\}$; if χ is the *Dirichlet function* and I is the set of irrational numbers, then $\chi | I$ is a constant, namely zero. Let us point out that $\chi | I$ is not defined for the value 3, for example.

Let us further note that $\chi | I$, in contrast to χ, is continuous, since $\chi | I$ is constant. Thus $f | M$ can be discontinuous while $f | X$ is continuous. Here we say that the function $f | X$ is continuous at $a \in X$ if for every sequence of points x_n, $n = 1, 2, 3, \ldots$, in X converging to a, the sequence of points fx_n also converges to fa. On the other hand, if there exists at least one sequence of points x_n, $n = 1, 2, 3, \ldots$, in X which converges to $a \in X$ but for which the sequence fx_n does not converge to fa, then $f | X$ is said to be discontinuous at the point a in X. The function $f | X$ is said to be continuous if it is continuous at every point $a \in X$. It is obvious that the functions $\chi | R$ and $\chi | I$ are, respectively, nowhere continuous and everywhere continuous in their domains of definition.

1.2. *Density*

If M is a subset[15] of the Euclidean space R^n, by a neighborhood of a point $x \in M$ we mean the intersection $K_x \cap M$ of M with an open hyperball K_x of R^n containing the point x; in order to emphasize that we are dealing with a neighborhood of x in M we shall sometimes speak of an "M-neighborhood of x."

In the following subsections of §1 *all points and sets of points*, unless otherwise described, *belong to a fixed subset E of R^n.*

1.2.1. *Isolated and Dense-in-Themselves Sets* Let M be a subset of E. A point $x \in M$ is said to be *isolated* in M provided there exists an E-neighborhood U_x of x containing only the point x in common with M (in other words, provided that there exists an M-neighborhood of x consisting of x alone). If every E-neighborhood of x contains at least one

[15] Let us emphasize that by a subset we do not necessarily mean a proper subset.

point of $M - \{x\}$, then the point x is not isolated. The set M is said to be *dense-in-itself* (isolated) provided that M contains no isolated (only isolated) points.

1.2.2. *Hull of a Set* It is customary to say that the point $x \in E$ *touches* the set M (in symbols, $x \in \overline{M}$) provided every neighborhood U_x of x in E intersects the set M: $U_x \cap M \neq \varnothing$. Then the set $\overline{M}$ of all points touching the set M is called the *hull* of M in E. For example, if M is the set of all rational numbers and E the set of all real numbers, then $\overline{M} = E$; for $E = R$ and $M := \{1, \frac{1}{2}, \frac{1}{3}, \ldots\}$ we have $\overline{M} = \{0\} \cup M$.

1.2.3. *Closed* (**F**-) *and Open* (**G**-)*sets* If $M \supseteqq \overline{M}$, then M is said to be *closed* or be a *closed set* (**F**-*set*). It is to be noted that the concept of a hull and consequently of closedness is defined with respect to E. Since it is always true that $M \subseteqq \overline{M}$, for a closed set M we must have $M = \overline{M}$. The complements of **F**-sets (with respect to E) are called *open sets* or **G**-*sets*.

Theorem 2. *A set $M \subseteqq E$ is open in E if and only if every $x \in M$ in E has a neighborhood U_x consisting entirely of points of M.*

The condition is necessary; for if M is open, its complement $\mathbf{C}M$ is closed. Now we are required to prove that every $x \in M$ is contained in an E-neighborhood U_x lying entirely in M. If there were no neighborhood U_x of this sort, every neighborhood of x in E would contain a point from the complement $\mathbf{C}M$ of M, so that we would have $x \in \overline{\mathbf{C}M}$. But since $\mathbf{C}M$ is closed (M is assumed to be open) we have $\overline{\mathbf{C}M} \subseteqq \mathbf{C}M$, i.e., $x \in \mathbf{C}M$, in contradiction to $x \in M$.

The condition is also sufficient: i.e., if every $x \in M$ contains an E-neighborhood $U_x \subseteqq M$, then M is open. In other words, we must prove that $\mathbf{C}M$ is closed: i.e., if $y \in \overline{\mathbf{C}M}$, then $y \in \mathbf{C}M$. But otherwise we would have $y \in M$ and, by the assumption, there would exist in E a neighborhood $U_y \subseteqq M$ of y, in contradiction to the fact that $y \in \overline{\mathbf{C}M}$ means that the neighborhood U_y must contain at least one point of $\mathbf{C}M$.

1.2.4. *Perfect Sets* Sets that are dense-in-themselves and closed are called *perfect* sets.[16]

If for E we take the set R of real numbers, then R itself and the ternary Cantor set T (with respect to R) are perfect. The set Q of rational numbers is not perfect in R, since it is not closed[17] in R, and the same remark holds for the half-closed interval $(0, 1]_R$ of real numbers $0 < x \leqq 1$. The set of integers is closed in R, but not perfect (in R), since it is not dense-in-itself.

[16] It is to be noted that this definition is relative to E, so that the property of being perfect depends on the choice of E.

[17] But if we take $E = Q$, then Q is perfect.

Theorem 3. *Every uncountable* **F**-*set* (**G**-*set*) *contains a perfect set.*

This statement is a very special case of the theorem that every uncountable analytic set contains a perfect set (Alexandroff-Hausdorff for **B**-sets, Suslin for **A**-sets; cf. §§3.2 and 3.3).[18]

1.2.5. *Everywhere Dense Sets, Nowhere Dense Sets* If two sets $A, B \subseteqq E$ are such that $\overline{A} \supseteqq B$, we say that A is *everywhere dense* with respect to B (if also $A \subseteqq B$, then A is everywhere dense *in* B). If there exists a point $b \in B$ and a neighborhood U_b of b in E such that A is dense with respect to $U_b \cap B$, then A is said to be *dense with respect to B in the neighborhood of b*; on the other hand, if there is no $b \in B$ such that A is dense with respect to B in a neighborhood of b, then A is said to be *nowhere dense* with respect to B. For example, the set Q of rational numbers is everywhere dense in R (here we have taken $E = R$), whereas the one-point set $\{x\}$, $x \in R$, is nowhere dense in R. Consequently, although the countable set Q is everywhere dense in R, it is certainly the union of countably many nowhere dense sets $\{x\}$, $x \in Q$. The ternary Cantor set T is also nowhere dense in R.

1.2.6. *Category of Sets* The following terminology is due to Baire. *A set is said to be of the first category (with respect to E) if it is a union of countably many nowhere dense (in E) sets; a set which is not of the first category is said to be of the second category.*

We now prove the following theorem.

Theorem 4. *The set Q of rational numbers is of first category (with respect to E = R). The sets I, Tr, R of irrational numbers, transcendental numbers, and real numbers are of second category (with respect to E = R).*

Proof. Let $x \in R$; then $\{x\}$ is nowhere dense in R; consequently, every countable set in R, in particular the set Q, is of first category. We now prove that every interval $[a, b]$ of the set R is of second category. Let $X = M_0 \cup M_1 \cup \cdots$ be the union of arbitrarily many nowhere dense sets $M_0, M_1, \ldots$ in $[a, b]$. In $[a, b]$ there exists a point x_0 and a neighborhood U_0 of x_0 such that $\overline{U_0} \cap M_0 = \varnothing$. Then U_0 contains a point x_1 with a neighborhood $U_1 \subseteqq U_0$ of x_1 such that $\overline{U_1} \cap M_1 = \varnothing$. By induction we obtain a decreasing sequence $\overline{U_0} \supset \overline{U_1} \supset \cdots$ of closed intervals $\overline{U_n}$, which are disjoint to $M_0, M_1, \ldots, M_n$. It follows that the nonempty subset $\cap \overline{U_n}$[19] of $[a, b]$ is disjoint to the union $\bigcup M_n$. But this means that $[a, b]$

[18] For the concepts of an **A**- (or analytic) set and a **B**-set see §§3.3 and 6 of this chapter.

[19] We have here made use of the so-called *Cantor intersection theorem*, which states: let $U_1, U_2, \ldots, U_k, \ldots$ be a decreasing sequence of non-empty closed bounded subsets of R^n: $U_k \supseteqq U_{k+1}$, $k = 1, 2, \ldots$. Then the intersection $\bigcap_{k=1}^{\infty} U_k$ of all the U_k is not empty.

cannot be exhausted by a sequence of nowhere dense sets: $[a, b]_R$[20] is of the second category and consequently so is R.

The same statement holds for the set $[a, b]_I$ of the irrational numbers in $[a, b]$; for if it were of the first category, say equal to the union $\bigcup X_n$ of nowhere dense sets X_n in I, we would have

$$(*) \qquad [a, b]_R = \{r_0\} \cup X_0 \cup \{r_1\} \cup X_2 \cup \{r_2\} \cup \cdots,$$

where $r_0, r_1, \ldots$ is a sequence containing all the rational numbers in $[a, b]_R$. But then by (*) it would follow that $[a, b]_R$ is also of first category, in contradiction to the above result.

In the same way we can show that $[a, b]_{Tr}$ is of second category, since the set of algebraic numbers is countable.

1.2.7. *Modules of Sets* Let us consider the system Y of all countable subsets of a set M. It is obvious that this system has the following two properties.

M_1: *Every subset of any element $X \in Y$ is an element of Y.*

M_2: *The union of countably many elements is again an element of Y.*

We also say, Y is a σ-module,[21] whereby we mean simply that Y has the two properties M_1 and M_2. The following theorem is worthy of note.

Theorem 5. *The sets of measure* 0 *form a σ-module, and the same statement is true for the sets of category* 1.

These two modules are of fundamental importance, as can be seen from the following theorems, stated here without proof.[22]

1. A bounded real function f, defined say on the unit interval $[0, 1]$, is integrable in the sense of Riemann if and only if the points of discontinuity of f form a set of measure 0, i.e., if f is continuous, apart from a set of measure 0, or as we shall say, if f is almost everywhere continuous (Lebesgue).

2. Every monotone real function defined on R is almost everywhere differentiable (Lebesgue).

3. Let $\frac{a_0}{2} + \sum_n (a_n \cos nx + b_n \sin nx)$ be a trigonometric series; if the

[20] By $[a, b]_M$ with $M \subseteq R$, $a, b \in R$ we mean the set of all $x \in M$ in $[a, b]$.

[21] We speak of a *module of sets* if the second σ-condition M_2 is replaced by the following module condition: the union of *finitely many* elements of Y is an element of Y. For example, the finite subsets of M form a module, but not a σ-module, if M is infinite.

[22] Our intention here is merely to give some impression of the various contexts in which these two modules, the sets of measure 0 and the sets of category 1, play a role. Since the sets of category 1 will occupy our attention in the following section, theorems 1–5 deal only with sets of measure 0, and since no subsequent use is made of these theorems, we do not give a detailed explanation of the concepts occurring in them. For these concepts see III 3, §2.

numerical series $\sum_n (a_n^2 + b_n^2) \log n$ converges, then the trigonometric series converges almost everywhere (Kolmogorov).

4. Let $p(x, y)$, $q(x, y)$ be continuous real functions defined in a subdomain of R^2; let the derivatives $\frac{\partial p}{\partial y}, \frac{\partial q}{\partial x}$ exist (and be finite at every point (x, y)). Then $p(x, y)dx + q(x, y)dy$ is a differential if and only if the equation $\frac{\partial p}{\partial y} = \frac{\partial q}{\partial x}$ holds almost everywhere (Montel).

5. The indefinite *Lebesgue integral* $c + \int_a^x f(t)dt$ is an (absolutely continuous) function, with the given function f as its derivative almost everywhere (Lebesgue).

In the following sections we shall make a closer examination of the sets of category 1. Our discussion of these two modules, the sets of measure 0 and the sets of category 1, will lead us to the metric and descriptive theory of real functions.

1.3. *Local and Global Properties*

A statement about a set or a function is called *global* or *local* according as it refers to the whole set (function) or to a subset (restriction of the function). For example, the property of countability is a global property. A topological space Y is said to be *locally countable* if for every point x there exists a countable neighborhood U_x of x. The concepts of *local compactness*, *local density*, and so forth, are defined in the same way. For example, R is locally compact but is not compact. The set I_{ω_1} of countable ordinal numbers[23] is locally countable, but is not countable. In a Euclidean space the countability of a set and its local countability are equivalent, as follows from the fact that in a Euclidean space the topology can be defined by a countable base.

Local and global properties of functions are defined in the same way. For example, a real function is said to be locally bounded if every point in the domain of definition of f is contained in a neighborhood in which f is bounded. For example, $\tan x$ is locally bounded, but is not bounded. This property is related to the fact that the domain of definition of $\tan x$ is not compact (i.e., in this case is not bounded and closed).

If the domain of definition $D \subseteq R^n$ of a locally bounded real function is compact, then f is bounded in D.

The proof is indirect: if f were bounded in D, there would exist a sequence of points $x_n \in D$, $n = 1, 2, \ldots$, such that $|fx_n| > n$ for every natural number n. But since D is compact, the sequence x_n contains a convergent subsequence $y_n \to y \in D$. Thus on the one hand $|fy_n| > n$, but on the other hand almost all the y_n are contained in every neighborhood

[23] The set I_{ω_1} can be interpreted in a natural way as a topological space.

U_y of y, so that, in particular, almost all the y_n are contained in the given neighborhood U of y in which f is bounded, in contradiction to the fact that $|fy_n| > n$ for almost all n.

It is equally easy to prove the following theorem.

If the domain of definition D of a locally constant function f is a connected topological space, then f is constant.

2. Continuous Functions

2.1. *Various Definitions of Continuity at a Point*

A function[24] *f on the set M, or in symbols $f|M$, is said to be continuous at $a \in M$ if for every sequence $x_n \to a$, $x_n \in M$, $n = 1, 2, \ldots$, the sequence fx_n approaches fa. The function $f|M$ is said to be discontinuous at $a \in M$ if f is not continuous at a.*

The concept of continuity can be defined in various ways, as is shown by the following theorem:

Theorem 6. *Let f be a function with domain of definition M and let a be a point in M. The following properties are equivalent:*

1. *The function f is continuous at a.*
2. *For every neighborhood Ufa of fa there exists a neighborhood Ua of a such that $fUa \subseteq Ufa$.*
3. *The variation of f at a is zero.*[25]
4. *For every subset $X \subseteq M$ we have: if $a \in \overline{X}$, then also $fa \in \overline{fX}$.*

We shall prove the chain of implications $1 \to 2 \to 3 \to 4 \to 1$, as follows.

$1 \to 2$: Let there be given a neighborhood Ufa of fa. If there were no neighborhood Ua of a such that $fUa \subseteq Ufa$, then in every sphere $K(a; n^{-1})$[26] there would exist a point x_n with $fx_n \notin Ufa$. But the x_n converge to a, whereas the fx_n lie outside Ufa and therefore cannot converge to fa, which is the desired contradiction.

$2 \to 3$: The proof is trivial; it is sufficient to consider a sequence $U^{(n)}a$ of neighborhoods of a for which $fU^{(n)}a \subseteq K(fa; n^{-1})$.

$3 \to 4$: Under hypothesis 3, let us assume that for some $X \subseteq M$ we have $a \in \overline{X}$ but not $fa \in \overline{fX}$. Since $\overline{fX}$ is closed, its complement $\mathbf{C}\overline{fX}$ is open and thus includes a sphere $K(fa; r)$. Since $a \in \overline{X}$, every neighborhood Ua

[24] By a function we shall always mean here a mapping of a subset of R^n into a subset of R^m.

[25] By the diameter dM of a set $M \subseteq R^n$ we mean the supremum of the distances between its points. The variation of a function f at the point a is the infimum of the $dfUa$ where Ua runs through the system of neighborhoods of a.

[26] Let M be the subset of R^m. Then by $K(a; \epsilon)$, $a \in N$, ϵ real and > 0, we mean the (open) set of all points $b \in M$ for which $d(a, b) < \epsilon$, where $d(a, b)$ is the Euclidean distance from b to a.

of a contains a point $x \in X$. Thus we can choose a sequence of neighborhoods $U^{(n)}a$ contracting to a and a sequence of points $x_n \in U^{(n)}a \cap X$. The fx_n lie in fX. But since $|fx_n - fa| \geqq r$, it follows that inf $|fx_n - fa| \geqq r$, so that f cannot have zero variation at a, which is the desired contradiction. $4 \to 1$: Under assumption 4, let $x_n \in M$, $n = 1, 2, \ldots$ be a sequence converging to a. Then by 4 it necessarily follows that fa lies in $\overline{fX}$ if for X we take the set $\{x_1, x_2, \ldots, x_n, \ldots\}$. Thus there exists a subsequence x_{n_i} of the x_n with $\lim_{x \to \infty} fx_{n_i} = fa$. But also $\lim_{x \to \infty} fx_n = fa$, since otherwise there would exist a subsequence fx_{m_i} with $fa \notin \overline{fY}$, where

$$Y := \{x_{m_1}, x_{m_2}, \ldots, x_{m_n}, \ldots\},$$

which is impossible in view of 4, since $x_{m_n} \to a$.

2.2. *Continuous Mappings; Open Mappings*

A function is said to be continuous in (on) a set M if it is continuous at every point of M. One then speaks of a continuous function in (on) M.

Lemma 1. *Let f be a continuous mapping of the set M into the set N. If D is an open subset of N, then $f^{-1}D$ is open in M.*

Proof. In view of theorem 2 in §1.2.3 we must prove that every $x \in f^{-1}D$ is contained in an M-neighborhood Ux of x that lies entirely in $f^{-1}D$. Now let $y = fx \in D$ be the image of x. Since D is open in N, the set D also contains a neighborhood V_y of y in N. But since f is continuous at x, there exists a neighborhood U_x of x in M which is mapped by f into V_y; thus U_x lies in $f^{-1}D$, and therefore $f^{-1}D$ is open.

The converse of this lemma is also true:

If f is a mapping of the set M into the set N and if the f-preimages of open sets in N are open in M, then f is continuous.

We must prove that for every point $a \in M$ and every neighborhood Ufa of fa in N there exists a neighborhood Ua of a in M such that fUa lies in Ufa. But since Ufa is open, $f^{-1}(Ufa)$ is also open. Thus in particular there exists a neighborhood Ua of a which is contained in $f^{-1}(Ufa)$ and for which therefore $fUa \subseteqq Ufa$, which means that f is continuous at a.

Theorem 7. *A mapping f of a set M into a set N is continuous if and only if the f-preimages of open sets in N are open in M.*

Remark. By taking the complements of the open sets, namely the closed sets, we see that in the above theorem the word "open" may be replaced by "closed."

If for the set N in theorem 7 we take in particular the real numbers R, we can express the theorem in a somewhat different way. Let us first introduce the notation: for every real number $y \in R$ let $[f > y]$ be the set of all $x \in M$ with $fx > y$, and define the sets $[f \geqq y]$, $[f < y]$, $[f \leqq y]$ similarly. Then by definition $[f > y] = f^{-1}(y, .)$, where by $(y, .)$ we mean the set of all $x \in R$ greater than y. Then the set $(y, .)$ is open in R. Now if f is continuous, $[f > y]$ is open in M and $[f \geqq y]$ is closed, since $[f \geqq y]$ is the complement of the open set $[f < y]$ in M. Conversely, it is easy to show that if f is a real function on the set M and if the sets $[f > y]$, $[f \geqq y]$ are respectively open and closed in M for all $x \in R$, then f is continuous. Let us state this situation in the following theorem.

Theorem 8. *A real function on a set M is continuous if and only if for all real numbers y the sets* $[f > y]$, $[f \geqq y]$ *are, respectively, open and closed.*

3. $\mathbf{F}_\sigma$ and $\mathbf{G}_\delta$-sets

3.1. *Points of Continuity and of Discontinuity of a Real Function*

Let f be a real function of the set M. For $X \subseteqq M$, the diameter of the image set fX is called the *variation* of the function f in X (see footnote 28).

Now let ϵ be an arbitrary real positive number, and let $S(\epsilon)$ denote the set of all $x \in M$ for which there exists a neighborhood U_x of x in M in which the variation of f is smaller than ϵ.

Lemma. *$S(\epsilon)$ is open in M, i.e., for every $x \in S(\epsilon)$ there exists an open neighborhood $U_x \subseteqq S(\epsilon)$ of x in M.*

Proof. Let x be a point in $S(\epsilon)$ and let U_x be an open neighborhood of x in M with $|fx' - fx''| < \epsilon$ for all $x', x'' \in U_x$. We assert that then $U_x \subseteqq S(\epsilon)$, i.e., that $y \in S(\epsilon)$ for every $y \in U_x$. But since U_x is open in M, there exists in M an open neighborhood U_y of y with $U_y \subseteqq U_x$; so the variation of y in U_y is not greater than that of f in U_x and is thus $< \epsilon$, so that y lies in $S(\epsilon)$, as was to be proved.

We now have a sequence of open sets

$$S(1), S(2^{-1}), S(3^{-1}), \ldots.$$

Let S be their intersection $S := \bigcap_{n=1}^{\infty} S(n^{-1})$. Then S is a $\mathbf{G}_\delta$-set if we define a $\mathbf{G}_\delta$-set (in M) as the intersection of countably many open sets (also called $\mathbf{G}$-sets) in M. Dually, the union of countably many closed sets (also called $\mathbf{F}$-sets) in M is called an $\mathbf{F}_\sigma$-set.[27]

Lemma. *The $\mathbf{G}_\sigma$-set S is precisely the set of points of continuity of f.*

For if a is a point of S, then $a \in S(n^{-1})$ for every natural number n. Thus for every natural number n the variation of f at a is smaller than n^{-1}, i.e., is equal to zero. Thus by theorem 6 in §2.1 the function f is continuous at a.

On the other hand, if f is continuous at a, then for every natural number n the variation of f at a is smaller than n^{-1}: thus $a \in S(n^{-1})$ for every natural number, i.e., $a \in S$.

[27] The letters **F**, **G**, σ and δ are intended to suggest the words "fermé" (closed) "Gebiet" (domain), "Summe" (union), and "Durchschnitt" (intersection).

The complement **C**S of the set S of points of continuity of f is the set of points of discontinuity of f. Thus

$$\mathbf{C}S = \mathbf{C}\bigcap_n S(n^{-1}) = \bigcup_n \mathbf{C}S(n^{-1}) = \bigcup_n F(n^{-1}),$$

where $F(n^{-1})$ denotes the complement of $S(n^{-1})$. The set $F(n^{-1})$ is closed (is an **F**-set), since it is the complement of an open set. Thus **C**S is the union of a countable set of **F**-sets, i.e., **C**S is an $\mathbf{F}_\sigma$-set. Let us combine these statements into the following theorem.

Theorem 9. *The set of points of continuity and the set of points of discontinuity of a real function f are respectively a $\mathbf{G}_\delta$-set and an $\mathbf{F}_\sigma$-set* (Lebesgue 1905).

3.2. *Further Investigation of the $\mathbf{G}_\delta$-Sets and $\mathbf{F}_\sigma$-Sets*

Every closed interval $[a, b] \subseteqq R$ is a $\mathbf{G}_\delta$-set, since

$$[a, b] = \bigcap_n (a - n^{-1}, b + n^{-1}).$$

In fact, we even have the following theorem.

Theorem 10. *Every* **F**-*set is a* $\mathbf{G}_\delta$-*set and, dually, every* **G**-*set is an* $\mathbf{F}_\sigma$-*set.*

For let $X \subseteqq M$ be closed and let $K(X; r)$ be the union of the

$$K(x; r), \qquad x \in X; \qquad K(X; r) := \bigcup_{x\in X} K(x; r).$$

For every $r > 0$, the set $K(X; r)$ is open in M. It is easy to prove that

$$X = \bigcap_{n=1} K(X; n^{-1}).$$

Thus X is a $\mathbf{G}_\delta$-set. Dually, we obtain

$$\mathbf{C}X = \bigcup_{n=1}^{\infty} \mathbf{C}K(X; n^{-1}).$$

Since the $\mathbf{C}K(X; n^{-1})$ are closed, as complements of open sets, it follows that $\mathbf{C}X$ is an $\mathbf{F}_\sigma$-set.

Theorem 10 is of basic importance. The concepts of $\mathbf{G}_\delta$-sets and $\mathbf{F}_\sigma$-sets play a great role, for example, in the theory of measure (every **L**-measurable set is a subset of a $\mathbf{G}_\delta$-set with the same measure (Lebesgue)) and in topology (for example, if $h: M \to M'$ is a topological mapping, there exist two $\mathbf{G}_\delta$-sets 0, 0′ and a topological mapping $k: 0 \to 0'$ such that $M \subseteqq 0$, $M' \subseteqq 0'$ and $k|M = h|M$ (Lavrentiev)).

The $\mathbf{F}_\sigma$-sets and $\mathbf{G}_\delta$-sets may be considered, after the **F**-sets and **G**-sets, as the simplest of all sets.

Following this pattern we will now attempt to construct further sets: namely, $\mathbf{F}_{\sigma\delta}$-sets as δ-sets of $\mathbf{F}_\sigma$-sets and, dually, $\mathbf{G}_{\delta\sigma}$-sets as σ-sets of $\mathbf{G}_\delta$-sets, and so forth. These sets (they are in fact the first classes of *Borel sets*[28] or, as we shall often say, of **B**-sets) have been thoroughly investigated.

3.3. *σ-Systems and δ-Systems*

Let S be a system of sets; then by S_σ we shall mean the system of all sets that can be represented as the union of countably many sets from S. Dually, S_δ denotes the system consisting of intersections of countably many sets from S. We now make the following definition.

A *Borel system* (also called a *B-system*) $\mathbf{B}(S)$ over S is the smallest system X of sets with the following properties: $X \supseteqq S$ and $X_\sigma \subseteqq X$, $X_\delta \subseteqq X$.

For a given S it is easy to construct a *Borel system* by a process which is of fundamental importance in measure theory and in the theory of functions.

By S^1 we denote the system S_σ, and by S_1 the system S_δ, so that for every ordinal number $\alpha > 1$ we can define inductively[29]

$$S^\alpha = \left(\bigcup_{\beta<\alpha} (S^\beta \cup S_\beta)\right)_\sigma, \qquad S_\alpha = \left(\bigcap_{\beta<\alpha} (S^\beta \cup S_\beta)\right)_\delta.$$

Then $S^{\omega_1} = S_{\omega_1}$, so that S^{ω_1} is both a σ-system and a δ-system and, since it is the smallest system of this sort over S, it is itself the *Borel system* $\mathbf{B}(S)$ over S. The elements $M \in \mathbf{B}(S)$ are called *Borel sets* (or **B**-*sets*) (of order α, if $M \in S^\alpha \cup S_\alpha$).

In many cases S will be the set of **G**-sets and **F**-sets in the Euclidean space R^n.

It is worth noting that although the $\mathbf{G}_\delta$-sets in the plane R^2 thus appear to be very simple in structure, their projections already produce the most general so-called **A**-sets[30] in R; we obtain in this way not only all **B**-sets in R but even certain *non-Borel sets*.

4. Construction of More Complicated Sets and Functions by Limit Processes

Our general purpose is now as follows: beginning with intervals, rectangles, and so forth, and with such simple functions as polynomials, we wish to construct more complicated mathematical objects.

All the functions considered in this section are real and are defined on a fixed interval $[a, b] \subseteqq R$.

[28] For the concept of a *Borel set* see §3.3.

[29] For the concept of an ordering and of an ordinal number see IA, §7.4.

[30] For the concept of an **A**-*set* (or *analytic set*) see §6.

4.1. *Baire Functions, i.e., Analytically Representable Functions*

From simple functions, say polynomials $a_0 + a_1x + a_2x^2 + \cdots + a_nx^n$, we can proceed by passage to the limit to more complicated functions, say power series $a_0 + a_1x + \cdots + a_nx^n + \cdots$ such as $\cos x$. Similarly, we can proceed from continuous functions f_n, $n = 1, 2, \ldots$, by passing to the limit (in case it exists)[31]

$$\lim_{n\to\infty} f_n$$

to discontinuous functions. Thus with respect to this passage to the limit the set of continuous functions is not closed. But the set of all functions on $[a, b]$ is certainly closed with respect to such limit processes (we say that this set is lim-closed). The question arises whether there exists a smaller lim-closed set of functions which contains all the continuous functions. The answer is affirmative, as we now proceed to show.

We first note that the intersection of (arbitrarily many) lim-closed sets of functions is again lim-closed. In particular, if we denote by St the set of all continuous functions, then the intersection of all lim-closed sets of functions including St is again lim-closed and is the smallest lim-closed set $\mathbf{B}(St)$ containing St. It is called the *Baire system of functions* and the functions in $\mathbf{B}(St)$ are called *Baire functions*. We can construct the system $\mathbf{B}(St)$ step by step as follows.

By $\mathbf{B}_0St$ we denote the set St itself and by $\mathbf{B}_1St$ the set of all functions obtainable from the f_n in $\mathbf{B}_0St$ by passage to the limit:

$$\mathbf{B}_1St := \{\lim_n f_n \mid f_n \in \mathbf{B}_0St\}.$$

Then we define

$$\mathbf{B}_2St := \{\lim_n f_n \mid f_n \in \mathbf{B}_0St \cup \mathbf{B}_1St\},$$

and for every ordinal number $\alpha > 0$,

$$\mathbf{B}_\alpha St := \left\{\lim_n f_n \mid f_n \in \bigcup_{\alpha' < \alpha} \mathbf{B}_{\alpha'}St\right\}$$

and thus obtain for every countable ordinal (i.e., $\alpha < \omega_1$)

$$\mathbf{B}_0St \subset \mathbf{B}_1St \subset \mathbf{B}_2St \subset \cdots \subset \mathbf{B}_\alpha St \subset \cdots.$$

For example, the *Dirichlet function* χ defined by $\chi(x) = 1$ for rational $x \in R$ and $\chi(x) = 0$ for irrational $x \in R$) is in $\mathbf{B}_2St - \mathbf{B}_1St$. On the other hand, the function equal to 0 on $R - \{1\}$ and equal to 1 on $\{1\}$ is discontinuous but is the limit of continuous functions: $f \in \mathbf{B}_1St - \mathbf{B}_0St$. The systems

$$\mathbf{B}_0St, \mathbf{B}_1St, \ldots, \mathbf{B}_\alpha St, \ldots, \alpha < \omega_1$$

[31] $f = \lim_{n\to\infty} f_n$ is defined by $f(x) = \lim_{n\to\infty} f_n(x)$ for every $x \in [a, b]$.

are *Baire classes*. Their union makes up the system of analytically representable functions or the *Baire system of functions* $\mathbf{B}(St)$, which is the smallest lim-closed set of functions containing the set St of all continuous functions.

4.2. *Almost-Continuity of the Baire Functions*

The functions f of the Baire classes appear to have a very complicated structure and to be very general. So it is interesting to note that these functions are "almost-continuous" in the sense that to every f there corresponds a $\mathbf{G}_\delta$-set on which $f|\mathbf{G}_\delta$ is continuous and whose complement $[a, b] - \mathbf{G}_\delta$ is of first category (with respect to $[a, b]$). Let us consider the *first Baire class* $\mathbf{B}_1 St$.

4.3. *Functions of the First Baire Class; A Typical Baire Theorem*

If the derivative $f'(x)$ of the function f exists at every point $x \in [a, b]$, then f' is in the first Baire class. The same is true if the set of points of discontinuity of the function f is finite. Let us examine the general case. Let

$$f(x) = \lim_{n\to\infty} f_n(x), \qquad f_n \in \mathbf{B}_0 St, \qquad n = 1, 2, \ldots$$

be a function in the first Baire class. What can we say about the set U of points of discontinuity of f? This set cannot have any interior points, since every perfect set $P \subseteqq [a, b]$ contains a point of continuity of the function $f|P$. This statement is precisely the first part of the famous theorem of Baire.

Theorem 11 (Theorem of Baire). *A real function on* $[a, b]$ *can be represented as the limit of real continuous functions if and only if every perfect set* P *in* $[a, b]$ *contains a point of continuity of the restricted function* $f|P$ (Baire, 1899).

The condition is necessary. For let P be a perfect set in $[a, b]$ and let the function $f = \lim_n f_n$, with f_n continuous on $[a, b]$, $n = 1, 2, \ldots$. Then for every real number $\epsilon > 0$ the set

$$F_{n,m}(\epsilon) = \{x \mid x \in [a, b] \wedge |f_n(x) - f_{n+m}(x)| \leqq \epsilon\}$$

is closed, and the same is true for the set

$$F_n(\epsilon) = \bigcap_{m=1}^{\infty} F_{n,m}(\epsilon).$$

On the other hand, the sequence $f_n(x)$ is convergent for every $x \in [a, b]$, which means that for sufficiently large n the inequality $|f_n(x) - f_{n+m}(x)| \leqq \epsilon$ holds for every m, i.e., $x \in F_n(\epsilon)$. Thus

$$(*) \qquad [a, b] = F_1(\epsilon) \cup F_2(\epsilon) \cup F_3(\epsilon) \cup \cdots.$$

Now let $p \in F_n(\epsilon)$, i.e., $|f_n(p) - f_{n+m}(p)| \leqq \epsilon$ for all m, from which it follows that $|f_n(p) - f(p)| \leqq \epsilon$ for $m \to \infty$. In particular, if p is an interior point of $F_n(\epsilon)$, it follows from the definition of an interior point that $|f_n(q) - f(q)| \leqq \epsilon$ for all points q in a neighborhood V_p of p. But an interior point p exists in at least one $F_n(\epsilon)$. For if $F_n(\epsilon)$ has no interior point, the set $F_n(\epsilon)$, since it is closed, is nowhere dense in $[a, b]$, and thus if there were no $F_n(\epsilon)$ with an interior point, the relation (*) would mean that the set $[a, b]$ would be of first category, in contradiction to theorem 4 in §1.2.6.

Now let $G(\epsilon)$ be the set of all $p \in [a, b]$ with a neighborhood V_p of p such that for a certain n we have

$$|f_n x - fx| \leqq \epsilon$$

for all $x \in V_p$. For the same reason as before, as the reader may easily see, the set $G(\epsilon)$ is everywhere dense; it is only necessary to bear in mind that every subinterval of $[a, b]$ is also of the second category. In particular, we obtain the sequence

$$G\left(\frac{1}{m}\right), \qquad m = 1, 2, \ldots,$$

of open everywhere dense sets. Let X be its intersection:

$$X = \bigcap_{m=1}^{\infty} G\left(\frac{1}{m}\right);$$

then X consists of exactly those points in $[a, b]$ at which f is continuous, as we shall now show.

We first show: for every $p \in X$ and every natural number $m > 0$ there exists a neighborhood U_p of p such that $|fx - fp| \leqq \dfrac{1}{m}$ for all $x \in U_p$.

Since p is in X, and thus in particular in $G\left(\dfrac{1}{3m}\right)$, there exists a natural number n and a neighborhood V_p of p such that for all $x \in V_p$ we have $|f_n x - fx| \leqq \dfrac{1}{3m}$. Since f_n is continuous at p (in fact, f_n is continuous on the whole of $[a, b]$), there exists a neighborhood $\tilde{U}_p$ of p such that $|f_n x - f_n p| \leqq \dfrac{1}{3m}$ for all $x \in \tilde{U}_p$. If we now set $U_p = \tilde{U}_p \cap V_p$, we have for all $x \in U_p$

$$|fx - fp| \leqq |fx - f_n x| + |f_n x - f_n p| + |f_n p - fp| \leqq \frac{1}{m},$$

i.e., f is continuous at p.

Conversely, it is easy to see that every point of continuity of f lies in X. Thus X is precisely the set of all points of continuity of f in $[a, b]$. As the following lemma shows, X is also dense in $[a, b]$.

Lemma. *The intersection of countably many open everywhere dense sets* G_n, $n = 1, 2, \ldots$, *in* $[a, b]$ *is everywhere dense in* $[a, b]$.

Proof. We must show that every open neighborhood U_x of a point $x \in [a, b]$ has a nonempty intersection with $X = \bigcap_{n=1}^{\infty} G_n$. Let us consider the open set $D_1 := U_x \cap G_1$; it contains a closed interval I_1. Let $\mathring{I}_1$ be the open kernel of I_1 and the set $\mathring{I}_1 \cap G_2 =: D_2$; then D_2 is open, since it is the intersection of two open sets. Let I_2 be a closed interval in D_2, let $\mathring{I}_2$ be its open kernel, and let $D_3 := \mathring{I}_2 \cap G_3$ be the intersection of $\mathring{I}_2$ with G_3; then D_3 is open. Let I_3 be a closed interval in D_3, etc. The intersection $I := \bigcap_{n=1}^{\infty} I_n$ is not empty, and $I \subseteqq X \cap U_x$.

Thus we have so far proved:

Theorem 12. *The set of points of continuity of every real function of the first Baire class on* $[a, b]$ *is everywhere dense on* $[a, b]$.

The set of points of discontinuity of a function of the first *Baire class* has no interior points. Since the set of points of continuity of the Dirichlet function χ is empty, χ cannot belong to the first Baire class.

The same argument can now be applied to P and $f|P$ in place of $[a, b]$ and $f|[a, b]$. In particular, P contains a point of continuity of the restricted function $f|P$. Thus the first part of the *Baire theorem is* proved.

For the second half of the *Baire theorem* the reader is referred to the literature. We confine ourselves here to the remark that the proof is simplified by the following theorem.

Theorem 13 (Lebesgue). *A real bounded function* $f|[a, b]$ *is in the first Baire class if and only if for any real number r the set* $[f > r] := \{x \mid x \in [a, b]$ *and* $fx > r\}$ *and* $[f < r] := \{x \mid x \in [a, b]$ *and* $fx < r\}$ *are* $\mathbf{F}_\sigma$*-sets* (Lebesgue, 1905).

4.4. *Remark.* The proof of the second half of the *Baire theorem*, not given here, depends on a certain characteristic procedure that was also used by Denjoy to invert the process of differentiation.

With respect to the Denjoy process let us consider the following function $f|[0, 1]$:

$$fx = x^2 \cos x^{-2} \qquad \text{for} \qquad 0 < x \leqq 1,$$
$$f0 = 0.$$

The derivative f' exists and is finite in $[0, 1]$, but f' is not Lebesgue-integrable in $[0, 1]$. For if f' were Lebesgue-integrable, then for

$$0 < x_0 < x_1 \leqq 1$$

we would have

$$\int_{x_0}^{x_1} f'(x)dx = x_1^2 \cos x_1^{-2} - x_0^2 \cos x_0^{-2}.$$

For $a_n = (2(4n + 1)^{-1})^{1/2}$, $b_n = (2n)^{-1/2}$ this would mean

$$\int_{a_n}^{b_n} f'(x)dx = \frac{1}{2n}.$$

Since the closed intervals $[a_n, b_n]$ are pairwise disjoint, for $E = \bigcup [a_n, b_n]$ we have

$$\int_E f'(x)dx \geqq \sum_{n=1}^{\infty} \frac{1}{2n} = \infty.$$

Thus the necessity arises of seeking the inversion of differentiation not by means of Lebesgue integration but in some new way, as was done by Denjoy and Perron.

5. Some Interesting Types of Functions

5.1. *Simple Functions of the First Baire Class*

5.1.1. Derivatives: if $f|[a, b]$ is everywhere differentiable, then since

$$f'x = \lim_{n\to\infty} \frac{f(x + n^{-1}) - f(x)}{n^{-1}}, \qquad x \in [a, b], \qquad n = 1, 2, \ldots,$$

the function $f'x$ is in the first *Baire class.*

5.1.2. *If the set of points of discontinuity of $f|[a, b]$ is countable, then f is in the first Baire class.*

For the proof we apply theorem 11. If $P \subseteqq [a, b]$ is a perfect set, then P is uncountable and therefore contains points of continuity of f and, in particular, points of continuity of $f|P$.

5.1.3. A real monotone function $f|[a, b]$ is necessarily in the first *Baire class*, since the set of points of discontinuity of f is countable.

5.1.4. A real function $f|[a, b]$ of bounded variation is necessarily in the first *Baire class,* since it is the difference of two monotone functions.

5.1.5. The semicontinuous functions are in the first *Baire class.* Here we have the definition: *a real function $f|M$ is upper-semicontinuous in $a \in M$ if for every $\epsilon > 0$ there exists a neighborhood $U_a \subseteqq M$ of a such that $Ua \subseteqq [f < (fa) + \epsilon]$, i.e., $fx < (fa) + \epsilon$ for $x \in Ua$.*

The following properties are equivalent.

1. *f is upper-semicontinuous at a.*
2. $\overline{\lim_{x\to a}}\, fx = fa.$

The function f is said to be upper-semicontinuous on a set $N \subseteqq M$ if f is upper-semicontinuous at every point of N.

It can be shown that *a real function defined on* $[a, b]$ *is upper-semicontinuous on this interval if and only if the set*

$$[f \geqq r] := \{x \mid x \in [a, b] \text{ and } fx \geqq r\}$$

is closed for every real number r.

The concept of *lower semicontinuity* can be defined dually to upper semicontinuity.

It can be shown that the semicontinuous (upper or lower) real functions on $[a, b]$ also belong to the first Baire class.

5.2. *General Baire Functions and Borel Sets*

Let us recall once again that M is a given subset of a Euclidean space; for example, M may be the set R of all real numbers, or the set I of irrational numbers (the so-called *Baire space*), or one of the spaces R^2, R^3, ... On M the system S of open and closed sets is defined by §1.2. In accordance with §3.3 we construct the *Borel system of sets* $\mathbf{B}(S)$ corresponding to S, namely the smallest system of sets containing the elements of S and closed with respect to the union and intersection of countably many elements. The elements of $\mathbf{B}(S)$ are also called *Borel sets* (or **B**-*sets*) in M.

A connection between the *Baire functions* and the *Borel sets* is given by a theorem of Lebesgue (we recall that the *Baire functions* arise from the continuous functions by an iterated lim-procedure; cf. §4).

Theorem 14 (Lebesgue): *For every real Baire function* $f \mid [a, b]$ *and for every real number r the sets*

$$(*) \qquad [f > r], \qquad [f \geqq r]$$

are Borel sets.

There are further connections between the sets (*) and the functions f: let us recall theorem 8 in §2.2: $f \mid [a, b]$ is continuous if and only if the sets (*) are open or closed respectively. Moreover, theorem 13 in §4.3 represents a connection between the functions $f \mid [a, b]$ of the first *Baire class* with the $\mathbf{F}_\sigma$- and the $\mathbf{G}_\delta$-sets in $[a, b]$.

5.3. *The Baire Condition for Sets and for Functions*

In the descriptive theory of functions the following two concepts are of great importance.

Let E be a subset of a Euclidean space (e.g., the real line R), and let $M \subseteqq E$. We say that M satisfies the *Baire condition* or is a β-set provided that M differs from an **F**-set or a **G**-set in E by a set of first category (with respect to E). In other words, M is congruent to an **F**-set or a **G**-set modulo a set of the first category.

Analogously, in the metric theory of functions we consider sets that differ from an **F**-set or a **G**-set by a set of measure zero. As was mentioned above, every Lebesgue-measurable set can be embedded in a $\mathbf{G}_\delta$-set with the same measure.

The above definition gives rise to the following terminology.

A real function $f|E$ is said to satisfy the *Baire* condition (or to be a β-function) if f is continuous in E up to a set of first category or, in other words, if there exists a set P of first category in E such that the restricted function $f|E - P$ is continuous.

These two concepts (β-set, β-function) are closely related, as is shown by the following theorems.

Theorem 15. *A real function $f|E$ is a β-function if and only if the set of preimages $f^{-1}(G)$ of every open set G in R*[32] *is a β-set.*

Theorem 16. *The Borel sets in E (the Baire functions in E) satisfy the Baire condition for sets or for functions, respectively* (Lebesgue).

6. Analytic Sets (A-Sets)

6.1. *The Error of Lebesgue and the Discovery of Suslin*

A faulty argument[33] of Lebesgue led to the discovery of the analytic sets. The error consisted in the assertion that the projection of a **B**-set is again a **B**-set; Lebesgue had wrongly assumed that

$$\text{Proj} \bigcap = \bigcap \text{Proj}.^{34}$$

This mistake was discovered in 1917 by Suslin who then proceeded, together with Luzin, to create the theory of **A**-sets. Suslin defined the **A**-sets by means of a special set-theoretic operation (the so-called **A**-operation, which we shall consider below). Luzin defined the **A**-sets as continuous images of the set I of irrational numbers. The **A**-sets arise in many investigations; even Lebesgue himself, in the article just cited, defined an **A**-set which is not a **B**-set. The following theorems are intended, before

[32] By footnote 4, a real function $f|E$ is a mapping from E into R.

[33] H. Lebesgue, Sur les fonctions représentables analytiquement. J. de Math. (6) 1 (1905).

[34] Counterexample: construct an (x, y)-coordinate system in the plane R^2 and let E_n, $n = 1, 2, \ldots$, be the open interval $(0, 1/n)$ of the y-axis. Then $\bigcap_{n=1}^{\infty} E_n$ is empty, so that $\text{Proj}_x \bigcap_{n=1}^{\infty} E_n$ is also empty, but since $\text{Proj}_x E_n = \{0\}$ we have

$$\bigcap_{n=1}^{\infty} \text{Proj}_x E_n = \{0\} \neq \varnothing.$$

we examine the concept of an **A**-set more closely, to give a preliminary idea of the ways in which *A*-sets arise:

1. The *linear*[35] **A**-*sets* are the images of continuous mappings of I (Luzin). This statement should be compared with the statement that the one-to-one continuous mappings of linear **B**-sets are again **B**-sets.
2. Every non-empty analytic set in R is the range of a real upper-semicontinuous (or lower-semicontinuous) function $f|R$.
3. A linear **A**-set is the range of the derivative of a continuous real function $f(x)$ which is almost everywhere differentiable (Popruženko).
4. The linear **A**-sets are orthogonal projections of the plane $\mathbf{G}_\delta$-sets (Suslin).
5. A linear **A**-set is the range of a one-valued implicit function $y(x)$, where $y(x)$ is defined by $f(x, y) = 0$, with continuous f.
6. Let X be a set in the plane R^2 and let aX be the set of all end points of open line-segments of the plane lying in the complement $R^2 - X$. Then if X is an $\mathbf{F}_\sigma$-set, aX is an **A**-set (Urysohn).

6.2. *The Suslin Definition of* **A**-*Sets; The* **A**-*Operation*

The most fruitful definition of the **A**-sets is probably the one due to Suslin. In order to arrive at this definition, we proceed in about the same way as Suslin himself, when he was attempting to construct a plane **B**-set whose projection on the real line R is not a **B**-set.

6.2.1. *The simplest sets on the real line R* The simplest sets on the real line are certainly the intervals (including the one-point intervals); as the simplest figures on the plane R^2 we should certainly take the rectangles and line-segments parallel to the axes. Then by σ-operations (unions of countably many sets) we construct more complicated sets and project them on the real line R. By δ-operations (intersections of countably many sets) we proceed from the sets already constructed in R^2 to sets of the form

$$X = E_1 \cap E_2 \cap E_3 \cap \cdots, \tag{1}$$

where the E_n are unions of rectangles E_{nk}:

$$E_n = E_{n1} \cup E_{n2} \cup E_{n3} \cup \cdots. \tag{2}$$

Thus

$$X = \bigcap_n \bigcup_k E_{nk},$$

[35] By a linear set we mean a subset of $R = R^1$, and correspondingly a planar set is a subset of R^2.

and by the distributive law

(3) $$X = \bigcup_{k_1, k_2, \ldots} \bigcap_n E_{nk_n},$$

where the $k_1, k_2, \ldots$ run independently through the natural numbers. If we project the set X into the first coordinate axis ($= R$), we obtain the following result (where $p_1 \colon R^2 = R \times R \to R$ is the projection on the first coordinate axis)

(4) $$p_1 X = p_1\left(\bigcup_{k_1, k_2, \ldots} \bigcap_n E_{nk_n}\right) = \bigcup_{k_1, k_2, \ldots} p_1\left(\bigcap_n E_{nk_n}\right).$$

But we cannot proceed any further without additional assumptions; for in general

(5) $$p_1\left(\bigcap_n E_{nk}\right) \neq \bigcap_n p_1 E_{nk_n}.$$

But if the rectangles E_{nk} are such that

$$E_{1k_1} \supset E_{2k_2} \supset E_{3k_3} \supset \cdots,$$

then the signs of equality hold in (5). Now we can represent the intersection

$$\bigcap_n E_{nk_n}$$

as the intersection of a decreasing sequence of sets; it is sufficient to consider the sequence

(6) $$E_{k_1, k_2, \ldots, k_n} := \bigcap_{\nu=1}^{n} E_{\nu k_\nu}. \qquad n = 1, 2, \ldots$$

We then obtain

(7) $$\bigcap_n E_{nk_n} = \bigcap_n E_{k_1, k_2, \ldots, k_n}$$

and

(8) $$p_1\left(\bigcap_n E_{nk_n}\right) = p_1\left(\bigcap_n E_{k_1, k_2, \ldots, k_n}\right)$$
$$= \bigcap_n p_1 E_{k_1, k_2, \ldots, k_n} = \bigcap_n \delta_{k_1, k_2, \ldots, k_n},$$

where $\delta_{k_1, k_2, \ldots, k_n}$ denotes the projection of $E_{k_1, k_2, \ldots, k_n}$ onto the first coordinate axis. Formula (4) then gives (we set $E = p_1 X$)

(9) $$E = \bigcup_{\delta} \bigcap_n \delta_{k_1, k_2, \ldots, k_n},$$

where δ runs through the system of all sequences $k_1, k_2, \ldots$ of natural numbers.

6.2.2. *The operation* $\mathbf{A}S$ *on a system of sets* S The principle of this construction for producing sets of real numbers of the form (9) is of fundamental importance. In particular, the representation (9) leads us to the concept of the analytic operation $\mathbf{A}S$ on the system of sets S:

To every k-tuple $(n_1, n_2, \ldots, n_k)$, $k = 1, 2, \ldots$, of natural numbers we assign a set $\delta_{n_1 n_2 \cdots n_k} \in S$. Then for every sequence $\nu = n_1, n_2, \ldots$ we form the intersection

$$E^\nu = \bigcap_{k=1}^{\infty} \delta_{n_1 \cdots n_k}$$

and then the union for all sequences

$$E = \bigcup_{\nu} E^\nu. \tag{9'}$$

We say that *E is an analytic set* (*or* **A**-*set*) *over S or that E is the result of an analytic operation* (*or an* **A**-*operation*) *over the system of sets S, and we write*

$$E \in \mathbf{A}S.$$

Here $\mathbf{A}S$ or $S_{\mathbf{A}}$ denotes the system of analytic sets E over S. Then the following theorem holds.

Theorem 17. *For every system of sets S we have* $\mathbf{AA}S = \mathbf{A}S$ (Suslin, 1917).

If S is the system of open and closed sets[36] of a subset T in R^n, the analytic sets E over S are also called analytic sets (or **A**-sets) in T.

6.2.3. *Relationship between the σ- and δ-operations and the* **A**-*operations* The δ-operations over a system of sets S are special **A**-operations. For if E is the result of a δ-operation over M,

$$E = \bigcap_{k=1}^{\infty} E_k, \qquad E_k \in M,$$

and if we define the $\delta_{n_1 \cdots n_k}$ by

$$\delta_{n_1 \cdots n_k} = E_k,$$

we obtain the form (9′)

$$E = \bigcap_{k=1}^{\infty} \delta_{n_1 \cdots n_k} = \bigcup_{\nu} \bigcap_{k=1}^{\infty} \delta_{n_1 \cdots n_k},$$

where ν runs through the set of all sequences of natural numbers.

Analogous remarks hold for the σ-operation. Thus we have proved:

Theorem 18. *The σ- and δ-operations are special* **A**-*operations.*

[36] For the definition of open and closed sets in T see §1.2 ff.

6.3. *Relationship between the Concepts of a* **B**-*Set and an* **A**-*Set*

Let us recall the concept of the **B**-set **B**(S) over the system of sets S: the system **B**(S) is the *smallest* system S of sets with $S \subseteqq X$, $X_\sigma \subseteqq X$, $X_\delta \subseteqq X$(§3.3). But by theorem 18 we have $(\mathbf{A}S)_\delta \subseteqq \mathbf{AA}S$, $(\mathbf{A}S)_\sigma \subseteqq \mathbf{AA}S$, and by theorem 17 $\mathbf{AA}S = \mathbf{A}S$, and thus $(\mathbf{A}S)_\delta \subseteqq \mathbf{A}S$, $(\mathbf{A}S)_\sigma \subseteqq \mathbf{A}S$; and it is trivial that $S \subseteqq \mathbf{A}S$. Thus $\mathbf{B}(S) \subseteqq \mathbf{A}S$. In other words, we have the theorem:

Theorem 19. *Every* **B**-*set is an* **A**-*set.*

The converse is false, as follows.

Theorem 20. *An* **A**-*set* E *in a set* T *in* R^n *is a* **B**-*set if and only if the complement* $\mathbf{C}E = T - E$ *is also an* **A**-*set* (Suslin) *or if* E *can be represented in the form* (9′) *with disjoint summands* (Luzin).

6.4. **A**-*Sets as Images under Projection*

Suslin introduced the linear **A**-sets as analytic sets over the closed linear intervals; but this is no restriction, since the analytic sets in R in the sense of §6.2.2 are already analytic sets over the closed linear intervals in R. Thus the sets $\delta_{n_1 \cdots n_k}$ in (9′) are, in this case, closed linear intervals. Let us now show how Suslin obtained every linear **A**-set as the projection of a plane $\mathbf{F}_{\sigma\delta}$-set.

Let

$$E = \bigcup_{\nu} \bigcap_{k=1}^{\infty} \delta_{n_1 n_2 \cdots n_k}$$

be a linear analytic set (in the x-axis of the plane $R^2 = R \times R$); the $\delta_{n_1 n_2 \cdots n_k}$ may be considered as closed intervals. On the y-axis of the plane R^2 we consider, for every k-tuple $(n_1, \ldots, n_k)$ of natural numbers, the closed interval

$$\alpha_{n_1 n_2 \cdots n_k} = \left[\sum_{l=1}^{k} 3^{-(n_1 + n_2 + \cdots + n_l)} \leqq y \leqq 3^{-(n_1 + \cdots + n_k)} + \sum_{l=1}^{k} 3^{-(n_1 + \cdots + n_l)}\right]$$

and set

$$D_{n_1 n_2 \cdots n_k} = \delta_{n_1 n_2 \cdots n_k} \times \alpha_{n_1 n_2 \cdots n_k},$$

so that $D_{n_1 n_2 \cdots n_k}$ is the rectangle with the projections $\delta_{n_1 n_2 \cdots n_k}$ and $\alpha_{n_1 n_2 \cdots n_k}$. If we further set

$$Q_k = \bigcup_{(n_1, \ldots, n_k)} D_{n_1 n_2 \cdots n_k},$$

where the summation is taken over all k-tuples of natural numbers, and

$$Q = \bigcap_{k=1}^{\infty} Q_k,$$

then Q_k is an $\mathbf{F}_\sigma$-set, and consequently Q is an $\mathbf{F}_{\sigma\delta}$-set, and for E we have $E = p_1 Q$ (where p_1 is the projection onto the x-axis).[37] Thus the following theorem holds.

Theorem 21. *Every linear* **A**-*set is the projection of a plane* **B**-*set* (*which may even be taken to be a* $\mathbf{G}_\delta$-*set or an* $F_{\sigma\delta}$-*set*) (Suslin, 1917).

6.5. *First Example of an* **A**-*Set That is Not a* **B**-*Set* (Lebesgue, 1905)

Let $0 \leqq x \leqq 1$ be a number with the dyadic expansion

$$x = \sum_{n=1}^{\infty} x_n 2^{-n},$$

in which infinitely many of the x_n are equal to 1. Let $r_1, r_2, r_3, \ldots$ be an enumeration of all the rational numbers between 0 and 1. Let Rx be the set of all r_n with $x_n = 1$, so that

$$Rx = \{r_n \mid x_n = 1\}.$$

The set Rx is either well ordered or not well ordered.[38] Let E be the set of all $x \in [0, 1]$ for which Rx is not well ordered. It turns out that E is an **A**-set but not a **B**-set.[39] The characteristic function of E is not **B**-measurable,[40] for if for $x \in [a, b]$ we define $\varphi x = 1$ or $\varphi x = 0$ according as $x \in E$ or $x \notin E$; then φ is not **B**-measurable, i.e., $\varphi^{-1}(0)$ and $\varphi^{-1}(1)$ are not both Borel sets.

6.6. *Projection and Subtraction as Methods for Generating Sets;* **C***A-Sets; Projective sets*

In §§6.4 and 6.5 we proved:

1. *Every linear* **A**-*set is the projection of a plane* **B**-*set.*
2. *There exists a linear* **A**-*set that is not a* **B**-*set.*

Thus we see that the operation of projection is very powerful for generating sets. Suslin also proved (Theorem 20):

3. *For an* **A**-*set E to be a* **B**-*set it is necessary and sufficient that the complementary* **C***E is also an* **A**-*set.*

[37] Outline of the proof. Here $\alpha_{n_1 \cdots n_k} = \bar{\alpha}_{n_1 \cdots n_k}$, and for every sequence $\nu_1 = \{n_1^{(1)}, \ldots, n_k^{(1)}, \ldots\}$ we have $\alpha_{n_1^{(1)} \cdots n_k^{(1)}} \supseteqq \alpha_{n_1^{(1)} \cdots n_k^{(1)} n_{k+1}^{(1)}}$. By the Cantor intersection theorem (see footnote 22) it follows that $D_{\nu_1} = \bigcap_{k=1}^{\infty} \alpha_{n_1^{(1)} \cdots n_k^{(1)}} \neq \varnothing$. Now let $P \in E = \bigcup_\nu \bigcap_{k=1}^{\infty} \delta_{n_1 \cdots n_k}$, i.e., let there exist a sequence $\nu_1 = \{n_1^{(1)}, \ldots, n_k^{(1)}, \ldots\}$ with $P \in \bigcap_{k=1}^{\infty} \delta_{n_1^{(1)} \cdots n_k^{(1)}}$. For some $\alpha \in D_{\nu_1}$ we have

$$(P, \alpha) \in \delta_{n_1^{(1)} \cdots n_k^{(1)}} \times \alpha_{n_1^{(1)} \cdots n_k^{(1)}} \subset \bigcup_{(n_1, \ldots, n_k)} \delta_{n_1 \cdots n_k} \times \alpha_{n_1 \cdots n_k}$$

for all k, i.e., $(P, \alpha) \in Q$. Thus the projection of Q onto the x-axis includes the set E: $Pr_1 Q \supset E$. The relation $E \supset Pr_1 Q$ is equally easy to prove.

[38] For the concept of well ordering, see IA, §7.4.

[39] The proof of this statement will not be given here.

[40] See footnotes 32 and 50.

It follows that taking the complement is also an important method of generating sets. Thus we first consider the complements of **A**-sets (the so-called **CA**-sets) and then their projections (**PCA**-sets). Then we consider the complements of these sets, the **CPCA**-sets, and then the **PCPCA**-sets, and so forth.

The sets thus defined are called *projective sets* by Luzin. In this way we obtain the sequence of classes of sets (the classical **A**-, **CA**-, **PCA**, ... -sets are denoted simply by **A**, **CA**, **PCA**, ...)

$$\mathbf{A}, \mathbf{CA}, \mathbf{PCA}, \mathbf{CPCA}, \ldots$$

or

$$\mathbf{A}_1, \mathbf{CA}_1, \mathbf{A}_2, \mathbf{CA}_2, \mathbf{A}_3, \mathbf{CA}_3, \ldots, \tag{1}$$

where

$$\mathbf{A} = \mathbf{A}_1,\ \mathbf{A}_2 = \mathbf{PCA}_1,\ \mathbf{A}_3 = \mathbf{PCA}_2, \ldots, \mathbf{A}_{n+1} = \mathbf{PCA}_n.$$

Of special interest are the systems of sets $\mathbf{B}_1, \mathbf{B}_2, \mathbf{B}_3, \ldots, \mathbf{B}_n, \ldots$, where by $\mathbf{B}_n$ we mean the system of all sets M that are both $\mathbf{A}_n$-sets and also $\mathbf{CA}_n$-sets; by Suslin's theorem (theorem 20 in §6.3) the $\mathbf{B}_1$-sets are identical with the **B**-sets.

The union of these sets is the class of projective or **P**-sets. For a given space R^n the class of **P**-sets can be characterized as the smallest system of sets that includes the **G**-sets and is closed with respect to complementation and to continuous mappings into R^n (i.e., if $X \in M$, then $\mathbf{C}X \in M$ and $fX \in M$ for every continuous mapping of X into the space R^n).

These classes of projective sets were introduced by Luzin. He raised the problem of determining the power of **CA**-sets and of deciding whether every uncountable **CA**-set contains a perfect set.[41] Luzin emphasized the unsolvability of this problem; in his opinion there was no contradiction in assuming either the existence or the nonexistence of such a perfect set. In 1951 Novikov succeeded in proving the following theorem.

Theorem 22. *The assumption that there exists a* **CA**-*set of power* $2^{\aleph_0}$ *without perfect subsets is consistent with the system of axioms* Σ, *provided that the system* Σ *itself is consistent. Here* Σ *denotes the system that was used by Gödel in 1940 to prove an analogous theorem with respect to the continuum hypothesis and the axiom of choice.* (See §7.)

Theorem 23. *Every* **A**-*set is measurable in the sense of Lebesgue and has the Baire property.*

From the results of Gödel and Novikov, we have

Theorem 24. *The existence of a linear* $\mathbf{B}_2$-*set that is not* **L**-*measurable is consistent with the system of axioms* Σ, *provided that* Σ *itself is consistent.*

We remark that there exist $2^{\aleph_0}$ projective sets; this number is "small" in comparison with the number $2^{2^{\aleph_0}}$ of all linear sets. Thus nonprojective

[41] For **A**-sets we have (Suslin): *Every uncountable* **A**-*set contains a perfect set.*

sets exist. Accordingly to Luzin the above classes (1) are distinct from one another.

6.7. *Separability*

It is frequently necessary to embed two given disjoint sets in disjoint sets of a simpler nature. For example, if X, Y are two closed disjoint sets in R^n, there exist disjoint **G**-sets, G_X, G_Y such that $X \subseteqq G_X$, $Y \subseteqq G_Y$.

In topology there are several separation axioms, e.g., the axiom of Hausdorff to the effect that every pair of points can be separated by disjoint open sets, and it is a remarkable fact that these topological separation axioms of various kinds are closely related to the existence of various *real* functions.

In the descriptive theory of functions, Luzin and others have investigated the problems of separation. Here the following theorems have been obtained.

Theorem 25. *Two disjoint* **A**-*sets can be separated by disjoint* **B**-*sets* (Luzin).

Theorem 26. *If* X_1, X_2 *are two* **A**-*sets, then the sets* $X_1 - (X_1 \cap X_2)$, $X_2 - (X_1 \cap X_2)$ *can be separated by disjoint* **CA**-*sets*[42] (Luzin).

Theorem 27. *If* X_1, X_2 *are two* **CA**-*sets, then the sets* $X_1 - (X_1 \cap X_2)$, $X_2 - (X_1 \cap X_2)$ *can be separated by disjoint* **CA**-*sets* (Luzin).

Theorem 28. *There exist two disjoint* **CA**-*sets that cannot be separated by disjoint* **B**-*sets* (Novikov).

Novikov also points out that theorems 25–28 provide us with new theorems if we rewrite them as follows:

$\mathbf{CA}_2$ in place of **A**,
$\mathbf{A}_2$ " " " **CA**,
$\mathbf{B}_2$ " " " **B**.

Theorem 29. *For higher classes* $\mathbf{A}_n$ *of projective sets it is consistent, according to Novikov, to assume in the system* Σ *that for almost all* n *the same separation theorems hold for the class* $\mathbf{A}_n$ *as for the class* $\mathbf{A}_2$.

In concluding this section, we give an application of theorem 25. Let E and $\mathbf{C}E$ both be **A**-sets; as such they are disjoint and by theorem 25 they are **B**-separable; then if X, Y are two disjoint **B**-sets with $X \supseteqq E$, $Y \supseteqq \mathbf{C}E$, it necessarily follows that $X = E$, $Y = \mathbf{C}E$, i.e., E is a *Borel set*. In this way we have deduced Suslin's theorem 20 (see §6.3) from theorem 25.

The *Novikov theorem* 28 plays an important role in the theory of implicit functions (see §6.9).

[42] I.e., there exist **CA**-sets N_1 and N_2 such that

$$N_1 \cap N_2 = \varnothing, \qquad X_1 - (X_1 \cap X_2) \subset N_1, \qquad X_2 - (X_1 \cap X_2) \subset N_2.$$

6.8. *Uniformization*

Let us consider as an example the "function" $y(x)$ with $y^2 - x = 0$. It is two-valued for every $x \neq 0$. There exist infinitely many one-valued functions[43] f which have the same domain of definition as $y(x)$ and for which $fx \in y(x)$ for every x, but only two of them are continuous.

The general uniformization problem in the descriptive theory of functions can be formulated as follows.

For a given set of points M with a property E (e.g., of being a Borel set), does there always exist in p_1M a (one-valued) function f such that the corresponding curve

$$y - fx = 0, \qquad x \in p_1M$$

lies in M and has a given property E' (e.g., of being of type **CA**)?

Theorem 30. *Every plane* **B**-*set or* **CA**-*set* M *can be uniformized by a* **CA**-*set, i.e., there exists a (one-valued) real function* f *such that the curve*

$$y - fx = 0, \qquad x \in p_1M$$

lies in M *and is of type* **CA** (Luzin and Kondo).

It is interesting to note that not every plane **A**-set can be uniformized by **A**-sets or **CA**-sets, i.e., there exists a plane **A**-set M such that, for every (one-valued) function $f|p_1M$, the curve $y - fx = 0$, $x \in p_1M$ is neither an **A**-set nor a **CA**-set.

6.9. *Implicit Functions*

If $\varphi(x, y)$ is a real Baire function, we can ask for the function (in general, many-valued) $y = fx$ for which the equation

$$\varphi(x, y) = 0 \tag{1}$$

holds (by a many-valued real function we now mean, in precise terms, a mapping into the power set of the real numbers).

What are the properties of this implicit function f?

Theorem 31. *If* f *is one-valued (i.e., if* fx *is a one-point set, so that* f *is a real function in the sense of* §1.1.1), *then the domain of definition of* f *is necessarily a* **B**-*set, and* $y = y(x) = f(x)$ *is a Baire function* (Lebesgue).

In general, the domain of definition of f is a linear **A**-set of the most general kind, in other words not a **B**-set (Suslin).

Proof. If E is a linear **A**-set, then E is the orthogonal projection of a plane **B**-set X; if we define $\varphi(x, y) = 0$ or 1, according as $(x, y) \in X$ or $(x, y) \notin X$, then φ is a *Baire function* with $\varphi^{-1}(0) = X$ and $p_1X = E$. Thus $y(x)$ is an implicit "function" whose "domain of definition" is E.

[43] By a function we have always meant, in the preceding sections, a one-valued function.

Theorem 32. *If $y = fx$ is a one-to-one* **B**-*measurable function,*[44] *then the function f^{-1}, where the domain of definition A of f is an* **A**-*set, is* **B**-*measurable in fA* (Lebesgue).

7. Questions of Constructibility and Existence

1. If a mathematical object can be constructed, the question of its existence is thereby settled. For example, it follows from $ax^2 + bx + c = 0$ $(a \neq 0)$ that $x = \frac{1}{2a}(-b \pm (b^2 - 4ac)^{1/2})$; yet the importance of this useful numerical fact is far greater than the mere existence of solutions of the equation $ax^2 + bx + c = 0$. On the other hand, the statement that, since the algebraic and the real numbers are respectively countable and uncountable, there must exist at least one transcendental number does not solve the problem of actually finding such a number. Clearly the statement that, e.g., $2^{2^{1/2}}$, e, π are transcendental numbers is of a quite different sort.

Now consider the precisely analogous situation for functions. The set of all *real Baire functions* in R has the same power as the set of all continuous real functions in R, namely, the power of the continuum, and similarly for the class of **A**-sets, of **CA**-sets, and the whole class of projective sets. But the system of all linear sets, or of all real functions in R, is of higher power. Consequently, there must exist nonprojective sets, *non-Baire functions* and so forth. But how are such functions to be specified or even constructed? (See the *Lebesgue example* in §6.5.)

2. The question of existence or constructibility depends on the methods to be used. For example, the classical problem of the quadrature of the circle is unsolvable as long as we are restricted to ruler and compass; but it becomes immediately "solvable" if we are allowed more complicated instruments.

3. Such questions are usually connected with the existence of the following sets: N (set of all natural numbers), R (set of all real numbers), $I\omega_1$ (set of all countable ordinals), and finally, for every set S, the system of all subsets of S, the so-called power set $\mathbf{P}S$ of S. A particularly important role is played by the existence, for every non-empty system of sets M, of a one-valued function f_M which to every non-empty $X \in M$ assigns an $f_M X \in X$. The existence of f_M for every M is equivalent to the axiom of choice.

[44] A mapping f of a set M' into a set M'', $M' \subseteq R^{m'}$, $M'' \subseteq R^{m''}$, is said to be **B**-measurable if $f^{-1}(F)$ is a Borel set for every closed set F in M''. The **B**-measurable functions are precisely the *Baire functions* (Lebesgue). The proof given by Lebesgue in 1905 is not quite correct, and it was the criticism of this proof that led Suslin to the discovery of the **A**-sets.

4. For several decades efforts have been made to prove[45] the *Cantor continuum hypothesis* to the effect that there exist in R only two distinct infinite cardinal numbers: one of them is the number of elements in the set N of natural numbers and the other, the number of elements in the set R. In 1900, Hilbert placed this problem first in his famous collection of problems and gave a sketch of a supposed proof. Various mathematicians (among them Luzin) have doubted the possibility of a proof of any kind for the equivalent statement $2^{\aleph_0} = \aleph_1$, and Luzin expressed the same doubt about his own question on the existence of a perfect set in every uncountable **CA**-set.

5. In 1939, Gödel took an essential step forward, in proving that certain statements H (see below) when adjoined to a system of axioms Σ for the theory of sets do not lead to a contradiction provided that Σ itself is consistent.[46] In other words, H cannot be disproved in the system Σ. Further progress was made by Novikov in 1951 and by Esenin in 1954.

The results obtained up to that point were as follows.

Theorem 33. *The statement H (see below) does not lead to a contradiction when adjoined to the above-mentioned system of axioms Σ provided that Σ itself is consistent; here H can be any one of the following statements:*

1. *The continuum hypothesis is valid; more generally, if a is an infinite cardinal, then 2^a is the next higher cardinal.*
2. *The axiom of choice is valid.*
3. *There exists an uncountable* **CA**-*set that does not contain any perfect subset.*
4. *There exists a* $\mathbf{B}_2$-*set that is not Lebesgue-measurable.*

6. In this connection we also mention the following conjecture: every totally ordered continuous set in which every disjoint system of open intervals is countable contains a countable everywhere dense subset (and is thus isomorphic to an interval of the real numbers) (Kurepa, 1934, in connection with the problem stated by Suslin).

Or in equivalent form:

If every disjoint system of open sets of points in a totally ordered dense set S is countable, then every system of open disjoint sets in the square $S^2 = S \times S$ is also countable (Kurepa, 1950).

Finally, in 1963, P. Cohen[47] succeeded in proving the independence of the continuum hypothesis by constructing a model of a theory of sets in

[45] For the continuum problem and for the definition of a cardinal number, see IA, §§4.7 and 7.

[46] For this concept, see IA, §4.7.

[47] *The independence of the continuum hypothesis*, Proceedings of the National Academy of Sciences 50 (1963), 1143–1148; 51 (1964), 105–110.

which Cantor's hypothesis does not hold. In other words, H cannot be proved in the system Σ.

Cohen's result brings to an end the century-long debate about Cantor's hypothesis. Just as Euclid's parallel postulate was at last shown by Lobachevsky and others to be independent of the other axioms for Euclidean geometry, the Cantor hypothesis is now seen to be independent of a set of axioms ordinarily regarded as suitable for a theory of sets.

Exercises

1. Consider a set S; e.g., $S = \{0, 1, 2\}$ and a function f defined on S; e.g., $f0 = 1, f1 = 2, f2 = 3$. Determine: (a) the restriction $f|\{0,1\}$; (b) the set $\mathbf{P}' = \mathbf{P}\{0, 1, 2\}$ and the corresponding induced function $f|\mathbf{P}'$. What is the range of this induced function?
2. **A universal G-set.** Let $I_1, I_2 \cdots$ be an enumeration of all the open intervals of R with rational end-points, where $I_1 = \varnothing$. For any irrational number $x \in R(0,1)$ let $x = \cfrac{1}{x_1 + \cfrac{1}{x_2} + \cdots}$ be the continued fraction for x and let $G(x) = I_{x_1} \cup I_{x_2} \cup \cdots$. Let M_2 be the set of all points (x, y) satisfying $y \notin G(x)$ and let $U_2 = R^2 \backslash \overline{M}_2$. Then prove that M_2 is a unversal set for open sets $\subseteqq R_1$ in the sense that the family of all open sets $\subseteqq R$ coincides with the family of all intersections of U_2 with parallels to y-axis. (cf. W. Siepiński, Fundamenta Mathematicae 7 (19) 198).
3. Let $0 < r_1 < r_2 < \cdots$ be a strictly increasing sequence of real positive numbers; let O be a fixed point and let F_n be the closed sphere with O as center and r_n as radius. Under what conditions on r_n is the union of the sets $F_1, F_2 \cdots$ closed?
4. Let $r_1 > r_2 > \cdots$ be a strictly decreasing sequence of positive real numbers; let O be a point and let G_n be the open sphere with O as center and r_n as radius. Under what conditions on r_n is the intersection of the sets G_n open?
5. Let r be a real number with $0 \leqq r < 1$. Is there a nowhere dense perfect subset $S(r)$ of $R[0, 1]$ such that mes $S(r) = r$?
6. Let $\sum_{n=1}^{\infty} a_n$ be any convergent series with nonnegative terms; consider the set $S(a)$ of all the sums $S(M) = \sum_x a_x$, where $x \in M$ and $M \subseteqq \{1, 2, 3, \ldots\} \equiv N$; in particular, let $S(\varnothing) = 0$. Prove the following: (1) The set $S(a)$ is finite or perfect according as the set $\{a_1, a_2, \ldots\}$ is finite or transfinite. (2) For $a_n = 2.3^{-n}$, $S(a)$ is the triadic set T.
7. Let $d_1, d_2, \ldots d_k$ (not all zeros and not all nines) be a fixed sequence of the decimal digits $0, 1, \ldots, 9$. Let $S(d_1, d_2, \ldots d_k)$ be the set of all

real $x \in [0, 1]$ such that no segment of the decimal representation of $x = \sum_n x_n 10^{-n}$ coincides with the sequence $d_1, d_2, \ldots, d_k$. Prove that $S(d)$ is measurable, perfect, and nowhere dense.

8. **Plane triadic set.** Consider any square K_2, divide it into 3^2 equal squares by means of lines parallel to the sides of K_2, delete the interior of the central square. With each of the remaining squares perform the analogous subdivision and deletion. By infinite iteration in this way a set G_2 is deleted (the union of the deleted central squares), and there remains a set $T_2 = K_2 \backslash G_2$. Determine the 2-dimensional measure of T_2 and prove that T_2 is a closed, nowhere-dense set in K_2.

   Generalize for R^3, K_3 (cube), G_3, T_3 as well as for R^n, K_n (hypercube), G_n, T_n.
9. If a set S is nowhere dense, so also $\bar{S}$.
10. The union of a finite number of nowhere-dense sets is nowhere dense.
11. The set S of all complex numbers Z satisfying $Z^{2^{1/2}} = 1$ is everywhere dense on the unit circle.
12. The set Q of rational numbers is an F_σ set but is not a G_δ-set.
13. Prove that the set T of transcendental numbers is a G_δ-set in the space R of real numbers.
14. Consider any Cartesian space R^n and the set $Q^n \subset R^n$ of all points with rational coordinates. Prove that $R^n \backslash Q^n$ is a G_δ-set.
15. Prove that the orthogonal projection on the x-axis of any bounded open set $X \subseteqq R^2$ is open but that the projection of a bounded closed set is not necessarily closed.
16. The set of points of continuity of a continuous function $f{:}R \to D$ is open.
17. If M is a bounded closed set and if $f|M$ is continuous, then fM is an F_σ-set but not necessarily an F-set.
18. Let $S = \{s_1, s_2, \ldots\}$ be any set of real numbers of cardinality $\leqq \aleph_0$. Is there a real-valued function $f|R$ such that S is the set of all points of continuity of f?
19. Is there a real-valued function $f|R$ such that

    (1) f is continuous on the set Q of rationals and discontinuous on $R \backslash Q$;

    (2) f is continuous on $R \backslash Q$ and discontinuous on Q?
20. Prove that every F_δ-set E of real numbers is the set of points of continuity of some function $f{:}R \to R$.
21. Let (1) $f_n|R \to R$ be any sequence of continuous real functions. Prove that the set X of convergence of the sequence (1) is an $F_{\sigma\delta}$-set.
22. Let S be any system of sets; consider S_σ, $S_{\sigma\delta}$, S_δ, $S_{\delta\sigma}$ (of §3.3); let $x \in \{r, \delta, \sigma\delta, \delta\sigma\}$; represent any number $E \in S_x$ by means of the **A**-operation (cf. §6.2.3).

Bibliography

[1] AUMANN, G., Reelle Funktionen. Springer, Berlin-Göttingen-Heidelberg 1954.

[2] HAUSDORFF, F.: Set theory. (Translated from the German by J. R. Aumann *et al.*) Chelsea, New York 1957.

[3] KURATOWSKI, C.: Topologie I. (Polska Akademia Nauk, Monogr. matem. 20) Warsaw 1958.

[4] LUZIN, N.: Ensembles analytiques. Gauthier-Villars, Paris 1930.

[5] NATANSON, I. P.: Theory of functions of a real variable. (Translated from the Russian by L. F. Boron *et al.*) Ungar, New York 1955.

[6] SIERPIŃSKI, W.: Les ensembles projectifs et analytiques (Mémorial des sciences math. 122). Paris 1950.

Bibliography added in Translation

BOAS, R. P.: A primer of real functions. (The Carus mathematical monographs, no. 13). Wiley, New York 1960.

KAMKE, E.: Theory of sets. (Translated from the German by F. Bagemihl). Dover, New York 1950.

ROYDEN, H. L.: Real analysis. Macmillan, New York 1963.

CHAPTER 13

Analysis and Theory of Numbers

Introduction

Of basic importance in analysis is the concept of convergence, defined as meaning that a given property concerning the set N of natural numbers holds for almost every $n \in N$, i.e., for all $n \in N$ with a finite number of exceptions.

Thus in contrast to the language of elementary mathematics, we find expressions in analysis such as *almost all*, *almost equal*, *approximately equal*, *asymptotically equal*, *equal in the limit*, and so forth. It is instructive to consider how these relations of equality and order have found many different applications in mathematics and have led to important advances. As further examples we may mention the well-known epsilon-language and the *Landau symbols* o and O (see §1.3). These expressions also play a role in the theory of numbers. As an example here let us give the following theorem from the modern theory of numbers.

Almost all odd natural numbers $2n + 1$ *can be represented as the sum of three primes.* This theorem was proved in 1937 by Vinogradov; according to a conjecture made by Goldbach in 1742, the assertion holds for all $n \geqq 4$. The theorem of Vinogradov is of an asymptotic character, whereas the Goldbach conjecture involves only the language of elementary mathematics.

Progress in number theory by the methods of analysis, or in other words progress in analytic number theory, began with the *Euler identity* (1737):

$$\sum_{n \in N} n^{-s} = \prod_{p \in P} (1 - p^{-s})^{-1}$$

(where P is the set of all primes) and the proof of the *Dirichlet theorem:*

If a, b are coprime, then the set M of numbers $ax + b$ ($x \in N$) contains infinitely many primes; in other words, the intersection $M \cap P$ is infinite.

The central theorems in analytic number theory proper are concerned with the asymptotic equality of the following functions[1]

$$\pi(x), \quad \frac{x}{\log x}, \quad \int_2^x \frac{dt}{\log t}$$

(here $\pi(x)$ denotes the number of prime numbers $p \leqq x$), i.e., with the fact that the quotient of any two of these three functions approaches 1 as $x \to \infty$ (see §1.3, theorem 3 and §1.9, theorem 14).

The analytic theory of numbers is made up of theorems on the relationships among integers that can be proved by the methods of analysis or that have the asymptotic form characteristic of analysis.

The present chapter is divided into three sections. In §1 we describe the invasion of the theory of numbers by analysis and attempt thereby to make clear the nature of analytic number theory. §2 deals with analysis as related to the transcendence properties of numbers. Here we investigate the arithmetic nature (irrationality and transcendence) of numbers and of the values of functions. The final section §3 gives some discussion of the interrelations between analysis and the theory of numbers in general.

Our chief purpose is to present some of the more important results in the theory of numbers in readily understood form, referring, for the most part, to the relevant literature for the proofs. Finally, it should be emphasized that at the present time many mathematicians are actively engaged in research in the problems of the analytic theory of numbers.

1. Invasion of the Theory of Numbers by Analysis

Analytic Number Theory

1.1. *The Euler Proof That the Set of Primes Is Infinite*

Let N be the set of all natural numbers and let P be the set of all primes. Then we have the following theorem:

Theorem 1. *The set P is infinite* (Euclid).

For our purposes it is important to show how this theorem was proved by Euler. For every prime p we have

$$(1-p^{-1})^{-1} = 1 + p^{-1} + p^{-2} + p^{-3} + \cdots. \tag{1}$$

If we assume that P is finite and if in (1) we let p run through the set P and multiply the resulting equations together, we obtain (by multiplying out the right sides) the formal relation:

$$\prod_{p \in P} (1-p^{-1})^{-1} = \sum_{n \in N} n^{-1}. \tag{2}$$

[1] By log here and below we will always mean the natural logarithm.

For if n is any natural number and $n = p_{n_1}^{\alpha_1} p_{n_2}^{\alpha_2} \cdots p_{n_k}^{\alpha_{n_k}}$ is its canonical factorization (see IB6, §2.6), then the term n^{-1} is obtained in (2) if for p in (1) we successively set $p = p_{n_\lambda}$ ($\lambda = 1, 2, \ldots, k$) and multiply together the terms $p_{n_\lambda}^{\alpha_\lambda}$ on the right-hand side. Since we have assumed that P is finite, the product on the left-hand side in (2) is also finite. But $\sum_{n \in N} n^{-1}$ is divergent, in contradiction to the assumption that P is finite. Thus there are infinitely many primes. In this way the theorem has been proved analytically, i.e., by an argument involving limits. Historically, this Euler proof of the theorem of Euclid represents the beginning of the analytic theory of numbers. From the divergence of the product in (2) Euler deduced the theorem:

Theorem 2. *The series* $\sum_{p \in P} p^{-1}$ *is divergent, from which it again follows that* P *is infinite.*

1.2. *The Euler-Riemann ζ-Function*

The relation (2) is a special case of the equation:

(3)
$$\prod_{p \in P} (1 - p^{-s})^{-1} = \sum_{n \in N} n^{-s},$$

which remains valid for every number $s > 1$, in fact for every complex number s with real part > 1. The series in (3) is usually denoted by $\zeta(s)$, and is called the *Euler-Riemann ζ-function*; thus

$$\zeta(s) = \sum_{n \in N} n^{-s}.$$

Then equation (3) reads

(4)
$$\prod_{p \in P} (1 - p^{-s})^{-1} = \zeta(s)$$

and is satisfied for every complex number $s = \sigma + it$ with $\sigma > 1$.

In the analytic theory of numbers it is customary to let s denote an arbitrary complex number and to set $\mathrm{Re}(s) = \sigma$ and $\mathrm{Im}(s) = t$.

Formula (4) states a relation between the sets P of primes and the ζ-function. Thus it is not surprising that the various analytic properties of the ζ-function enable us to make statements about the set P. By analytic continuation we define ζ for the whole complex plane. Except for $s = 1$ the function is everywhere analytic. The ζ-function is meromorphic with its only pole at 1, where its residue is 1. Thus the function $f(s) = \zeta(s) - \dfrac{1}{s - 1}$ is an entire function, i.e., $f(s)$ is a well-defined number for every complex number s.

1.3. *The Function $\pi(x)$; The Prime Number Theorem* (cf. IB6, §9)

For every real number $x \geqq 2$, let $\pi(x)$ denote the number of primes $\leqq x$. By theorem 1 we have $\lim \pi(x) = \infty$. But Euclid's theorem does not give any information about the rate of growth of $\pi(x)$. Thus we are left with the fundamental question of finding some measure and some asymptotic estimate for the growth of $\pi(x)$. Protracted efforts on the part of famous mathematicians (Legendre, Gauss, Čebyšev, Hadamard, de la Vallée-Poussin, and so forth) led finally to the following theorem, which is basic for the whole of the analytic theory of numbers:

Theorem 3 (The Prime Number Theorem). *The function $\pi(x)$ is asymptotically equal to $\frac{x}{\log x}$, i.e.,*

$$\lim_{x\to\infty} \frac{\pi(x)}{x/\log x} = 1; \tag{5}$$

in other words, for large values of x the function $\pi(x)$ is approximately equal to $\frac{x}{\log x}$. Thus we have the relation $\frac{\pi(x)}{x/\log x} = 1 + \epsilon(x)$, with $\lim\limits_{x\to\infty} \epsilon(x) = 0$.

In other words, if we divide the difference $\pi(x) - \frac{x}{\log x}$ by $\frac{x}{\log x}$, the quotient approaches 0. This relation is also expressed in the following way:

$$\pi(x) - \frac{x}{\log x} = o\left(\frac{x}{\log x}\right) \quad \text{and} \quad \pi(x) = \frac{x}{\log x} + o\left(\frac{x}{\log x}\right). \tag{6}$$

Let us explain the notation further: let f and g be functions defined on a final segment of the set R of real numbers, and let g be positive; then $f = o(g)$ means that $\lim\limits_{x\to\infty} \frac{f(x)}{g(x)} = 0$. Similarly, $f = O(g)$ (Landau) or $f \ll g$ (Vinogradov) means that there exist numbers $C > 0$ and $x_0 > 0$ such that $|f(x)| \leqq Cg(x)$ for every $x > x_0$.

The same notation is also used for sequences. For example, $f = O(1)$ or $f \ll 1$ means that f is bounded. Instead of $x \to \infty$ we may also consider some other approach to a limit, say $x \to 0$, and use the corresponding notation. The symbols o, $\ll$ and O have proved to be very clear and useful.

1.4. *Some Consequences of the Prime Number Theory*

Let us emphasize again some of the most important facts:

1. Since $\frac{x}{\log x}$ approaches ∞ with x, it follows from (5) that $\pi(x)$ also approaches ∞ with x. The function $\pi(x)$ becomes infinitely great with increasing x and thus is not bounded, so that in the notation just introduced: $\pi(x) \neq O(1)$ (theorem 1).

2. The function $\pi(x)$ increases more slowly than x (Legendre), a fact which we can express in the form $\pi(x) = o(x)$. For by (6) in §1.3 we have

$$\frac{\pi(x)}{x} = \frac{1}{\log x} + o\left(\frac{1}{\log x}\right),$$

from which it follows that $\lim_{x \to \infty} \frac{\pi(x)}{x} = 0$. This result corresponds to the empirical fact that at first the prime numbers occur very frequently, but then more and more rarely (*Sieve of Eratosthenes*). Let us mention that there exist arbitrarily long connected sections of N containing only composite numbers; e.g., the numbers $n! + r$ $(r = 2, 3, 4, \ldots, n)$ for $n > 2$ form such sets.

Closely related to the prime number theorem are other theorems in the analytic theory of numbers; we shall mention some of them here, and give a few proofs. First of all let us cite the following theorem.

Theorem 4. *For every $\epsilon > 0$ and every $n \in N$ there exists a number x_1 such that for every $x > x_1$ there are at least n primes between x and $(1 + \epsilon)x$, and consequently $\pi((1 + \epsilon)x) - \pi(x) \geqq n$.*

For the proof we make use of the prime number theorem, according to which for every $\eta > 0$ there exists a number $x_0(\eta)$ such that $1 - \eta < \frac{\pi(x)}{x/\log x} < 1 + \eta$ for every $x > x_0$, from which it follows that $(1 - \eta)\frac{x}{\log x} < \pi(x) < (1 + \eta)\frac{x}{\log x}$. Setting $(1 + \epsilon)x$ in place of x we obtain

$$(1 - \eta)\frac{(1 + \epsilon)x}{\log(1 + \epsilon) + \log x} < \pi((1 + \epsilon)x) < (1 + \eta)\frac{(1 + \epsilon)x}{\log(1 + \epsilon) + \log x}.$$

From these two relations we have

$$\pi((1 + \epsilon)x) - \pi(x) > \frac{x}{\log x + \log(1 + \epsilon)}[\epsilon - \eta(\epsilon + \epsilon' + 2) - \epsilon'],$$

where $\epsilon' = \frac{\log(1 + \epsilon)}{\log x}$, Since the positive numbers η and ϵ' can be made arbitrarily small, the expression in brackets $[\cdots]$ is positive for all sufficiently large values of x. Since the factor $\frac{x}{\log((1 + \epsilon)x)}$ becomes arbitrarily large with increasing x, there must exist an x_0 such that the difference $\pi((1 + \epsilon)x) - \pi(x)$ is greater than n, for all $x > x_0$, as was to be proved.

As a special case of theorem 4 we may consider part of the following statement (Bertrand-Čebyšev).

Theorem 5. *For every $x > 1$ there exists at least one prime between x and $2x$.*

If we set $\epsilon = 1$ and $n = 1$ in theorem 4, it follows at once that theorem 5 is valid for almost all x in N. Let us note that theorem 5 is much simpler to prove than theorem 4.

Proceeding more deeply into the subject, we can now prove the following theorem.

Theorem 6. *On the set* $(0, \infty)$ *of positive real numbers there exists a continuous real function* $\rho(x)$ *satisfying the three conditions:*

$$\lim_{x \to \infty} \frac{\rho(x)}{x} = 1, \tag{7}$$

$$\pi(\rho(x)) - \pi(x) > 1 \quad \textit{for every } x > 0. \tag{8}$$

$$\lim_{x \to \infty} [\pi(\rho(x)) - \pi(x)] = \infty. \tag{9}$$

For let $\epsilon_1 > \epsilon_2 > \cdots \to 0$ be a strictly decreasing sequence converging to zero, and to every ϵ_n let there be assigned a natural number a_n with the property

$$\pi((1 + \epsilon_n)x) - \pi(x) > n \quad \text{for every } x \geqq a_n. \tag{10}$$

Without loss of generality we may assume that the sequence a_n $(n \in N)$ is strictly increasing and approaches infinity. By A_n we denote the intersection of the straight lines $y = (1 + \epsilon_n)x$ and $x = a_n$ $(n \in N)$. If we now define the function $\rho(x)$ in such a way that in the domain (a_1, ∞) it coincides with the broken line $A_1, A_2, A_3, \ldots$ and $\rho(x) = (1 + \epsilon_1)a_1$ for $0 < x < a_1$, then it satisfies the desired conditions. For by (10) the continuous function $\rho(x)$ satisfies (8) and (9), so that only (7) remains to be proved. Let x_n $(n \in N)$ be a sequence of real numbers $n \geqq a_1$ approaching ∞; also let $v(n)$ be defined for every n in such a way that $a_{v(n)} \leqq x_n \leqq a_{v(n+1)}$. Then $v(n)$ increases with n beyond all bounds. From

$$(1 + \epsilon_{v(n)})x_n \geqq \rho(x_n) \geqq (1 + \epsilon_{v(n+1)})x_n$$

it follows that

$$1 + \epsilon_{v(n)} \geqq \frac{\rho(x_n)}{x_n} \geqq 1 + \epsilon_{v(n+1)}. \tag{11}$$

If we let $n \to \infty$ in (11), then, since $v(n) \to \infty$ also, we obtain the desired relation (7).

It is now natural to ask whether theorem 6 will enable us to make definite statements about the distribution of the primes, e.g., whether there is at least one prime between x^n and $(x + 1)^n$ for large x. This statement is true for every $n > 5$, as can be deduced from the Čudakov theorem below.

Let us now consider the sequence of differences $\Delta n = p_{n+1} - p_n$. By the Čebyšev theorem there exists a prime (theorem 5, §1.4) between n

and $2n$ and thus, in particular, between p_n and $2p_n$. Consequently $p_n < p_{n+1} < 2p_n$, so that $\Delta n = p_{n+1} - p_n < p_n$, i.e., $p_{n+1} - p_n = O(p_n)$, which may be regarded as the simplest estimate of Δn.

A much better estimate is that of Čudakov:

Theorem 7. $p_{n+1} - p_n = O(p_n^{(3/4)+\epsilon})$.

These remarks on prime numbers offer a good opportunity to discuss an interesting theorem on the binomial coefficients. From the equation

$$2n = (n + r)\left(1 + \frac{n - r}{n + r}\right) \quad \text{and} \quad \lim_{n\to\infty} \frac{n - r}{n + r} = 1$$

(r is a fixed natural number) it follows from theorem 4 that for sufficiently large n there are arbitrarily many primes between $n + r$ and $2n$. Together with $\binom{2n}{n - r} = \frac{(2n)!}{(n - r)!(n + r)!}$ this means that the number $\binom{2n}{n - r}$ contains arbitrarily many primes to the first power, provided that n is sufficiently large, and the same statement holds for $\binom{2n + 1}{n - r}$. Thus we have the following theorem.

Theorem 8. *For every integer* $r \geqq 0$ *we can find a natural number* r_0 *such that for every* $n > r_0$ *the number* $\binom{n}{n' - r}$, *where* $n' = \left[\frac{n}{2}\right]$, *contains at least one prime to exactly the first power; thus the number* $\binom{n}{n' - r}$ *cannot be expressed as a higher power than the first* (Kurepa, 1953).

According to a conjecture of Erdös (1939) the number $\binom{n}{k}\left(1 < k \leqq \frac{n}{2}\right)$ is never an l-th power ($l > 2$). This conjecture has already been proved for $l = 3, 4, 5$.

1.5. *The "Greatest" Prime* (see IB6, §2.12, theorem 5)

The essential content of theorem 1 (§1.1) is the statement that no greatest prime exists (Euclid). However, it is meaningful to speak of the greatest prime number known at the present time. This is the number $p = 2^{4423} - 1$ (cf. A. Hurwitz: New Mersenne Primes; Math. of Computation 16 (1962), p. 249).

1.6. *Riemann Conjecture and Theorems on the Nontrivial Zeros of the ζ-Function*

It is a remarkable fact that estimation of $p_{n+1} - p_n$ is closely related to the set of zeros of the *Riemann ζ-function*. The trivial zeros are -2, -3, $-4, \ldots$, as can be seen from the functional equation

$$\zeta(1 - s) = \frac{2}{(2\pi)^s} \cos \frac{\pi s}{2}\, \Gamma(s)\zeta(s). \tag{12}$$

The other zeros lie in the strip $0 < \sigma < 1$. If we let $N(T)$ be the number of these zeros in the rectangle $0 < \sigma < 1$, $0 < t \leqq T$, then:

Theorem 9.

$$N(T) = \frac{1}{2\pi} T \log T - \frac{1 + \log 2\pi}{2\pi} T + O(\log T).$$

This theorem was conjectured by Riemann in 1859 and proved by Mangold in 1905.

According to the famous Riemann conjecture (1859), every nontrivial zero of the ζ-function is of the form $\frac{1}{2} + it$. Up to the present time this much-discussed conjecture has been neither proved nor refuted, although many of its consequences have been proved by other means.

In 1942 the following remarkable theorem was proved by Selberg:

Theorem 10. *The number $N_0(T)$ of solutions of $\zeta(\frac{1}{2} + it) = 0$ for $0 < t \leqq T$ has the property that $\dfrac{N_0(T)}{T \log T}$ lies between two positive constants; in particular, there exists a number $A > 0$ and a $T_0 > 0$ such that*

$$N_0(T) > AT \log T \quad \text{for } T \geqq T_0.$$

If for the infinitely many zeros $\frac{1}{2} + ic_n$ of the function ζ we arrange the positive c_n in order of magnitude ($c_1 = 14.134725\ldots$, $c_2 = 21.022040\ldots$), then $\lim_{n\to\infty} (c_{n+1} - c_n) = 0$. By a theorem of Bohr-Landau, the nonreal zeros of ζ lie chiefly in an arbitrary ϵ-strip around $\sigma = \frac{1}{2}$ (see the hatched strip in Figure 1). If we denote by $N_1(\epsilon, T)$ the number of these zeros ρ with $|\text{Re}\, \rho - \frac{1}{2}| < \epsilon$, $|\text{Im}\, \rho| < T$, and by $N_2(\epsilon, T)$ the number of the remaining nontrivial zeros with $|\text{Im}\, \rho| < T$, then:

Theorem 11 (Bohr-Landau, 1913).

$$\lim_{T\to\infty} \frac{N_2(\epsilon, T)}{N_1(\epsilon, T)} = 0,$$

for every $\epsilon > 0$.

The best estimate known at the present time for the above function $N_0(T)$ is given by the formula $\zeta(\frac{1}{2} + it) = O(t^{(15/92)+\epsilon})$ for every $\epsilon > 0$ (Min, 1949).

The close connection between the ζ-function and the set P of prime numbers can also be seen from the following theorem:

Theorem 12. *From $\zeta(\frac{1}{2} + it) = O(t^c)$ it follows that*

$$p_{n+1} - p_n = O(p_n^{[(1+4c)/(2+4c)]+\eta})$$

for every $\eta > 0$ (Ingham, 1937).

The determination of $g = \lim_{n\to\infty} (p_{n+1} - p_n)$ is still an open problem. Is $g < \infty$? Or is $g = 2$, so that there are infinitely many twin primes? Although $\sum p^{-1}$ is divergent, the subseries over the twin primes is convergent (Brun).

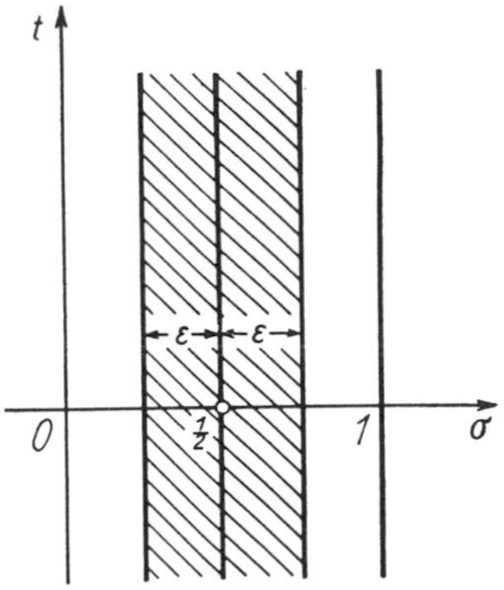

Fig. 1

1.7. *The Function $N(\sigma, T)$*

Fix $\sigma > \frac{1}{2}$, and denote by $N(\sigma, T)$ the number of $\zeta(\sigma + it) = 0$ for $0 < t < T$. By the Riemann conjecture, we would have $N(\sigma, T) = 0$. The difficulties in estimating $N(\sigma, T)$ are clear from the following theorem.

Theorem 13. *The relation $N(\sigma, T) = O(x(\sigma, T))$ is satisfied by the following expressions for $x(\sigma, T)$:*

$T \log T$	(Bohr-Landau, 1914),
$T^{1-4\delta^2+\epsilon}$	(Carlson, 1921),
$T^{1-2\delta}(\log T)^6$	(Turán, 1951),
$T(\log T)^{\delta/(\delta-1)}$	(Wenneberg, 1920),
$T^{1-4\delta^2}(\log T)^6$	(Hoheisel, 1930),
$T^{15/92}(\log T)^{1/58}$	(Titchmarsh-Min, 1949),

and so forth, where $\delta = \sigma - \frac{1}{2}$ and $\epsilon > 0$ is arbitrary.

1.8 *Conjecture of Lindelöf*

The Riemann conjecture implies the conjecture of Lindelöf (Littlewood, 1912) that $\zeta(\frac{1}{2} + it) = O(t^{\epsilon})$ for every $\epsilon > 0$. The converse question whether the Lindelöf conjecture also implies the Riemann conjecture is still open. If the Lindelöf conjecture is correct, then in theorem 12 (Ingham) we could let c approach 0 and in this way obtain the still unproved asymptotic estimate:

$$p_{n+1} - p_n = O(p_n^{(1/2)+\eta}) \quad \text{for every } \eta > 0.$$

1.9. *Second Prime Number Theorem*

Theorem 14. *The function $\pi(x)$ is asymptotically equal to $\int_2^x \frac{dt}{\log t}$, i.e., the quotient of these two functions converges to 1 as x approaches ∞.*

Moreover, we have

$$\pi(x) = \int_2^x \frac{dt}{\log t} + o\left(\frac{x}{\log^r x}\right) \quad \textit{for every } r > 0.$$

The proof of the second prime number theorem is not essentially different from that of the first (§1.3, theorem 3). This theorem has been expressed in many different ways; for example, the relation between the functions $\pi(x)$ and $\int_2^x \frac{dt}{\log t}$ is written in the form:

$$\pi(x) = \int_2^x \frac{dt}{\log t} + o\left(\frac{x}{g(x)}\right),$$

$$g(x) = [-C(\log x)^{4/7}(\log\log x)^{-3/7}] \quad \text{with } C > 0.$$

1.10. *Giant Numbers; Zeros of the Function*

$$P(x) = \pi(x) - \int_2^x \frac{dt}{\log t}.$$

On the basis of the tabulated values, it had been believed that $P(x) \leqq 0$, but in 1914 Littlewood proved that $P(x)$ takes both positive and negative values. The zeros of $P(x)$ are gigantically large; for example, the first of them is of the order of magnitude of $10^{10^{10^{34}}}$ (Skewes, 1955), which is certainly the largest definite number that has ever played a role in science.

1.11. *Certain Sets of Primes*

In connection with the set P of prime numbers let us mention the following infinite subset of P: $\pi_1 = 2 = p_1$, $\pi_2 = 3 = p_2$, $\pi_3 = f(\pi_1\pi_2 + 1), \ldots,$ $\pi_n = f(\pi_1\pi_2 \cdots \pi_n + 1), \ldots,$ where $f(x)$ is the smallest prime factor of the natural number x. It is still an open question how often $\pi_{n+1} < \pi_n$.

Among the *Fermat numbers* $F_n = 2^{2^n} + 1$, the first five, F_0, F_1, F_2, F_3, F_4 are primes, but not $F_5, F_6, \ldots, F_{12}, F_{15}, F_{16}, F_{18}, F_{23}, F_{36}, F_{38}, F_{73}$. It is still unknown whether the numbers F_{13}, F_{14} and F_k for $k > 73$ are prime or not. Since F_{73} has more than 10^{21} decimal places, it can hardly be written out in the usual decimal form, in view of the fact that the volumes required to hold it would fill several billions of libraries.

1.12. *The Dirichlet Theorem on Prime Numbers* (see IB6, §9)

The Euclidean theorem on prime numbers was generalized by Dirichlet in the following way:

Theorem 15. *If the natural numbers a and b are coprime, then the arithmetic progression $an + b$ $(n \in N)$ contains infinitely many prime numbers* (Dirichlet, 1837). *If $p_1(a, b)$ is the smallest of these primes and if $2 \leqq a$, $1 \leqq b < a$, there exists a constant c independent of a such that $p_1 < a^c$* (Linnik, 1947). *For $a = b = 1$ we have the theorem of Euclid.*

With his proof of the first half of this theorem Dirichlet laid the foundations for the analytic theory of numbers in the proper sense of the term. The other theorems on prime numbers have also been generalized to arithmetic progressions.

If we let $\pi(x; a, b)$ be the number of primes p for which $p \leqq x$; $p \equiv b \pmod{a}$, then the following theorem is easy to prove.

Theorem 16. *For fixed coprimes a and b the functions* $\pi(x; a, b)$ *and* $\frac{x}{\varphi(a) \log x}$ *are asymptotically equal; here* $\varphi(a)$ (*the Euler function*) *is the number of terms in* $1, 2, 3, \ldots, a$ *that are prime to a.*

1.13. *Dirichlet Series*

A very useful instrument in the whole of analytic number theory is provided by the so-called *Dirichlet series* $\sum_{n=1}^{\infty} a_n n^{-s}$, which are generalizations of the ζ-function. There is an extensive literature on these series (see the bibliography).

1.14. *Elementary Proofs of the Prime Number Theorems*

The prime number theorems were proved at the end of the 19th century by arguments on the ζ-function regarded as a function of a complex variable. But repeated attempts were made to prove these theorems by elementary means in the domain of the real numbers alone. In these efforts Selberg has been especially successful. He recognized the basic importance of the following theorem.

Theorem 17. $\vartheta(x) \log x + \sum_{p \geqq x} \vartheta\left(\frac{x}{p}\right) \log p = 2x \log x + o(x)$ *with the Čebyšev function* $\vartheta(x) = \sum_{p \leqq x} \log p$.

This equation is a simple consequence of the prime number theorem (§1.3, theorem 3), but can also be proved, independently of that theorem, by "elementary" means. Now Selberg recognized that conversely the prime number theorem can also be deduced by elementary means from theorem 17. Thus the prime number theorem can be proved in an elementary way, i.e., independently of the theory of the functions of a complex variable. These proofs (Selberg, Erdös) depend chiefly on the properties of the logarithmic functions, but it must be admitted that although they are "elementary" they are by no means simple.

1.15. *Goldbach Conjecture* (see IB6, §9)

In a letter to Euler in 1742 Goldbach expressed the conjecture that *every odd number* $\geqq 9$ *can be expressed as a sum of three primes.* For the proof it is sufficient to show that every even number $\geqq 6$ is the sum of two primes. In 1937 Vinogradov proved the following theorem.

Theorem 18. *There exists a number* x_0 *such that every odd number* $> x_0$ *can be expressed as a sum of three primes.*

The proof is quite lengthy and depends on an ingenious method of estimating trigonometric sums; however, the result can also be deduced from the properties of the ζ-function. But it is still an open question whether every "large" even number can be expressed as the sum of two primes. However, we have the following theorem.

Theorem 19. *The number of even numbers* $\leqq x$ *not contained in* $P + P$ (see IB6, §9) *is equal to* $O\left(\frac{x}{(\log x)^r}\right)$, *where* $r > 0$ *can be chosen arbitrarily* (Čudakov).

Although the Goldbach conjecture on even and odd numbers as sums of primes can be stated in very elementary terms, none of the proofs that have been found for parts of it are at all elementary. This example shows clearly that relatively simple statements in the theory of numbers can be dealt with at the present time only by the methods of analysis.

1.16. *Waring Problem: the Functions G(k), g(k)* (see IB6, §9)

First let us introduce certain concepts and notation that will simplify our discussion. Frequent use is made of the following sets of numbers:

$$N = \{1, 2, 3, \ldots\}, \qquad N_0 = \{0, 1, 2, 3, \ldots\} = \{0\} \cup N,$$
$$P = \{2, 3, 5, \ldots\}, \qquad P_0 = \{0, 1, 2, 3, 5, \ldots\} = \{0, 1\} \cup P.$$

If A, B are sets of numbers, we define (see IB6, §9):

$$A + B = \{a + b, a \in A, b \in B\},$$
$$AB = \{ab, a \in A, b \in B\},$$

and in particular

$$1A = A, A^1 = A, \ldots, (n + 1)A = nA + A, A^{n+1} = A^nA.$$

We say that two sets A and B are asymptotically equal, in symbols $A \sim B$, if they have a common final segment, i.e., if there exists a number m such that

$$A \cap [m, \infty] = B \cap [m, \infty] = [m, \infty]_N.$$

Then by theorem 18 we have $4P_0 \sim N_0$ (Vinogradov, 1937), and in the notation of the present section, the still unproved *Goldbach conjecture* is written $3P_0 \sim N_0$.

Waring conjectured in 1770 (see IB6, §9) that the set

$$N^{[k]} = \{0, 1^k, 2^k, 3^k, \ldots\}$$

is a basis, i.e., that there exists a number $r = r(k)$ with the property that every number of the set N_0 can be represented as the sum of r numbers of the form x^k with $x \in N_0$, so that $N_0 = r(k)N^{[k]}$.

If $G(k)$ is the smallest number r such that the sets N_0 and $rN^{[k]}$ are asymptotically identical, i.e., such that almost every natural number can be represented as the sum of r numbers x^k with $x \in N_0$, then the following relations, proved only after considerable effort, are found to hold.

Theorem 20. 1. $G(2) = 4$ (Lagrange),
2. $G(k) \geqq k, k \geqq 2$,
3. $G(k) < \infty$ (Hilbert, 1909),
4. $G(3) \leqq 8$ (Landau),
5. $G(4) \leqq 17$ (Estermann, Davenport, Heilbronn, 1936),
6. $G(k) \leqq 6k (\log k + 1)$ für $k \geqq 16$ (Vinogradov, 1937),
6a. $G(k) < 3k \log k + 11k$.

In conclusion let us mention the following relation between the functions $G(k)$ and $g(k)$.

Theorem 21. *If we denote by $\bar{g}(k)$ the smallest number n such that every natural number can be expressed as the sum of n numbers x^k with $x \in N_0$, then*

$$\lim_{k \to \infty} \frac{G(k)}{\bar{g}(k)} = 0.$$

2. Analysis and Questions of Transcendence

2.1. *Quadrature of the Circle and the Number π*

The classical Greek problem of the quadrature of the circle is closely related to the nature of the number π.

It is well known that if a number x can be constructed from a given unit length by means of ruler and compasses, it must be expressible by a finite number of repetitions of the algebraic operations $+$, $-$, $\cdot$, $:$ and $\sqrt{}$; thus x is necessarily an algebraic number. If quadrature of the circle were possible, i.e., if with ruler and compasses alone we could construct a square with area equal to that of a given circle, we could take its radius equal to unity and thereby construct the side of the square with length equal to $\pi^{1/2}$. Thus the number $\pi^{1/2}$, and therefore also $\pi = \pi^{1/2} \cdot \pi^{1/2}$, would be algebraic. But Lindemann proved in 1882 that this is not the case, so that quadrature of the circle is impossible. Nonalgebraic numbers are usually called *transcendental*. Thus we have

Theorem 22. *π is a transcendental number.*

This result provides an example to settle the long and wearisome controversy whether transcendental numbers exist or not.

As a remark of theorem 22 let us note that the circumference of an ellipse with axes of algebraic length is transcendental.

2.2. *First Appearance of Transcendental Numbers; Liouville Property of Algebraic Numbers*

The number π was followed by the appearance of many other transcendental numbers; moreover, it is not the first number whose transcendence was recognized. In 1873 Hermite had already proved the following theorem.

Theorem 22a. *The number e is transcendental.*

At the present time many other transcendental numbers are known.

Liouville gave a very interesting property of the algebraic numbers of degree n, in connection with his investigation into the accuracy and rapidity with which irrational numbers can be approximated by rational numbers; a great deal of the present knowledge about transcendental numbers depends on his result, which runs as follows.

If α is a real root of $f(x) = 0$ of degree $n \geqq 2$, then the mean-value theorem shows that, for arbitrary $\frac{x}{y}$:

$$f\left(\frac{x}{y}\right) = f(\alpha) + \left(\frac{x}{y} - \alpha\right)f'(\xi), \qquad \xi = \alpha + \vartheta\left(\frac{x}{y} - \alpha\right), \qquad 0 < \vartheta < 1.$$

Thus, if the coefficients of f are integers and $f\left(\frac{x}{y}\right) \neq 0$, $y > 0$, it follows from $f(\alpha) = 0$, that

$$y^{-n} \leqq \left|f\left(\frac{x}{y}\right)\right| = \left|\frac{x}{y} - \alpha\right| \cdot |f'(\xi)|,$$

and therefore

$$\left|\frac{x}{y} - \alpha\right| \geqq C \cdot y^{-n} \quad \text{with } C > 0.$$

So we obtain the following theorem of Liouville (1844).

Theorem 23. For every algebraic number α of degree $n > 1$ there exists a number $C > 0$ such that for all rational $\frac{x}{y}$:

$$\left|\alpha - \frac{x}{y}\right| \geqq C \cdot y^{-n}.$$

The inequality $0 < \left|\alpha - \frac{x}{y}\right| < y^{-n-1}$ has finitely many solutions, or no solutions at all, in integers x, y (integer solutions). The same remarks even holds for $\left|\alpha - \frac{x}{y}\right| < y^{-n}$, $n > 2$ (Roth, 1955).

It follows that the algebraic numbers are "not very approximable" by rational numbers. Thus if a real number α is "very approximable," it is certainly transcendental. For example, the number $\alpha = \sum 10^{-n!}$ is obviously transcendental, since the partial sums are rational and converge very rapidly to α with increasing n. Thus we have the following theorem.

Theorem 24. *Every number* $\sum_n a_n \cdot 10^{-n!}$ *with* $a_n = 0$ *or* 1 *and with infinitely many* $a_n = 1$ *is transcendental.*

2.3. *Power of the Set of Transcendental Numbers*

It was shown by G. Cantor that the algebraic numbers are countable but the real numbers are not, from which he concluded that the transcendental numbers are uncountable and therefore have the power of the continuum (Cantor, 1874). Thus

Theorem 25. *The real numbers are predominantly transcendental and not algebraic, just as they are predominantly irrational and not rational.*

Let us remark that there exists a measure of transcendence according to which one number is more or less transcendental than another. In this connection we may speak of *hypertranscendental* numbers.

It is a different, and often very difficult, problem to establish the arithmetical character (rationality, transcendence, and the like) of a given number α, i.e., to decide whether a number is rational or irrational, algebraic or transcendental. Consider, for example, the following facts.

The problem of the transcendence of π remained unsolved for thousands of years. In 1958 Erdös showed that the numbers $\sum_n \frac{p_n^k}{n!}$ $(k = 0, 1, 2, \ldots)$ are irrational. The question whether the Euler constant

$$\gamma_0 = \lim_{n \to \infty} \left(-\log n + 1 + \frac{1}{2} + \frac{1}{3} + \cdots + \frac{1}{n} \right)$$

is rational or irrational has remained unanswered up to the present time.

The numbers $0.123456789101112\cdots$ and

$$\cfrac{1}{1 + \cfrac{1}{2 + \cfrac{1}{3 + \cdot_{\cdot_{\cdot}}}}}$$

are transcendental, and so is the number $\sum_{n=1}^{\infty} \frac{\alpha^n}{n!n}$ for every algebraic $\alpha \neq 0$ (Šidlovskiĭ, 1954).

It should be emphasized that the set A of algebraic numbers forms a field; in particular, it follows from $a, b \in A$ that also $a + b$, $a - b$, $a \cdot b \in A$.

If we expand the ζ-function about the point 1,

$$\zeta(s) = \frac{1}{s-1} + \sum_{n=0}^{\infty} \frac{(-1)^n \gamma_n}{n!} (s-1)^n,$$

then for every γ_n we have the estimate

$$\sum_{k=1}^{r} \frac{(\log k)^n}{n} = \frac{(\log r)^{n+1}}{n+1} + \gamma_n + o(1)$$

(conjectured by Hardy in 1912 and proved by Chowla and Briggs in 1955). The arithmetical nature of the numbers γ_n is unknown; the numbers γ_{2n} (or γ_{2n+1}) include infinitely many positive and infinitely many negative (D. Mitrović, 1957).

2.4. *Arithmetical Nature of a Number; Solution of Hilbert's Seventh Problem*

Theorem 26. *Hilbert's seventh problem* (1900), *namely, whether the numbers* α^β (α *and* β *algebraic*, $0 \neq \alpha \neq 1$, β *irrational*) *are transcendental, was answered in the affirmative* (Gelfond, 1929–34, Schneider, 1935).

The proof makes use of some of the results of the theory of analytic functions of a complex variable. For example, e^π $(= i^{-2i})$ is transcendental, but it remains unknown whether $e + \pi$, $e - \pi$, $e \cdot \pi$ are transcendental, although the transcendence of e and π has been established. But we do have the following theorem.

Theorem 27. *For every integer* $n \geqq 0$ *at least one of the numbers* e^{e^n}, $e^{e^{2n}}$, $e^{e^{3n}}$ *is transcendental* (Gelfand; the proof is very complicated).

2.5. *Transcendence of the Values of Certain Functions*

As early as 1748, Euler raised the question of the transcendence of the numbers $\log_b a$ for rational a, b; this problem has been completely solved (see §2.4, theorem 26).

At the present time much study is devoted to the transcendence of values of $f(x)$ for analytic functions f and algebraic values x, in view of the close connection between the growth of an analytic function and the arithmetical character of its values in an algebraic field. Hermite was the first (1873) to recognize this situation for a particular function. We are dealing here with theorem 26 in §2.4.

Theorem 28. *The functions* e^x, $\sin x$, $\cos x$, $\tan x$, $\sinh x$, $\cosh x$, $\tanh x$ *have transcendental values for every algebraic number* $x \neq 0$. *Their inverse functions also assume transcendental values for every algebraic number* $x \neq 0$.

Theorem 27 is now a simple consequence of the *Lindemann-Weierstrass* theorem:

Theorem 29. *For every natural number n and every sequence of distinct algebraic numbers* $\alpha_1, \alpha_2, \ldots, \alpha_n$, *the numbers* $e^{\alpha_1}, e^{\alpha_2}, \ldots, e^{\alpha_n}$ *are linearly independent over the field A of algebraic numbers, i.e., from* $\sum_{i=1}^{n} x_i e^{\alpha_i} = 0$ $(i = 1, 2, \ldots, n)$ *it follows that* $x_i = 0$ $(i = 1, 2, \ldots, n)$.

For example, if e were not transcendental, we would have $p(e) = 0$, where $p(x)$ is a polynomial. But in fact, by the theorem just quoted, $p(e)=0$ is not possible. The transcendence of the number e^{α} for every $\alpha \in A$, $\alpha \neq 0$ is proved in the same way. As a further example, let us consider the function sin: if $\sin \alpha$ $(\alpha \in A, \alpha \neq 0)$ were an algebraic number x, $x = \dfrac{e^{2i\alpha} - e^{-2i\alpha}}{2i}$, we would have $e^{2i\alpha} - e^{-2i\alpha} - 2xie^0 = 0$, in contradiction to the linear independence of the numbers $e^{2i\alpha}$ $e^{-2i\alpha}$ and e^0 over A.

In conclusion, we give two further results about the transcendence of certain numbers.

Theorem 30. *The Euler function* $B(p, q) = \int_0^1 x^{p-1}(1 - x)^{q-1}\,dx$ *has transcendental values for all rational fractions* p, q (Schneider).

Theorem 31. *Let* ω_1 and ω_2 *be two numbers with nonreal quotient, let* t_1 *and* t_2 *be two numbers not both zero, and finally let* z_1 *be such that* $z_1 \neq k_1\omega_1 + k_2\omega_2$ (k_1, k_2 *rational integers*). *If* $\wp(z)$ *is the Weierstrass elliptic function with periods* $2\omega_1$, $2\omega_2$ *and if* g_2, g_3 *are the corresponding invariants, then at least one of the numbers*

$$t_1, t_2, g_2, g_3, \wp(z_1), t_1 z_1 + t_2\zeta(z_1)$$

is transcendental, where

$$\zeta(z) = \tfrac{1}{2} - \int_0^z (\wp(z) - z^{-2})\,dz.$$

This Weierstrass elliptic ζ-function is not the same as the Euler-Riemann ζ-function in §1. The number-theoretic nature of the numbers $\zeta(2n + 1)$ is unknown. If the invariants g_2, g_3 are algebraic, the number $k_1\omega_1 + k_2\omega_2$ is transcendental.

2.6. *Diophantine Approximation* (see IB6, §7)

1. We say that an equation is diophantine if we are looking for integer solutions of it; and if we are looking for integer solutions of corresponding inequalities, we speak of diophantine inequalities or diophantine approximation.

For example, the inequality

$$(1) \qquad 0 < \left|\frac{a}{b}x - y\right| < \frac{1}{z},$$

where a, b are integers, $b > 0$, has integer solutions y and x for only finitely many integer z, or for no integer z.

In fact, the number $\left|\frac{a}{b}x - y\right|$ is rational and of the form $\frac{c}{b}$ so that

(2) $$\left|\frac{a}{b}x - y\right| > \frac{1}{2b}.$$

Comparing (1) and (2), we see that $\frac{1}{z} > \frac{1}{2b}$, so that $0 < z < 2b$.

On the other hand, we can prove that for every irrational number α the inequality

(3) $$0 < |\alpha x - y| < \epsilon$$

for every $\epsilon > 0$ has infinitely many integer solutions x, y.

Thus we have the following theorem, which shows how the arithmetical nature of a number is related to its approximability by rationals.

Theorem 32. *For a real number α to be irrational it is necessary and sufficient that for every $\epsilon > 0$ the diophantine inequality $0 < |\alpha x - y| < \epsilon$ has infinitely many solutions.*

2. In connection with the *Liouville property* of the algebraic numbers (see §2.2) we have already considered diophantine approximations of the form

$$\left|\alpha - \frac{y}{x}\right| < x^{-n-1}$$

or

(4) $$|\alpha x - y| < x^{-n}.$$

Comparison of (3) and (4) indicates the possibility of a connection between ϵ and x.

According to Liouville the inequality

(5) $$0 < |\alpha x - y| < cx^{-1}, \quad \alpha \text{ irrational}$$

for certain $c > 0$ has infinitely many integer solutions x, y. Now it is a remarkable fact that such values of c have an infimum:

Theorem 33. *For every irrational real number α the inequality*

$$0 < |\alpha x - y| < 5^{-1/2}x^{-1}$$

has infinitely many integer solutions. The number $5^{-1/2}$ cannot be replaced (as can be shown, in particular, for the case $\alpha = \frac{1}{2}(\sqrt{5} - 1)$) by any smaller number (Hurwitz, 1891).

Thus we may speak of the best approximation of an irrational α by means of rational numbers in the form (5).

As for the accuracy of the approximation to α by means of $\frac{y}{x}$ we may consider either $\left|\alpha - \frac{y}{x}\right|$ or $|\alpha x - y|$ as a measure for the approximation. The two points of view are not equivalent.

Theorem 34. *For every natural number z and every real number α, the inequality $|\alpha x - y| < z^{-1}$ has at least one integer solution x, y with* $0 < x \leqq z$ (Dirichlet).

Moreover, theorem 34 is only a special case of the following theorem:

Theorem 35. *If $\alpha_1, \alpha_2, \ldots, \alpha_n$ are real and z is a natural number, there exist integers $x_1, x_2, \ldots, x_n, y$ such that*

$$\left|\sum_{i=1}^{n} \alpha_i x_i - y\right| < z^{-n}, \quad |x_i| < z\ (1 \leqq i \leqq n), \quad \sum_{i=1}^{n} x_i = 0 \quad \text{(Dirichlet).}$$

3. As for the nonhomogeneous case, we can prove the following theorem:

Theorem 36. *If α is irrational and β real, the inequality $|\alpha x - y - \beta| < \frac{3}{|x|}$ has infinitely many integer solutions* (Čebyšev).

This theorem can be strengthened.

The study of diophantine approximations has been carried much further, but lack of space prevents our giving any more attention to them here. For results and for the literature, see J. K. Koksma.

2.7. *Uniform Distribution of Numbers and Functions*

Let $\alpha_1, \alpha_2, \ldots$ be a sequence of numbers in the interval [0, 1], let $[x, y]$ be a closed subinterval of [0, 1] and let n be a natural number. Then the number of terms of $\alpha_1, \alpha_2, \ldots, \alpha_n$ contained in $[x, y]$ is denoted by $a_n(x, y)$. The number

$$\frac{a_n(x, y)}{n} \tag{6}$$

is called the *mean-density* of the numbers $\alpha_1, \alpha_2, \ldots, \alpha_n$ with respect to $[x, y]$. We now consider

$$\lim_{n\to\infty} \frac{a_n(x, y)}{n}. \tag{7}$$

If for all x, y with $0 \leqq x \leqq y < 1$

$$\lim_{n\to\infty} \frac{a_n(x, y)}{n} = |x - y|,$$

the sequence $\alpha_1, \alpha_2, \ldots$ is said to be *uniformly distributed* in [0, 1].

Correspondingly we define:

A sequence of real numbers $\alpha_1, \alpha_2, \ldots$ *is said to be uniformly distributed* modulo 1 *if the sequence* $\{\alpha_n - [\alpha_n]\}$ *is uniformly distributed in the unit interval* $[0, 1]$. *A function* $f(x)$ *is said to be uniformly distributed* modulo 1 *if the sequence* $f(1), f(2), f(3), \ldots$ *is uniformly distributed* modulo 1.

If for every number α we consider the sequence $\alpha, 2\alpha, 3\alpha, \ldots$, we can express the most important results in the following set of theorems, stated here without proof.

Theorem 37. *For every irrational number* α *the sequence* $\{n\alpha\}$ *is uniformly distributed* modulo 1.

Theorem 38. *For a real number* α *to be irrational, it is necessary and sufficient that the sequence* $\{n\alpha - [n\alpha]\}$ *forms an everywhere dense set in the unit interval* $[0, 1]$ *or, in other words, that the sequence* $\{n\alpha\}$ *forms an everywhere dense set on the number circle.*

The number circle, in contrast to the number line, serves to interpret and represent the real numbers modulo 1.

Theorem 39. *If the function* $f(x)$ *is uniformly distributed* modulo 1 (see above), *then*

$$\int_0^1 g(x)dx = \lim_{n \to \infty} \frac{1}{n} \sum_{x=1}^{n} g(f(x))$$

for every function $g(x)$ *Riemann-integrable in* $[0, 1]$.

In particular, if $g(x) = e^{2\pi hix}$ (h an integer $\neq 0$), then

$$\lim_{n \to \infty} \frac{1}{n} \sum_{x=1}^{n} e^{2\pi ihf(x)} = 0. \tag{8}$$

The converse of this theorem was proved by H. Weyl in 1914:

Theorem 40. *The real function* $f(x)$ *is uniformly distributed* modulo 1 *if and only if* (8) *holds for every integer* $h \neq 0$.

This theorem and its generalizations have proved the starting point for applications of many different kinds (Hardy, Littlewood, Vinogradov, and others).

2.8. *Asymptotic Distribution of the Digits in Real Numbers*

If $x = 0, x_1x_2x_3, \ldots = \sum_{n=1}^{\infty} 10^{-n}x_n$, ($x_n \in \{0, 1, 2, \ldots, 9\}$) is an irrational number, we may ask how often a given digit, e.g., 6, occurs in its decimal expansion.

Are some of the digits predominant or are they all on an equal footing? This question was answered by Borel to the effect that for almost all real numbers all the digits behave alike; they all occur with asymptotically equal frequency.

The situation can be expressed more precisely as follows:

Theorem 41. *Let* z *be a digit in the decimal system* ($z \in \{0, 1, 2, \ldots, 10-1\}$), *let* $x = \sum_{n=1}^{\infty} 10^{-n}x_n$, *and let* z_n *denote the number of occurrences of* z *in the sequence* $x_1, x_2, x_3, \ldots, x_n$. *Then* $\lim_{n\to\infty} \frac{z_n}{n} = 10^{-1}$ *for every x apart from a set of measure* 0 (Borel, 1909).

Of course the base 10 could be replaced here by any other natural number $r > 1$.

2.9. *Representation of Real Numbers by Continued Fractions* (see IB6, §3)

For every irrational number x in [0, 1] the sequence of partial quotients $x_1, x_2, \ldots, x_n$ ($x_i \in N$) in the representation

$$(9) \qquad x = \cfrac{1}{x_1 + \cfrac{1}{x_2 + \cdots + \cfrac{1}{x_n} \cdots}} = \lim_{n\to\infty} [x_1, x_2, \ldots, x_n]$$

is uniquely determined (see IB6, §3).

In order to represent the real numbers by their continued fractions (9), the entire set of natural numbers is indispensable, in contrast to the g-adic representation ($g > 1; g \in N$), where finitely many numbers (i.e., "digits"; in the decimal system, ten digits) are sufficient. The following theorem gives further information:

Theorem 42. *The natural numbers are not uniformly distributed in the continued fractions. The larger a natural number is, the more rarely it appears as an* x_n *in* (9) (Hinčin).

Equally interesting is the following theorem (also due to Hinčin, 1935):

Theorem 43. *For almost all irrational numbers* (9)

$$(10) \qquad \lim_{n\to\infty} \sqrt[n]{x_1 x_2 \cdots x_n} = \prod_{k=1}^{\infty} \left(1 + \frac{1}{k(k+2)}\right)^{\log k/\log 2} = 2.6\cdots.$$

Thus the number $2.6\cdots$ in (10) plays the role of a universal constant, since (10) shows how a given number can be brought into relationship with the entire set of irrational numbers, a sign that we are on the track of some law.

2.10. *Invasion of Number Theory by Arguments from Probability*

The quotient (6) actually represents the probability that one of the numbers $\alpha_1, \alpha_2, \ldots, \alpha_n$ is contained in the interval $[x, y]$. Uniform distribution is thus equivalent to equality of probability of the appearance of the numbers $\alpha_1, \alpha_2, \ldots, \alpha_n$ everywhere on the number circle (with circumference 1). Similarly, theorem 41 can be interpreted in probability terms and in fact was first discovered in this way (Borel, 1909).

3. General Remarks on the Relation between Analysis and Number Theory

3.1. *Invasion of Number Theory by Analysis*

From the above remarks it is clear that analytic arguments play a great role in the theory of numbers.

3.2. *Invasion of Analysis by Number Theory*

However, the theory of numbers has not failed to repay the debt. Its role in modern analysis has been many-sided and fruitful, but here we can mention only the most important aspects.

1. The sequence of natural numbers 1, 2, 3, ..., the foundation of the whole of classical arithmetic, plays only a modest role in present-day analysis; nevertheless, it is the foundation on which the whole proud edifice has been constructed. The many generalizations of a sequence that have grown out of the original sequence of integers have gradually become fundamental concepts of analysis, and also of a great part of topology.

2. Principles that were first introduced in the theory of numbers (e.g., the principle of complete induction) have not only found application in analysis but have actually become basic methods there, and have spread out to other branches of mathematics.

3. Certain number-theoretic functions appear also in analysis, in generalized form. A typical example is the Γ-function (see III 10, §3). The number $P(n)$ of permutations of n distinct elements is easily stated:

$$P(n) = n! = 1 \cdot 2 \cdot 3 \cdots n.$$

Euler extrapolated the function $P(n)$ to the continuum and thus obtained his Γ-function

$$\Gamma(s) = \int_0^\infty e^{-t} t^{s-1} dt \quad \text{for } \operatorname{Re} s > 0.$$

The Γ-function can be continued analytically into the whole complex plane; at the points $s = 0, -1, -2, \ldots$, it has simple poles with the residues $\frac{(-1)^n}{n!}$. It satisfies the functional equation:

$$\Gamma(s + 1) = s\Gamma(s).$$

For every natural n, as is easily seen from the preceding remarks, we have

$$n! = \Gamma(n + 1).$$

Moreover, the Γ-function is closely related to many other functions of extremely varied kinds. For example

$$\Gamma(s)\Gamma(1-s) = \frac{\pi}{\sin \pi s},$$

$$e^{-z} = \frac{1}{2\pi i}\int_{c-i\infty}^{c+i\infty} \Gamma(s)z^{-s}ds \quad \{c > 0, \operatorname{Re}(z) > 0\}$$
(Cahen, Mellin),

$$\zeta(s) = \frac{1}{\Gamma(s)}\int_0^\infty \frac{x^{s-1}}{e^x - 1}\,dx,$$

$$\zeta(1-s) = \frac{2}{(2\pi)^s}\cos\frac{\pi s}{2}\,\Gamma(s)\zeta(s) \quad \text{(Riemann)},$$

$$\frac{\Gamma'(s)}{\Gamma(s)} = -\frac{1}{s} - \gamma_0 + \sum_{n=1}^{\infty}\left(\frac{1}{n} - \frac{1}{n+s}\right).$$

Of great importance in practice and also in theory are the Stirling approximation to the Γ-function

$$\Gamma(n+1) = n^n e^{-n}(2\pi n)^{1/2} \quad \text{(Stirling, 1730)}$$

and the estimate

$$\log \Gamma(s+b) = (s+b-\tfrac{1}{2})\log s - s + \tfrac{1}{2}\log 2\pi + O(|s^{-1}|)$$

for $s \to \infty$.

4. Arithmetico-algebraic structures and concepts (e.g., groups and fields) are becoming more and more important in almost every part of pure and applied mathematics. Let us mention here the modern arithmetic function-theory, particularly for algebraic functions (L. Hensel, Enzyklopädie der math. Wiss., Bd. II 3/1, Leipzig 1921, p. 533–650).

The name "transcendental number" is to be explained by the fact that such a number cannot be calculated by algebraic methods ($\cdots$*quod algebrae vires transcendit*). In modern times the zeros of algebraic polynomials $p(x)$ with integral coefficients have been called algebraic numbers, and the nonalgebraic numbers are transcendental (see the discussion above in §2). An algebraic number α is of degree n, or of order n, if α is a zero of a polynomial $p(x)$ of degree n but not of a polynomial of lower degree.

Exercises

1. For every integer $n > 0$ at least one of the numbers $n, n+1, n+2, n+3, n+4, n+5$ has at least two prime divisors.

2. Using the formula $\pi(x) \sim \frac{x}{\ln x}$, calculate the approximate value of $(\pi)x$ and the corresponding relative error w: (1) $\pi(50)$; (2) $\pi(100)$; (3) $\pi(500)$; (4) $\pi(1000)$.
3. (1) Check the following table:

n	10	50	10^2	200	500	10^3	10^4	10^8	10^9	10^{10}
$\pi(r)$	4	15	25	46	95	169	1229	5761455	50847534	455052512

(2) Prove that $\frac{\pi(n-1)}{n-1} < \frac{\pi(n)}{n}$ or $\frac{\pi(n-1)}{n-1} > \frac{\pi(n)}{n}$ according as n is prime or composite.
4. (Unsolved problem). It is not known whether there exist infinitely many pairs of consecutive integers n, $n + 1$ each of which is a prime power. Examples are $\{2, 3\}$, $\{3, 4\}$, $\{4, 5\}$, $\{7, 8\}$, $\{8, 9\}$, $\{16, 17\}$, $\{31, 32\}$. Twenty-six such pairs are known, the greatest being $\{2^{4423} - 1, 2^{4423}\}$.
5. (Unsolved problem). It is not known whether there exist infinitely many triples of consecutive integers n, $n + 1$, $n + 2$ each of which is the product of exactly two distinct primes. Two examples are $33 = 3\cdot 11$, $34 = 2\cdot 17$, $35 = 5\cdot 7$; $93 = 3\cdot 31$, $94 = 2\cdot 47$, $95 = 5\cdot 19$.
6. Given that $\sum n^{-2} = \frac{\pi^2}{6}$ is irrational, prove that there exist infinitely many primes.
Hint: Apply the Euler's formula

$$\zeta(s) \equiv \sum n^{-s} = \prod_p (1 - p^{-s})^{-1} \quad \text{for } s = 2.$$

7. Prove that every initial segment $\sum_{k=1}^{n} k^{-1}$ of the harmonic series $1 + \frac{1}{2} + \frac{1}{3} + \cdots$ is nonintegral.
8. Prove that every initial segment of the series $1 + \frac{1}{3} + \frac{1}{5} + \frac{1}{7} + \cdots$ is nonintegral.
9. For natural numbers b, n let $b(n)$ denote the number of digits $(0, 1, \ldots, b - 1)$ in the representation of n with base b; e.g. $10(25) = 2$.

Determine (1) $\lim_{n\to\infty} \frac{2(n)}{10(n)}$; (2) $\lim_{n\to\infty} \frac{8(n)}{10(n)}$; (3) $\lim_{n\to\infty} \frac{b(n)}{c(n)}$ for $b < c$.

10–22. Various problems, solved and unsolved.
10. It is not known whether there are infinitely many primes of the form $a_n = 9^{-1}(10^n - 1)$; e.g., a_n is prime for $n = 2, 3$.
11. Does the Fibonacci series

$$1, 1, 2, 3, 5, 8, 13, \ldots, f_{n+2} = f_n + f_{n+1}$$

contain an infinite number of primes?

12. Is there a trinomial $f(x) = ax^2 + bx + c$ such that $f(x) = p \in P$ implies $p + 2 \in P$ (P is the set of prime numbers)?
13. What is the number of solutions of
$$x^3 + y^3 + z^3 \in P$$
in positive integers x, y, z?
14. Is there a $p \in P$ between n and n^2?
15. Is there a constant $n \in N$ such that p and $p + n$ are both prime for infinitely many values of p? Is $n = 2$ such a number?
16. Let $\pi_1(x)$ denote the number of primes $p \leq x$ of the form $4h + 1$, and let $\pi_3(x)$ denote the number of the form $4h + 3$. Check that $\pi_1(10) = 1 < \pi_3(10)2$, $\pi_4 10^2 < \pi_3(10^2)$. It has been proved that $\pi_1 x \leq \pi x$ for $x < 26861 = a$, but that $\pi_1 a = 1473 > \pi_3 a = 1472$. There are infinitely many x with $\pi_1 > \pi_3 x$ and also infinitely many with $\pi_1 x < \pi_3 x$ (Littlewood, 1914).
17. For each of the numbers
$$(1)\ \sqrt{3} + \sqrt{5};\ (2)\ 2^{\sqrt{3}};\ (3)\ 3^{\sqrt{5}};\ (4)\ 5^{\sqrt{3}+\sqrt{5}};\ (5)\ 2^{i\sqrt{2}}$$
determine whether it is algebraic or transcendental.
18. Let $\mathfrak{T}(n)$ be the number of representations of the old integer n as the sum of three primes (cf. theorem 18); then
$$\mathfrak{T}(n) \sim \frac{n}{2 \ln n} \prod_{p} \left(1 + \frac{1}{(p-1)^3}\right) \prod_{p/n} \left(1 - \frac{1}{p^2 - 3p + 3}\right).$$
(Vinogradov).
19. An an upper bound for the number x_0 in theorem 18 note the result of Borozdin (1956): $\ln \ln x_0 \leq 16.038$.
20. Let $d_0 n$ denote the number of positive integers dividing n, and let $d_r n$ denote the number of sequences $m_1, m_2, \ldots, m_r$ of positive integers such that $m_1 m_2 \cdots m_r = n$. Then, (1) $d_0 n = d_2 n$; (2) If $q_1, q_2, \ldots, q_k$ are pairwise distinct prime numbers, then $d_r(q_1, q_2, \ldots, q_k) = r^k$. (3) The number of solutions in natural numbers of the inequality $x_1 x_2 \cdots x_r \leq n$ is given by $\sum_{0<a\leq n} d_r(a)$.
(4) Prove by induction on r that
$$[\zeta(s)]^r = \sum_{n=1}^{\infty} \frac{d_r(n)}{n^s}.$$
21. Find some numbers satisfying $\psi n = \psi(n + 1)$. It is not known whether there are infinitely many such numbers. Examples are:
$$1, 3, 15, 104, 164, 194, 255, 495, 584, 975.$$
22. Conjecture. (R. D. Carmichael). There is no positive integer m such that $\psi(n) = m$ has a unique solution $n \in N$ (the conjecture is known to hold for $m \leq 10^{400}$).

Bibliography

[1] GEL'FOND, A. O.: Transcendental and algebraic numbers. (Translated from the Russian,) Dover, New York 1960.

[2] KOKSMA, I. F.: Diophantische Approximationen. (Ergebnisse der Mathematik, Vol. 4, Berlin 1936.)

[3] LANDAU, E.: Vorlesungen über Zahlentheorie, I–III, S. Hirzel, Leipzig 1927. (Vol. I, part 1 was translated into English by J. E. Goodman as Elementary Number Theory, Chelsea, New York 1958.)

[4] NAGELL, T.: Introduction to number theory. Second edition, Chelsea, New York 1964.

[5] NIVEN, I.: Irrational numbers. (The Carus Math. Monographs 11, 1956.)

[6] OSTMANN, H.-H.: Additive Zahlentheorie (Ergebnisse der Mathematik, Heft 7 u. 11). Springer, Berlin-Göttingen-Heidelberg 1956.

[7] PRACHAR, K.: Primzahlverteilung. Springer, Berlin-Göttingen- Heidelberg 1957.

[8] SCHNEIDER, TH.: Einführung in die transzendenten Zahlen. Springer, Berlin-Göttingen-Heidelberg 1957. Also available in French translation: Introduction aux nombres transcendants (Tr. by P. Eymard.) Gauthier-Villars, Paris 1959.

Bibliography added in Translation

AYOUB, R. G.: An introduction to the analytic theory of numbers. American Mathematical Society, Providence 1963.

GEL'FOND, A. O.: Elementary methods in analytic number theory. (Translated from the Russian by A. Feinstein.) Rand McNally, Chicago 1965.

HARDY, G. H. and WRIGHT, E. M.: An introduction to the theory of numbers. Clarendon Press, Oxford 1960.

LE VEQUE, W. J.: Topics in number theory. Addison-Wesley, Reading, Mass. 1956.

NIVEN, I. M.: Numbers: rational and irrational. Random House, New York 1965.

NIVEN, I. M. and ZUCKERMAN, H. S.: An introduction to the theory of numbers. Wiley, New York 1960.

ROBINSON, A.: Numbers and ideals. Holden-Day, San Francisco 1965.

CHAPTER 14

The Changing Structure of Modern Mathematics

1. Introduction

A glance at the mathematical journals, which every year are increasing in number and size, or at the immense outpouring of mathematical books, dealing every year with numerous new subjects, shows at once that mathematics is a flourishing science, in a state of rapid development and ramification. In certain branches of mathematics, for example, topology and algebraic geometry, some of the topics are expanding so rapidly that only by constant personal contact with the leaders in the field can a mathematician take an active share in the work. A printed article, published a year after it is written, no longer represents the latest research in the field.

Of course, this phenomenon is by no means confined to mathematics. The acceleration of progress in the various branches of natural science has been at least as great. However, the reasons can hardly be the same in mathematics, which is a purely intellectual science. The immense resources of modern technology, its refined methods of experimentation and the consequent flood of new empirical data, are not at the disposal of mathematics. However, they have a kind of analogue. If we compare the methods of modern mathematical research with those of the preceding generation, we are struck by the greater ease and boldness with which nowadays a whole apparatus of concepts is developed for the attack on a given problem. The methods have become much more abstract, the same concepts are applied to problems of widely varying kinds, and unexpected connections are continually found among mathematical disciplines that appear to be widely separated from one another. The purely computational arguments of former days, involving masses of formulae, have to a great extent disappeared.

Of course, these remarks are true in such an emphatic form only for parts of present-day mathematics, and our task in the following section will be to describe these parts more precisely. We shall be interested in those branches of mathematics where the changes have been most far-reaching.

The change in the structure of mathematics can be traced to a very simple cause, namely, to the fact that the axiomatic way of thinking has penetrated more and more deeply into each of the individual subjects of mathematics and has brought with it a greater freedom and flexibility in dealing with them from an abstract point of view. We shall trace this process by various examples and shall attempt to understand it to its full extent.

A group of French mathematicians, who are publishing their huge work "Eléments de Mathématique"[1] under the collective pseudonym of Bourbaki, deserve the credit for systematically thinking through the consequences of the axiomatic method for the whole of mathematics and for undertaking, in their theory of the structures of mathematics, to reconstruct the entire subject.

The results of this attempt are surprising; they imply a complete rearrangement of mathematics. The hitherto unquestioned division of mathematics into the four great fields of algebra, theory of numbers, analysis and geometry appears suddenly to be out of date. It is now realized that these fields, whose framework has been strongly influenced by the accidents of history and by practical needs, are by no means uniform in content. The modern reconstruction of mathematics, based on the axiomatic method, often leads to a complete shift of emphasis regarding the importance of classical theorems and to the total destruction of many traditional contexts, an activity which at first sight may make a very unpleasant impression. On the other hand, the subsequent result is an unexpected unification of the entire structure of mathematics. Suddenly we seem to be able again to look at mathematics as a whole and to see that its various parts are much more closely united with one another than could have been suspected before the change. Of course, this new insight into mathematics is in many respects one-sided, with varying influence from one branch of mathematics to another; but it shows very clearly the direction in which modern mathematics as a whole is developing. The ideas of Bourbaki have found a place in many passages of the present work, though naturally it has not been possible to construct the entire work on the *Bourbaki system*. In particular, in chapter 10 of the first volume we gave an introduction to the concept of a structure, which forms the theoretical basis of the "Eléments."

[1] Paris, Hermann et Cie.

2. The Development of the Axiomatic Method in Geometry

The mathematics of the nineteenth century, as described in such lively fashion in the famous lectures of Felix Klein,[2] completed and firmly established the system of mathematics begun by Descartes, Leibniz and Newton, where the central role is played by the infinitesimal calculus and the other parts of the great discipline of analysis that spring from it. Geometry underwent a vigorous development in the newly discovered projective and differential geometries. In another direction, the work of many outstanding mathematicians contributed to the theory of functions of a complex variable, also a creation of the nineteenth century. The flood of ideas developed in these fields continues almost undiminished; the ideas are refined and extended and even today such questions make up an essential part of the whole of mathematical research. The mathematics of the nineteenth century was strongly influenced by the contemporary world-picture of the physicists. Not only did the requirements of physical application, always a stimulus to mathematical research, determine the character of the mathematics of the time, but the view of the physicists that the actual space of our experience is adequately described by the three-dimensional space of Euclid led the majority of mathematicians to regard non-Euclidean geometry as mere abstract play. A pronounced aversion to excessively abstract ideas was everywhere to be seen. The complex numbers, and with them the whole theory of functions of a complex variable, were accepted only after these numbers were represented as points in the complex plane. A hundred years ago no one was as well satisfied as we are today with the representation, now quite commonplace, of complex numbers as pairs of real numbers.

The situation is exactly the same for the concept of n-dimensional space. Even Riemann avoided the word "space" and preferred to speak of a "manifold," a term which today is used in a different sense. The realization that *more than one* geometry is conceivable, and indeed the whole axiomatic conception of geometry, was the result of long and strenuous efforts. It was the study of this question that gave rise to the modern concept of a system of axioms. In his "Foundations of Geometry," Hilbert expressed the idea in unmistakable terms, although he did not make use of the concepts of mathematical logic current today. The points, lines, and planes of Hilbert are precisely the variables of modern logic, for which certain relations are defined, such as "lies on," "between," and so forth, and the axioms determine the properties of these relations. The system of axioms for Euclidean geometry obtained in this way defines a structure (cf. IB10), namely, that of Euclidean geometry.

[2] Vorlesungen über die Entwicklung der Mathematik im 19. Jahrhundert. Springer-Verlag, Berlin 1926/27.

For such a complicated axiomatic system it is natural to ask whether we cannot deduce a contradiction from the given axioms. Hilbert answered this question by constructing a model (or a configuration, in the terminology of IB10) which satisfies all the axioms, namely, the model consisting of analytic geometry. Thus he reduced the question of the consistency of Euclidean geometry to that of the consistency of the real number system.

The structure consisting of Euclidean geometry is monomorphic, i.e., any two models can be mapped onto each other in such a way that the points, lines and planes of one model correspond in one-to-one fashion with the points, lines and planes of the other and the relations "lies on," "between," and so forth, hold for the elements of the one model if and only if they also hold for the corresponding elements of the other model. In this sense there is only one Euclidean geometry. But the situation changes if we discard the parallel axiom. The resulting system of axioms, called absolute geometry, is satisfied not only by Euclidean geometry but also by non-Euclidean (hyperbolic) geometry, and these two geometries cannot be mapped onto each other in the above sense. Thus the structure of absolute geometry is no longer monomorphic.

Other changes in the system of axioms for geometry were soon undertaken, and in this way whole theories of new geometric structures were obtained. At first these structures were regarded as "pathological" and they were thought to be interesting only because they made it possible to analyze more precisely the various simple concepts at the basis of geometry; but this opinion changed as soon as it was realized that these geometric structures also give rise to interesting problems in algebra, in view of the fact that the models for them could only be constructed by means of algebraic theories. This aspect of the axiomatic theory of geometry is already to be found in Hilbert.

Not quite so important for further development, but still an immense achievement of the nineteenth century, was the axiomatic foundation of the natural numbers by Dedekind and Peano. Here too we are dealing with a monomorphic structure; there is only one system of natural numbers.

3. The Theory of Sets and Mathematical Logic

The discovery of the nineteenth century most important for our modern attitude toward mathematics was the theory of sets by Georg Cantor, in view of the fact that the paradoxes in it led to the well-known crisis in the foundations of mathematics. Neither the theory of sets itself nor the logical difficulties arising from it were properly recognized by the mathematicians of the nineteenth century. On the one hand, it was the Lebesgue theory of measure and integration, in the early twentieth, that first made clear the importance of the theory for the whole of analysis, where it is now too firmly rooted ever to be dislodged; and on the other hand, the efforts to overcome the paradoxes gave rise both to mathematical logic and to study of the foundations of mathematics; and the mutual influence of these two subjects, now largely clarified, continued to claim the absorbed interest of mathematicians throughout the first half of the present century. The results were: first, the venerable discipline of logic was mathematicized, and has now become a science surpassing mathematics in precision of thought and, second, mathematics itself has become more rigorous and better understood. Of particular interest to us here are the repercussions of this development on the methods of mathematics. The precise, logical formulation and analysis of the axiomatic method, as provided by mathematical logic, and the recognition that, in the framework of mathematical logic, a theory can be investigated only after it has been axiomatized, combined to persuade mathematicians of the importance of the axiomatic procedure even in the so-called "inner-mathematical" study of a given branch of mathematics itself.

Yet they hesitated to draw the logical consequences. For a long time they regarded the axiomatization of a theorem as merely the last step in its development; only when the theory lay before them in complete form were they willing to give it the final polish by stating it in axiomatic terms. They did not expect that early axiomatization of a theory would give them essential guidance for further research, although certainly the example of geometry spoke against this attitude. It required further successes of the axiomatic method in another central field of mathematics to complete the revolution in this respect.

4. Change in the Attitude Toward Algebra

Up to the middle of the nineteenth century, algebra consisted of the solution of equations or systems of equations. The concept of a group, whose decisive importance for equations of higher degree was discovered by Galois, was first defined in abstract form by Cayley. However, it was not until the final twenty years of the century that research was undertaken on abstract groups, and at first only on finite groups. But by this time the concept of a group had proved extremely fruitful in widely differing fields of mathematics; one has only to think, for example, of Jordan's already well-developed theory of substitution groups, or of Felix Klein's Erlanger Programm. Perhaps it was precisely because of this phenomenon, namely, that the structure of a group turned up in so many widely separated disciplines, that the creation of an autonomous theory of groups was so long delayed, since such a theory fitted so badly into the traditional division of mathematics into various subjects.

The concept of a field (or, in the earlier terminology, of a domain of rationality) underwent the same slow development. It was only in the second decade of the present century that Steinitz created an axiomatic theory of fields and began to look for all possible nonisomorphic fields. The breakthrough of axiomatic algebra, also called "abstract" algebra, came finally in 1920, particularly through the efforts of E. Artin and Emmy Noether, and brought with it a complete reorganization of algebra, with consequent rapid advances. These developments in algebra show most clearly the change in the structure of mathematics as a whole, and it is here that the reorganization may be said to be more or less complete. In classical algebra the problem of the solution of an equation of the nth degree $p(x) = x^n + a_{n-1}x^{n-1} + \cdots + a_0 = 0$ was in the foreground. Here the coefficients a_i are rational numbers, and the question of greatest interest was whether by repeated extraction of roots of the coefficients a formula for the n roots of the equation could be found. Galois discovered that this question can be answered by investigating the group of permutations of the roots of the polynomial (cf. IB7, §7). The situation can also be interpreted as follows: to the field $\mathbf{P}$ of rational numbers we adjoin the roots of the polynomial $p(x)$ and thus obtain a larger field $\mathbf{\Sigma}$, the so-called root field of the polynomial. Then $\mathbf{\Sigma}$ is of finite rank over $\mathbf{P}$, i.e., there exist finitely many elements α_i, $i = 1, \ldots, m$ in $\mathbf{\Sigma}$ such that every element β of $\mathbf{\Sigma}$ can be written in the form $\beta = \sum_{i=1}^{m} b_i\alpha_i$ with rational b_i. Certain permutations of the roots produce in $\mathbf{\Sigma}$ an automorphism over $\mathbf{P}$, i.e., a one-to-one mapping of $\mathbf{\Sigma}$ onto itself in which the sum and product of two elements in $\mathbf{\Sigma}$ are mapped onto the sum and product of the corresponding elements and each element of $\mathbf{P}$ corresponds to itself. Conversely, every automorphism of $\mathbf{\Sigma}$ over $\mathbf{P}$ generates a permutation of the roots of $p(x)$, which uniquely determines the given automorphism. Under the operation of successive application these automorphisms form a group called the Galois group of $\mathbf{\Sigma}$ over $\mathbf{P}$. Not only does the Galois group determine the nature of the roots of $p(x)$, but from it we can also read off the number of fields existing between $\mathbf{P}$ and $\mathbf{\Sigma}$ and their relations to one another.

These results were at first regarded only as subsidiary details in the theory of equations, but gradually the point of view was completely changed. For it is clear that by means of a substitution we can change the polynomial $p(x)$ into another polynomial $q(x)$, and thereby also change the roots, without changing the field $\mathbf{\Sigma}$. Thus the decisive feature in the situation is the root field itself and its various automorphisms, quite apart from the way in which these automorphisms are represented by permutations of the roots of $p(x)$ or $q(x)$. If we follow this line of thought to its logical conclusion, the problem of solving $p(x) = 0$ is changed into another problem, at first sight quite different: namely, to investigate the group of automorphisms of $\mathbf{\Sigma}$ over P for two given fields P and $\mathbf{\Sigma}$, where $\mathbf{\Sigma}$ is of finite rank over P. Since this problem deals only with fields and certain groups of automorphisms of these fields, it is natural to ask whether we can replace the field P by an arbitrary field $\mathbf{K}_1$ and $\mathbf{\Sigma}$ by an arbitrary $\mathbf{K}_2$ of finite rank over $\mathbf{K}_1$. Moreover, since the polynomial $p(x)$ now plays no role in the formulation of the problem, we are tempted to discard the hypothesis that $\mathbf{K}_2$ is of finite rank over $\mathbf{K}_1$, although we will find that the problem then becomes more difficult.

At any rate we can see that this reformulation of the original problem in terms of the general theory of fields at once gives rise to a large number of new questions and that in its new form the problem is simpler and more clear-cut.

As soon as we have given an axiomatic formulation for the structure of a field and have recognized its importance for algebra, it is natural, as in the case of geometry, to define new structures by making changes in the axioms and to study them in the same way as for fields. Thus if we discard the commutative law of multiplication, we obtain a skew field. For a long time the quaternions were the only known example of a skew field, but a systematic study of the structure of skew fields brought others to light and thereby gave rise to a new and interesting theory. For example, the abstract form of the Galois theory has been taken over for skew fields. If we also abandon division, we arrive at the concept of a ring, which may be commutative or not. Many examples of this structure were already known; the integers form a commutative ring; the n-rowed matrices, under the well-known operations for matrices, form a noncommutative ring, and so forth.

If we discard the associative law for multiplication in rings, we obtain the nonassociative rings, whose theory is quite different, but whose importance for other parts of mathematics is constantly growing. For example, a particular class of such rings, the *Lie rings*, has long been in the foreground of the picture because of its well-known importance in the theory of *Lie groups*.

These structures, each of which requires two operations, namely, addition and multiplication, are related in a natural way to the structure of a group, in which there is only one operation. An important extension of the concept of a group was given by Krull, who introduced groups with operators.

These examples will serve to indicate that the new view of algebra as a theory of structures with one or two operations gave rise to many new questions. But we must now turn our attention to the closely related change in methods, indispensable for such a rapid development.

The new methods developed for the study of structures were suggested by experience with the oldest theory of structures, namely, the theory of groups. The most important concepts there had proved to be those of a subgroup, a normal divisor, a factor group, and a homomorphism, with its special cases of isomorphism and automorphism.

The analogues to these concepts become the central notions for the other structures, namely, the subrings and subfields of a ring or a field, and the homomorphisms, isomorphisms and automorphisms of a ring, where we now require that the mapping be consistent with the two operations of the ring. To the concept of a normal subgroup in a group there corresponds the concept of an ideal[3] in a ring. Ideals were already recognized in the nineteenth century by Dedekind as a central concept in the theory of algebraic numbers, but their fundamental role in the structure of a general ring was first recognized by Krull and Emmy Noether. Corresponding to a factor group is the residue class ring with respect to a two-sided ideal.

The systematic use of these new concepts and of the methods depending on them led to the almost total disappearance of the kind of calculation that had been customary in algebra, namely, calculation with certain chosen elements; because of random properties connected with the choice of such elements, calculation of this kind had often concealed the simple abstract relations involved in the problem.

Of course, every new algebraic structure produced new difficulties; however, the methods developed for well-known structures usually required only rather obvious changes in order to be successful in the new setting. For example, the theory of lattices, described in IB9, was developed with surprising rapidity in the thirties. In IB9 we have already discussed the peculiar double nature of a lattice, which as a dual group is an algebraic structure and as a lattice is an ordered structure.

The first comprehensive treatment of algebra in its new form was given in the well-known book by van der Waerden, Modern Algebra.[4]

5. The Axiomatic Method in Analysis

In algebra it may be said that the changes are relatively complete, at least as far as methods are concerned, but the situation in analysis is quite different. Here we are still in the midst of the change and cannot yet foresee whether it will ever be as complete as in algebra.

The first great achievements of the axiomatic method in analysis are the exact definition of the concept of a limit and the consequent establish-

[3] An ideal in a commutative ring is a subring whose elements are transformed by multiplication with an arbitrary element of the ring into elements of the same subring. In a noncommutative ring there exist right ideals, left ideals and two-sided ideals, according to whether we multiply only on the right, only on the left, or on both sides with an arbitrary element of the ring (cf. IB5, §3).

[4] Two volumes. Springer, Berlin-Göttingen-Heidelberg, fourth edition 1955/1959.

ment by Cantor and Dedekind of the system of real numbers. The field of real numbers cannot be regarded as a purely algebraic structure, since its definition involves properties of order and continuity.

Part of the Cantor theory of sets consisted of the theory of sets of points, with concepts such as "closed," "open," "connected," "multiply connected," and basic theorems such as the covering theorem of Heine-Borel, all of which contributed greatly to the increasing precision of analysis.

But we were already far into the present century before the penetrating analysis of many mathematicians, of whom we may mention Fréchet and Hausdorff, led to the realization that the concept of a topological space is the simplest and most general structure embracing the properties of continuity.

This concept was developed from that of a metric space, whose importance had been realized much earlier; in its definition of distance a metric space depends on the real numbers, whereas a topological space is defined in purely set-theoretic terms. The natural attempt to transfer at least parts of analysis from n-dimensional Euclidean space to general topological spaces met with difficulties of various kinds. The first question, namely, whether a sufficient number of continuous functions can be defined in such spaces, was answered by the lemma of Uryson, which gives a precise statement of the properties of a space that are necessary from this point of view. The absence in a general topological space of the concept of a distance, defined by means of real numbers, gave rise to difficulties in defining uniform convergence for a sequence of functions. These difficulties were solved by means of the uniform structures introduced on topological spaces by A. Weil (see III 1, §6). The Weil theory and the *Cartan concept of a filter* (see III 1, §5), which is the generalization of the limit concept made necessary by the absence of axioms of countability, transformed general topology (theory of topological structures) into the effective tool of analysis which it now represents.

The necessity of developing more general notions of space first became noticeable in the theory of functions of a complex variable. It was this theory that created, in the form of *Riemann surfaces*, the general notion of a two-dimensional manifold, which is the suitable framework for the theory of analytic functions of one variable. At the present time we are witnessing an analogous development of the theory of analytic functions of several variables, which is attempting to find, in the complex manifolds of n dimensions of various types, the concept of a space in which the theory can be axiomatized in the simplest way. (For a detailed discussion of these constructions see III 6, §4 and III 7, §6). Another early development was the differentiable manifold, which is the natural concept of space for the arguments of differential geometry. And as a last example let us mention the theory of integration. In the search for the most general assumptions under which the Lebesgue theory could be carried through, it was discovered that this theory is possible in all locally compact topological spaces.

What conclusion do we draw from this brief list? Obviously that a different concept of space is necessary for each of the different branches of analysis. As soon as this conclusion was formulated in detail, classical analysis, carried on solely in Euclidean space, branched out into many individual theories, each with its own point of departure.

Thus the picture of analysis given in the ordinary introductory lectures on differential and integral calculus by no means corresponds to the present-day state of the subject. Of course, from a practical point of view it is obviously impossible to begin such a course with an introduction to general topology; yet in a systematic presentation of the calculus from a modern point of view this is exactly what should be done.

In fact, the situation is still more complicated. It is not only the topological structures that must be introduced if we are to have deeper understanding of analysis. We must also examine the structures that have been developed in functional analysis, since they are now taking a more and more central place in large parts of analysis. Let us look more closely at this latest development, where a particularly radical change is taking place at the present time.

Purely algebraic methods have for a considerable time taken a quite modest place in analysis. For example, in the theory of linear differential equations of higher degree an important role is played by the linearly independent fundamental solutions and the Wronskian determinant, in other words, by an application of linear algebra, and the theory of groups is indispensable in the study of automorphic functions. But of much greater importance is the development in analysis arising from the study of integral equations and linear equations with infinitely many unknowns (cf. III 11, §4). Poincaré and Fredholm discovered that an integral equation

$$\varphi(x) + \int_a^b K(x, y)\varphi(y)dy = f(x)$$

with known kernel $K(x, y)$, known right-hand side $f(x)$ and unknown function $\varphi(x)$ behaves in very much the same way as a system of linear equations with n equations and n unknowns.

If we expand $\varphi(x)$, $f(x)$, and $K(x, y)$ each in a series of functions $h_n(x)$, $n = 1, 2, \ldots$, of a complete orthogonal system and substitute these series into the integral equation, we obtain a system of infinitely many equations in infinitely many unknowns, where the unknowns are the coefficients of the expansion of the desired function $\varphi(x)$ in the orthogonal system $h_n(x)$.

Thus it is natural to attempt to solve such an integral equation by developing a theory of equations with infinitely many unknowns after the pattern of the theory for finitely many unknowns. Since the only method in common use at the time for solving a finite system of equations was the method of determinants, it is not surprising that the first attempt consisted in setting up a theory of infinite determinants (von Koch, Fredholm). But it was soon realized that only a narrow range of problems could be solved in this way.

It was Hilbert who created the new methods for solving these problems, but E. Schmidt was the first to introduce in a thoroughgoing way the geometric terminology which throws into clear relief the analogy with questions of analytic geometry.

The fundamental concept proved to be that of a Hilbert space H, namely, the set of all vectors $\mathfrak{x} = (\xi_1, \xi_2, \ldots)$ of real numbers ξ_n with convergent sum of squares $\sum_{n=1}^{\infty} \xi_n^2$, for which the usual vector operations $\mathfrak{x} + \mathfrak{y}$ and $\alpha\mathfrak{x}$ are defined, and also a scalar product $(\mathfrak{x}, \mathfrak{y}) = \sum_{i=1}^{\infty} \xi_i\eta_i$. This concept is obviously an infinite-dimensional generalization of n-dimensional Euclidean space. After choice of an orthogonal system $h_n(x)$, we see that the function $\varphi(x) = \sum_{n=1}^{\infty} \xi_n h_n(x)$ corresponds to the vector $\mathfrak{x} = (\xi_1, \xi_2, \ldots)$ of the Hilbert space, and the operation $\int_b^a K(x, y)\varphi(y)dy$ corresponds to multiplication of the vector $\mathfrak{x}$ by a matrix $\mathfrak{A}$ with infinitely many rows and columns. In this way we obtain a geometric interpretation of $\mathfrak{A}$ as a linear mapping of the space H into itself, and then the generalization of such geometric concepts as orthogonality, projection, orthogonal transformation, and so forth, provides a conceptual framework in which problems of the theory of integral equations can be solved in close analogy with the methods of Euclidean geometry.

The axiomatization of the structure called Hilbert space was first given by J. von Neumann (see III 11, §2), who characterized it as a linear space with a scalar product, and it is this axiomatic description that has proved most convenient as a starting point for modern investigations. The space H introduced above, whose elements are vectors with infinitely many coordinates, is a model for this axiomatic structure.

The structure of a Hilbert space is recognized today as basic in the theory of linear partial differential equations and their corresponding eigenvalue problems, and large parts of this theory can now be built up in a uniform way.

Of some interest too are the reverse effects of this development on the elementary fields of linear algebra and analytic geometry. The inadequacy of the classical method of determinants in the infinite-dimensional case led to the recognition that the most important facts concerning the solution of systems of equations can be expressed without determinants in the finite case also. Then the relations between rank and number of solutions and the connection with the transposed system of equations were extended by Toeplitz from the finite case to a general type of infinite system of equations in a very simple determinant-free manner. His arguments could also be extended at once to skew fields, where the very concept of determinants is not available. But even this thoroughgoing use of coordinates and matrices proved inadequate in the case of a Hilbert space; only the abstract definition of a linear mapping as a mapping Ax of a Hilbert space into itself under the linearity condition $A(\alpha_1 x_1 + \alpha_2 x_2) = \alpha_1(Ax_1) + \alpha_2(Ax_2)$ was suitable for dealing with all the problems. As a result the abstract formulation is now common in analytical geometry as well, and the notation in coordinates and matrices is used only where it serves the purposes of explicit calculation.

Hilbert space was the first of a class of structures that became important for analysis. The significance of the structure of a normed space was recognized by S. Banach. A normed space (cf. III 11, §2) is a linear space, or a vector space, for whose elements x there is defined an absolute value or a norm $\|x\|$ with the three properties:

1. $\|x\| \geqq 0$ and $\|x\| = 0$ only for $x = 0$;
2. $\|\alpha x\| = |\alpha| \cdot \|x\|$ for every real α;
3. The triangle inequality holds: $\|x + y\| \leqq \|x\| + \|y\|$.

The Hilbert space H is an example of a normed space; we have only to set $\|\mathfrak{x}\| = \sqrt{(\mathfrak{x}, \mathfrak{x})}$.

The structure of a normed space is no longer a purely algebraic structure, but a mixed algebraic-topological structure. For on the one hand a normed space R is certainly a metric space, since $\|x - y\|$ defines a distance between x and y, and on the other hand R is a linear space and therefore an algebraic structure. The two structures are connected by the fact that the algebraic operations $x + y$ and αx are continuous in the two variables with respect to the given metric.

Just as metric spaces were a preliminary step toward a general topological space, so normed spaces are the first step toward a topological vector space, which is defined as a topological space that is at the same time a vector space with operations continuous in the topology. The theory of these topological vector spaces, made possible only by the development of general topology, is of very recent date; its most important result up to now is the *Schwartz theory of distributions* (see III 3, §3), which is proving to be more and more fruitful in the unification of analysis, especially for differential equations. It appears that large parts of analysis can be included and systematically developed in the theory of these mixed algebraic-topological structures; let us give two more, particularly important examples.

If for the elements of a normed space there is also defined a multiplication xy related to the norm by the inequality $\|xy\| \leqq \|x\| \cdot \|y\|$, we have a normed algebra. For example, the continuous functions on the interval

[0, 1] with the norm $\|f(t)\| = \max_{t\in[0,1]} |f(t)|$ form a normed algebra with respect to the operations $f(t) + g(t)$, $\alpha f(t)$, $f(t)\cdot g(t)$. The theory of these normed algebras, which is due to Gelfand, shows an interesting mixture of algebraic and analytic methods. Important roles are played, on the one hand, by the concept of a prime ideal and by other concepts of the theory of rings, and on the other hand by the theory of analytic functions.

The theory of these normed algebras stands in close connection not only with Hilbert space but also with the theory of topological groups. A topological group is a topological space whose elements form a group with respect to a multiplication such that the operation xy^{-1} is continuous in both variables (cf. III 1, §2). Thus a topological group is a topological-algebraic structure. Examples have been known for a long time, e.g., the group of motions or of projective mappings. In fact, these groups were already used by Klein in his Erlanger Programm to characterize the various types of geometry. The first systematic investigation of certain special topological groups was undertaken by Sophus Lie in his theory of groups of continuous transformations, later called *Lie groups*.

The elements of a Lie group are the transformations of an n-dimensional space into itself carrying every point $x = (x_1, \ldots, x_n)$ into a point $x' = (x'_1, \ldots, x'_n)$. The coordinates of the image x' are given by a formula $x'_i = f_i(x_1, \ldots, x_n; a_1, \ldots, a_m)$, where the $a_1, \ldots, a_m$ are parameters determining the transformation T. Thus every transformation T can be assigned to a point $a = (a_1, \ldots, a_m)$ in R^m and can thus be denoted by T_a. The point c representing the transformation $T_c = T_a T_b$ must depend continuously on a and b, i.e., on the parameters of the two factors.

By this correspondence $T_a \rightarrow a$ we can then map a suitably chosen neighborhood of the identical transformation topologically onto a neighborhood of the origin in R^m; for this reason a Lie group is said to be locally Euclidean.

But in addition to this local Euclideanism other requirements on the differentiability of c as a function of the $a_1, \ldots, a_m$ and $b_1, \ldots, b_m$ were included in the Lie theory, so that the structure of a Lie group was determined by a rather complicated system of axioms. Then in 1900, in his famous lecture "Mathematical Problems" at the international congress of mathematicians in Paris, Hilbert proposed the question whether these differentiability requirements cannot be dispensed with; in other words, is a locally Euclidean topological group always a Lie group?

This problem, the fifth in Hilbert's lecture, proved intractable for a long time, in spite of the most strenuous efforts of many mathematicians, until it was recently answered in the affirmative by Gleason, Montgomery, and Zippin.

Here we have an example of an unusually profound and productive question that arose as a direct result of the axiomatic attitude toward mathematics, namely, the question whether the additional axiom on differentiability is already implied by the other hypotheses.

Important parts of analysis can be built up on a topological group as the underlying space. Thus in recent times great advances have been made in so-called harmonic analysis on topological groups, i.e., in the questions which correspond in classical analysis to the theory of Fourier series and the Fourier integral.

On the other hand, many questions in the theory of finite groups, in particular questions relating to irreducible representations by means of matrices, can be extended to topological groups, where they give rise to questions of great importance for the present-day theory of quantum mechanics.

With the concept of a topological group we bring to a close our examples of axiomatic theories developed from classical analysis. It is already clear that the methods and even the very questions themselves have undergone rapid and important changes from their origins in classical analysis, and that completely new fields have been developed, in which the axiomatic method has led to vigorous growth.

Although it can be said with some justification that in algebra the change is already complete, this is by no means the case in analysis. Although the study of linear partial differential equations has received important new stimulus from the theory of Hilbert space and of distributions, nevertheless the classical methods have continued to assert themselves and have not been in any sense absorbed in the new concepts. For the difficult subject of nonlinear differential equations the axiomatic method has provided almost no new ideas. Here we must simply wait until we see how far the new concepts will prevail.

6. Development in Other Mathematical Disciplines

Least affected by these general developments has been the theory of numbers, at least in its simplest form, the theory of integers. No doubt, the explanation lies partly in the fact that this theory was already in the form of a definite structure, described by a small number of axioms, e.g., the axioms of Peano. But here also the algebraic structures have found some entrance, for example, in the arguments based on residue classes modulo an integer or in the realization that Fermat's lesser theorem is a special case of a simple theorem in group theory. However, it must be admitted that the astonishing success of function-theoretic methods in the analytic theory of the distribution of primes still remains quite puzzling; it cannot be said that we understand the situation here as well as in the application of group theory to the Galois theory for the solution of algebraic equations.

Of particular interest from our present point of view is a remarkable invasion of topological methods into the theory of numbers, namely, Hensel's discovery of the p-adic numbers, which we now describe briefly on account of its simplicity.

If in the field **P** of rational numbers we define the absolute value $|x|$ in the usual way, **P** becomes a metric space with the distance $|x - y|$ between the rational numbers x and y. Then by the well-known Cantor process of completion by fundamental sequences we can complete **P** to the field of real numbers (cf. IB1, §4.4). By a fundamental sequence we mean a sequence x_n of rational numbers with $\lim_{n,m\to\infty} |x_n - x_m| = 0$. Thus P is provided with a special topological structure, namely, the metric generated by the absolute value $|x - y|$.

But Hensel introduced a different absolute value on **P**. If p is a fixed prime and x is a rational number, then x can be uniquely represented as a (positive or negative) $\nu_p(x)$th power of p, multiplied by certain (positive or negative) powers of other primes. If we now set $|x|_p = p^{-\nu_p(x)}$ for $x \neq 0$ and $|0|_p = 0$, we obtain a new "absolute value," which is easily seen to satisfy the rules: $|x|_p \geqq 0$, $|x|_p = 0$ only for $x = 0$; $|xy|_p = |x|_p|y|_p$ and $|x + y|_p \leqq \max(|x|_p, |y|_p) \leqq |x|_p + |y|_p$. Consequently, on the pattern of the ordinary definition of the absolute value we can introduce a distance defined by $|x - y|_p$, thereby making **P** into a metric space, and then we can complete this space by adjoining to it the fundamental sequences. In this way we obtain the field $\mathbf{P}_p$ of p-adic numbers, in which, e.g., the sequence $1 + p + p^2 + \cdots$ converges and represents a p-adic number not in **P**. Thus **P** has been embedded in a field that is quite different from the field of real numbers.

It is obvious that this discussion is very closely related to the theory of normed spaces in the preceding section; when further developed into the theory of valuations, it has proved to be of fundamental importance for profound questions in algebraic number fields and in algebraic geometry.

Thus not only analysis but also algebra and the theory of numbers make increasing use of composite algebraic-topological structures, which to a greater and greater extent are seen to be the essential kernel of many theories.

Let us make at least a few remarks about a part of mathematics which by now has become a huge and complicated edifice and is still rapidly expanding, namely, combinatorial topology. The earliest beginnings go back to the nineteenth century, but the decisive developments took place at about the same time as those of algebra and were much influenced by them. Again it has been algebraic methods which, when applied in a thoroughgoing way, have led to rapid changes. Group theory and the theory of modules have been particularly important in this algebraization of combinatorial topology, and it has even been necessary to construct a whole series of new algebraic theories, known today as "homological algebra," which have turned out to be important for the investigation of algebraic structures as well. Let us merely mention two theories of central interest, namely, cohomology and the theory of sheaves, without making any attempt to state exactly what they mean. The most attractive feature of these subjects is that, by means of theorems about very abstract and quite complicated algebraic structures, it is possible again and again to obtain concrete geometric results which up to now have not been obtainable in any simpler way.

7. Mathematics as a Hierarchy of Structures

In 1900, at the end of his lecture in Paris, Hilbert expressed his hopes for the development of mathematics in the coming century:

> The problems I have mentioned are only examples; yet they are enough to show how rich, how various and extensive the science of mathematics has already become; so the question arises whether mathematics is now faced with the same fate as other sciences have met long ago, namely, of being split up into numerous subsciences, almost incomprehensible to scientists in other branches, and more and more loosely connected with one another. I believe and hope that this will not happen; in my opinion mathematics is an inseparable whole, an organism whose very ability to remain alive depends on the close relationship of its parts. Although these parts are different in nature, yet in all of them we can see the same logical procedures, the same methods for the construction of new ideas, with numerous analogies throughout. Moreover, we realize that the more extensively a mathematical theory is developed, the more harmonious and uniform is its structure, and the more numerous the unsuspected relationships between parts of the science that were up to now quite separated from one another. Thus it happens that with the growth of mathematics the uniformity of its character is not lost but becomes clearer than before.
>
> But then we ask whether with this extension of mathematical knowledge it will not at last become quite impossible for the individual mathematician to embrace all parts of the subject? In answer I would say this: it lies at the very heart of mathematics that every advance goes hand in hand with the discovery of sharper instruments and simpler methods, which replace the older, more complicated arguments and thus make the earlier theories easier to understand; as a result, the individual mathematician, with these new instruments and methods at his command, finds it easier to orient himself in the various branches of mathematics, to a far greater extent than can be the case in any other science.
>
> The uniform character of mathematics is at the very heart of the subject; for mathematics is the foundation of all knowledge of nature. In order that it may not fall short of this high calling, let us hope that in the coming century there will arise leaders of genius surrounded by eager disciples.

It is well known that Hilbert's optimistic expectations regarding a proof of the consistency of classical analysis have not been fulfilled, at least not to their full extent, and yet it can be said that he has been completely justified in the hopes he expressed for greater uniformity in mathematics. By means of a few striking examples we have tried to follow the process whereby mathematics has become more uniform as a result of the axiomatic method, a method which Hilbert, its most outstanding representative at the beginning of the century, certainly had in mind when he spoke the above words. We saw how strongly this process has affected the classical division of mathematics into various disciplines, how this division has been seen to be out of date and how an ever increasing number of new theories have arisen from this realization. But if the over-all view of mathematics is not again to be lost, an attempt must be made to understand the consequences of the new order in mathematics and to describe it in a systematic way.

It is precisely this attempt that has been made in the great work "Eléments de Mathématique" of N. Bourbaki, of which more than thirty volumes have now appeared, all of them belonging to the first part of the work, "Les structures fondamentales de l'analyse."

Before we describe the Bourbaki construction in detail, let us examine its underlying ideas. From what has been said, and also from the subtitle of the book, it is clear that Bourbaki regards mathematics as the theory of structures and that for him the reorganization of mathematics must consist in selecting those structures that have proved particularly fruitful for mathematics, in arranging the various structures according to their mutual relationships and in incorporating them in a natural way into the edifice as a whole, which then may be called a hierarchy of structures.

As a particularly simple type of structure we have already studied the algebraic structures in §4; they are characterized by one or more operations, e.g., the structure of a group or the structure of a ring.

A second, different type consists of the topological structures. A topological space is defined as a set in which certain subsets are distinguished and are called neighborhoods; these subsets must satisfy certain relations, the axioms of the topological space.

A third type of structure, not mentioned explicitly above, consists of the ordered sets. An ordered set (cf. IB9, §1) is a set on which a two-place relation, called inclusion, is defined in such a way that: (1) $a \subseteqq a$; (2) from $a \subseteqq b$ and $b \subseteqq c$ follows $a \subseteqq c$; (3) from $a \subseteqq b$ and $b \subseteqq a$ follows $a = b$.

These three types of structures stand on equal footing with one another as the basic structure of the whole edifice.

If to the axioms for a group we adjoin the further axiom of commutativity of multiplication, we obtain the structure of an Abelian group, which is therefore obtained by specialization of the general structure of a group. Analogously by adjoining further axioms to the structure of a topological space we obtain special topological structures, e.g., the locally compact spaces. By specialization of an ordered set, as discussed in IB9, §1, we obtain, e.g., the structure of a lattice.

The whole edifice of algebra, of general topology and of the theory of ordered sets will therefore be constructed by first investigating the most general of these structures, and then proceeding to more special structures by the stepwise adjunction of further axioms.

In the above remarks on analysis we have seen that in addition to these very simple structures a particularly important role is played by mixed structures, e.g., the structure of a topological group or of a topological vector space. Here we are dealing with sets that simultaneously carry two, or even more than two, structures of the three fundamental types. Of great importance here is the way in which these structures are interrelated. In a topological group, for example, the topological structure and the group structure cannot be completely unrelated but must be such that the group operation xy^{-1} is continuous in both variables. The purpose of this requirement is to avoid structures of a pronounced pathological character. The situation is the same when we superimpose an order structure on a field; the laws of monotonicity must be satisfied, i.e., $a \leqq b$ implies $a + c \leqq b + c$ and $0 \leqq a$, $0 \leqq b$ implies $0 \leqq ab$.

Requirements of this sort are called compatibility conditions for a composite structure. The choice of compatibility conditions will depend in each case on the existence of a significant and productive theory for the structures in question.

A look at the historical development will show what long and laborious paths had to be traversed before the proper structures were found for the various sets of problem, which could then be formulated in the simplest way, free of unnecessary hypotheses. But the solution is not always unique; in many cases there are various equivalent axiomatic formulations of a theory, resting on different fundamental concepts, and the decision in favor of one or the other of these formulations may well depend on the extent to which it harmonizes with the closely related structures. But even here a great deal is still a matter of taste.

The great importance of the mixed structures makes it clear that from the standpoint of theory of structures we should consider mathematics as a unified edifice, an interwoven hierarchy of structures. If we wish to carry out the construction of mathematics according to this principle, as is done in Bourbaki, we will put the three general types of structures at the head, and then the mixed types of structures, in an order which is to some extent a matter of choice, and finally the various structures that are obtained by specialization from their predecessors and are therefore richer in hypotheses. It is clear that the classical problems, dealing say with rational or real numbers or with n-dimensional space as their basic structure, will appear far down the list, either in the theory of special structures or perhaps only as examples or exercises (the "Exemples" and "Exercices" in Bourbaki are not always very simple). Many readers may find that this adherence to the basic principle is somewhat exaggerated, since it is necessary to learn a great deal about the general theory before arriving at elementary facts that could have been proved in a much simpler way; of course, this is a disadvantage, but it is far more than outweighed by the impressive uniformity and clarity of the new edifice. Mathematics has suddenly become perspicuous again, as Hilbert had hoped in the lecture quoted at the beginning of this section. The edifice erected in the "Eléments" has already brought greater uniformity into the terminology of mathematics, and the immense number of methods and results contained in it are now available in definitive form as an effective tool for further research. In recent years the consequences have been unmistakable; many a new stimulus to research has come from this work.

8. The Bourbaki Construction of Mathematics

The part of the Bourbaki work printed up to the time of writing of the present chapter consists of six books, in some of which a few "fascicules" are still missing. Book I contains the theory of sets and order structures; Book II, algebra; Book III, general topology; Book IV, the elementary theory of functions of one variable; Book V, the theory of topological vector spaces; and Book VI, the theory of integration.

The first to appear was a slender fascicule in 1939, called "Fascicule de résultats," of the book on the theory of sets. It contains the basic concepts of the theory of sets and a number of theorems without proofs, which serve as the foundation and as a guide to the terminology for the later parts of the work. Here the point of view is the naive one, the concept of a set being regarded as given. Included are operations with sets, the concept of a mapping, ordered sets, cardinal and ordinal numbers, and transfinite induction in the form of the Zorn lemma, and finally a purely set-theoretic formulation of the concept of a structure and of isomorphism of structures. The treatment here is somewhat more general than in IB10, but the discussion given there and the above examples are enough for a clear understanding of the subject.

The program announced in the "Fascicule de résultats" has been carried out completely; the four chapters of the book on set theory are in three fascicules: I. Description of Formalized Mathematics; II. Set Theory; III. Ordered Sets, Cardinal Numbers, Natural Numbers; IV. Structures.

Chapter I contains the elements of mathematical logic, and in Chapter II, in contrast to the naive standpoint of the "Fascicule de résultats," the theory of sets is presented in formalized axiomatic form. Questions of metamathematics, in particular, freedom from contradiction, are not discussed. The formalized language is abandoned in later fascicules, but it is a guiding principle of the authors that the theories are always presented in such a way that, at least in principle, they could easily be formalized.

We have no space to construct the theory of sets in detail. Let us merely remark that in Chapter III the natural numbers are introduced as those cardinal numbers $\mathfrak{a}$ for which $\mathfrak{a} \neq \mathfrak{a} + 1$, and the rules for calculation with natural numbers are deduced from the arithmetic of the cardinal numbers.

The second book contains algebra. Chapter I deals with algebraic structures in general. If on a set E a rule is given which to every pair of elements x, y in E assigns a third element $z = x \tau y$ in E, we speak of a law of inner composition, and the operation τ is called a composition. If the composition is associative, i.e., if $(x \tau y) \tau z = x \tau (y \tau z)$, we obtain the simplest algebraic structure, the monoid. A very general discussion is given of the problems related to the existence of a unit element and of inverses, and the discussion is illustrated by the construction of the rational numbers from the integers.

Next comes the study of outer composition. Let there be given a set E and a second set $\mathbf{\Omega}$, the domain of operators, such that to every α in $\mathbf{\Omega}$ and every x in E there corresponds an element αx in E. If the operators are written on the left, we speak of a left outer composition, and correspondingly on the right. Then an algebraic structure is defined, with complete generality, as a set E on which there are given a number of inner compositions, and possibly also a number of outer compositions with respect to certain domains of operators $\mathbf{\Omega}, \mathbf{\Omega}', \ldots$. There follows a discussion of homomorphisms, isomorphisms, the formation of quotient structures, products of structures, and the possible compatibility conditions for the various compositions, e.g., the left-distributive law $\alpha(x \tau y) = (\alpha x) \tau (\alpha y)$.

The first structure to be investigated in detail is the group, and together with it the structure of a group with operators, where the operator domain is assumed to be left-distributive in the above sense. Only the basic parts of the theory of groups are discussed in detail, up to and including the Jordan-Hölder theorem. Then come the basic parts of ring theory and field theory, in each case without the assumption of commutativity.

The second chapter deals with linear algebra. Here the basic concept is the left module, i.e., an Abelian group M with a ring R as left-operator domain and with the following compatibility conditions: $\alpha(x + y) = \alpha x + \alpha y$, $(\alpha + \beta)x = \alpha x + \beta x$, $\alpha(\beta x) = (\alpha\beta)x$, where x, y are arbitrary in M and α, β are arbitrary in R. If R has a unit element ϵ for multiplication, then $\epsilon x = x$ is also required. In particular, if R is a commutative field K, then M is also called a vector space or a linear space over K. The concepts of basis, linear mapping, dual spaces of a vector space, and the theory of equations and matrices find their place here.

The next important structures are the algebras; an algebra is a ring R with a field K as a left-operator domain. An algebra is also called a hypercomplex system over K. Examples are the quaternions and the group algebras.

Chapter III contains multilinear algebra. The basic structure here is the tensor product of two modules, on which the theory of tensors and the exterior algebra of Grassman are constructed; a discussion of these topics is impossible here for lack of space. The theory of determinants is included in this chapter.

The modern axiomatic form of the Galois theory, i.e., the theory of algebraic extension of a ground field, forms the content of Chapter IV (Polynomials and Rational Functions) and of Chapter V (Commutative Fields). Chapter VI deals with ordered groups and ordered fields, the first mixed structure to appear in the work. Chapter VII discusses modules over principal ideal rings; here we find, for example, the theory of divisibility of integers and of elementary divisors for matrices. Chapter VIII deals with modules and semisimple rings, Chapter IX with semilinear and quadratic forms.

The picture of algebra sketched out by "Bourbaki" is thus seen to be not very different from that of van der Waerden, as is not surprising in view of what was said above in §4. However van der Waerden has no mention of multilinear algebra, and certain other subjects, such as duality and the detailed theory of modules, receive their proper emphasis only in the systematic construction of Bourbaki.

Book III deals with the third group of basic structures, the topological structures and, since the two other basic structures have now been dealt with in detail, it also contains a large number of composite structures. This book has now been printed in full, in five fascicules and a "Fascicule de résultats." Chapter I deals with topological structures; the axioms for a topological space are given both in terms of open sets and in terms of neighborhoods.

Systematic use is made of the concept of a filter, with a detailed study of continuous mapping, topological products, quotient spaces, and the basic notions of topology. Compact and locally compact spaces are the first special topological structures to be investigated. The concept of connectedness also finds its place here.

Chapter II deals with uniform structures, defined above in III 1, §6. Every space that carries a uniform structure can be completed. This is the most general formulation of the familiar passage from rational numbers to real numbers or, more generally, from a metric space to its completion. Instead of the fundamental or Cauchy sequences we now have

Cauchy filters or fundamental sequences in the sense of *Moore-Smith convergence* (cf. III 1, §5). The possibility of defining uniform continuity in these structures was mentioned above.

Chapter III deals with composite structures, first the topological groups, then topological rings and topological fields, and Chapter IV with the structure of the real numbers, which is a three-fold structure (an ordered topological field). The discussion of series, and of functions on arbitrary sets but with real values, is carried out only to the extent to which it appears to be necessary for later applications. Chapter V has the title "Groups with a Parameter." Various axiomatic characterizations are presented for the additive group of real numbers, and some attention is given to the power function and the exponential function. In Chapter VI the n-dimensional space R^n is introduced as the topological product of n spaces isomorphic to the line R, and some of its simple geometric-topological properties are investigated. The n-dimensional projective space P^n is constructed from the $(n + 1)$-dimensional space R^{n+1} by forming a quotient space, which amounts to regarding the lines through the origin in R^{n+1} as the points of P^n. Chapter VII deals with the additive groups in R^n, in other words with a special type of topological Abelian groups, and Chapter VIII with the complex numbers, the complex n-dimensional space and the complex n-dimensional projective space.

Thus Chapters V through VIII are concerned with special mixed structures, which it was necessary to investigate at this place because of their importance for the construction of the whole work.

Under the title "Application of Real Numbers in General Topology," Chapter IX takes up the study of general topological structures, beginning essentially with the metric spaces, for whose definition (cf. §5) the real numbers are necessary. Here we find various composite structures, groups with a metric, normed spaces, fields with a valuation (i.e., fields on which an absolute value is defined; e.g., the p-adic fields in §6), normed algebras and various classes of topological spaces, e.g., the normal spaces.

Chapter X also studies the special class of structures consisting of the function spaces, i.e., spaces of mappings of one topological space into another, equipped with suitable topologies. Important theorems of classical analysis are formulated here in full generality, e.g., the theorems of Ascoli and Dini, and the theorem of Weierstrass on the approximation of continuous functions by polynomials.

Book III, of which we have now given a fairly detailed description, shows rather clearly the extent to which analysis has been influenced by the innovations of Bourbaki. Consider, for example, how much material already precedes the systematic investigation of the real numbers; e.g., essential use is already being made of the theorems on topological groups, to say nothing of all the assumptions borrowed from algebra. Theorems like the Weierstrass approximation theorem do not appear until the end of Book III, where they are special cases of very general theorems requiring the whole equipment of general topology.

The authors themselves feel that this arrangement of the material is somewhat hard on the reader, so that in Book IV they make a certain compromise. In order not to keep him waiting too long for classical analysis, they deal in this book with the elementary theory of functions of a real variable, with differentiation, the Riemann integral, ordinary differential equations and even, in the last chapter, the gamma function. Nevertheless, it is in general assumed that the values of these functions may be not only real or complex numbers but elements of a topological vector space or of a normed space, since in most cases the theorems and their proofs can be stated just as easily for the general case.

As we have tried to show in §5, the structure of a normed space and its generalization, the structure of a topological vector space, has claimed a more and more fundamental position in analysis. Book V, now in complete form with two fascicules and a "Fascicule de résultats," gives the first systematic discussion of the theory of topological vector spaces and contains many new results. Lack of space prevents us from discussing them here in detail but let us mention at least the following fact: it was not until mathematics was considered as a systematic arrangement of structures that the concept of a topological vector space gained its outstanding position, which led to the elaboration of its theory by Dieudonné, Grothendieck and Schwartz.

The sixth book of the "Eléments" deals with integration as a theory of measure and integration on local compact spaces. In the system as a whole, it appears as the study of a special class of topological vector spaces, namely, of continuous linear functionals on vector spaces consisting of functions defined and continuous on a locally compact space. The basic ideas here are to found in the third chapter of the present volume.

Further fascicules of Bourbaki, recently published, deal with the theory of Lie algebras and commutative algebra.

Many questions are left unanswered by the "Eléments" as published up to now. For example, it is not easy to see where the classical theory of functions of a complex variable will find its place, or how its various parts will be distributed. The answer to these questions depends entirely on the structures that will be considered of greatest importance in the remaining part of the work. But as is shown by the books already published, the decision here does not depend exclusively on the general hierarchy of structures; from time to time there appear certain special structures, which take a very central position, e.g., the real and complex numbers, and which for this reason must undergo an early and detailed investigation. In spite of his devotion to systematization, Bourbaki never forgets that his purpose is to present only what is actually utilized in present-day mathematics as a whole, of which it is his intention to present only the foundations. He realizes quite well that the structures to which he has given a leading position deserve this honor only in the present-day state of mathematics; the rapid development of the subject will not only change the emphasis but will continually bring new structures into view.

Index